Sustainable and Innovative Vegetable Production in times of Climate Change

Zafar Iqbal
Muhammad Akbar Anjum
Sezai Ercisli • Saeed Rauf
Editors

Sustainable and Innovative Vegetable Production in times of Climate Change

Concepts of Olericulture

Springer

Editors
Zafar Iqbal
Faculty of Agriculture
University of Sargodha
Sargodha, Pakistan

Muhammad Akbar Anjum
Department of Horticulture
Bahauddin Zakariya University
Multan, Pakistan

Sezai Ercisli
Department of Horticulture, Faculty of Agriculture
Atatürk University
Erzurum, Türkiye

Saeed Rauf
Department of Plant Breeding & Genetics
College of Agriculture, University of Sargodha
Sargodha, Pakistan

ISBN 978-981-96-6282-1 ISBN 978-981-96-6283-8 (eBook)
https://doi.org/10.1007/978-981-96-6283-8

This Springer imprint is published by the registered company Springer Nature Singapore Pte Ltd.
The registered company address is: 152 Beach Road, #21-01/04 Gateway East, Singapore 189721, Singapore

We dedicate this humble effort to our parents and beloved family members who inspired us to higher ideals of life. Their deep love is evident from every page of this book and life as well.
We also pay gratitude to our true friends who shared insightful ideas and always prayed for our success and happiness. The words in this edition glisten with their brilliance and warmth of deep affection.

Foreword

Vegetables play a significant role in our daily diet as they are rich in essential minerals, nutrients, vitamins, phytochemical compounds, antioxidants, and fiber. Daily intake of vegetables enhances immunity and saves from risk of acute and chronic disease conditions. Keeping in view the importance of vegetables, there was a dire need to provide concise and beneficial information on vegetables in one book for all the professionals and vegetable growers craving friendship with Mother Nature.

I have the honor to look through contents of the book "Sustainable and Innovative Vegetable Production in Times of Climate Change: Concepts of Olericulture." The editors and contributors of this book are well-renowned and famous scientists having expertise in their respective fields. My professional relationship particularly with the editors spans over years. Their deep commitment is reflected from the impressive write-up of this book.

This book contains both basic and advanced knowledge required for successful raising of vegetables. It beautifully integrates chapters not only for researchers and scientists but also for home gardeners. This book provides valuable information on sustainable and eco-friendly methods for growing vegetables. It is also a practical guide for establishing an edible garden with diverse local, exotic, underutilized, traditional, and heirloom vegetables.

The book structure splits into six significant parts each of which unveils unique features related to vegetable cultivation. The chapters in each part spark inquisitiveness of readers. Part I covers origin, evolution, and classification of vegetables, their therapeutic value, and color specificity. Part II details precise and rigorous agricultural practices. It makes aware of cultivation techniques, production systems, nursery management, vegetable grafting, urban edible landscaping, indoor production techniques, protected olericulture, hydroponic and organic vegetable production, and hybrid seed production.

Part III focusses on nutrition, irrigation, and weed management and growth regulators' application. Part IV discusses constraints in the wake of climate change, drought stress, and toxic impacts of heavy metals. Part V delves into biotechnological methods and genetics including CRISPR-based gene tweaking. Part VI provides an insight into postharvest technology, value addition, advances in fresh-cut vegetables, innovative use of nanotechnology in vegetable processing, and preparation of juices and blends.

I hope this book will be a substantial resource and real scientific asset for vegetable lovers and broad range of readers including agriculturists, nutritionists, novice gardeners, plant breeders, botanists, climatologists, researchers, food industry professionals, academicians, and students of related fields. It suggests a roadmap to transform existing agricultural systems in future to achieve the diverse targets of climate-resilient farming, food security, sustainability, off-season production, and self-sufficiency.

Maqsood Ahmed
Ex-Vice Chancellor, Mirpur University of Science and Technology
Mirpur, Azad Jammu & Kashmir, Pakistan

Preface

The theme of this book was perceived years back. It was our dream to write an advanced book on olericulture to provide precise and comprehensive information on all aspects of vegetable cultivation and management. Before proceeding, we reviewed contents of many published books but variety of information desired by stakeholders was not available at a single source and in one book. So, this endeavor led us to complete a book of 28 chapters with succinct style covering diverse topics. This book responds to queries that strike minds of professionals, agriculturists, researchers, outreach consultants, nursery owners, and gardeners.

We have written this book because overpopulation alarm, natural calamities, and changing climatic patterns force to produce more food for hunger hotspots and food insecure countries of the world. Securing food by raising conventional crops and traditional means cannot achieve this noble target. In addition to field cultivation, growing diverse vegetables by space maximization in urban areas and utilizing even a small area of available land for edible landscaping can provide not only safe vegetables but also stave off adverse effects of climate change.

This book provides a thorough analysis of the diverse challenges and approaches associated with vegetable production in the face of environmental modifications, stress tolerance, soil health, safe and protective farming, postharvest handling, sustainable practices, value addition, and the pressing issue of food insecurity. Its contents facilitate an in-depth understanding of the current climatic requirements, technical prerequisites, better crop management, and eco-friendly techniques necessary for successful cultivation of vegetables. Most of the vegetables can happily be grown in different climates by modifying the existing methods keeping in view the local conditions. Concerns about pesticides' hazards and environmental impacts drive a growing demand for safe, residue-free food in a world that is increasingly embracing organic agriculture.

The contributors of this book are seasoned experts in their profession. Their rich experience in basic and applied research in olericulture has helped to make this book much beneficial for all the stakeholders. It provides policymakers, experimenters, avid gardeners, educators, and agricultural professionals with state-of-the-art knowledge of various methods and principles for innovative vegetable production. This book will be equally useful for industrialists, horticulturists, nutritionists, extension workers, students, faculty members of universities, and gardeners who have passion to pique their curiosity for seeking unique knowledge and technological contrivance.

We greatly acknowledge the technical assistance and kind support provided by Dr. Muhammad Nadeem, Associate Professor, Institute of Food Science and Nutrition, University of Sargodha, Pakistan; Dr. Muhammad Kabir, Head, Department of Biological Sciences, Thal University, Bhakkar, Pakistan; Muhammad Waqar Alam, Assistant Professor, Department of Botany, University of Okara, Pakistan; and Muhammad Abdullah Asif (IT Expert), Faisalabad, Pakistan.

We also thank Springer for offering the opportunity to present this scientific work in valuable form.

Sargodha, Pakistan — Zafar Iqbal
Multan, Pakistan — Muhammad Akbar Anjum
Erzurum, Türkiye — Sezai Ercisli
Sargodha, Pakistan — Saeed Rauf

Contents

Part I Origin, Classification and Importance

1 **Origin, Evolution, and Classification of Vegetables** 3
Zafar Iqbal, Abdul Rehman, Muhammad Waqar Alam, Muhammad Asim, and Muhammad Qasim Shahid

2 **Therapeutic Value and Health Benefits of Vegetables** 15
Zafar Iqbal, Shaheer Zafar, Muhammad Kabir, Muhammad Nadeem, and Syed Nassar Abbas

3 **Specificity of Vegetables' Natural Coloration** 31
Arslan Akbar, Shaheer Zafar, Zafar Iqbal, Muhammad Waqar Alam, and Ozair Chaudhry

Part II Cultivation Techniques and Production Systems

4 **Vegetable Nursery Management: Foundation of Successful Production System** 45
Mateen Sajid, Zahoor Hussain, Muhammad Akbar Anjum, Shuhua Zhu, and Raheel Anwar

5 **Vegetable Grafting: An Innovative Approach in Olericulture** 69
Saddia Galani, Maria Babar, Muhammad Anees, and Muhammad Omar Kaseb

6 **Methods of Open Field Cultivation of Vegetables** 89
Muhammad Saqib, Aqsa Haram, Arslan Akbar, and Muhammad Akbar Anjum

7 **Edible Landscaping: Planning and Designing** 101
Zafar Iqbal, Muhammad Akbar Anjum, Shahid J. Butt, Alaiha Asif, and Rashid Mukhtar Balal

8 **Urban and Indoor Vegetables' Production Techniques** 121
Zafar Iqbal, Muhammad Akbar Anjum, Shahid J. Butt, Muhammad Adnan Shahid, and Muhammad Safeer Abbas Jafrey

9 **Protected Olericulture** 151
Anam Noor, Sakeena Tul Ain Haider, Arslan Akbar, Khurram Ziaf, and Muhammad Akbar Anjum

10 **Hydroponic Vegetable Production** 183
Shahid Javed Butt, Zafar Iqbal, Servet Varis, and Kamran Shehzad

11 **Organic and Sustainable Vegetable Production** 203
Safina Naz, Muhammad Akbar Anjum, Hafiza Muniba Din Muhammad, Riaz Ahmad, and Muhammad Ahsan Altaf

12 **Hybrid Seed Production in Vegetable Crops** 221
Muhammad Abu Bakar Saddique, Muhammad Ahmad Tariq, Tanveer Ahmad, Sadia Hakeem, and Cai Linlin

Part III Management and Care

13 **Mineral Nutrition of Vegetable Crops** 245
Zahra Batool, Muhammad Baqir Hussain, Niaz Ahmad, and Waseem Hassan

14 **Irrigation Management in Vegetable Crops** 269
Atique-ur-Rehman, Rafi Qamar, Axel Garcia y Garcia, Muhammad Shoaib, and Shakeel Ahmad

15 **Weed Management in Vegetable Crops** 289
Omer Farooq, Rao Muhammad Ikram, Imran Shabbir, Hulya Dogan, and Shakeel Ahmad

16 **Applications of Plant Growth Regulators in Vegetables** 311
Ishtiaq Ahmad, Mehmet Ramazan Bozhuyuk, Maryam, Riaz Ahmad, and Muhammad Akbar Anjum

Part IV Stresses and Constraints

17 **Constraints in Vegetables Production Due to Climate Change** 327
Muhammad Saqib, Muhammad Akbar Anjum, Ishtiaq Ahmad, Sami Abou Fayssal, and Riaz Ahmad

18 **Understanding Drought Stress, Its Effects, and Tolerance Mechanism for Sustainable Vegetable Production** 345
Ali Asad Bahar, Hafiz Nazar Faried, Sami Ullah, and Ahmad Ali

19 **Toxicity of Heavy Metals in Vegetables and Their Impacts on Human Health** 365
Muhammad Kabir, Natasha Aziz, Muhammad Hammad, Muhammad Zafar Iqbal, and Muhammad Athar Tariq

Part V Genetics and Biotechnology

20 Applications of Biotechnology in Vegetable Crops 385
Riaz Ahmad, Sezai Ercisli, Arslan Akbar, and Muhammad Akbar Anjum

21 Hybrid Breeding in Vegetable Crops 403
Saeed Rauf, Taiyyibah Basharat, and Zafar Iqbal

22 CRISPR-Based Gene Tweaking for Improvement in Vegetable Crops 421
Muhammad Faheem, Anam Niamat, Sameer H. Qari, Nayla Munawar, and Aftab Ahmad

Part VI Postharvest Technology, Processing and Value Addition

23 Postharvest Storage Life Extension and Quality Management of Vegetables 445
Sajid Ali, Muhammad Hassan, Muhammad Akbar Anjum, Safina Naz, and Shaghef Ejaz

24 Postharvest Management of Vegetable Seeds 463
Tanveer Ahmad, Muhammad Anwar-ul-Haq, Anam Noor, Sajjad Akhtar, and Taki Demir

25 Prospects of Value Addition in Veggies: A Tool to Minimize Postharvest Losses 483
Muhammad Nadeem, Zafar Iqbal, Tahir Mahmood Qureshi, Mian Nadeem Riaz, and Muhammad Adnan Shahid

26 Recent Advances in Fresh-Cut Vegetables 501
Anwar Ali, Joanna Trafialek, and Muhammad Faisal Manzoor

27 Innovative Use of Nanotechnology in Vegetable Processing 521
Rimsha Gulzar, Amna Shehbaz, Tuba Mustafa, Muhammad Afzaal, and Hanisah Kamilah

28 Vegetables' Juices, Blends, and Smoothies: Preparation and Processing 537
Hafsa Ashraf, Shakeela Khalid, Waseem Khalid, Muhammad Nadeem, and Muhammad Usman Tahir

Editors and Contributors

About the Editors

Zafar Iqbal has served in different positions including Chairman, Principal, Director Sub Campus, and Dean, Faculty of Agriculture, University of Sargodha, Pakistan. He has vast research and academic experience with awards from different national and international organizations. He has 200 publications to his credit including 12 Monographs/Lab Manuals and 02 book chapters. He has delivered many keynote/invited talks in national and international conferences. He is a recipient of research productivity award from Pakistan Council for Science and Technology. He was awarded exquisite trophy on his excellent work on mango malformation disease. He has expertise in postharvest problems of horticultural crops, mycological issues, and disease management. He is member of Australia Global Alumni, Asian Council of Science Editors, and many other professional societies. He has designed a mobile app to disseminate latest technologies to global farming community. He won first prize in gender equality competition arranged by an NGO in Pakistan.

Muhammad Akbar Anjum served as Chairman, Department of Horticulture, Principal College of Agriculture, and Director, Office of Research, Innovation and Commercialization (ORIC) at Bahauddin Zakariya University, Multan (Pakistan). With a strong academic background, he has received numerous distinctions. He secured first position in B.Sc. (Hons.) Agriculture and M.Sc. (Hons.) Horticulture and was awarded a Gold and a Silver Medal, respectively. During his Ph.D. at University of Salford, Manchester (UK), he was awarded ORS (Overseas Research

Students) Award. In 2020, he won University level Best Teacher Award. He also received accolades several times for his research productivity from the Pakistan Council for Science and Technology. He is author/co-author of 03 books and 24 book chapters. He has published more than 200 research articles in journals of international repute. He has served as Chief Editor of two journals and is currently Editor/Member Advisory Board of five journals. He is member of several professional bodies and has served as Vice President and President of Pakistan Society for Horticultural Science.

Sezai Ercişli is currently serving as the Head of Department of Horticulture and Faculty of Agriculture, and Advisor to the Rector at Atatürk University. He also served as Vice Rector of Atatürk University between 2016 and 2020. He was a Visiting Professor at the University of Nebraska between 1997 and 1998. He has been the President of the World Apricot Association and the Turkish representative of the European Fruit Producers Association since 2018. He conducted scientific studies on Horticulture Plant Biotechnology in many laboratories in different countries of the world. He is ranked among the 100,000 most successful scientists in the world, according to the Stanford University classification published for 2021. With 655 articles published, his h-index is 52. Currently, he is Chief Editor of one journal of international repute since 2014. Previously, he also served as Editor-in-Chief of another journal.

Saeed Rauf is a plant scientist at the Department of Plant Breeding & Genetics, College of Agriculture, University of Sargodha, Pakistan. He was awarded a Silver Medal for securing second position in his M.Sc. (Hons.) program. With approximately 50 research articles published in internationally recognized journals and eight book chapters to his credit, he has made significant contributions to the field. He has successfully developed several inbred lines that are now commercially utilized in sunflower hybrid development. His expertise spans genetic tools and protocols applicable to all field crops, with a focus on expanding their yield potential. Additionally, he specializes in biometrical genetics, cytogenetics, and tissue culture.

As a dedicated mentor, he has supervised numerous M.Phil. and Ph.D. students. He also serves as the Editor of a research journal recognized by the Higher Education Commission (HEC) of Pakistan.

Contributors

Syed Nassar Abbas Curtin University, Perth, Australia

Muhammad Afzaal Safety & Biotechnology Lab, Department of Food Science, Government College University, Faisalabad, Pakistan

Aftab Ahmad Natural and Medical Science Research Center (NMSRC), University of Nizwa, Nizwa, Oman

Ishtiaq Ahmad Department of Horticultural Sciences, The Islamia University of Bahawalpur, Bahawalpur, Pakistan

Niaz Ahmad Department of Soil Science, Bahauddin Zakariya University, Multan, Pakistan

Ahmad Department of Horticulture, The University of Agriculture, Dera Ismail Khan, Pakistan

Shakeel Ahmad Institute of Agronomy, Bahauddin Zakariya University, Multan, Pakistan

Tanveer Ahmad Department of Horticulture, MNS University of Agriculture, Multan, Pakistan

Arslan Akbar Department of Horticulture, Bahauddin Zakariya University, Multan, Pakistan

Sajjad Akhtar Rachna Agribusiness, Gujranwala, Pakistan

Muhammad Waqar Alam Department of Botany, University of Okara, Okara, Punjab, Pakistan

Ahmad Ali Department of Soil, Water and Ecosystem Sciences, Institute of Food and Agricultural Sciences, University of Florida, Gainesville, FL, USA

Anwar Ali Institute of Human Nutrition Sciences, Warsaw University of Life Sciences–SGGW, Warsaw, Poland

Sajid Ali Department of Horticulture, Bahauddin Zakariya University, Multan, Pakistan

Muhammad Ahsan Altaf College of Horticulture, Hainan University, Haikou, China

Muhammad Anees Zhengzhou Fruit Research Institute, Chinese Academy of Agricultural Sciences, Zhengzhou, China

Muhammad Akbar Anjum Department of Horticulture, Bahauddin Zakariya University, Multan, Pakistan

Raheel Anwar Institute of Horticultural Sciences, University of Agriculture, Faisalabad, Pakistan

Muhammad Anwar-ul-Haq Department of Horticulture, MNS University of Agriculture, Multan, Pakistan

Department of Horticulture, Bahauddin Zakariya University, Multan, Pakistan

Department of Life Science, University of Modena and Reggio Emilia, Reggio Emilia, Italy

Hafsa Ashraf Department of Food Science, Government College University, Faisalabad, Pakistan

Alaiha Asif Horticultural Sciences Department, North Florida Research and Education Centre, University of Florida/IFAS, Quincy, FL, USA

Muhammad Asim Department of Plant Pathology, College of Agriculture, University of Sargodha, Sargodha, Pakistan

Atique-ur-Rehman Institute of Agronomy, Bahauddin Zakariya University, Multan, Pakistan

Natasha Aziz Department of Biological Sciences, Thal University, Bhakkar, Pakistan

Maria Babar Dr. A.Q. Khan Institute of Biotechnology and Genetic Engineering (KIBGE), University of Karachi, Karachi, Pakistan

Ali Asad Bahar Department of Horticulture, MNS University of Agriculture, Multan, Pakistan

Rashid Mukhtar Balal Department of Horticulture, College of Agriculture, University of Sargodha, Sargodha, Pakistan

Taiyyibah Basharat Department of Plant Breeding & Genetics, College of Agriculture, University of Sargodha, Sargodha, Pakistan

Zahra Batool Department of Soil and Environmental Sciences, MNS University of Agriculture, Multan, Pakistan

Mehmet Ramazan Bozhuyuk Horticulture Department, Igdir University, Igdir, Türkiye

Shahid Javed Butt Department of Horticulture, Faculty of Agriculture, PMAS-Arid Agriculture University, Rawalpindi, Pakistan

Ozair Chaudhry Durham Environment and Climate Advisory Committee (DECAC), Planning and Economic Development Department, Whitby, Ontario, Canada

Taki Demir Department of Horticulture, Sakarya University of Applied Sciences, Sakarya, Türkiye

Hulya Dogan Department of Agriculture and Food, Hemp Research Institute, Yozgat Bozok University, Yozgat, Türkiye

Shaghef Ejaz Department of Horticulture, Bahauddin Zakariya University, Multan, Pakistan

Sezai Ercisli Department of Horticulture, Faculty of Agriculture, Atatürk University, Erzurum, Türkiye

Muhammad Faheem Department of Biochemistry, Faculty of Sciences, University of Agriculture, Faisalabad, Pakistan

Hafiz Nazar Faried Department of Horticulture, MNS University of Agriculture, Multan, Pakistan

Omer Farooq Institute of Agronomy, Bahauddin Zakariya University, Multan, Pakistan

Sami Abou Fayssal Faculty of Agronomy, Department of Agronomy, University of Forestry, Sofia, Bulgaria

Department of Plant Production, Faculty of Agriculture, Lebanese University, Beirut, Lebanon

Saddia Galani Dr. A.Q. Khan Institute of Biotechnology and Genetic Engineering (KIBGE), University of Karachi, Karachi, Pakistan

Axel Garcia y Garcia Department of Plant Sciences, University of Wyoming, Powell, WY, USA

Rimsha Gulzar Safety & Biotechnology Lab, Department of Food Science, Government College University, Faisalabad, Pakistan

Sakeena Tul Ain Haider Department of Horticulture, Bahauddin Zakariya University, Multan, Pakistan

Sadia Hakeem Institute of Plant Breeding and Biotechnology, MNS University of Agriculture, Multan, Pakistan

Vegetable Research Substation, Multan, Ayub Agricultural Research Institute, Faisalabad, Pakistan

Muhammad Hammad Department of Biological Sciences, Thal University, Bhakkar, Pakistan

Aqsa Haram Department of Horticulture, Bahauddin Zakariya University, Multan, Pakistan

Muhammad Hassan Department of Horticulture, Bahauddin Zakariya University, Multan, Pakistan

Waseem Hassan Department of Soil and Environmental Sciences, MNS University of Agriculture, Multan, Pakistan

Muhammad Baqir Hussain Department of Soil and Environmental Sciences, MNS University of Agriculture, Multan, Pakistan

Zahoor Hussain Department of Horticulture, Ghazi University, Dera Ghazi Khan, Pakistan

Rao Muhammad Ikram Department of Agronomy, MNS University of Agriculture, Multan, Pakistan

Muhammad Zafar Iqbal Department of Botany, University of Karachi, Karachi, Pakistan

Zafar Iqbal Faculty of Agriculture, University of Sargodha, Sargodha, Pakistan

Muhammad Safeer Abbas Jafrey BinQain Agro Services, Lahore, Pakistan

Muhammad Kabir Department of Biological Sciences, Thal University, Bhakkar, Pakistan

Hanisah Kamilah Department of Crop Science, Faculty of Agricultural and Forestry Sciences, Universiti Putra Malaysia, Bintulu Saralu Sarawak Campus, Bintulu, Malaysia

Muhammad Omar Kaseb Horticulture Research Institute, Agriculture Research Centre, Giza, Egypt

Shakeela Khalid Department of Zoology, University of Sargodha, Sargodha, Pakistan

Waseem Khalid Faculty of Chemical Sciences and Technologies, Department of Organic Chemistry, University of Castilla La Mancha, Ciudad Real, Spain

University Institute of Food Science and Technology, The University of Lahore, Lahore, Pakistan

Cai Linlin School of Life Science and Engineering, Southwest University of Science and Technology, Mianyang, China

Muhammad Faisal Manzoor Guangdong Provincial Key Laboratory of Intelligent Food Manufacturing, School of Food Science and Engineering, Foshan University, Foshan, China

Faculty of Sciences and Technology, ILMA University, Karachi, Pakistan

Maryam Department of Botany, The Govt. Sadiq College for Women University, Bahawalpur, Pakistan

Hafiza Muniba Din Muhammad Department of Horticulture, Bahauddin Zakariya University, Multan, Pakistan

Nayla Munawar Department of Chemistry, United Arab Emirates University, Al-Ain, United Arab Emirates

Tuba Mustafa Safety & Biotechnology Lab, Department of Food Science, Government College University, Faisalabad, Pakistan

Muhammad Nadeem Institute of Food Science and Nutrition, University of Sargodha, Sargodha, Pakistan

Horticultural Sciences Department, North Florida Research and Education Center, Institute of Food and Agricultural Sciences, University of Florida, Quincy, FL, USA

Safina Naz Department of Horticulture, Bahauddin Zakariya University, Multan, Pakistan

Anam Niamat Department of Biochemistry, Faculty of Sciences, University of Agriculture, Faisalabad, Pakistan

Anam Noor Department of Horticulture, Bahauddin Zakariya University, Multan, Pakistan

Rafi Qamar Department of Agronomy, College of Agriculture, University of Sargodha, Sargodha, Pakistan

Sameer H. Qari Department of Biology, AlJumum University College, Umm Al-Qura University, Makkah, Saudi Arabia

Tahir Mahmood Qureshi Department of Food Sciences, Cholistan University of Veterinary and Animal Sciences, Bahawalpur, Pakistan

Saeed Rauf Department of Plant Breeding & Genetics, College of Agriculture, University of Sargodha, Sargodha, Pakistan

Abdul Rehman Department of Agronomy, College of Agriculture, University of Sargodha, Sargodha, Pakistan

Mian Nadeem Riaz Department of Food Science and Technology, Texas A&M University, College Station, TX, USA

Muhammad Abu Bakar Saddique Institute of Plant Breeding and Biotechnology, MNS University of Agriculture, Multan, Pakistan

Mateen Sajid Department of Horticulture, Ghazi University, Dera Ghazi Khan, Pakistan

Muhammad Saqib Department of Horticulture, The University of Agriculture, Dera Ismail Khan, Pakistan

Imran Shabbir Institute of Agronomy, Bahauddin Zakariya University, Multan, Pakistan

Muhammad Qasim Shahid South China Agricultural University, Guangzhou, China

Muhammad Adnan Shahid Horticultural Sciences Department, North Florida Research and Education Center, Institute of Food and Agricultural Sciences, University of Florida, Quincy, FL, USA

Amna Shehbaz Safety & Biotechnology Lab, Department of Food Science, Government College University, Faisalabad, Pakistan

Kamran Shehzad National Agricultural Research Center, Islamabad, Pakistan

Muhammad Shoaib Department of Agricultural Engineering, Bahauddin Zakariya University, Multan, Pakistan

Muhammad Usman Tahir Department of Food Science, Government College University, Faisalabad, Pakistan

Muhammad Ahmad Tariq Institute of Plant Breeding and Biotechnology, MNS University of Agriculture, Multan, Pakistan

Muhammad Athar Tariq California Department of Food and Agriculture, Sacramento, CA, USA

Joanna Trafialek Institute of Human Nutrition Sciences, Warsaw University of Life Sciences–SGGW, Warsaw, Poland

Sami Ullah Department of Horticulture, MNS University of Agriculture, Multan, Pakistan

Servet Varis Department of Horticulture, Namik Kemal University, Tekirdag, Türkiye

Shaheer Zafar Hamdard College of Medicine and Dentistry, Hamdard University, Karachi, Pakistan

Shuhua Zhu Department of Horticulture, Shihezi University, Shihezi, Xinjiang, China

Khurram Ziaf Institute of Horticultural Sciences, University of Agriculture, Faisalabad, Pakistan

Part I

Origin, Classification and Importance

Origin, Evolution, and Classification of Vegetables

1

Zafar Iqbal, Abdul Rehman, Muhammad Waqar Alam, Muhammad Asim, and Muhammad Qasim Shahid

Abstract

Vegetables include edible plant parts which are essential for maintaining human health. These contain vitamins, minerals, phytochemical compounds, antioxidants, and dietary fiber. Vegetables attract with diverse colors, shapes, flavors, and textures. Adding vegetables to our daily diet keeps us healthy and safe and boosts our immune system. This chapter provides detailed information on origin, etymology, historical development, classification, and benefits of common vegetables. It apprises us of informative description of different vegetable groups viz. green, bulb, inflorescence, root and sea vegetables and legumes.

Keywords

Etymology · Classification · Fruit vegetables · Inflorescence vegetables · Sea veggies · Legumes

Z. Iqbal (✉)
Faculty of Agriculture, University of Sargodha, Sargodha, Pakistan
e-mail: zafarrohils@yahoo.com

A. Rehman
Department of Agronomy, College of Agriculture, University of Sargodha, Sargodha, Pakistan

M. W. Alam
Department of Botany, University of Okara, Okara, Punjab, Pakistan

M. Asim
Department of Plant Pathology, College of Agriculture, University of Sargodha, Sargodha, Pakistan

M. Q. Shahid
South China Agricultural University, Guangzhou, China

Z. Iqbal et al. (eds.), *Sustainable and Innovative Vegetable Production in times of Climate Change*, https://doi.org/10.1007/978-981-96-6283-8_1

1.1 Introduction

Vegetable is called as an edible part of roots, stems, leaves, flowers, fruit, or seeds of any herbaceous plant (Wikipedia). These plant parts may be consumed fresh (Lotha 2020) or cooked in diverse ways. According to Merriam-Webster, it is a herbaceous plant like bean, cabbage, or potato, which is raised for an edible part to be eaten in meal. Vegetables are annual or perennial horticultural crops with sections like roots, stalks, flowers, fruits, and leaves that can be consumed wholly or partially (Welbaum 2015).

1.2 Origin and Etymology

In English, the word vegetable was noted for the first time in the early fifteenth century. It originated from Old French, and was actually applied to all plants (Swedenborg 2003). It has been derived from the Latin word vegetabilis, which means "growing, flourishing" (Harper 2000). Before eighteenth century, the meaning of vegetable as 'a plant grown for food' was not introduced (John 1993). In 1767, this word was particularly used to describe a "plant grown for food". The term "veggie" was used for the first time in 1955 (Hai 2015; Harper 2000).

The exact definition of vegetable varies as different parts viz. roots, stems, leaves, flowers, fruits,and seeds are consumed as food globally. The broadest definition refers to "matter of plant origin". Vegetable may be more precisely defined as: "any plant part utilized for food that is not a fruit or seed, but includes mature fruits that are eaten as part of a main meal" (Sinha et al. 2010; Vainio and Bianchini 2003; Wikipedia). It excludes foods obtained from some plants such as nuts, fruits, and flowers but includes savory fruits like tomatoes and courgettis, flowers like broccoli, and some seeds like pulses.

1.3 Historical Development

Vegetables were picked from the wild by hunter-gatherers and their farming initiated in different countries on the globe with gradual agricultural development during 10,000 BC to 7000 BC. Naturally occurring vegetables were cultivated locally and confined to specific areas but with the passage of time, international trade collected exotic crops from other countries to add to local types. At present, most of the local or exotic vegetables are raised in almost all the countries of the world and in case of less suitable conditions, varieties are acclimatized or grown in protected environment.

1.3.1 History of Common Vegetables

1.3.1.1 Potato

The first domestication of potato has been reported from the region of southern Peru and extreme western Bolivia (Spooner et al. 2005) between 8000 and 5000 BC. Potato cultivation in South America goes back 10,000 years (David and Gordon 2014) but identification becomes difficult as tubers do not preserve well in the archeological record. European explorers visualized the potential of potato and gradually expanding, potato became the most favorite vegetable in the world (Anonymous 2020).

1.3.1.2 Tomato

French botanist Tournefort suggested the Latin botanical name, *Lycopersicon esculentum,* for the tomato. The English word *tomato* comes from the Spanish word, *tomate*, derived from Nahuatl (Aztec language) word, *tomatl (Filippone* 2019b*)*. Tomato is being cultivated for the last 2500 years. People of ancient Aztec empire included it into their lives and cuisine keeping in view its importance and value (Anonymous 2020). Tomato was brought from Andean region to Europe in the sixteenth century (Bergougnoux 2014). It has now essential, beautiful, and colorful presentation in salad all over the world. It has also culinary value while cooking other vegetables or meat. Etc.

1.3.1.3 Carrot

The carrot is a root vegetable, available in orange color but purple, white, red, black, and yellow varieties are also available (Lorrizo et al. 2020, 2013; Sifferlin 2018). History of carrot emerges for the last five thousand years. The plant, originated in Persia, was cultivated for its leaves and seeds. Domestication of carrots is reported from Iran and Afghanistan and expanding later to China and Egypt with prominence in medieval Europe (Anonymous 2020).

1.3.1.4 Cucumber

It is a member of the gourd family Cucurbitaceae and it has witnessed a journey of 5000 years of human growth, development, and expansion. It was reportedly brought to different parts of Europe by the Romans or Greeks. Cultivation of cucumber was done in the ninth century in France, in the fourteenth century in England, and by the mid-sixteenth century in North America (Doijode 2001; Zhuang et al. 2006; Renner et al. 2007). It is now used as part of fresh salad globally.

1.3.1.5 Cabbage

It's difficult to explore its specific origin due to some cultivars of leafy greens identified as Brassicas (Ingram 2000*)*. According to the most common theory, West Cabbage was domesticated in Europe 3000 years ago from its wild predecessors. It was introduced into Europe by the Romans and the plant was later bred into the present popular form. It was also grown in the cooler parts of northern Europe (Rose 2008). In the East, cabbage is being used since 4000 BC. The headed cabbage was called as "krambe" by Greeks while the Romans named it "brassica" or "olus". The use of cabbage was thought to be a luxury in Rome. It was also preferred due to medicinal properties as it provided relief from gout, headaches, and the symptoms caused by poisonous mushrooms. The first round-headed cabbages were introduced in fourteenth century in England which became famous cuisine in Europe. Cultivated variants of cabbage spread to Asia and Americas. Between fourteenth and seventeenth century, cabbage was brought to India by colonizing traders from Portugal (Anonymous 2020).

1.3.1.6 Garlic

Garlic is considered as one of the oldest cultivated plants in the world having history of several thousand years of human consumption (Block 2010). It originated from central Asia as it domesticated during Neolithic times. This plant was recognized by the ancient Indians who domesticated it about 6000 years back. In 3000 BC, it was introduced into the Middle East and northern Africa enabling later to reach Europe. Sumerians (2600–2100 BC) carried garlic to China, from where it spread to Korea and Japan. Garlic expansion has been reported to occur first in the old world, and then in the new world. Some historians still claim its origin from China (Petrovska and Cekovska 2010).

Egyptian royals used to feed garlic to their slaves or laborers who built famous pyramids in order to strengthen their bodies. In other cultures, it was used to increase working potential of laborers. The wild plant called *Allium longicuspis* evolved into the form of the modern *Allium sativum*, or garlic (Rivlin 2001).

1.3.1.7 Onion

Many botanists, archaeologists, and food historians consider central Asia as the onion's place of origin. Another report finds its first cultivation in Iran and West Pakistan. However, majority of people agree that onion has been cultivated for 5000 years or more. Onions were grown in Sumeria 4500 years ago, in India and China 5000 years ago and in ancient Egypt, 5500 years back. In Egypt, onions were considered to be symbols of eternity and endless life. Egyptians used to make sketch of onions on the walls of burial chambers and tombs.

1.3.1.8 Capsicum

The origin of *Capsicum* is Western hemisphere where it is used as food since 7500 BC. It was cultivated for the first time in South America, but later introduced to Central America between 5200 and 3400 BC (Anonymous 2020). From America, it was introduced to rest of the world by Columbus (Srivastava and Mangal 2019).

1.3.1.9 Bean

This term is used for a variety of plants belonging to family Leguminosae. The oldest proofs for the use of bean as food in Thailand are 9000 years old. People native to Mexico and Peru used to grow this crop since 7000 BC. The large-seeded broad beans were reported to be cultivated in Iberia, Aegean, and transalpine Europe 4000 years ago (Ehler 2020a). The genetic studies of the common bean *Phaseolus* confirm its origin in Mesoamerica (Bitocchi et al. 2012).

1.3.1.10 Radish

It most likely originated in the area between the Mediterranean and the Caspian Sea (Crisp 1995). Historical record also relates it to the wild radish growing in southwest China (Cheo et al. 1987). Archaeological evidence suggested that they were among the first crops on which Europeans used to survive (Garrison 2016).

1.3.1.11 Spinach

Spinach was first cultivated in Persia almost 2000 years back (Williams and Long 1997). The Saracens brought spinach to Sicily in AD 827 (Rolland and Sherman 2006). Its cultivation started in India and China in seventh century. Spinach gained popularity in America in the 1880s.

1.3.1.12 Pumpkins

The word ‘pumpkin’ has been derived from the Greek word *pepon*, which means “large melon” (Paris 1989). The pumpkin seeds have an old record between 7000 and 5550 BC. Pumpkins have been reported to be originated in Central America over 7500 years ago. Pre-Columbian natives used to cultivate pumpkins for their flesh (Avey 2014). These were considered a significant food item for native Americans.

1.3.1.13 Bitter Gourd

Bitter gourd has originated from the tropical areas of Eastern India. It has a cultivation history of more than 600 years and is popular in China as a vegetable and ornament plant. It is now widely cultivated in different zones viz. tropical, subtropical, and temperate, especially in Pakistan, India, Indonesia, Malaysia, Japan, Southeast Asia, Tropical Africa, and China.

1.3.1.14 Okra

Okra has been reported to be cultivated near Ethiopia. Egyptians used to grow okra during twelfth century BC. Later its cultivation expanded to the Middle East and North Africa. People used to eat cooked seed pods, and the seeds were also toasted and ground to substitute coffee (Ehler 2020b).

1.3.1.15 Ginger

Ginger has been greatly admired in the world as it possesses strong medicinal properties (Drewnicki 2020). Ginger roots have been used in herbal medicines for the last 4500 years (Kurcina 2021). The first record of ginger dates back to 475–221 BC. The Arabs introduced ginger to the Mediterranean. Ginger was also a part of trade items for Europe during the Middle Ages (Pickersgill 2005; Wikipedia).

1.3.1.16 Eggplant (Brinjal)

India has been reported to be pioneer in eggplant cultivation followed by Africa (Trujilo 2003; Matthee 2016). Ancient Chinese documents reveal the first record of eggplant in 544 CE (Dunlop 2006). Its cultivation by the Arabs in the early middle ages has been reported from the Mediterranean region wherefrom it was introduced into Spain in eighth century (Sanderson and Renfrew 2005).

1.3.1.17 Sugar Beet

Cultivation of fodder beets started in Europe in 1500 AD. Later it gained popularity due to presence of sugar content. Germany initiated its cultivation in the 1700s. The expansion of beet industry occurred in France and other European countries. The suspension of raw cane sugar supply to France forced Napoleon to consider it as good sucrose source (Yamane 2020).

1.3.1.18 Coriander

Coriander is a very old crop as its seeds have been found in 5000 BC (Filippone 2019a). It was considered as spice of happiness by Egyptians as it possessed medicinal properties in addition to its use in kitchen. Coriander seeds were imported from Egypt and introduced into Britain by Romans. The leaves of coriander emit strong smell when they are rubbed and bruised. Its oil has been used as food item and aromatic ingredient since the 1900s.

1.3.1.19 Fenugreek

Fenugreek has been reported to be cultivated since 4000 BC. It is considered native to India and Southeastern Europe. Its medicinal use has been reported for different disorders and ailments. It has also been grown as forage crop in many countries of the world. Use of its seeds as spice is common in some Asian countries, and leaves are used in Pakistani and Indian cuisines.

1.3.1.20 Broccoli

Broccoli is a nutrient-rich cruciferous vegetable which is a prominent part of French, Italian, and Asian cuisines. It's a type of cabbage which has long history of cultivation in Mediterranean since Roman times. It was brought to Britain by Italians in the mid-1700s. Broccoli originated from old cultivars raised in the Roman Empire, however, it bred through artificial selection in Italy. It was grown commercially in the 1500s (Filippone 2019c), but its commercial cultivation in California, USA started in 1923.

1.3.1.21 Sea Vegetables

Sea vegetables include algae and plants that grow in or close to ocean. According to archaeological evidence, Japan has an old tradition of using sea vegetables for over 10,000 years. These are also widely consumed in Korea, Vietnam, China, Malaysia, and some other countries which border seas (GMF 2020). They generally contain fiber, iron, calcium, potassium, phosphorus, folic acid, and vitamins A and C. Value-added products of sea veggies are becoming popular which include snacks, soups, sauces, jellies, flakes, pills, etc.

1.4 Classification of Vegetables

Globally, 10,000 plant species are used as vegetables. They may be classified on the basis of some common features. Classification is necessarily required for food scientists, agrarians, nutritionists, academicians, and other stakeholders to subclassify vegetables on the basis of some scientific and desired parameters. If this subdivision relates to similar food composition, then it may be more useful (Pennington and Fisher 2010; Ülger et al. 2018). Classification may be based on plant part used for nutrition and the specific nutritive value (Lintas 1992; Ülger et al. 2018). There are various types and varieties of vegetables which are considered powerhouse of nutrients, minerals, and vitamins (Hai 2015).

1.4.1 Green Vegetables

1.4.1.1 Leafy Vegetables

Leafy vegetables are actually plant leaves which are consumed as vegetables. This group includes cabbage, Chinese cabbage, cress, collards, Brussels sprouts, arugula, endive, curly lettuce, savoy, sea kale, spinach, purslane, radicchio, nasturtium, sorrel, chicory, and chad. These are considered good source of important minerals (iron and calcium), vitamins (A, C, and riboflavin), and fiber. There is more vitamin C in young fresh leaves than in mature plants. Greener and thinner leaves have more nutritive value with lower calories (Hai 2015; Ülger et al. 2018).

1.4.1.2 Stalk Vegetables

Celery and asparagus are included in stalk vegetables. They contain minerals and vitamins in proportion to the green color. Asparagus is a rich source of folic acid (Ülger et al. 2018).

1.4.1.3 Fruit and Flower Vegetables

Broccoli, cauliflower, and artichoke are included in the list of flowering vegetables. Broccoli is a good source of iron, phosphorus, vitamins A and C, and riboflavin. Cauliflower is also a good source of vitamin C. They can be cooked or eaten as salads. Artichoke contains potassium, calcium, and phosphorus and high dietary fiber content.

Fruit vegetables are placed in both fruits and vegetables. Peppers and tomatoes are the most common fruit vegetables which are rich in vitamin C. Tomato is a fruit by scientific name but from culinary point of view, it is named as vegetable due to its use in savory. Other fruit vegetables include avocados, chayote, eggplant, cucumbers, okra, squash, olives, zucchini, and tomatillos (Hai 2015; Ülger et al. 2018).

1.4.2 Bulb Vegetables

Bulb vegetables are fragrant vegetables which add flavor to broths, casseroles, soups, etc. People eat their bulbs so the name bulb vegetables has been given. Bulb vegetables include chives, garlic, leeks, onions, scallions, shallots, and water chestnuts (Hai 2015). Onion is generally called as magic bulb while garlic referred as wonder drug.

1.4.3 Inflorescence Vegetables

These are actually flowers, floral buds, leaves, and stems of a plant which are consumed as vegetables. These vegetables, being nutrient-dense, are recommended to be included in diet plan of people due to less calories. Some inflorescent vegetables are: Broccoli, artichokes, cauliflower, and banana flower (Hai 2015).

1.4.4 Root Vegetables

Root vegetables are natural bulbs that flourish under the soil and absorb nutrients from the rhizosphere (Denenberg 2019). These are enlarged enough for storage of energy in the form of carbohydrates. Root vegetables include carrots, celeriac, radish, burdock, parsnips, malanga, turnips, rutabaga, salsify, stalk vegetable, asparagus, bamboo, cardoon, chard, celery, fiddlehead, fennel, kohlrabi, cassava, Jerusalem artichoke, jicama, potato, sweet potato, taro, and yam (Hai 2015).

1.4.5 Sea Veggies (Sea Weeds)

The term sea vegetables refers to edible marine algae, which includes some of the most primitive life forms on earth. Seaweed includes a subset of this enormous category of thousands of species (Tanabe and Gandhi 2020). Sea veggies are a good source of dietary fiber and chlorophyll. Due to natural presence of sodium, potassium, calcium, phosphorus, magnesium, iron, and other trace minerals in the ocean, they have a pleasant salty flavor. They are so versatile to be used in any dish (Angle 2016). This group includes vegetables like aonori, arame, carola, dabberlocks or badderlocks, dulse or dillisk, hijiki, kombu, laver (or gim), mozuku, nori, ogonori, sea grape, sea lettuce, and wakame (Hai 2015).

1.4.6 Legumes

The word “legumes” originates from the Latin legumen which means vegetable. Legumes are considered as a class of vegetables that includes beans, peas, soybeans, and lentils. Legumes contain protein, iron, and zinc typically found in animal products like chicken or beef. They are high in fiber, folate, magnesium, and potassium (Grieger 2020). This group also contains saponin and soluble fiber (Lintas 1992; Ülger et al. 2018).

References

Angle S (2016) Are Sea Veggies the Superfood Missing from Your Kitchen? https://wwwshape-com/healthy-eating/cooking-ideas/are-sea-veggies-superfood-missing-your-kitchen. Accessed 02 Sept 2020

Anonymous (2020) Vegetable facts. http://wwwvegetablefactsnet/vegetable-history/. Accessed 20 June 2020

Avey T (2014) History of pumpkins and recipe round-up. https://wwwpbsorg/food/the-history-kitchen/history-pumpkins-recipes/#. Accessed 28 Aug 2020

Bergougnoux V (2014) The history of tomato: from domestication to biopharming. Biotechnol Adv 32(1):170–189

Bitocchi E, Nanni L, Bellucci E, Rossi M, Giardini A, Zeuli PS, Logozzo G, Stougaard J, McClean P, Attene G, Papa R (2012) Mesoamerican origin of the common bean (*Phaseolus vulgaris* L) is revealed by sequence data. Proc Natl Acad Sci 109(14):E788–E796. https://doi.org/10.1073/pnas1108973109. PMC 3325731 PMID 22393017

Block E (2010) Garlic and other alliums. the lore and the science royal society of chemistry, pp 5–6. ISBN 9780854041909

Cheo TY, Guo RL, Lan YZ, Lou LL, Kuan KC, An ZX (1987) Angiospermae dicotyledoneae cruciferae. In: Cheo TY (ed) Flora Reipublicae Popularis Sinicae 33. Sci Press, Beijing (China), pp 1–483

Crisp P (1995) Radish (*Raphanus sativus*) (Cruciferae). In: Smartt J, Simmonds NW (eds) Evolution of crop plants, 2nd edn. Longman Scientific & Technical UK, pp 86–89

David RH, Gordon CH (2014) Foraging and farming: the evolution of plant exploitation. Routledge. ISBN 1317598296, p 496

Denenberg Z (2019) Everything you need to know about root vegetables. https://wwwsouthernliv-ingcom/side-dishes/vegetables/what-are-root-vegetables. Accessed 03 Sept 2020

Doijode SD (2001) Seed storage of horticultural crops. Haworth Press. ISBN 1-56022-901-2, p 281

Drewnicki A (2020) The amazing history and health benefits of ginger. https://wwwhonestly-goodco/the-amazing-history-and-health-benefits-of-ginger/. Accessed 28 Aug 2020

Dunlop F (2006) Revolutionary Chinese cookbook: recipes from Hunan Province. Ebury Press, p 202

Ehler JT (2020a) Beans: one of the first cultivated crops. http://wwwfoodreferencecom/html/art-beanshtml. Accessed 23 Aug 2020

Ehler JT (2020b) Okra history and facts. http://wwwfoodreferencecom/html/artokrahtml#:~:text. Okra. Accessed 28 Aug 2020

Filippone PT (2019a) The long history and many uses of coriander. https://wwwthespruceeatscom/what-is-coriander-1807009#. Accessed 31 Aug 2020

Filippone PT (2019b) The history of tomatoes as food. https://wwwthespruceeatscom/history-of-tomatoes-as-food-1807678. Accessed 26 Aug 2020

Filippone PT (2019c) The history of Broccoli. https://wwwthespruceeatscom/broccoli-history-1807573#:. Accessed 31 Aug 2020

Garrison H (2016) The Radish: a history rooted in ancient times. https://wwwbulbolightcom/blogs/digest/the-radish. Accessed 23 Aug 2020

GMF (2020) Sea vegetables George Mateljan Foundation. http://wwwwhfoodscom/genpagephp?tname=foodspice&dbid=135. Accessed 02 Sept 2020

Grieger L (2020) Legumes: a nutritious no fat food. https://wwwsummitmedicalgroupcom/news/nutrition/legumes/. Accessed 03 Sept 2020

Hai HD (2015) The worldwide vegetables. http://theworldwidevegetablesweeblycom/introduction-to-vegetableshtml
Harper D (2000) "Vegetable" Online etymology dictionary. https://www.etymonlinecom/. Accessed 21 Aug 2020
Ingram C (2000) The cook's guide to vegetables. Hermes House, pp 64–66. ISBN 978-1-84038-842-8
John A (1993) Dictionary of word origins. Arcade Publishing, New York. ISBN 1-55970-214-1
Kurcina K (2021) A brief history of ginger and its benefits to your body. https://spotterup.com/a-brief-history-of-ginger-and-its-benefits-to-your-body/. Accessed 18 April 2025
Lintas C (1992) Nutritional aspects of fruit and vegetable consumption. In: Lauret F (ed) Les fruits et légumes dans les économies méditerranéennes: actes du colloque de Chania. CIHEAM, Montpellier, pp 79–87
Lorizzo M, Senalik DA, Ellison SL, Grzebelus D, Cavagnaro PF, Allender C, Brunet J, Spooner DM, Van Deynze A, Simon PW (2013) Genetic structure and domestication of carrot (*Daucus carota* subsp *sativus*) (Apiaceae). Am J Bot 100(5):930–938. https://doi.org/10.3732/ajb1300055. PMID 23594914
Lorizzo M, Curaba J, Pottorff M, Ferruzzi MG, Simon P, Cavagnaro PF (2020) Carrot anthocyanins genetics and genomics: status and perspectives to improve its application for the food colorant industry. Genes 11(8):906. https://doi.org/10.3390/genes11080906. ISSN 2073-4425 PMID 32784714
Lotha G (2020) Vegetable. https://wwwbritannicacom/topic/vegetable
Matthee R (2016) Patterns of Food consumption in early modern Iran. https://doi.org/10.1093/oxfordhb/9780199935369013l3
Paris HS (1989) Historical records origins and development of the edible cultivar groups of *Cucurbita pepo* (Cucurbitaceae). Econ Bot New York Botanical Garden Press 43(4):423–443. https://doi.org/10.1007/bf02935916. JSTOR 4255187
Pennington JAT, Fisher RA (2010) Food component profiles for fruit and vegetable subgroups. J Food Compos Anal 23:411–418
Petrovska BB, Cekovska S (2010) Extracts from the history and medical properties of garlic. Pharm Rev 4(7):106–110
Pickersgill B (2005) In: Ghillean P, Mark N (eds) The cultural history of plants. Routledge, pp 163–164. ISBN 0415927463
Renner SS, Schaefer H, Kocyan A (2007) Phylogenetics of Cucumis (Cucurbitaceae): Cucumber (*C. sativus*) belongs in an Asian/Australian clade far from melon (*C. melo*). BMC Evol Biol 7:58. https://doi.org/10.1186/1471-2148-7-58. PMC 3225884 PMID 17425784
Rivlin RS (2001) Historical perspective on the use of garlic. J Nutr 131(3):951–954
Rolland JL, Sherman C (2006) The food encyclopedia. Robert Rose, Toronto, pp 335–338. ISBN 9780778801504
Rose M (2008) Cabbage: Its history uses and culture. https://davesgardencom/guides/articles/view/753. Accessed 27 Aug 2020
Sanderson H, Renfrew JM (2005) Ghillean P, Mark N (eds) The cultural history of plants. Routledge. p 118. ISBN 0415927463
Sifferlin A (2018) Eat this now: rainbow carrots. Time Archived from the original on 23 Jan 2018. Retrieved 27 Jan 2018
Sinha N, Hui YH, Evranuz EÖ, Siddiq M, Ahmed J (2010) Handbook of vegetables and vegetable processing. Wiley, pp 192–352. ISBN 978-0-470-95844-5
Spooner DM, McLean K, Ramsay G, Waugh R, Bryan GJ (2005) A single domestication for potato based on multi locus amplified fragment length polymorphism genotyping PNAS 102 (41):1469499 Bibcode:PNAS10214694 S https://doi.org/10.1073/pnas0507400102 PMC 1253605 PMID 16203994 Lay summary
Srivastava A, Mangal M (2019) Capsicum breeding: history and development. In: Ramchiary N, Kole C (eds) The capsicum genome compendium of plant genomes. Springer, Cham. https://doi.org/10.1007/978-3-319-97217-6_3

Swedenborg E (2003) Swedenborg concordance 1888. Kessinger Publishing, p 502. ISBN 0-7661-3728-7

Tanabe L, Gandhi B (2020) Everything you need to know about sea vegetables. https://wwwmindbodygreencom/articles/what-are-sea-vegetables. Accessed 02 Sept 2020

Trujilo L (2003) The Elegant Eggplant. Master Gardener Journal archived from the original on 4 Dec 2016. Retrieved 16 Sep2018

Ülger TG, Songur AN, Cirak O, Çakıroğlu FP (2018) Role of vegetables in human nutrition and disease prevention. https://doi.org/10.5772/intechopen77038

Vainio H, Bianchini F (2003) Fruits and vegetables international agency for research on cancer, World Health Organization. IARC Press, Lyon, p 293

Welbaum GE (2015) Vegetable production and practices; IARC handbooks of cancer prevention: fruit and vegetables. In: Vegetable History Nomenclature and Classification, vol 8, pp 1–15

Wikipedia Vegetable. https://enwikipediaorg/wiki/Vegetable. Accessed on 21 Aug 2020

Williams SR, Long S (1997) Nutrition and diet therapy. p 229 ISBN 978-0-8151-9273-2

Yamane T (2020) Sugar beet plant. https://wwwbritannicacom/plant/sugar-beet. Accessed 28 Aug 2020

Zhuang FY, Chen J, Staub JE, Qian C (2006) Taxonomic relationships of a rare cucumis species (*C. hystrix* Chakr) and its interspecific hybrid with cucumber. Hortic Sci 41(3):571–574. https://doi.org/10.21273/HORTSCI413571

Therapeutic Value and Health Benefits of Vegetables

2

Zafar Iqbal, Shaheer Zafar, Muhammad Kabir, Muhammad Nadeem, and Syed Nassar Abbas

Abstract

Inclusion of vegetables in daily diet offers numerous health benefits. Vegetables provide important nutrients for better health and maintenance of our body. The risk of some chronic diseases is likely to be reduced in people who use fresh fruits and vegetables. Vegetables and fruits have held a place in nutritional management due to concentrations of minerals, vitamins, dietary fiber, and more recently phytochemicals, especially antioxidants. This chapter reveals therapeutic properties of veggies and protection they provide against numerous disease conditions like diabetes, cardiovascular problems, osteoporosis, paralysis, cancer, obesity, eye diseases, gastrointestinal ailments, blood pressure, arthritis, and prostate/bladder and immunity issues.

Z. Iqbal
Faculty of Agriculture, University of Sargodha, Sargodha, Pakistan

S. Zafar
Hamdard College of Medicine and Dentistry, Hamdard University, Karachi, Pakistan

M. Kabir
Department of Biological Sciences, Thal University, Bhakkar, Pakistan

M. Nadeem (✉)
Institute of Food Science and Nutrition, University of Sargodha, Sargodha, Pakistan

Horticultural Sciences Department, North Florida Research and Education Center, Institute of Food and Agricultural Sciences, University of Florida, Quincy, FL, USA
e-mail: nadeem.abdul@uos.edu.pk; nadeem.muhammad@ufl.edu

S. N. Abbas
Curtin University, Perth, Australia

Z. Iqbal et al. (eds.), *Sustainable and Innovative Vegetable Production in times of Climate Change*, https://doi.org/10.1007/978-981-96-6283-8_2

Keywords

Beneficial compounds of vegetables · Cardiovascular · Osteoporosis · Cellular rejuvenation · Immunity booster

2.1 Health Benefits and Protection Against Devastating Diseases

2.1.1 Diabetes

Diabetes mellitus (DM) and metabolic syndrome (MS) have increased to an alarming level due to unhealthy lifestyle. In addition to increasing body vigor, vegetables reduce risk of DM by provision of fiber, antioxidants, and some other micronutrients that support to manage blood sugar regulation, and reduce weight. Garlic possesses antihyperglycemic and lipid-lowering properties. It also helps prevent some chronic complications (Eidi et al. 2006; Madkor et al. 2011; Kumar et al. 2013; Ülger et al. 2018). Adding vegetables like spinach, tomatoes, kale, arugula, onion, broccoli, green beans, zucchini, and carrots to daily meal can keep blood sugar levels lower.

2.1.2 Cardiovascular Diseases

These diseases are considered as the key basis of ailments and demise in the world. Some prophylactic and positive activities may eliminate the life-threatening factors. Enhanced use of vegetables reduces CVD-related mortality rates (Wang et al. 2014; Roth et al. 2015). Some vegetables provide protection against CVDs as having trans-fat, low saturated fat, and cholesterol content and good levels of bioactive fusions (Kris-Etherton et al. 2002; Ülger et al. 2018).

Epidemiological studies reveal that regular use of garlic reduces cholesterol level, lowers high blood pressure and checks advancement of CVD (Yeh and Liu 2001; Chang et al. 2005; Ried et al. 2013). Green leafy vegetables intensify antioxidant capacity through vitamins, minerals, pulp, and phytochemical fusions and defend against oxidative stress which plays a significant chunk in the pathogenesis of CVD. Generally, green leafy vegetables including lettuce, spinach, mustard greens, and Swiss chard, and cruciferous including Brussels sprouts, cauliflower, broccoli, cabbage, Bok choy, and kale provide more protection to vegetarians against CVDs (Joshipura et al. 2001; Hung et al. 2004; Huang et al. 2012; Lidder and Webb 2013).

2.1.3 Osteoporosis

Osteoporosis is a health problem in which bones become thin and weak. This disease makes a person prone to bone fracture. Although osteoporosis is considered to be a disease of older persons but susceptibility may start at an early age. Vegetables provide important nutrients to improve bone strength. Vegetables which prevent osteoporosis include curly kale, collard greens, spinach, turnip greens, broccoli, mustard greens, Chinese cabbage, green and white cabbage, and bok choy.

2.1.4 Paralysis

In paralysis, muscles don't perform their normal function in the human body. The persons who consume green leafy vegetables regularly remain safe from the attack of paralysis (Ülger et al. 2018).

2.1.5 Cancer

Cancer occurs when cells proliferate without control (Insel et al. 2007; Heim and Spinalla 2016). Globally, ten million deaths have been reported due to cancer in the year 2020 (WHO 2022). Avoidance of risk factors can prevent between 30 and 50% of cancers (WHO 2022). Important factors include smoking, foods containing synthetic chemicals, unhealthy lifestyle, and carcinogens present in the environment (Heim and Spinalla 2016; Ülger et al. 2018).

Increased vegetable consumption reduces the risk of cancer. Some vegetables' fibers have been described to protect against colorectal cancer (Insel et al. 2007). Vegetables possess anticancer properties due to the existence of antioxidants, minerals, vitamins, pulp, and phytochemicals (Soh et al. 2003; AICR 2015). Due to inadequate vegetable consumption, about 14% of deaths are caused by gastrointestinal cancers (Arusha 2011; Ülger et al. 2018). American Cancer Society reports that energetic food including vegetables and fruits reduces the danger of different types of cancer, i.e., lung, stomach, esophageal, oral pharyngeal, and colon cancer. A study confirms that tomatoes provide protection to men against aggressive forms of prostate cancer and a pigment named as lycopene which gives red color, possesses protective effect (Wiseman 2008). Root, tuber, and bulb vegetables also provide protection against cancer (Ülger et al. 2018). Green leafy vegetables minimize the possibility of cancer due to natural presence of vitamins C, E, K, and A (AICR 2015; Pollock 2016; Tao et al. 2018).

2.1.6 Gastrointestinal Health

Vegetables and fruits include complex fiber, which inhabits water and expands as it passes through the gastrointestinal system. This can ease signs of a short-tempered bowel, regulate bowel activities, and relieve constipation (Aldoori et al. 1998; Lembo and Camilleri 2003).

2.1.7 Eye Diseases

Eating fruits and vegetables keeps our eyes fresh and vigorous. Nutrients like lutein, zinc, omega-3 fatty acids, and vitamins C and E support age-related visualization difficulties including cataracts and macular degeneration. These diseases distress millions of Americans over age 65 (Brown et al. 1999; Cho et al. 2004; Christen et al. 2008). Consuming green leafy vegetables like collards, spinach, kale, and also beans helps fight vision problems. Vitamin A also plays an essential role in vision as retina needs this vitamin to support its function to turn light rays into image (Sajovic et al. 2022). Beta carotene helps to reduce the oxidative stress. Carrots are rich source of beta carotene and vitamin. Like carrots, sweet potatoes are a good source of beta carotene and vitamin E.

2.1.8 Blood Pressure/Hypertension

According to recent estimates, 1.28 billion adults belonging to the age group of 29 to 79 years suffer from hypertension. Surprisingly, 46% of the adult victims are unaware that they suffer from this disease condition (WHO 2023). A healthy, nutritious, and balanced food is important for lowering blood pressure and sustaining it at optimum levels. Recent investigation has revealed that various foods rich in specific nutrients like potassium, magnesium, calcium, and nitrates reduce blood pressure levels (Yokoyama et al. 2014). Leafy green vegetables like cabbage, kale, lettuce, collard greens, fennel, mustard greens, spinach, and Swiss chard are rich in nitrates and consuming 1–2 servings of these vegetables daily can decrease high blood pressure for up to 24 hours. Garlic is a typical antifungal antibiotic food and it improves the body's manufacturing of nitric oxide, which assists the blood vessels to dilate and smooth muscles to relax. Garlic extracts also decrease both systolic and diastolic blood pressure in hypertensive individuals.

2.1.9 Arthritis

Arthritis damages the joints and reduces mobility due to severe symptoms like swelling and stiffness. People of different age groups and racial backgrounds are affected by this disease. Out of various types of arthritis, osteoarthritis affects the joints with overuse while in rheumatoid arthritis (RA), immune system affects joints (Sen and Hurley 2023). Harmful byproducts like free radicals are produced by metabolic processes in the body. These free radicals cause injury to cells that are associated with rheumatoid arthritis and inflammation. Green leafy vegetables such as spinach, Brussels sprouts, broccoli, kale, Bok choy, and Swiss chard contain antioxidants like vitamins A, C, and K, which defend cells from free-radical injury and decrease the potentials of arthritis. These foods also contain high levels of calcium which imparts strength to bones. Kidney beans, red beans, and pinto beans, being good sources of magnesium, zinc, iron, folic acid, and potassium, boost immune system. Ginger extract is also well known to reduce knee pain (Alman and Marcussen 2001).

2.1.10 Improved Memory and Brain Function

Eating a brain-boosting diet can support diverse brain functions. Omega-3 fatty acids provide protection against neurodegeneration while antioxidants reduce cellular stress and inflammation (Wang et al. 2018; Dighriri et al. 2022). Brussels sprouts are rich in omega-3 fatty acids, vitamins K, C, and fiber. Regular use of vegetables including broccoli, cabbage, spinach, kale, sprouts, bell peppers, asparagus, eggplant, and onions may protect from memory loss. Although pumpkin seeds are small in size, they are full of important nutrients and antioxidants like magnesium, copper, iron, potassium, zinc, Vitamin B2, and folate. They provide protection against free radical damage.

2.1.11 Improvement of Prostate and Bladder Health

According to the American Cancer Society, prostate cancer is the utmost common cancer among males, distressing 01 in 08 males in the United States (ACS 2024). Risk of prostate cancer may be minimized if beneficial veggies are added to the daily menu. Pumpkin seeds help to give relief from symptoms of benign prostatic hyperplasia (BPH) (Fruhwirth and Hermeter 2007). Meals containing pumpkin seeds provide relief as natural treatment against overactive bladder. Green leafy and cruciferous vegetables minimize the risk of BPH and prostate cancer. Brussels sprouts, Broccoli, bok choy, cauliflower, and cabbage contain "sulforaphane" which is much beneficial for a healthy prostrate. Use of onion and garlic in daily diet may reduce the risk of BPH. The bright-colored carotenoid lycopene in tomato also reduces the risk of prostate cancer.

2.1.12 Improved Sleep Habits

The chemicals, nutrients, minerals, and compounds which control the sleep cycle with coordination include gamma-aminobutyric acid (GABA), melatonin, omega-3 fatty acids, calcium, tryptophan, potassium, magnesium, pyridoxine, L-ornithine, histamine, serotonin, folate, acetylcholine, antioxidants, vitamin B and D, copper, and zinc. In addition to some other factors, falling asleep may occur due to the non-inclusion of fruits and vegetables in the daily diet. Leafy greens like kale and spinach contain fiber which ensures deeper and more restorative sleep (Grandner et al. 2014; Nisar et al. 2019). Mediterranean diet has been found associated with better vigor and sleep value (Muscogiuri et al. 2020). Short sleep duration has been found related to lower consumption of vegetables and fruits and an improved eating of fruits and veggies in children is associated with satisfactory and adequate sleep (Dashti et al. 2015; Noorwali et al. 2019).

2.1.13 Immunity Booster

Presence of sulfur-containing compounds, such as allicin in garlic and vitamin C in red bell peppers, helps to strengthen the immune system (Arreola et al. 2015). Turmeric has various traditional medicinal uses in Asian countries. It enhances the body's antibody response. Curcumin in turmeric has been found to boost healthy immune system which leads to a fresh and lively lifestyle. The persons and athletes suffering from Exercise-Induced Muscle Damage (EIMD) get relief if they use turmeric or curcumin as oral treatment (McFarlin et al. 2016). Spinach, broccoli, ginger, sweet potatoes, carrots, kale, lettuce, arugula, artichokes, and collard greens are also well-known veggies and powerhouse of essential nutrients which improve immune functions.

2.1.14 Anti-aging, Cellular Rejuvenation, and Skin Glowing Properties

Watercress is a nutrient-dense and hydrating dark leafy green vegetable. It is a virtuous foundation of calcium, potassium, manganese, zinc, iron, copper, phosphorus, selenium, folate, niacin, thiamin, riboflavin, protein, and vitamins A, C, E, K, and B-6. The antioxidants like vitamin A and C in watercress may be beneficial for neutralization of harmful free radicals, helping to protect against skin problems. Broccoli is an anti-aging vegetable containing vitamins C, K, B-6, antioxidants, fiber, folate, iron, magnesium, lutein, and calcium.

Spinach contains high vitamin C content and it is helpful to keep skin firm, fresh, and smooth. Kale has occupied its space as a super food. This cruciferous veggie contains skin's favorite anti-aging nutrients including A, C, E, K, B1, B2, B3, B6, lutein, carotenoids, iron, folate, magnesium, calcium, phosphorus, and omega-3 s. It possesses anticancer properties and also increases detoxification. Red bell peppers have high content of vitamin C and prominent antioxidants called as carotenoids. They provide protection against sun damage and toxic effects of environmental toxins. Beta carotene in sweet potato and carrots acts as a natural sun protectant. Ginger reduces scars, wrinkles, and fine lines on skin so it is used as face mask and scrub. Cucumber contains important minerals, nutrients, and antioxidants. It has high water content which hydrates skin and keeps it fresh and glowing. Its slices are also placed under eyes by people to reduce dark circles. Beetroot juice contains antioxidants and vitamin C which keep safe from skin damage and early signs of aging like wrinkles and fine lines. Face of people who use beetroot regularly in their diet becomes pink and glowing. Purslane is considered anti-aging as it is rich in vitamin A, zinc, phosphorus, calcium, iron, magnesium, copper, potassium, and omega-3 fatty acids. Garlic protects against sun damage, reduces blackheads, and soothes human skin. It contains allicin which possesses antifungal, antibacterial, and anti-aging properties. Presence of sulfur in garlic prevents infections.

Complete detail of all beneficial compounds naturally present in vegetables is given in Table 2.1.

Table 2.1 Beneficial compounds of vegetables

Sr. #	Compound	Rich Source	Functions/Disease Prevention	Citation
1.	Carotenoids Including vitamin A, lycopene, and beta-carotene	Red, orange, and yellow veggies like beets, squash, carrots, sweet potato, orange bell peppers Dark green like kale, collard greens, spinach, broccoli Tomato	Potent antioxidants necessary for immune function. They possess an important antioxidant function of deactivating free radicals. Carotenoid-rich food fights against many diseases like cancers, cardiovascular diseases, age-related macular degeneration. Addition of adequate carotenoids may significantly reduce the risk of several disorders mediated by reactive oxygen species.	Perera and Yen (2007), Fiedor and Burda (2014) and Saini et al. (2022)
2.	Diallyl sulfide	Allium vegetables including garlic, onion, chives, shallots, scallions, leeks	Reduced risk of cardiovascular diseases. Garlic has protective effect against colorectal cancer and also possesses anti-hypertensive, anti-diabetic, and anti-microbial properties.	Midde et al. (2015)
3.	Dithiolethiones	Cruciferous vegetables like broccoli, collard greens, kale, cabbage, Brussels sprouts, watercress, arugula, radishes, Bok choy.	Dithiolethiones are a well-known class of potentchemo-preventive, cytoprotective, and neuroprotective agents	Zhang and Munday (2008) and Ansari et al. (2018)
4.	Flavonoids	Parsley, onions, kale, tomatoes, lettuce, broccoli, asparagus, leafy greens.	These compounds reduce inflammation and protect against cancer and cardiovascular diseases. They are rich in antioxidant activity and avert effects of toxins. They also possess antibacterial properties.	Liga et al. (2023)
5.	Plant sterols and stanols	Legumes, broccoli, Brussels sprouts, cauliflower	These substances occur naturally in small amounts in vegetables. These compounds reduce cholesterol's absorption in the gut.	Barkas et al. (2023)

(continued)

Table 2.1 (continued)

Sr. #	Compound	Rich Source	Functions/Disease Prevention	Citation
6.	Polyphenols	White beans, black beans, artichokes, chicory, red onions, spinach, potatoes, red lettuce, shallots	They possess antioxidant properties. They prevent cell damage caused by free radicals and improve brain health. They help against digestive and weight management issues, diabetes, neurodegenerative disease, and cardiovascular diseases. They minimize the chances of developing a blood clot.	Bertelli et al. (2021)
7.	Isothiocyanates	Cruciferous vegetables like broccoli, watercress, cabbage, cauliflower, Brussels sprouts, kale, collard green, turnip, and Japanese radish	These are sulfur-containing plant chemicals having good anticancer properties. They reduce risk of colorectal and lung cancer.	Higdon et al. (2007) and Olayanju et al. (2024)
8.	Indoles	Cruciferous vegetables like broccoli, cabbage, and kale	Indoles are organic compounds having various health benefits. They reduce risk of breast, prostate, and colon cancer.	Qian et al. (2024)
9.	Vitamin A	Dark green leafy vegetables like pumpkins, carrots, spinach, amaranth, chard, sweet potatoes, squashes, kale, collards, turnips, sweet red pepper, and lettuce	It helps maintain the integrity and function of all surface tissue. It keeps our skin and eyes healthy and protects against infection. It supports the daily replacement of skin cells.	Menezes et al. (2024)

10.	Vitamin B	Red vegetables (beets, carrots, sweet potatoes), green vegetables (artichoke, green pepper, okra, broccoli, asparagus), green leafy vegetables, cauliflower, and legumes (beans).	**B1 (thiamin)**: It is important for nerve, heart, and muscle function. It helps in muscle contraction. It supports body cells to convert carbohydrates into energy. **B2 (riboflavin)**: It produces red blood cells. It plays role for functioning of immune system. It is also required for healthy skin and hair. **B3 (niacin)**: Niacin helps in food digestion, and supports nervous system. **B5 (pantothenic acid)**: It helps in converting food into energy. It plays a vital role for production of red blood cells. **B6 (pyridoxine):** It supports immune and nervous system function and protects the body from infection by helping to make antibodies. **B7 (biotin)**: It helps to maintain healthy nails, hair, and skin. **B9 (folate)**: It displays its role in cell growth and metabolism. It helps in the production of body's genetic material. It works with vitamin B12 for making of red blood cells. **B12 (cobalamin):** It supports the nervous system to function properly.	Ali et al. (2022), Ibrahim et al. (2023) and Gharibzahedi et al. (2023)
11.	Vitamin C	Artichoke, asparagus, carrots, Brussels sprouts, broccoli, cauliflower, cabbage, turnip greens, parsley, celery, cucumber, yellow peppers green, red and sweet, bell peppers, kale, onions, potatoes, sweet potato, spinach, squash, tomatoes.	The human body cannot yield or stock vitamin C. So, it's important to munch it frequently in adequate amounts. It supports immune system function. It is essential for the biosynthesis of collagen which is needed for healthy connective tissue and wound healing. It also helps in iron absorption.	Kim (2020) and See et al. (2024)

(continued)

Table 2.1 (continued)

Sr. #	Compound	Rich Source	Functions/Disease Prevention	Citation
12.	Vitamin K	Cruciferous vegetables (broccoli, cauliflower, cabbage, Brussels sprouts, turnip greens), green leafy vegetables (spinach, kale, Swiss chard, mustard greens, collard greens, parsley, lettuce, cabbage), pumpkin and asparagus.	It protects immune system. It is essentially required for healthy bones. It plays a role in preventing Alzheimer's disease. It regulates blood clotting and prevents osteoporosis.	Aaseth et al. (2024)
13.	Calcium	Spinach, kale, turnips, collard greens, sweet potatoes, okra, arugula, legumes, bok choy, cress, rhubarb, beet greens, and tomatoes.	Calcium is important for strong bones and teeth, and for sustaining typical function of muscles, nerves, and some glands. Its use is recommended for muscle cramps and osteoporosis.	Ahmed et al. (2023)
14.	Chromium	Broccoli, tomatoes, onions, garlic, cabbage, carrots, peppers, radishes, potatoes, parsnip, asparagus, peas, green leafy vegetables, string beans.	Human body needs chromium to break down food efficiently. It helps keep blood sugar levels normal. It also boosts immune system.	Gossa Al-Saadde et al. (2023)
15.	Copper	Artichokes, greens, beet greens, spinach, asparagus, winter squash, sweet potatoes, lima beans, turnip, green peas, pumpkin seeds, and green leafy vegetables.	It is essential for producing blood cells. It strengthens bones and plays a significant part in immune system function. It can help reduce wrinkles and increase wound healing. It boosts skin health by helping build collagen.	Burkhead and Collins (2022)
16.	Magnesium	Artichokes, asparagus, squash, Brussels sprouts, carrots, sweet potatoes, leafy greens (turnip greens, kale, spinach, collard greens, mustard greens), legumes (black beans, chickpeas, kidney beans), green bell peppers, and pumpkin seeds.	It is required for important functions of muscles and nerves and also maintains heart rhythm. It helps maintain blood glucose level. It helps in prevention of osteoporosis, diabetes, bronchial asthma, migraine, and cardiovascular diseases.	Ashique et al. (2023) and Al Alawi et al. (2018)

17.	Manganese	Sweet potatoes, leeks, eggplant, beets, cruciferous vegetables, dark green leafy vegetables, lima beans, chickpeas	It is an important trace inorganic obligatory by the body for calcium absorption, bone growth, blood sugar regulation, and healthy skin. It is needed for brain and memory function and intracellular activities.	Bowman et al. (2011) and Chen et al. (2018)
18.	Potassium	Cauliflower, broccoli, sprouts, cabbage, kale, Brussels, turnip greens, carrots, squash, sweet potato, tomato, white potato, beet greens, lima beans, eggplant, spinach, dark green leafy vegetables, and parsnips.	It helps to keep healthy blood pressure. It supports nerves, muscles, and heart for their better function.	Weaver (2013) and D'Elia (2024)
19.	Folate/folic acid	Asparagus, spinach, broccoli, romaine lettuces, turnip greens, dried or fresh beans, peas, beetroots, Brussels sprouts, tomato juice, mustard greens, artichokes, okra, parsley, parsnips, and red bell peppers.	Terms viz. folate and folic acid are used for vitamin B9. Folic acid is synthetic folate. It is required for DNA biosynthesis, cell division, and cell growth. It helps our body form red blood cells.	NIH (2020) and Tola (2024)
20.	Sulfur	Cruciferous vegetables (cauliflower, broccoli, cabbage, arugula, Brussels sprouts, kale, radish, watercress) Allium (garlic, leeks, onions, scallions, and shallots) Legumes (faba beans, kidney beans, peas)	It is used in biochemical processes. It assists in maintaining the integrity of connective tissues, such as tendons, ligaments, and skin. It provides strength to hair in the form of disulfide bonds. It plays a role in the inhibition of cancer cells.	Nimni et al. (2007), Ingenbleek and Kimura (2013), Doleman et al. (2017) and Hill et al. (2022)

Table 2.1 (continued)

Sr. #	Compound	Rich Source	Functions/Disease Prevention	Citation
21.	Zinc	Spinach, green peas, asparagus, beet greens, broccoli, okra, beans, pumpkin seeds, and garlic	Zinc is needed from daily diet as human body does not store it. It is an essential mineral which helps in healing wound, supporting immune system, and building proteins and DNA. Zinc is also required for cell growth, cell division, and enhancing action of insulin. It is needed by more than 300 enzymes and plays important role in cellular metabolic processes.	Roohani et al. (2013) and Mai et al. (2024)
22.	Iron	Spinach, curly kale, water cress, Swiss chard, collard, beet greens, turnip greens, tomato paste, potatoes, asparagus, leeks, acorn squash, and green beans	It is needed for normal functions of all the cells in human body. It helps in formation of red blood cells and hemoglobin. Deficiency of iron causes anemia and its chronic deficiency leads to organ failure.	Welk et al. (2023)
23.	Fiber	Artichokes, acorn squash, butternut squash, green peas, collard greens, kale, parsnips, broccoli, cabbage, cauliflower, carrots, spinach, legumes, potatoes, sweet potatoes, beet greens, Brussels sprouts, okra, Swiss chard, asparagus, turnips, eggplant, green bell peppers, celery, leeks, tomato, and lettuce.	It is a carbohydrate present in plant cell walls. There are two types of fiber viz. insoluble and soluble. Insoluble fiber does not liquify in water and prevents constipation. Soluble fiber absorbs water and may help lower cholesterol levels. It helps in feeding beneficial bacteria in our digestive tracts. Fiber-containing diets decrease the risk of cardiac disease and several cancerous conditions. A high-fiber diet reduces the chances of mortality in patients suffering from kidney problem and diabetes.	Viuda-Martos et al. (2010) and Khorasaniha et al. (2023)

References

Aaseth JO, Finnes TE, Askim M, Alexander J (2024) The importance of vitamin K and the combination of vitamins K and D for calcium metabolism and bone health: a review. Nutrients 16(15):2420. https://doi.org/10.3390/nu16152420

Ahmed A, Saleem MA, Saeed F, Afzaal M, Imran A, Akram S, Hussain M, Khan A, Jbawi EA (2023) A comprehensive review on the impact of calcium and vitamin D insufficiency and allied metabolic disorders in females. Food Sci Nutr 11(9):5004. https://doi.org/10.1002/fsn3.3519

AICR American Institute for Cancer Research (2015) Available from: http://www.aicr.org/. Accessed 3 Mar 2018

Al Alawi AM, Majoni SW, Falhammar H (2018) Magnesium and human health: perspectives and research directions. Int J Endocrinol 9041694:17. https://doi.org/10.1155/2018/9041694

Aldoori WH, Giovannucci EL, Rockett HR, Sampson L, Rimm EB, Willett AW (1998) A prospective study of dietary fiber types and symptomatic diverticular disease in men. J Nutr 128(4):714–719

Ali MA, Hafez HA, Kamel MA, Ghamry HI, Shukry M, Farag MA (2022) Dietary vitamin B complex: orchestration in human nutrition throughout life with sex differences. Nutrients 14(19):3940. https://doi.org/10.3390/nu14193940

Alman RD, Marcussen KC (2001) Effects of a ginger extract on knee pain in patients with osteoarthritis. Arthritis Rheum 44(11):2531–2538

American Cancer Society (2024) Key statistics for prostate cancer. https://www.cancer.org/cancer/types/prostate-cancer/about/key-statistics.html. Accessed 19 Oct 2024

Ansari MI, Khan MM, Saquib M, Khatoon S, Hussain MK (2018) Dithiolethiones: a privileged pharmacophore for anticancer therapy and chemoprevention. Future Med Chem 10(10):1241–1260. https://doi.org/10.4155/fmc-2017-0281

Arreola R, Quintero-Fabián S, López-Roa RI, Flores-Gutiérrez EO, Reyes-Grajeda JP, Carrera-Quintanar L, Ortuño-Sahagún D (2015) Immunomodulation and anti-inflammatory effects of garlic compounds. J Imm Res 2015:1–13. https://doi.org/10.1155/2015/401630

Arusha (2011) Fruits and vegetables importance for public health. Joint FAO/WHO Workshop on promotion of the production and consumption of fruits and vegetables. https://www.fao.org/fileadmin/templates/agphome/documents/horticulture/WHO/arusha/F_V_public_health_XUEREB.pdf. Accessed April 18 2025.

Ashique S, Kumar S, Hussain A, Mishra N, Garg A, Gowda BHJ, Farid A, Gupta G, Dua K, Taghizadeh-Hesary F (2023) A narrative review on the role of magnesium in immune regulation, inflammation, infectious diseases, and cancer. J Health Popul Nutr 42:74. https://doi.org/10.1186/s41043-023-00423-0

Barkas F, Bathrellou E, Nomikos T, Panagiotakos D, Liberopoulos E, Kontogianni MD (2023) Plant sterols and plant Stanols in cholesterol management and cardiovascular prevention. Nutrients 15(13):2845. https://doi.org/10.3390/nu15132845

Bertelli A, Biagi M, Corsini M, Baini G, Cappellucci G, Miraldi E (2021) Polyphenols: from theory to practice. Food Secur 10(11):2595. https://doi.org/10.3390/foods10112595

Bowman AB, Kwakye GF, Hernandez EH, Aschner M (2011) Role of manganese in neurodegenerative diseases. J Trace Elem Med Biol 25(4):191–203

Brown L, Rimm EB, Seddon JM, Giovannucci EL, Chasan-Taber L, Spiegelman D, Willett WC, Hankinson SE (1999) A prospective study of carotenoid intake and risk of cataract extraction in US men. Am J Clinic Nutr 70(4):517–524

Burkhead JL, Collins JF (2022) Nutrition information brief-copper. Adv Nutr 13(2):681–683. https://doi.org/10.1093/advances/nmab157

Chang HS, Yamato O, Yamasaki M, Maede Y (2005) Modulatory influence of sodium 2-propenyl thiosulfate from garlic on cyclooxygenase activity in canine platelets: possible mechanism for the anti-aggregatory effect. Prostaglandins Leukot Essent Fat Acids 72:351–355. https://doi.org/10.1016/j.plefa.2005.01.003

Chen P, Bornhorst J, Aschner M (2018) Manganese metabolism in humans. Front Biosci 23:1655–1679

Cho E, Seddon JM, Rosner B, Willett WC, Hankinson SE (2004) Prospective study of intake of fruits, vegetables, vitamins, and carotenoids and risk of age-related maculopathy. Arch Opthalmol 122(6):883–892

Christen WG, Liu S, Glynn RJ, Gaziano JM, Buring JE (2008) Dietary carotenoids, vitamins C and E, and risk of cataract in women: a prospective study. Arch Ophthalmol 126(1):102–109

Dashti HS, Scheer FA, Jacques PF, Lamon-Fava S, Ordovas JM (2015) Short sleep duration and dietary intake: epidemiologic evidence, mechanisms, and health implications. Adv Nutr 6:648–659

D'Elia L (2024) Potassium intake and human health. Nutrients 16(6):833. https://doi.org/10.3390/nu16060833

Dighriri IM, Alsubaie AM, Hakami FM, Hamithi DM, Alshekh MM, Khobrani FA, Dalak FE, Hakami AA, Alsueaadi EH, Alsaawi LS, Alshammari SF, Alqahtani AS, Alawi IA, Aljuaid AA, Tawhari MQ (2022) Effects of Omega-3 polyunsaturated fatty acids on brain functions: a systematic review. Cureus 14(10):e30091. https://doi.org/10.7759/cureus.30091

Doleman JF, Grisar K, Liedekerke LV, Saha S, Roe M, Tapp HS, Mithen RF (2017) The contribution of alliaceous and cruciferous vegetables to dietary Sulphur intake. Food Chem 234:38–45

Eidi A, Eidi M, Esmaeili E (2006) Antidiabetic effect of garlic (*Allium sativum* L.) in normal and streptozotocin-induced diabetic rats. Phytomedicine 13:624–629. https://doi.org/10.1016/j.phymed.2005.09.010

Fiedor J, Burda K (2014) Potential role of carotenoids as antioxidants in human health and disease. Nutrients 6(2):466–488

Fruhwirth GO, Hermeter A (2007) Seeds and oil of the Styrian oil pumpkin: components and biological activities. Eur J Lipid Sci Technol 109(11):1128–1140

Gharibzahedi SMT, Moghadam M, Amft J, Tolun A, Hasabnis G, Altintas Z (2023) Recent advances in dietary sources, health benefits, emerging encapsulation methods, food fortification, and new sensor-based monitoring of vitamin B12: a critical review. Molecules 28(22):7469

Gossa Al-Saadde DL, Haider AM, Ali A, Abdu Musad Saleh E, Turki Jalil A, Abdulelah FM, Romero-Parra RM, Tayyib NA, Ramírez-Coronel AA, Alkhayyat AS (2023) The role of chromium supplementation in cardiovascular risk factors: a comprehensive reviews of putative molecular mechanisms. Heliyon 9(9):e19826. https://doi.org/10.1016/j.heliyon.2023.e19826

Grandner MA, Jackson N, Gerstner JR, Knutson KL (2014) Sleep symptoms associated with intake of specific dietary nutrients. J Sleep Res 23(1):22–34

Heim KC, Spinalla MJ. In: Bagchi D, Preuss H, Swaropp A (2016) Editors. Nutraceuticals and functional foods in human health and disease prevention. 1st ed. CRP Press, Taylor Francis Group, Las Vegas. pp 361–390

Higdon JV, Delage B, Williams DE, Dashwood RH (2007) Cruciferous vegetables and human cancer risk: epidemiologic evidence and mechanistic basis. Pharmacol Res 55(3):224–236

Hill CR, Shafaei A, Balmer L, Lewis JR, Hodgson JM, Millar AH, Blekkenhorst LC (2022) Sulfur compounds: from plants to humans and their role in chronic disease prevention. Crit Rev Food Sci Nutr 63(27):8616–8638. https://doi.org/10.1080/10408398.2022.2057915

Huang T, Yang B, Zheng J, Li G, Wahlqvist ML, Li D (2012) Cardiovascular disease mortality and cancer incidence in vegetarians: a meta-analysis and systematic review. Ann Nutr Metab 60:233–240. https://doi.org/10.1159/000337301

Hung HC, Joshipura KJ, Jiang R, Hu FB, Hunter D, Smith-Warner SA, Colditz GA, Rosner B, Spiegelman D, Willett WC (2004) Fruit and vegetable intake and risk of major chronic disease. J Natl Canc Inst 96(21):1577–1584

Ibrahim M, Khan S, Pathak S, Mazhar M, Singh S (2023) Vitamin B-complex and its relationship with the health of vegetarian people. Nat Resour Human Health 3(3):342–354

Ingenbleek Y, Kimura H (2013) Nutritional essentiality of sulfur in health and disease. Nutr Rev 71(7):413–432

Insel P, Turner RE, Ross D (2007) Diet and health. In: Insel P, Turner RE, Ross D (eds) Nutrition, 3rd edn. Jones and Barlett Publishers, Judbury, pp 598–640

Joshipura KJ, Hu FB, Manson JE, Stampfer MJ, Rimm EB, Speizer FE, Willett WC (2001) The effect of fruit and vegetable intake on risk for coronary heart disease. Ann Int Med 134:1106–1114. https://doi.org/10.7326/0003-4819-134-12-200106190-00010

Khorasaniha R, Olof H, Voisin A, Armstrong K, Wine E, Vasanthan T, Armstrong H (2023) Diversity of fibers in common foods: key to advancing dietary research 139: https://doi.org/10.1016/j.foodhyd.2023.108495

Kim EE (2020) Vitamin C in human health and disease: effects, mechanisms of action, and new guidance on intake. J Nucl Med 61(4):623. https://doi.org/10.2967/jnumed.119.240952

Kris-Etherton PM, Hecker KD, Bonanome A, Coval SM, Binkoski AE, Hilpert KF, Etherton TD (2002) Bioactive compounds in foods: their role in the prevention of cardiovascular disease and cancer. Am J Med 113:71–88. https://doi.org/10.1016/S0002-9343(01)00995-0

Kumar R, Chhatwal S, Arora S, Sharma S, Singh J, Singh N, Bhandari V, Khurana A (2013) Antihyperglycemic, antihyperlipidemic, anti-inflammatory and adenosine deaminase-lowering effects of garlic in patients with type 2 diabetes mellitus with obesity. Diabetes Metab Syndr Obes 6:49–56. https://doi.org/10.2147/DMSO.S38888

Lembo A, Camilleri M (2003) Chronic constipation. New England J Med 349(14):1360–1368

Lidder S, Webb AJ (2013) Vascular effects of dietary nitrate (as found in green leafy vegetables and beetroot) via the nitrate-nitrite-nitric oxide pathway. Br J Clinic Pharmacol 75:677–696. https://doi.org/10.1111/j.1365-2125.2012.04420.x

Liga S, Paul C, Péter F (2023) Flavonoids: overview of biosynthesis, biological activity, and current extraction techniques. Plan Theory 12(14):2732. https://doi.org/10.3390/plants12142732

Madkor HR, Mansour SW, Ramadan G (2011) Modulatory effects of garlic, ginger, turmeric and their mixture on hyperglycemia, dyslipidemia and oxidative stress in streptozotocin-nicotinamide diabetic rats. Br J Nutr 105:1210–1217. https://doi.org/10.1017/S0007114510004927

Mai W, Wang F, He S, Wen Y, You G, Zhang L, Dong H (2024) Zinc contents in foods and estimates of dietary intakes in Guangzhou, Guangdong Province, China. Front Nutr 11. https://doi.org/10.3389/fnut.2024.1364033

McFarlin BK, Venable AS, Henning AL, Sampson JN, Pennel K, Vingren JL, Hill DW (2016) Reduced inflammatory and muscle damage biomarkers following oral supplementation with bioavailable curcumin. BBA Clin 18(5):72–78. https://doi.org/10.1016/j.bbacli.2016.02.003

Menezes MSS, Almeida CMM (2024) Structural, functional, nutritional and clinical aspects of vitamin A: a review. PharmaNutrition 27. https://doi.org/10.1016/j.phanu.2024.100383

Midde NM, Miller DD, Chauhan S, Kumar A, Kumar S (2015) Diallyl sulfide: potential use in novel therapeutic interventions in alcohol, drugs, and disease mediated cellular toxicity by targeting cytochrome P450 2E1. Curr Drug Metab 16(6):486–503

Muscogiuri G, Barrea L, Aprano S, Framondi L, Di Matteo R, Laudisio D, Pugliese G, Savastano S, Colao A (2020) Sleep quality in obesity: does adherence to the mediterranean diet matter? Nutrients 12(5):1364.1-11. https://doi.org/10.3390/nu12051364

NIH (2020) Folate. National Institutes of Health, US Department of Health and Human Services. https://ods.od.nih.gov/factsheets/Folate-HealthProfessional/. Accessed 13 Oct 2020

Nimni ME, Han B, Cordoba F (2007) Are we getting enough sulfur in our diet? Nutr Meta 4:24. https://doi.org/10.1186/1743-7075-4-24

Nisar M, Mohammad RM, Arshad A, Hashmi I, Yousuf SM, Bai S (2019) Influence of dietary intake on sleeping patterns of medical students. Cureus 11(2):e4106. https://doi.org/10.7759/cureus.4106

Noorwali E, Hardie L, Cade J (2019) Bridging the reciprocal gap between sleep and fruit and vegetable consumption: a review of the evidence, potential mechanisms, implications, and directions for future work. Nutrients 11(1382):1–45. https://doi.org/10.3390/nu11061382

Olayanju JB, Bozic D, Naidoo U, Sadik OA (2024) A comparative review of key Isothiocyanates and their health benefits. Nutrients 16(6):757. https://doi.org/10.3390/nu16060757

Perera CO, Yen GM (2007) Functional properties of carotenoids in human health. Int J Food Prop 10(2):201–230. https://doi.org/10.1080/10942910601045271

Pollock RL (2016) Comparing consumption of green leafy vegetables to cruciferous vegetables in relations to incidence of 17 different cancers: a meta-analysis. Glob J Med Res 16:30–39

Qian JC, Zhang HP, Wang Y, Liu D (2024) Heating conversion of indole-3-carbinol into N-substituted oligomers with anti-melanoma effect. Food Chem 22:101410. https://doi.org/10.1016/j.fochx.2024.101410

Ried K, Frank OR, Stocks NP (2013) Aged garlic extract reduces blood pressure in hypertensives: a dose–response trial. Eur J Clinic Nutr 67:64–70. https://doi.org/10.1038/ejcn.2012.178

Roohani N, Hurrell R, Kelishadi R, Schulin R (2013) Zinc and its importance for human health: an integrative review. J Res Med Sci 18:144–157
Roth GA, Forouzanfar MH, Moran AE, Barber R, Nguyen G, Feigin VL, Murray CJ (2015) Demographic and epidemiologic drivers of global cardiovascular mortality. New Engl J Med 372:1333–1341. https://doi.org/10.1056/NEJMoa1406656
Saini RK, Prasad P, Lokesh V, Shang X, Shin J, Keum YS, Lee JH (2022) Carotenoids: dietary sources, extraction, encapsulation, bioavailability, and health benefits – a review. Recent Advanc Antioxidants (Basel) 11(4):795. https://doi.org/10.3390/antiox11040795
Sajovic J, Meglič A, Glavač D, Markelj Š, Hawlina M, Fakin A (2022) The role of vitamin a in retinal diseases. Int J Mol Sci 23(3):1014. https://doi.org/10.3390/ijms23031014. PMID: 35162940
See XZ, Yeo WS, Saptoro A (2024) A comprehensive review and recent advances of vitamin C: overview, functions, sources, applications, market survey and processes. Chem Eng Res Des 206:108–129. https://doi.org/10.1016/j.cherd.2024.04.048
Sen R, Hurley JA (2023) Osteoarthritis. National Library of Medicine. https://www.ncbi.nlm.nih.gov/books/NBK482326/ Accessed 24 Oct 2024
Soh Y, Shin M, Lee J, Jang J, Kim OH, Kang H, Surh Y (2003) Oxidative DNA damage and glioma cell death induced by tetrahydropapaveroline. Mutat Res 544:129–142. https://doi.org/10.1016/j.mrrev.2003.06.023
Tao J, Li Y, Li S, Li HB (2018) Plant foods for the prevention and management of colon cancer. J Fun Food 42:95–110. https://doi.org/10.1016/j.jff.2017.12.064
Tola FS (2024) The concept of folic acid supplementation and its role in prevention of neural tube defect among pregnant women: PRISMA. Medicine (Baltimore) 103(19):e38154. https://doi.org/10.1097/MD.0000000000038154
Ülger TG, Bozdağ ANS, Çırak O, Cakiroglu FP (2018) Role of vegetables in human nutrition and disease prevention. In: Asaduzzaman, M, Asao, T (ed) Vegetables - importance of quality vegetables to human health https://doi.org/10.5772/intechopen.77038
Viuda-Martos M, López-Marcos MC, Fernández-López J, Sendra E, López-Vargas JH (2010) Role of fiber in cardiovascular diseases: a review. Comp Rev Food Sci Food Saf 9(2):240–258
Wang L, Fan H, He J, Wang L, Tian Z, Wang C (2018) Protective effects of omega-3 fatty acids against Alzheimer's disease in rat brain endothelial cells. Brain Behav 8:0. https://doi.org/10.1002/brb3.1037
Wang X, Ouyang Y, Liu J, Zhu M, Zhao G, Bao W, Hu FB (2014) Fruit and vegetable consumption and mortality from all causes, cardiovascular disease, and cancer: systematic review and dose-response meta-analysis of prospective cohort studies. BMJ. 349:g4490. https://doi.org/10.1136/bmj.g4490.
Weaver CM (2013) Potassium and health. Adv Nutr 4(3):368S–377S. https://doi.org/10.3945/an.112.003533
Welk AK, Mehlhose C, Daum D, Enneking U (2023) Vegetables with enhanced iron bioavailability-German consumers' perceptions of a new approach to improve dietary iron supply. Nutrients 15(10):2291. https://doi.org/10.3390/nu15102291
WHO (2022) Cancer. https://www.who.int/news-room/fact-sheets/detail/cancer. Accessed 29 Jan 2025
WHO (2023) Hypertension. https://www.who.int/news-room/fact-sheets/detail/hypertension. Accessed 16 Dec 2024
Wiseman M (2008) The Second World Cancer Research Fund/American Institute for Cancer Research Expert Report. Food, nutrition, physical activity, and the prevention of cancer: a global perspective: nutrition society and BAPEN medical symposium on 'nutrition support in cancer therapy'. Proc Nutr Soc 67(3):253–256
Yeh YY, Liu L (2001) Cholesterol-lowering effect of garlic extracts and organosulfur compounds: human and animal studies. J Nutr 131:989–993. https://doi.org/10.1093/jn/131.3.989S
Yokoyama Y, Nishimura K, Barnard ND, Takegami M, Watanabe M, Sekikawa A, Okamura T, Miyamoto Y (2014) Vegetarian diets and blood pressure: a meta-analysis. JAMA Int Med 174(4):577–587
Zhang Y, Munday R (2008) Dithiolethiones for cancer chemoprevention: where do we stand? Mol Canc Therap 7(11):3470–3479

Specificity of Vegetables' Natural Coloration

3

Arslan Akbar, Shaheer Zafar, Zafar Iqbal,
Muhammad Waqar Alam, and Ozair Chaudhry

Abstract

Colorful vegetables have much attraction worldwide as these vegetables contain different beneficial phytochemicals in their edible parts with positive health benefits and decorative value as well. Diverse vegetables of different colors are produced globally to consume in breakfast, lunch, brunch, and dinner. Color is related to antioxidant capacity in some fruits and vegetables. Vegetables with bright colors are pleasing to the eye and appear beautiful and nutritious in the market and on the table. This chapter describes in detail the categories of plant pigments, diversity of nutrients, specificity of natural colors, and beauty of variegated, multihued, and distinctive veggies.

Keywords

Plant pigments · Phytonutrients · Aromatic veggies · Heirloom · Color specificity

A. Akbar
Department of Horticulture, Bahauddin Zakariya University, Multan, Pakistan

S. Zafar
Hamdard College of Medicine and Dentistry, Hamdard University, Karachi, Pakistan

Z. Iqbal (✉)
Faculty of Agriculture, University of Sargodha, Sargodha, Pakistan
e-mail: zafarrohils@yahoo.com

M. W. Alam
Department of Botany, University of Okara, Okara, Punjab, Pakistan

O. Chaudhry
Durham Environment and Climate Advisory Committee (DECAC), Planning and Economic Development Department, Whitby, Ontario, Canada

Z. Iqbal et al. (eds.), *Sustainable and Innovative Vegetable Production in times of Climate Change*, https://doi.org/10.1007/978-981-96-6283-8_3

Plant pigments present in colored vegetables are naturally color markers. The term pigment is used to indicate colored molecules which impart unique color to vegetables, fruits, flowers, and leaves. Plant pigments are responsible for the shining color of plant tissue (Schoefs 2004). Pigmentation in plant parts also prevents disease and a pest attack (Singh 1996; Polturak et al. 2017) along with enhanced tolerance to abiotic stresses.

3.1 Categories of Plant Pigments

Plant pigments which color the foods are classified into four major groups. They include betalains (yellow, red), anthocyanins in flavonoids (blue, red, purple), carotenoids (orange, red, yellow), and chlorophylls (green). In each category of pigments, there are some compounds with special name, chemical structure, properties, and distinctive color. They have the ability to color due to their chemical structures. The atoms are specifically organized to absorb specific wavelengths.

3.1.1 Carotenoids

There are more than 600 types of carotenoids (Rutz et al. 2016). These plant pigments display yellow, orange, and bright red tint in different vegetables and fruits. Foods rich in carotenoids provide protective health benefits. Dietary carotenoids function as antioxidants and also reduce risk of many diseases especially cancers and eye diseases. Fruits and vegetables contribute most important source of carotenoids to human food (Scott et al. 1996; Krinsky and Johnson 2005; Su 2022). These are known to be disseminated in the cellular tissues and also found as colored pigments in plants (Simpson 1985; Minguez-Mosquera et al. 1990; Furr and Clark 1997). Isolation of carotenoids can be done from the granum (membranous structure made of stacks of thylakoids) of chloroplasts in the form of carotenoprotein complexes, which give varying colors to the external plant surfaces (Takyi 2001). The visibility of colors in the plants is due to the conjugated double bonds of carotenoids that absorb light. Presence of more double bonds ensures more absorbance of red color wavelength. When a fruit attains maturity, the chlorophyll content is decreased resulting in colored-carotenoid pigments (Marín et al. 2004). Improvement of carotenoids color retention during ripening has also been studied by various researchers (Markus et al. 1999; Khoo et al. 2011). Carotenoids are subdivided into two main groups viz. xanthophylls and carotenes both of which possess antioxidant properties. Some carotenoids can be converted into vitamin A.

3.1.1.1 Xanthophylls

Xanthophylls comprise oxygen and occasionally a yellow color. Xanthophyll carotenoids provide protection against too much sunshine. They are well known to support eye health (Jurja et al. 2023). Xanthophylls include lutein and zeaxanthin.

3.1.1.2 Carotenes

Carotenes having no oxygen are associated with orange pigment. Carotenes support growth of the plants. Carotenes include beta carotene and lycopene (Crupi et al. 2023; Gebregziabher et al. 2023).

3.1.2 Flavonoids

They possess antioxidant, anti-inflammatory, antimicrobial, anticancer, and cardio- and neuroprotective properties (Liga et al. 2023). This is a group of pigments producing color in various fruits and vegetables.

3.1.2.1 Classification of Flavonoids

Flavonoids are divided into 12 major subclasses on the basis of chemical structures. However, six among these viz. flavonols, flavones, flavanones, isoflavones, flavan-3-ols, and anthocyanidins occupy dietary significance.

3.1.2.1.1 Flavonols

This group possesses antioxidant properties and provides protection against vascular disease. Vegetables rich in flavonols include onion, kale, tomatoes, lettuce, parsley, oregano, spinach, etc.

3.1.2.1.2 Flavones

Flavones are much beneficial as they possess antioxidative, anti-inflammatory, and anticancer properties. Vegetables which contain flavones include celery, mint, red peppers, parsley, oregano, fenugreek seed, and sage (Hostetler et al. 2017). Most of the flavones are present in all parts of plants. White flowers contain flavones as primary pigments (Jiang et al. 2016; Liga et al. 2023).

3.1.2.1.3 Flavanones

They possess antioxidant, anti-inflammatory, and antimicrobial properties. Flavanone-rich foods include peppermint, tomato, and chickpea. Isolation of flavanones may be done from vegetative parts including stem, leaves, flowers, rhizomes, and fruits and generative ones, i.e., bark, branches, and roots (Mukherjee 2019; Liga et al. 2023).

3.1.2.1.4 Isoflavones

Isoflavones possess anti-cancer, antioxidant, antimicrobial, and anti-inflammatory properties. Isoflavones prevent bone loss. The dietary sources of isoflavones include chickpeas and beans (Křížová et al. 2019).

3.1.2.1.5 Flavan-3-Ols

They include many monomeric, oligomeric, and polymeric compounds (Godos et al. 2024). They are also called as flavanols. They act as antioxidant, anticarcinogen, antimicrobial, antiviral, and neuroprotective agents. Legumes contain flavan-3-ol compounds viz. catechins and epicatechins.

3.1.3 Anthocyanins

Anthocyanins have been added to the human nourishment for times and used as traditional herbal medications due to their various properties to treat symptoms such as pyrexia, dysentery, hypertension, liver disorders, urinary problems, and the common cold (Konczak and Zhang 2004; Shipp and Abdel-Aal 2010; Tang et al. 2023). Naturally, many purple-colored foods contain anthocyanins. These are widespread in food plants, occurring in 27 families (Ghosh 2005). These pigments show many health benefits and also give deep red, purple, or blue shade to fruit and veggies. Anthocyanin is observed as red pigment in acidic condition but becomes blue with increasing pH (Khoo et al. 2017).

3.1.3.1 Types of Anthocyanins in Plants

Pelargonidin, cyanidin, peonidin, malvidin, delphinidin, and petunidin are the most communal anthocyanidins found in plants. The distribution of these anthocyanidins in vegetables and fruits is 50, 12, 12, 12, 7, and 7%, respectively (Castañeda-Ovando et al. 2009; Khoo et al. 2017). Cyanidin is the main color in berries (Seeram et al. 2001) and some red-colored vegetables such as purple corn and red sweet potato (Cevallos-Casals and Cisneros-Zevallos 2003). Delphinidin manifests as a blue-reddish or purple pigment in the plant (Katsumoto et al. 2007) while pelargonidin gives an orange tint to flowers and red to some of the berries and fruits (Bąkowska-Barczak 2005; Jaakola 2013; Khoo et al. 2017).

3.1.4 Chlorophyll

Due to the presence of chlorophyll, plants look green. It is found in nearly every green plant part including leaves, stem, and chloroplast. Chloroplast possesses thylakoid membranes which comprise green chlorophyll color. It is placed under the distinct class of composites named tetrapyrroles because it comprises four pyrrole rings combined together with a covalent bond, as are heme molecule and vitamin-B12 (Willows 2004; Pareek et al. 2018). It fascinates light in the blue and red areas of the observable light spectrum but green light is not absorbed rather reproduced making the green appearance of the plant. Types found in plants are chlorophyll "a" and "b" (Gaur et al. 2006; Belitz et al. 2009). Chemical formula of chlorophyll is surprisingly similar to that of hemoglobin present in our blood. Difference is due to one substance, i.e., iron is replaced by magnesium in the chlorophyll (Inanc 2011). Besides making plants healthy and green, chlorophyll also has antioxidants, vitamins, and therapeutic properties which ultimately benefit our body (Larsen and Christensen 2005; Pareek et al. 2018). It is also used as a normal color ingredient in treated foods or food additive in food industry. In soaps and cosmetics products, chlorophyll "a" is used.

3.1.5 Betalains

Betalains are natural pigments having distinct antioxidant property. They are extracted from the tubers, leaves, flowers, roots, and fruits of various plants (Zannou et al. 2023). Due to their extraction from common beet (*Beta vulgaris*), they have been named as Betalains (Akbar Hussain et al. 2018). The water-soluble nitrogenous betalains are classified into yellow-orange betaxanthins, and red-violet betacyanins (Gandia-Herrero et al. 2016; Rodriguez-Amaya 2019). Betanin, probetanin, isobetanin, and neobetanin are the betacyanins found in plants. Miraxanthin, vulgaxanthin, indicaxanthin, and portulaxanthin are betaxanthins available in plants. Betalains are present in edible plant parts, bracts, stems, leaves, and flowers (Heuer et al. 1994; Schliemann et al. 1996; Kugler et al. 2004; Wang et al. 2007; Felker et al. 2008; Gandıa-Herrero et al. 2009; Osorio-Esquivel et al. 2011).

3.2 Choice of Multi-Hued and Nutrient-Dense Veggies

The natural colors of diverse vegetables unveil the presence of a variety of phytochemicals and phytonutrients. Globally, consumers use color of vegetables for identification of a specific food item and evaluation of quality, safety, and ripeness of the product (Schiffersteina et al. 2019). Slight color differences may have a significant impact on food choice (Bruhn 1995; Crisosto et al. 2003), because these color variations disclose meaningful information. A healthy eating style requires dense concentration of beneficial nutrients (Drewnowski and Fulgoni 2014). Those foods which have diversity of phytonutrients with greater color density are considered to be the finest choice for inclusion in the diet (Minich 2019).

Food neophobia and food pickiness (Pliner and Hobden 1992; Taylor et al. 2015) cause hurdle to children's sophisticated eating of vegetables and fruits (Birch and Fischer 1998; Galloway et al. 2003; Dovey et al. 2008; Lafraire et al. 2016). As this habit continues with growing age (Nicklaus et al. 2005), so it's wise to devise effective interventions to overwhelmed children's food concern and neophobia (Rioux et al. 2018) for better health and mental sharpness. Growing colorful and unique vegetable varieties infuses a sense of love, charm, fun, and adventure in both young and old age groups.

3.3 Specificity of Vegetables' Natural Colors

3.3.1 Red Vegetables

Red color seems to be much attractive and perceptible. Red vegetables pick their beautiful tint and nutrition enrichment from lycopene and anthocyanin. These vegetables contain phytochemicals like ellagitannins, ellagic acid, carotenoids, fisetin, flavones, lycopene, phloretin anthocyanins, and quercetin (Minich 2019). Natural coloring agents are especially attractive and beneficial for the food industry as an alternative to synthetic coloring compounds which have negative effects on human's health (Leong et al. 2018). Red vegetables include radishes, red jalapeño pepper, red cabbage, radicchio, red bell peppers, red beets, red chard, tomatoes, red potatoes, and red onion.

3.3.2 Orange Vegetables

The lively orange color represents strength and endurance. The attractive brightness of this color activates energizing mood. They are quite beneficial as they tackle harmful free radicals. Flavorful and warmly-colored orange vegetables include sweet potato, winter squash, carrots, orange bell peppers, turmeric, yams, tomatoes, and pumpkin.

3.3.3 Yellow Vegetables

Yellow fruits and vegetables are treasures of essential vitamins and minerals that protect against negative effects on skin cells. Phytochemicals available in yellow vegetables include bioflavonoids, bromelain, lutein, gingerol, nobiletin, prebiotic fibers, rutin, and zeaxanthin (Minich 2019; Ashfaq et al. 2021). Yellow vegetables include corn, ginger, potatoes (Yukon), squash (acorn, buttercup, butternut, summer, winter), yellow bell peppers, and yellow onions.

3.3.4 Green Vegetables

Green is a naturally scattered color in diversified vegetation on this planet having soothing effect on eyes and mind. The compounds which impart green color to vegetables are glucosinolates which are a large group of sulfur-containing amino acid derivatives (Garden-Robinson 2009; Minich 2019). Some glucosinolates and their derivatives are helpful to reduce certain types of cancer. Green vegetables include bean sprouts, bamboo sprouts, cucumbers, edamame bell peppers, bitter melon, bok choy, broccoli, cabbage, celery, artichokes, asparagus, green beans, green peas, greens (dandelion, kale, lettuce, mustard, beet, chard, collards, spinach, turnip), okra, rosemary, snow peas, and watercress.

3.3.5 Blue Purple Vegetables

Purple fruits and vegetables are highly liked by people due to their bright, attractive, and vibrant color and health benefits. Blue purple vegetables include purple kale, purple cauliflower, eggplant, purple cabbage, purple bell peppers, purple carrots, purple asparagus, purple onions, purple yam, purple radicchio, and purple potatoes. Purple peppers look much pretty because they initially show purple color, but when they ripen, they turn to light purple, light yellow, darker yellow, and then red. Radicchio is distinctive due to maroon leaves and unique slightly spicy flavor.

3.3.6 White and Neutral-Colored Veggies

Some vegetables have no color on their inside. It does not show deficiency of nutrients or phytochemicals. Some flavonoids are colorless and they have properties to combat the free radical formation. The white fruits and vegetables contain a water-soluble pigment named as anthoxanthin. Phytochemicals like allyl sulfides, allicin, magnesium, and potassium are also present in white vegetables. White vegetables include garlic, leek, cauliflower, chicory, cucumber, russet potatoes, plantain, turnips, white onions, Idaho potatoes, parsnips, jicama, white asparagus, and yuca (Oude Griep et al. 2011).

3.3.7 Distinctive Veggies

3.3.7.1 Aromatic Vegetables

These vegetables bring flavor to diverse dishes like soups, stews, sauces, and favorite cuisines and contribute to good health through beneficial nutrients. Quality of food splits into sensory and hidden quality (Kramer 1973). In vegetables and fruits, the characteristics imparting typical superiority may be defined by four diverse qualities including appearance and color, flavor (aroma and taste), nutritional value, and texture (Barrett et al. 2010). Colors of vegetables attract on first look. Aroma is considerably more intricate and has numerous proportions (Gormley 1981). Aroma emanating from aromatic vegetables attracts and serves as appetizer.

Based on flavor characteristics, vegetables may be divided into two groups: (1) Vegetables with strong flavor which may be ascribed to a single compound or group of associated compounds, e.g., sulfide compounds in onions and phthalides in celery. (2) A number of volatiles determine the flavor, e.g., snap beans and tomatoes (Stevens 1985; Barrett et al. 2010). *Carrots, onions, parsley, bay leaves, celery (aka mirepoix), thyme, garlic, ginger, fennel, leeks, shallots, bell peppers, and scallions are well-known aromatic vegetables*. These vegetables turn simple dishes into flavorful meals.

3.3.7.2 Baby or Miniature Vegetables

The term mini-vegetables first appeared in the 1980s. They are called with different names like teenage, personal-sized, mini-sized, baby, or miniature veggies. Miniature vegetables are consumed by people at homes and restaurants as salads, dessert bowls, and garnishes. Due to their short root system, miniature vegetables are best suited for container growing, small spaced yards, and small gardens. Some baby vegetables are bred with specific and desired traits and unique phenotypic features. These smaller versions of veggies add color and beauty to home gardens and dwarf landscapes.

Miniature varieties of different vegetables are available in global markets. These include beans, beets, cabbage, cauliflower, cantaloupe, bell peppers, carrots, cucumber, eggplant, leeks, Tom Thumb lettuce, onions, kale, basil, peas, Chinese broccoli, pumpkin, turnips, endive, arugula, artichokes, celery, squash-winter, scallopini (hybrid of scallop and zucchini), Brussels sprouts, and cherry tomato.

3.3.7.3 Quirky, Heirloom, and Weird-Colored Veggies

Unusual, quirky, and weird-colored veggies are much liked by zealous and ambitious gardeners. These may be identified with common or distinct colors and varying shapes, flavors, and textures. Unlike cross-bred vegetable crops, some heirlooms give production during the whole season. In an effort to preserve biodiversity, some species of heirlooms are saved and cultivated by individuals, groups, or families from one generation to another in different countries of the world. Heirloom varieties are mostly centuries old and open-pollinated by bees, insects, birds, or breeze without human interference. Out of various naturally occurring species, only few have been explored yet.

Many organic consumers wish to purchase quirky and heirloom veggies from some super and specialty stores along with other groceries. The weird colored or heirloom veggies include purple sweet potato, kohlrabi, purple carrot, celeriac, amaranth, snake gourd, yard long beans, serpent cucumber, purple peas, sunchoke, tomatillo, sea beans, black turmeric, black garlic, black radish, blue pumpkin, white asparagus, black tomato, golden beets, purple asparagus, green eggplant, rainbow chard, purple cauliflower, and orange cauliflower (Veteto 2008; Mimura 2024).

References

Akbar Hussain E, Sadiq Z, Zia-Ul-Haq M (2018) Introduction. In: Betalains: biomolecular aspects. Springer, Cham, pp 1–14. https://doi.org/10.1007/978-3-319-95624-4_1

Ashfaq M, Tariq M, Parveen I, Khalid S, Faridi TA, Nazir S (2021) Bio-active potential of naturally colored foods and their nutraceutical significance. Nat J Bio Sci 2(I):13–29

Bąkowska-Barczak A (2005) Acylated anthocyanins as stable, natural food colorants – a review. Pol J Food Nutr Sci 55(2):107–116

Barrett DM, Beaulieu JC, Shewfelt R (2010) Color, flavor, texture, and nutritional quality of fresh-cut fruits and vegetables: desirable levels, instrumental and sensory measurement, and the effects of processing. Crit Rev Food Sci Nutr 50:369–389

Belitz HD, Grosch W, Schieberle P (2009) Food chemistry, 4th edn. Springer-Verlag, Berlin/Heidelberg, p 1070

Birch LL, Fisher JO (1998) Development of eating behaviors among children and adolescents. Pediatrics 101(Supplement 2):539–549

Bruhn CM (1995) Consumer and retailer satisfaction with the quality and size of California peaches and nectarines. J Food Qual 18(3):241–256

Castañeda-Ovando A, Pacheco-Hernández ML, Páez-Hernández E, Rodríguez JA, Galán-Vidal A (2009) Chemical studies of anthocyanins: a review. Food Chem 113(4):859–871

Cevallos-Casals BA, Cisneros-Zevallos L (2003) Stoichiometric and kinetic studies of phenolic antioxidants from Andean purple corn and red-fleshed sweet potato. J Agric Food Chem 51(11):3313–3319

Crisosto CH, Crisosto GM, Metheney P (2003) Consumer acceptance of 'brooks' and 'Bing' cherries is mainly dependent on fruit SSC and visual skin color. Postharvest Biol Technol 28(1):159–167

Crupi P, Faienza MF, Naeem MY, Corbo F, Clodoveo ML, Muraglia M (2023) Overview of the potential beneficial effects of carotenoids on consumer health and well-being. Antioxidants (Basel) 12(5):1069. https://doi.org/10.3390/antiox12051069

Dovey TM, Staples PA, Gibson EL, Halford JCG (2008) Food neophobia and "picky/fussy" eating in children: a review. Appetite 50(2–3):181–193

Drewnowski A, Fulgoni VL III (2014) Nutrient density: principles and evaluation tools. Am J Clin Nutr 99(5):1223S–1228S

Felker P, Stintzing FC, Mussig E, Leitenberger M, Carle R, Vogt T, Bunch R (2008) Color inheritance in cactus pear (Opuntia *Ficus indica*) fruits. Ann Appl Biol 152:307–318

Furr HC, Clark RM (1997) Intestinal absorption and tissue distribution of carotenoids. J Nutr Biochem 8:364–377

Galloway AT, Lee Y, Birch LL (2003) Predictors and consequences of food neophobia and pickiness in young girls. J Am Diet Assoc 103(6):692–698

Gandía-Herrero F, Jimenez-Atienzar M, Cabanes J, Escribano J, García-Carmona F (2009) Fluorescence detection of tyrosinase activity on dopamine-betaxanthin purified from *Portulaca oleracea* (common purslane) flowers. J Agric Food Chem 57:2523–2528

Gandía-Herrero F, Escribano J, García-Carmona F (2016) Biological activities of plant pigments betalains. Crit Rev Food Sci Nutr 56:937–945

Garden-Robinson J (2009) What color is your food? Taste a rainbow of fruits and vegetables for better health. NDSU Extension Service, North Dakota State University, Fargo, ND

Gaur S, Shivhare U, Ahmed J (2006) Degradation of chlorophyll during processing of green vegetables: a review. Stewart Postharvest Rev 5(14):1–8

Gebregziabher BS, Gebremeskel H, Debesa B, Ayalneh D, Mitiku T, Wendwessen T, Habtemariam E, Nur S, Getachew T (2023) Carotenoids: dietary sources, health functions, biofortification, marketing trend and affecting factors–a review. J Agri Food Res 14. https://doi.org/10.1016/j.jafr.2023.100834

Ghosh D (2005) Anthocyanins and anthocyanin-rich extracts in biology and medicine: biochemical, cellular, and medicinal properties. Curr Top Nutraceutical Res 3(2):113–124

Godos J, Romano GL, Laudani S, Gozzo L, Guerrera I, Dominguez Azpíroz I, Martínez Diaz R, Quiles JL, Battino M, Drago F et al (2024) Flavan-3-ols and vascular health: clinical evidence and mechanisms of action. Nutrients 16(15):2471. https://doi.org/10.3390/nu16152471

Gormley TR (1981) Aroma in fruit and vegetables. In: Goodenough PW, Atkin RK (eds) Quality in stored and processed vegetables and fruit. Academic Press, New York, pp 35–51

Heuer S, Richter S, Metzger JW, Wray V, Nimtzt M, Strack D (1994) Betacyanins from bracts of *Bougainvillea glabra*. Phytochemistry 37:761–767

Hostetler GL, Ralston RA, Schwartz SJ (2017) Flavones: food sources, bioavailability, metabolism, and bioactivity. Adv Nutr 8(3):423–435. https://doi.org/10.3945/an.116.012948

İnanç AL (2011) Chlorophyll: structural properties, health benefits and its occurrence in virgin olive oils. Akadem Gida 9(2):26–32

Jaakola L (2013) New insights into the regulation of anthocyanin biosynthesis in fruits. Trends Plant Sci 18(9):477–483

Jiang N, Doseff AI, Grotewold E (2016) Flavones: from biosynthesis to health benefits. Plan Theory 5:27

Jurja S, Negreanu-Pirjol T, Vasile M, Hincu MM, Coviltir V, Negreanu-Pirjol B (2023) Xanthophyll pigments dietary supplements administration and retinal health in the context of increasing life expectancy trend. Front Nutr 10. https://doi.org/10.3389/fnut.2023.1226686

Katsumoto Y, Fukuchi-Mizutani M, Fukui Y, Brugliera F, Holton TA, Karan M, Nakamura N, Yonekura-Sakakibara K, Togami J, Pigeaire A, Tao G, Nehra NS, Lu C, Dyson BK, Tsuda S, Ashikari T, Kusumi T, Mason JG, Tanaka Y (2007) Engineering of the rose flavonoid biosynthetic pathway successfully generated blue-hued flowers accumulating delphinidin. Plant Cell Physiol 48(11):1589–1600

Khoo H, Prasad KN, Kong K, Jiang Y, Ismail A (2011) Carotenoids and their isomers: color pigments in fruits and vegetables. Molecules 16(2):1710–1738

Khoo HE, Azlan A, Tang ST, Lim SM (2017) Anthocyanidins and anthocyanins: colored pigments as food, pharmaceutical ingredients, and the potential health benefits. Food Nutr Res 61(1):1–21 101361779

Konczak I, Zhang W (2004) Anthocyanins–more than nature's colours. J Biomed Biotechnol 5:239–240

Kramer A (1973) Food and the consumer. AVI Publishing Co, Westport

Krinsky NI, Johnson EJ (2005) Carotenoid actions and their relation to health and disease. Mol Asp Med 26:459–516

Křížová L, Dadáková K, Kašparovská J, Kašparovský T (2019) Isoflavones. Molecules 24(6):1076. https://doi.org/10.3390/molecules24061076

Kugler F, Stintzing FC, Carle R (2004) Identification of betalains from petioles of differently colored Swiss chard (*Beta vulgaris* L. sp. *cicla* [L.] Alef. cv. Bright lights) by high-performance liquid chromatography–electrospray ionization mass spectrometry. J Agric Food Chem 52:2975–2981

Lafraire J, Rioux C, Roque J, Giboreau A, Picard D (2016) Rapid categorization of food and non-food items by 3- to 4-year-old children. Food Qual Prefer 49:87–91

Larsen E, Christensen LP (2005) Simple saponification method for the quantitative determination of carotenoids in green vegetables. J Agric Food Chem 53(17):6598–6602

Leong HY, Show PL, Lim MH, Ooi CW, Ling TC (2018) Natural red pigments from plants and their health benefits: a review. Food Rev Int 34(5):463–482

Liga S, Paul C, Péter F (2023) Flavonoids: overview of biosynthesis, biological activity, and current extraction techniques. Plan Theory 12(14):2732. https://doi.org/10.3390/plants12142732

Marín A, Ferreres F, Tomas-Barberan FA, Gil MI (2004) Characterization and quantitation of antioxidant constituents of sweet pepper (*Capsicum annuum* L.). J Agric Food Chem 52:3861–3869

Markus F, Daood HG, Kapitany J, Biacs PA (1999) Change in the carotenoid and antioxidant content of spice red pepper (paprika) as a function of ripening and some technological factors. J Agric Food Chem 47:100–107

Mimura Y (2024) Introduction to heirloom vegetables in Kyoto prefecture (*Kyo-yasai*). Japan J Ethn Food 11:26. https://doi.org/10.1186/s42779-024-00240-8

Minguez-Mosquera MI, Gandul-Rojas B, Garrido-Fernandez J, Gallardo-Guerrero L (1990) Pigments present in virgin olive oil. J Am Oil Chem Soc 67:192–196

Minich DM (2019) A review of the science of colorful, plant-based food and practical strategies for "eating the rainbow.". J Nutr Metab 2019:2125070

Mukherjee PK (2019) Bioactive phytocomponents and their analysis. In: Mukherjee PK (ed) Quality control and evaluation of herbal drugs. Elsevier, Amsterdam, pp 237–328

Nicklaus S, Boggio V, Chabanet C, Issanchou S (2005) A prospective study of food variety seeking in childhood, adolescence and early adult life. Appetite 44(3):289–297

Osorio-Esquivel O, Alicia-Ortiz-Moreno AVB, Dorantes-Alvarez L, Giusti MM (2011) Phenolics, betacyanins and antioxidant activity in opuntia joconostle fruits. Food Res Int 44:2160–2168

Oude Griep LM, Verschuren WM, Kromhout D, Ocke M, Geleijnse JM (2011) Colors of fruit and vegetables and 10-year incidence of stroke. Stroke 42(11):3190–3195

Pareek S, Sagar NA, Sharma S, Kumar V, Agarwal T, González-Aguilar GA, Yahia EM (2018) Chlorophylls: chemistry and biological functions. In: Yahia EM (ed) Fruit and vegetable phytochemicals: chemistry and human health, vol I, 2nd edn. Wiley, pp 269–284

Pliner P, Hobden K (1992) Development of a scale to measure the trait of food neophobia. Appetite 19:105–120

Polturak G, Grossman N, Vela-Corcia D, Dong Y, Nudel A, Pliner M, Levy M, Rogachev I, Aharoni A (2017) Engineered gray mold resistance, antioxidant capacity, and pigmentation in betalain-producing crops and ornamentals. Proc Natl Acad Sci USA 114(34):9062–9067

Rioux C, Lafraire J, Picard D (2018) Visual exposure and categorization performance positively influence 3- to 6-year-old children's willingness to taste unfamiliar vegetables. Appetite 120:32–42

Rodriguez-Amaya DB (2019) Betalains. In: Encyclopedia of food chemistry, pp 35–39

Rutz JK, Borges CD, Zambiazi RC, da Rosa CG, da Silva MM (2016) Elaboration of microparticles of carotenoids from natural and synthetic sources for applications in food. Food Chem 202:324–333

Schiffersteina HNJ, Wehrleb T, Carbon C (2019) Consumer expectations for vegetables with typical and atypical colors: the case of carrots. Food Qual Prefer 72:98–108

Schliemann W, Joy RW IV, Komamine A, Metzger JW, Nimtz M, Wray V, Strack D (1996) Betacyanins from plants and cell cultures of *Phytolacca americana*. Phytochemistry 42:1039–1046

Schoefs B (2004) Determination of pigments in vegetables. J Chromatogr A 1054(1–2):217–226

Scott KJ, Thurnham DI, Hart DJ, Bingham SA, Day K (1996) The correlation between the intake of lutein, lycopene and β-carotene from vegetables and fruits, and blood plasma concentrations in a group of women aged 50-65 years in the UK. Brit J Nutr 75:409–418

Seeram NP, Momin RA, Nair MG, Bourquin LD (2001) Cyclooxygenase inhibitory and antioxidant cyanidin glycosides in cherries and berries. Phytomedicine 8(5):362–369

Shipp J, Abdel-Aal EM (2010) Food applications and physiological effects of anthocyanins as functional food ingredients. Open Food Sci J 4:7–22

Simpson KL (1985) Chemical changes in natural food pigments. In: Richardson T, Finley JW (eds) Chemical changes in food during processing. Van Nostrand Reinhold Company, Inc, New York, pp 409–437

Singh P (1996) Essentials of plant breeding. Kalyani Publishers, p 206

Stevens MA (1985) Tomato flavor: effects of genotype, cultural practices and maturity at picking. In: Evaluation of quality of fruits and vegetables. AVI Publishing Company, Westport, p 410

Su X (2022) Note on important health benefits of carotenoids. Oxidants Antioxidants Med Sci 11(6):01–02

Takyi EEK (2001) Bioavailability of carotenoids from vegetables versus supplements. In: Watson RR (ed) Vegetables, fruits, and herbs in health promotion. CRC Press LCC, Danvers, pp 19–31

Tang R, He Y, Fan K (2023) Recent advances in stability improvement of anthocyanins by efficient methods and its application in food intelligent packaging: a review. Food Biosci 56:103164. https://doi.org/10.1016/j.fbio.2023.103164

Taylor CM, Wernimont SM, Northstone K, Emmett PM (2015) Picky/fussy eating in children: review of definitions, assessment, prevalence and dietary intakes. Appetite 95:349–359

Veteto JR (2008) The history and survival of traditional heirloom vegetable varieties in the southern Appalachian Mountains of western North Carolina. Agric Hum Values 25(1):121–134. https://doi.org/10.1007/s10460-007-9097-6

Wang CQ, Chen M, Wang BS (2007) Betacyanin accumulation in the leaves of C3 halophyte *Suaeda salsa* L. is induced by watering roots with H2O2. Plant Sci 172:1–7

Willows RD (2004) Chlorophylls. In: Davies K (ed) Annual plant reviews, vol. 12, plant pigments and their manipulation. CRC Press/Blackwell, Oxford, p 24

Zannou O, Oussou KF, Chabi IB, Odouaro OBO, Deli MGEP, Goksen G, Galanakis CM (2023) A comprehensive review of recent development in extraction and encapsulation techniques of betalains. Crit Rev Food Sci Nutr 64(30):11263–11280. https://doi.org/10.1080/10408398.2023.2235695

Part II

Cultivation Techniques and Production Systems

Vegetable Nursery Management: Foundation of Successful Production System

4

Mateen Sajid, Zahoor Hussain, Muhammad Akbar Anjum, Shuhua Zhu, and Raheel Anwar

Abstract

Vegetables are integral part of human diet as they contain carbohydrates, protein, dietary fiber, minerals, vitamins, and bioactive compounds. The expansion in cultivated area under vegetable crops with increased production is the dire need to meet food requirements of expanding population around the globe. Currently, unavailability of reliable, true to type vegetable seedlings is one of the major problems to achieve this goal. On the other hand, nursery industry in most of the developing countries is lacking modern facilities to meet the international standards for good quality and quantity of plant produce. Lack of awareness about modern technologies used in vegetable nursery production aggravates the existing problems. Nursery workers are still using conventional methods to grow plants. Vegetable quality entirely depends on health of seedlings which are shifted from nurseries to vegetable fields. Seedlings produced using conventional methods are mostly under the biotic and abiotic stresses. Therefore, to provide true to type, healthy, and disease-free seedlings to the growers, it is imperative that the nursery should be established by adopting modern and advanced technologies. Information on nursery types, growth media, nutrient management, and disease and insect-pest management will not only help in production of

M. Sajid · Z. Hussain (✉)
Department of Horticulture, Ghazi University, Dera Ghazi Khan, Pakistan
e-mail: zhussain@gudgk.edu.pk

M. A. Anjum
Department of Horticulture, Bahauddin Zakariya University, Multan, Pakistan

S. Zhu
Department of Horticulture, Shihezi University, Shihezi, Xinjiang, China

R. Anwar
Institute of Horticultural Sciences, University of Agriculture, Faisalabad, Pakistan

Z. Iqbal et al. (eds.), *Sustainable and Innovative Vegetable Production in times of Climate Change*, https://doi.org/10.1007/978-981-96-6283-8_4

healthy seedlings but also make the growers well-off. This chapter is aimed to provide guidelines to raise vegetable seedlings using modern techniques and will also serve as a valuable reference for anyone involved in nursery production.

Keywords

Automation in nurseries · Hardening · Hydroponics · Organic practices · Protected nursery · Protrays · Plug culture · Soil treatments

4.1 Introduction

A vegetable nursery is a place, area, or establishment for raising and management of young vegetable seedlings until these are sold or transplanted. The production of healthy and vigorous seedlings imparts a vital role in establishing a healthy crop in the main field. In nursery, seedlings receive ideal conditions for their growth and development. For vegetables, seeds are typically used to raise seedlings in the nursery. Nursery conditions also facilitate better seed germination of small, hybrid, and costly seeds. However, depending upon the kind of seed and its sowing method, germination of vegetables' seeds is very critical. Many seedlings are still grown in the soil beds, uprooted when attain a suitable size, sold as bare-root plants, or raised in plastic containers containing a mixture of soil, sand, and compost. When moved to the open field, such seedlings face transplanting shock and/or may suffer from soil-borne diseases. Most of the nursery growers in developing countries do not use standard seedling-raising technologies, for example, proper media preparation, seedling trays, hygienic conditions, quality insect mesh and shade net materials, or a double door system. Commercial vegetable production demands better control over plant growth and production, which begins in the nursery. Plant performance in the field is based on plant genetics, and seed and seedling health. The vegetable nursery is the cornerstone of a successful vegetable production system (Kumar et al. 2023). Under controlled environmental conditions, seeds are transformed into robust seedlings, which become ready in a short time for transplantation into the fields. Understanding nursery management is essential for anyone aspiring to be a successful vegetable grower. This chapter explores the critical aspects of establishing and maintaining a vegetable nursery, detailing best practices, techniques, and common pitfalls.

4.1.1 Raising Nursery and Classification of Vegetables

Vegetable crops are divided into three groups depending upon their seed sowing in the nursery or directly into the field. Vegetables such as cabbage, cauliflower, broccoli, Brussel's sprout, tomato, eggplant, sweet pepper, and onion cannot be grown by sowing their seeds directly into the field. Therefore, raising of nursery seedlings for these vegetable crops is necessary before transplanting into the main field

(Sharmila et al. 2014). The vegetables like okra, turnip, radish, carrot, peas, beans, cucurbits, and leafy vegetables (spinach, fenugreek) are propagated by seeds and directly planted in main fields (Arya 2003). However, there are also few vegetables which can be grown by both methods, i.e., directly sowing seeds in the field or through raising seedlings in the nursey and then transplanting in the field, e.g., chilli and lettuce.

4.1.2 The Importance of a Vegetable Nursery

A vegetable nursery is a vital component of successful crop production, offering several benefits that enhance the overall health and yield of vegetable crops, as described below:

4.1.2.1 Optimal Growth Conditions

A nursery allows growers to create an ideal environment for seed germination and seedling growth. Key environmental factors such as temperature, humidity, and light can be controlled to ensure optimal conditions for young plants. This is particularly beneficial in regions with variable climates, as it allows growers to extend the growing season and produce high-quality seedlings that are better prepared for transplantation.

4.1.2.2 Early Harvests

Starting seedlings in a nursery enables growers to get a head start on the growing season. This early planting leads to staggered harvesting, which can improve cash flow and provide a competitive edge in the market. Early harvests are especially valuable for high-value crops, allowing growers to take advantage of higher prices at the beginning of the season.

4.1.2.3 Disease Management

Seedlings are vulnerable to pests and diseases in their early stages. Growing them in a nursery reduces exposure to harmful field conditions, thereby lowering the risk of pest infestations and disease outbreaks. A controlled nursery environment allows for close monitoring, enabling growers to detect and address any issues swiftly before these spread.

4.1.2.4 Better Resource Utilization

Nurseries promote efficient use of resources like water and nutrients. By nurturing plants in a controlled setting, growers can apply precise irrigation and fertilization practices, minimizing waste and ensuring that the seedlings receive the exact amount of nutrients needed for their optimal growth. This leads to better growth rates and a more sustainable approach to vegetable production.

4.2 Types of Vegetable Nurseries

There are different types of nurseries available in the world. However, the types of these nurseries are classified on several bases as described below:

4.2.1 On the Basis of Structure

4.2.1.1 Open Field Nursery

This type of vegetable nursery is prepared in the open field after final preparation of soil. This nursery does not have any kind of protective structure (Ahmad et al. 2017). Moreover, this type of vegetable nursery is raised during the favorable conditions for establishment of seedlings. However, survival percentage of seedlings in open-field nursery conditions is poor (Singh 1998).

4.2.1.2 Protected Nursery

For better seed germination and seedling growth, seeds are grown in some structures with partial or full control on environmental conditions. These structures vary in the size, shape, design, construction material, and other facilities including control of environmental factors. In advanced countries, seedlings are usually raised in greenhouses or glasshouses. However, in developing countries, usually cheap alternatives are used, as described below:

4.2.1.2.1 Polyhouse

Polyhouse is a good and cheap alternative to greenhouse with partial control on environmental factors. It is a type of protected technology for raising plant nurseries, which is now becoming popular in developing countries. The polyhouse provides better environmental conditions for early and robust growth of plants, which subsequently facilitates better survival and performance of plants in the field (Rymbai and Reddy 2011). Nowadays, polyhouses are very common in Pakistan and other countries for production of summer vegetables during winter season.

4.2.1.2.2 Shade House

A shade house is a horticultural structure which is used to grow plants under artificial cover (shade or partial shade). It creates a mix of light and shade to provide favorable environment for shade-loving plants. In this type of structure, different types of frames are used for the construction such as iron or wooden poles, and usually chicks of sarkanda (reed) are placed over the frame to create shade. Nowadays, besides sarkanda chicks, different types of nets such as green, black, and white synthetic nets with various mesh sizes are used to cover the structure. The technique is low-cost to raise seedlings during summer season for planting early winter vegetables (Hannah 2002). This type of shade house also protects young seedlings from the entry of insects and pests (Fig. 4.1).

Fig. 4.1 Shade house

4.2.1.2.3 Thatch Roof

Thatch roofing is a very old traditional method in which dry vegetation is used to make the roof. In thatched roof nurseries, seedlings are grown on the beds that are protected by the roof, which is made up of dry grass. This type of nursery protects young seedlings from frost and harsh weather conditions.

4.2.1.2.4 Container/Plug Culture

In this type of nursery, different types of containers/plugs (trays) are used for vegetable nursery. Containers include plastic pots, plastic plugs, polythene bags, etc., which can be kept or shifted anytime in a protective structure. In this type of nursery, usually a single seed is sown in the individual pot/plug and a large number of seedlings can be produced in a little area under optimized and controlled conditions. Actually, this technology (Fig. 4.2) was developed to grow ornamental flowering and fruit plants such as guava (Usman et al. 2013). Nowadays this type of nursery has become an important nursery-raising technology and is commonly used to raise vegetable seedlings (Hannah 2002).

4.2.1.2.5 Natural Conservatory

It is a very simple technique for raising vegetables' nursery under the partial shade of trees such as mango, mulberry, and java plum. The technique is cheaper and good for growing nursery of early winter vegetables during the summer (Hannah 2002).

Moreover, in vitro techniques are also used to develop nurseries in the tissue culture laboratories on artificial media under aseptic and controlled conditions (Thorpe 1981). This technique is used to eliminate viruses from stock plants and propagate healthy and disease-free plants on a large scale. The plants produced

Fig. 4.2 Container/plugs

through tissue culture are highly delicate and, therefore, are hardened or acclimatized before transplanting to open field conditions (Thorpe 2007; Singh and Kumar 2020).

4.2.2 On the Basis of Sale Business

4.2.2.1 Wholesale Nursery

This type of nursery is well known, which sells seedlings in bulk on wholesale basis.

4.2.2.2 Retail Nursery

This is another type of nursery on the basis of sale. Such nurseries sell their seedlings on a retail basis.

4.2.3 On the Basis of Time Duration

4.2.3.1 Temporary Nursery

This type of nursery is set up on temporary basis for the growing of seedlings to fulfill the seasonal requirements of an area and then it also becomes the part of field as cultivated area. This type of nursery is also known as flying nursery.

4.2.3.2 Permanent Nursery

This type of nursery is set up for the growing of seedlings on permanent basis. This type of nursery has a permanent structure for growing seedlings, so that seedlings/saplings can be provided to the stakeholders on a regular basis. In permanent nurseries, the capacity of seedlings and saplings production varies from nursery to nursery. However, permanent nurseries are well equipped with irrigation and fertilization facilities, and plant protection measures.

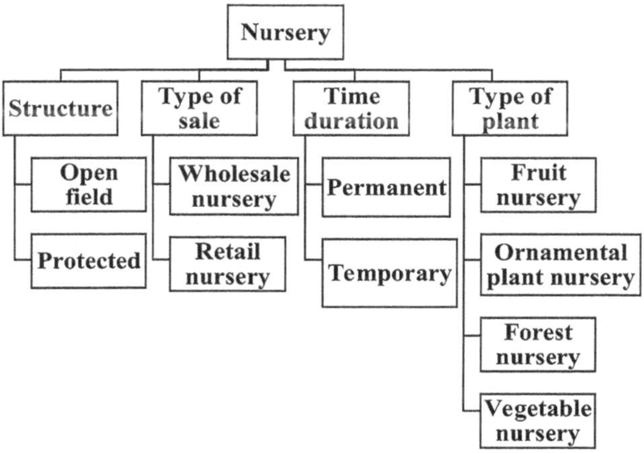

Fig. 4.3 Flow chart of nursery classification

4.2.4 On the Basis of Type of Plants

On the basis of type of plants produced in the nurseries, these can be classified as (a) Fruit plant nursery, (b) Ornamental nursery, (c) Forest nursery, and (d) Vegetable nursery (Fig. 4.3).

4.3 Establishment of Nursery

4.3.1 Selection of Location

A well-designed nursery is essential for producing healthy seedlings. Several key factors should be considered in its setup. Selecting the right location for a vegetable nursery is a crucial and first step. Key considerations include the following aspects:

4.3.1.1 Sunlight

A nursery should ideally receive 6–8 h of direct sunlight each day, which is critical for photosynthesis and robust plant growth (Palmer 1989; Smart 1989).

4.3.1.2 Drainage

The site should have good drainage to avoid waterlogging, which can cause root diseases. Conducting soil tests can help in assessing drainage capabilities and making necessary adjustments.

4.3.1.3 Accessibility

Proximity to water source and transport routes is essential for easy maintenance and distribution of seedlings. This ensures that watering, fertilization, and transport processes are efficient and manageable.

4.3.2 Preparation of Land and Soil

Soil is very important for raising vegetable seedlings, especially in open-field nurseries. The soil system consists of air, water, and organic matter as well as living organisms. The soil is the main source of nutrients and water for plant growth and development. It also supports the root system of plants to anchor. The ideal soil for vegetable nursery has a good mixture of sand, silt, and clay, and is rich in organic matter with good water drainage. It also has optimum pH 6.5–7.0 with a minimum of 2% organic matter (ICFRE 2019). For preparation of nursery soil, it should be deeply plowed 3–4 times. The soil should be free from all kinds of clods, stones, plant debris, and weed seeds. Adding 4–5 kg of well-decomposed farmyard manure (FYM) or 500 grams of vermicompost per square meter further enriches the soil (Vishwakarma et al. 2018).

4.3.3 Soil Treatment

Soil is often a host to insects and their eggs, fungal spores, bacteria, nematodes, and weed seeds, which can hinder seedlings' development. Different soil treatments are used to overcome these threats, which are described below:

4.3.3.1 Formalin Treatment

A solution of 2.5 mL of formaldehyde per liter of water is applied saturating the topsoil to a depth of 15–20 cm. The treated soil is covered with a 200-gauge polythene sheet to allow the fumes to work for 48 hours, and the soil is then loosened for 7–10 days before sowing.

4.3.3.2 Fungicides and Insecticides

Soil can be treated with suitable pesticides. Fungicides like Captan or Thiram at 5 g/m^2 are effective for fungal control. Insecticides like chlorpyrifos at 2 mL/L of water help in controlling soil-dwelling pests.

4.3.3.3 Hot Steam

Hot steam treatment, applied under an airtight polythene sheet, effectively sanitizes the soil over 4–6 h using steam-generating machines. This is mostly used on small scale, especially when soil is used as growing medium in containers.

4.3.3.4 Soil Solarization

Solarization is a soil treatment method that utilizes solar heat to eliminate soil-borne pests, pathogens, and weed seeds. It involves covering moist soil with a 200-gauge polythene sheet for 5–6 weeks between May and August. This method leverages the sun's heat to eliminate weed seeds, insect larvae, and pathogens in the soil.

For solarization, elevated beds are thoroughly moistened, as wet soil is more effective in conducting heat than dry soil, making soil organisms more vulnerable to heat exposure. A transparent 200-gauge sheet is laid over the moist soil. This is practiced ideally during the hot summer when temperatures are high. Edges of the polythene sheet are pressed down with compacted, damp soil to ensure an airtight seal. This step is crucial to prevent moisture loss and to block air from entering beneath the polythene cover, which could reduce the heat effect. The polythene sheet is kept in place for 5–6 weeks period. During this time, heat trapped beneath the cover will increase temperature high enough to kill weed seeds, insects (including their eggs, larvae, and pupae), and soil-borne pathogens like those causing damping off disease. After the treatment period, the sheet is carefully removed and the soil is allowed to cool down before proceeding to sow seeds. This solarization process effectively controls unwanted soil organisms within the top 12–15 cm of soil, ensuring a healthier environment for seedling growth.

4.3.4 Nursery Bed Preparation

Proper bed preparation is essential for successful seedling growth in a vegetable nursery. There are three types of nursery beds (flat beds, raised beds, and sunken beds) prepared for raising vegetable seedlings depending upon seasonal conditions and desired outcomes.

4.3.4.1 Flat Nursery Beds

Flat nursery beds are appropriate during non-rainy periods like summer and winter when waterlogging is less of a concern. These are ideal for areas where rainfall is minimal or controlled irrigation is practiced (Fig. 4.4). Important aspects include:

Fig. 4.4 Flat nursery beds

- Dimensions: Width of 1 meter (east to west) and length of 3–5 m (north to south), are adjusted as needed.
- Orientation: The length is typically arranged from north to south to ensure better sunlight exposure.
- Spacing: A 30–40 cm space is maintained between rows to create water channels for controlled irrigation. These channels also help to drain excessive water during heavy rains.
- Use: These are best suited for dry seasons, as they avoid the risk of excessive moisture and facilitate easier management during watering routines.

4.3.4.2 Raised Nursery Bed

Raised nursery beds are highly suitable for all seasons, especially during the rainy season. Their elevation ensures adequate drainage, reducing the risk of waterlogging (Fig. 4.5). Here are the key details:

- Height: 15–20 cm above ground level.
- Dimensions: Typically sized at 3 × 1 m or 5 × 1 m.
- Orientation: Beds are aligned in the same direction as flat beds for better sunlight.
- Drainage: Proper elevation allows water to flow away easily during heavy rainfall.
- Spacing: A 30–40 cm gap is maintained between adjacent beds to facilitate accessible cultural practices like weeding, watering, and harvesting.

Fig. 4.5 Raised nursery beds

4.3.4.3 Sunken Nursery Bed

This type of nursery bed is prepared 10–15 cm downwards from the soil surface. This bed is very suitable during the winter months, especially when the air blows around the nursery area. Hence, the seedlings in sunken bed are not hit by the cool breeze. Further, to protect the seedlings from frost and cool air, sunken bed can be easily covered with polyethylene sheets.

4.4 Raising of Vegetables Nursery

After seed bed preparation, the seed sowing is a very important step for growing vegetables' nursery. However, the success of vegetable nursery depends upon the quality of seed, sowing time, methods, and season or prevailing environmental conditions.

4.4.1 Choosing Quality Seeds

Selecting high-quality seeds is fundamental to nursery success. Important factors to be considered while purchasing seeds are given below:

4.4.1.1 Varietal Suitability

Choose cultivars that are well-suited for local climatic conditions and have proven market demand. Local agricultural extension services can provide valuable insights. However, nowadays hybrid cultivars are preferred over conventional ones to exploit potential of hybrid vigor.

4.4.1.2 Disease Resistance

Only seeds of those cultivars should be sown which have resistance to common insect-pests and diseases in the area.

4.4.1.3 Germination Rates

High germination rates indicate seed quality, thus the seeds with optimum germination percentage should be planted. Further, seeds should be purchased only from reputable suppliers to ensure quality.

4.4.2 Seed Sowing Methods

Different methods used for sowing of vegetable seeds in the nursery are given below.

4.4.2.1 Broadcast

This type of vegetable seed sowing method is very old and common which is used for all types of crops. In this method, the seeds are directly broadcasted on the bed surface and then covered with a fine layer of farm yard manure (FYM) or compost or silt. Some vegetables such as spinach, coriander, fenugreek, etc., are directly sown in the field by broadcasting the seeds. Seed broadcasting is done manually or mechanically. It is not possible to maintain proper planting distance in this method; hence line sowing method is preferred over broadcast method (Pandey et al. 2023) in both nursery raising as well as direct sowing of vegetables in the field.

4.4.2.2 Line Sowing

Line sowing is the most effective method for growing vegetable seedlings. This approach involves creating shallow furrows, approximately 0.5 cm deep, aligned width-wise (east to west) with 5–15 cm spacing between them. Seeds are placed individually or about 1–5 cm apart in these furrows depending upon the kind of vegetable. After sowing, a fine mixture of sand, soil, and well-decomposed FYM or leaf compost in a 1:1:2 ratio is used to cover the seeds. A gentle watering with a sprinkler is done, till seed germination is completed.

4.4.2.3 Sowing in Protrays or Plugtrays

High-quality seeds and vigorous seedlings are crucial for achieving maximum yield potential in every vegetable crop. Seed sowing and raising nursery seedlings can be challenging without proper techniques. However, germination trays make it possible to grow strong and vigorous seedlings. In recent decades, seeds with robust properties or vigorous seedlings have become increasingly important to vegetable growers (Sharmila et al. 2014). For this purpose, use of protrays is a common practice for propagating vegetables through seeds (Fig. 4.6).

Due to their higher costs, the hybrid seeds are often started in controlled nursery conditions to maximize germination rates and produce healthy plants.

Vegetables such as tomatoes, eggplant, capsicum, and cucurbits are commonly grown this way. This method offers better results for both nurserymen and growers

Fig. 4.6 Protray method of nursery sowing

and has gained rapid popularity in the agro-industry worldwide (Pandiyaraj et al. 2017). Selection of protrays is crucial and depends on the size of the growing area and the number of plants needed for transplantation. Typically, trays consist of 72–800 cells and measure around 53.7 × 27.5 cm. The cell size is also important, as it determines the quantity of growing medium required and the water-holding capacity. Larger cells are often preferred because these provide better conditions for growth and produce more vigorous seedlings compared to smaller cells. The process of raising vegetable seedlings in protrays is described below:

Sowing of seeds: For effective germination, only one seed is planted in each cell of a tray. Seeds are treated with a suitable pesticide before sowing to reduce the risk of diseases and insect attacks. After sowing, light irrigation is applied using a sprinkler or spray machine. Maintaining optimal moisture level in the trays is critical, as over-irrigation can lead to root rot and damping off.

Light: Light is essential for seed germination and seedling growth. However, direct sunlight, which can scorch the seedlings, should be avoided. Proper light conditions help in acclimatizing the seedlings before they are transplanted into more challenging outdoor conditions.

Germination: After sowing the seeds, trays are covered with black polythene sheets to maintain optimal moisture and provide a warm environment conducive to germination. The cover is removed once seedlings begin to emerge. Ideal temperatures for germination vary depending on the crop. Tomatoes and brinjal thrive at 21–24 °C, while chillies and sweet pepper prefer 28–32 °C. During the first 4 weeks, root zone temperatures are kept between 26 and 29 °C, and between 20 and 26 °C in the fifth and sixth weeks (Reddy 2019).

Hardening: Hardening is the process of gradually acclimatizing seedlings to outdoor conditions. This ensures that these can withstand less favorable conditions after transplanting. Without this step, seedlings are more likely to experience high mortality rates. Hardening can be achieved by applying low-concentration salt sprays to limit water uptake, gradually increasing exposure to direct sunlight, or temporarily stopping irrigation (Pandiyaraj et al. 2017).

Transplanting: The appropriate transplant time depends on the crop type, plant age, and leaf stage. Typically, transplanting is done at 2–4 leaf stage or when plants are 21–30 days old. Transplanting is usually done in the morning or evening to avoid exposing the seedlings to high temperatures.

Advantages It offers the following advantages:

- The tray method reduces the seed rate compared to open-field nurseries due to minimal seed wastage.
- As trays occupy small areas, land preparation and field operations for transplanting can be done efficiently, freeing up space for the main crop.
- Trays are placed in protected conditions, reducing the risk of disease and insect attacks.

- Sterilized medium in trays minimizes the occurrence of soil-borne diseases.
- Seedlings grown using this method are more vigorous, with lower mortality rates after transplantation.
- This method results in uniform seed germination and seedling growth.
- The soft medium allows seedlings to be easily removed without damaging the roots.
- Irrigation and fertilizer applications are more accessible and more precise.
- This method is especially suitable for raising costly hybrid seedlings.

Disadvantages This method has the following disadvantages:

- Skilled workers are required for the process.
- Careful monitoring is needed throughout the stages, from sowing to transplanting.
- Despite the protective measures, there remains a risk of pest and disease incidence.

4.5 Care and Maintenance of Vegetable Nursery

4.5.1 Irrigation Management Practices

Nursery crops require a consistent supply of high-quality irrigation water to grow and develop into healthy plants. Water must be applied correctly and precisely where needed to optimize plant growth while minimizing runoff and conserving water and fertilizers (Schoene et al. 2006). Various irrigation systems are available for nursery use, ranging from simple hand-watering to complex sprinkler and drip systems.

4.5.1.1 Sprinkler Irrigation Systems

Overhead sprinkler systems in which the sprinkler head pivots on a bearing, commonly apply water over a wide area, irrigating many plants simultaneously. These systems typically use impact sprinkler heads with different discharge rates. In small nurseries, sprinklers with lower discharge rates are used, categorizing them as micro-irrigation systems. Due to their versatility, impact sprinklers or impulse sprinklers are commonly emphasized in discussions about nursery irrigation systems.

4.5.1.2 Micro-Irrigation Systems

Micro-irrigation, often called as drip irrigation, uses emitters to provide a low, precise discharge rate. These emitters can be built into drip tubing (in-line emitters), attached to spray stakes or configured as individual emitters with weights or stakes to hold them in place. Discharge rate for these devices also varies depending on crop requirements, row or drip line length and spacing between the emitters. While drip irrigation offers greater control over water distribution, it requires more complex maintenance and operation than other systems.

4.5.2 Fertilization

Proper fertilization is crucial for maintaining a profitable nursery business. Optimal plant growth rates can only be achieved when nutrients are supplied appropriately; not too little or too much. A significant challenge for nurserymen is to provide the seedlings with necessary nutrients without over-fertilizing, which can contribute to nitrogen runoff.

Insufficient fertilization, particularly of nitrogen, can result in poor plant growth. Typical growing media, such as sand and organic matter mixes, do not contain enough nutrients to sustain plant needs over extended periods. This issue is often exacerbated by over-irrigation, which causes leaching of essential nutrients such as nitrogen from the soil. Additionally, organic matter in potting mixes usually has a low pH, which can negatively impact nutrient uptake. To address these challenges, it is necessary to adjust the chemical properties of the growing medium using amendments that balance pH levels and provide good fertility. Additional fertilizers are needed throughout the growth period to maintain adequate fertility levels for plant development.

Excessive fertilizer application can also harm the seedlings, causing growth issues or direct injury. Moreover, over-fertilization can negatively impact water quality and pose risks to the health of plants, animals, and ecosystems relying on that water.

4.5.3 Integrated Pest Management

Due to the diversity of crops grown in greenhouses and container nurseries, they often face a range of pests and diseases. Pests can cause significant economic losses; therefore, a large portion of nursery budget is allocated to pest management. Pesticide use is often intensive, driven by the need for high-quality, blemish-free plants and regulations that mandate clean plants for shipping. However, frequent pesticide use can lead to resistance among pests, creating a cycle of increasing dependency on chemicals. At the same time, regulatory constraints on pesticide usage and the risk of pesticide runoff are increasing. Moreover, pesticides required for quarantine and shipping protocols may also be subject to restrictions due to water quality concerns.

Many growers adopt integrated pest management (IPM) programs to control pests and reduce pesticide runoff. The IPM emphasizes preventive measures, pest monitoring, and taking actions based on threshold levels. This approach minimizes pesticide use, lowering production costs and reducing the risk of liabilities related to runoff, pesticide drift, and workers safety. Additionally, IPM limits pest spread within the nursery and minimizes pest infestations in harvested plants sent to other locations. The core principles of IPM include pest identification, monitoring, and establishing action thresholds. Nursery staff must be skilled in diagnosing plant issues to ensure that appropriate pest management measures are implemented timely. Control measures include physical, mechanical, cultural, chemical, and biological approaches. Considerations like pesticide resistance management, proper application methods, and selection of pesticides to minimize runoff are also crucial.

4.5.4 Integrated Weed Management

Weeds persist in nurseries due to the year-round cropping system and the ease with which their seeds spread. Weeds in nursery compete with seedlings for water, nutrients, and light, sometimes blocking water and fertilizer from reaching the plants. Additionally, weeds can harbor insect-pests, pathogens, and viruses, which can decrease marketability of seedlings. Managing weeds can be time-consuming and costly, involving direct expenses for labor, equipment, herbicides, and mulches as well as the opportunity cost of diverting labor from other tasks.

An effective weed management program should be part of an overall IPM strategy. This includes using cultural control methods like sanitation, mechanical methods like mulching and hand weeding, physical processes such as heat-treating or steaming of growing medium, and chemical methods. Choosing a nursery site that minimizes weed problems and implementing runoff management practices to protect water quality is also crucial.

Specific herbicide recommendations are not provided here due to frequent changes in product labels. For guidance, growers can consult Department of Agriculture or Plant Protection or the other respective government departments.

4.5.5 Thinning

Thinning is a critical process that removes weak, unhealthy, diseased, insect-infested, and overly crowded plants from nursery beds. This also ensures uniform space between each plant, allowing equal light and air distribution to every seedling.

4.5.6 Hardening

Hardening is the process of acclimatizing seedlings to environmental stress, improving their resilience, and increasing their chances of survival and successful growth after transplantation. Before transplanting into the main field, irrigation should be stopped for 5–6 days. Light irrigation on the transplanting day allows for easier seedling removal with minimal root damage.

4.5.7 Uprooting and Packing of Nursery Plants

Once the seedlings attain a suitable size and have 4–6 leaves (typically around 25–30 days), these should be carefully uprooted for transplanting. Careful handling is crucial to prevent root damage and essential for subsequent successful establishment of seedlings. Open roots should be avoided during transplanting to minimize stress factors. Sometimes, a slurry of soil, water, fertilizers, and a suitable fungicide is prepared to treat the roots. Seedlings are bundled into groups of 50–100 and have their roots dipped into this slurry. Proper packing is essential to maintain firmness of the seedlings and ensure that these retain their vigor, allowing for better establishment at the new site.

4.6 Advanced Techniques for Nursery Management

4.6.1 Automation in Nurseries

Automation technologies, such as automated irrigation systems and climate control, can significantly enhance efficiency and reduce labor costs. Monitoring systems that provide real-time data on prevailing temperature, relative humidity, and soil moisture enable precise management of the nursery environment, resulting in better growth conditions and reduced manual efforts.

4.6.2 Organic Practices

Organic practices in nurseries can improve soil health and enhance plant resilience. Utilizing organic fertilizers, compost and biological pest control methods supports sustainable agriculture and reduces chemical inputs, leading to healthier, more sustainable plant production.

4.6.3 Crop Rotation and Diversity

Integrating diverse vegetable crops and practicing crop rotation can help to prevent build-up of diseases and improve soil health over time. This approach fosters biodiversity and enhances overall resilience of the nursery, leading to healthier plants and reduced pest pressures.

4.7 Nursery Raising Techniques for Some Important Vegetables

4.7.1 Tomato

To raise tomato seedlings for planting one-hectare area, 200–250 g seeds of hybrid cultivars and 350–400 g seeds of other cultivars are sufficient. Seeds are treated with antifungal agents like Thiram or Captan @ 2.5 g/kg of seeds. Additionally, seed treatment with *Trichoderma viride* 4 g/kg or *Pseudomonas fluorescens* 10 g/kg is also helpful in protecting the seedlings from soil-borne diseases. Depending on the available space, a tomato nursery can be grown using two methods viz. raised bed method and tray method. The optimum time for sowing seeds to raise seedlings for open-field transplantation is early to mid of October.

4.7.1.1 Raised Bed Method

The land is deep plowed, and clods, stubbles, and weeds are removed. About 20–25 kg of well-decomposed FYM is added per bed to ensure fine tilth. Furadan-3G 15 g per bed can be mixed into the soil to protect the seeds from ants and termites. Raised beds are prepared 5 m in length, 1 m wide, and 15–20 cm high from the ground. One bed is sufficient for planting 20 g seeds, while at least 10 beds are needed for transplanting one hectare. Furrows are created over the beds using a wooden stick or metal rod, spaced at 10–15 cm. Seeds are sown in these furrows 4 cm apart and 1 cm deep, covered with soil and lightly irrigated with a sprinkler. The beds are irrigated once or twice daily depending on soil moisture conditions. If there are chances of damping off disease, Copper oxychloride @ 2.5 g/L of water can be drenched and repeated after 15 days to prevent the disease. Watering should be withheld 1 week before transplanting to harden the seedlings.

4.7.1.2 Tray Method

Germination trays are filled with coco peat 1.2 kg per tray or with some other suitable growing medium. Seeds are sown in the germination trays ensuring 1 seed per cell. Seeds are covered with coco peat or the other growing medium. The trays are stacked and covered with a polythene sheet until germination starts, i.e., for about 5–6 days. Then the germination trays with the germinated seeds are moved onto raised beds inside a shade house (Fig. 4.7). The seedlings are ready for transplanting about 25–30 days after sowing, when these have developed 5–6 true leaves. A tomato plant which is 7.5–10 cm in height and has a sturdy stem is considered suitable for transplanting.

4.7.2 Chilli and Sweet Pepper

For outdoor planting, the best time for raising capsicum seedlings is mid of October to mid of November. For planting one hectare of land, 200–250 g seeds of hybrid cultivars are enough. For other cultivars, seed quantity requirements are little higher

Fig. 4.7 Young seedlings in germination trays

due to low germination rates. Seeds are treated with a systemic fungicide such as Mancozeb or Captan @ 3 g/kg seeds to prevent from soil-borne diseases. Depending on the available space, growing medium, and other facilities, capsicum seedlings can be produced through raised bed method or tray method.

4.7.2.1 Raised Bed Method

The land is prepared and beds are aligned in the same way as for tomato. The size of the bed may vary depending on the available area, but a reasonable size could be 1 × 3 m^2 or a larger bed could be 1 × 5 m^2. The beds are raised about 15–20 cm above the ground level. A small amount of a complete fertilizer (about 0.5 kg) is applied to each nursery bed to help the seedlings get a good start. After preparing the seed bed, a metal rod or a stick is used to make straight narrow furrows on the beds 10 cm apart. Seeds are placed side by side in the furrows, about 5 cm apart. Then, the seeds are covered and light irrigation is applied by using a hand sprinkler. Water is applied gently to prevent washing the seeds away. The seedlings usually become ready for transplanting after 25–30 days of seed sowing. The beds are irrigated a few hours before removing the seedlings. This will loose the soil and cause less harm to the roots when the seedlings are uprooted. A trowel or a small spade is used to lift the seedlings. Simply pulling the seedlings from the nursery bed is avoided because this causes damage to very fine, almost invisible root hairs.

4.7.2.2 Tray Method

Growing seeds in seedling trays is a modern technique for raising vegetable seedlings. Trays are made of plastic or some natural materials such as compressed peat moss. The trays have a number of individual cells, which contain a suitable growing medium and seeds can be planted separately in these cells. The benefit of this method is that roots of one seedling don't intermingle with those of adjoining seedlings, permitting them to be uplifted from the trays without any harm. Moreover, the seedlings can be uprooted with some soil or growing medium attached to the roots as a small ball and transplanted without stress. Nurserymen usually use sterilized soil or a commercially available growing medium to raise seedlings in these trays. The sterilization process kills microbes that cause soil-borne diseases in the seedlings, facilitating the plants to take a better and healthy start. Seedling trays can be placed in a greenhouse or on raised beds in the field or at any elevated and sunny location. Water is applied on a regular basis to the seedlings, as and when needed, to prevent the seedlings from moisture stress. After 25–35 days of planting seeds, the seedlings are ready to be uprooted and transplanted into the field. Seedlings can be removed from the trays easily by gently taking the plants out, gently bending or twisting the trays.

4.7.3 Eggplant

The seeds are planted during the month of February and March. About 375 g of seeds are enough to produce seedlings for planting one hectare of land. Before sowing, seeds are treated with Thiram or Captan @ 2.5 g/kg of seeds to prevent soil-borne diseases such as root rot and damping off.

Land is thoroughly plowed and a fine tilth is prepared. Well-rotten FYM is mixed into the soil @ 4–5 kg per m^2 before preparing nursery beds. The beds are prepared 3 m long, 1 m wide, and 15 cm high. The beds are drenched with a Captan solution (2.5 g/L water) 2–3 days prior to sowing the seeds to prevent damping off disease. Rows are made on the seed beds 10 cm apart with a wooden stick or metal rod. Seeds are sown in these rows at 4–5 cm distance and 1.0–1.5 cm deep, and then covered with silt, compost, or some other medium. A light irrigation is applied immediately after sowing the seeds and continued till seedlings are uprooted. The nursery beds are covered with black polythene sheets or paddy straw until germination occurs. Healthy seedlings that are 12–15 cm in height with 3–4 leaves are ready for transplanting.

4.7.4 Onions

For planting onions on one hectare of land, about 5–7 kg of seed with a good germination percentage is required to raise seedlings. If good quality seeds are used and proper nursery management practices are followed, seed rate can be reduced to 4–5 kg/ha. Seeds are treated with Thiram, Captan, or Carbendazim @ 2–3 g/kg of seeds to prevent soil and seed-borne diseases. For normal crop, seeds are sown during October–November.

The area is deep plowed and clods are broken using a cultivator. Debris, weeds, and stones are removed before preparing the beds. Approximately 0.05 hectares of nursery area is needed to produce enough seedlings for planting one hectare of field. About 500 kg of well-decomposed FYM with 12.5 kg *Trichoderma viride* is applied during the last plowing and thoroughly mixed it into the soil. Raised nursery beds (15 cm in height) are prepared which are 1.0–1.2 m in width; however, the length of the bed may be adjusted as per area available or according to convenience. Seeds may be broadcasted or sown in lines on the raised beds. However, line sowing is considered better as this allows proper growth of seedlings, and easier weeding and pesticide application. The lines are made 0.5–1.0 cm deep and 5 cm apart. After sowing, the seeds are covered with a fine mixture of soil and powdered FYM or vermicompost for uniform germination. Light watering is applied immediately after sowing and continued till the seedlings are uprooted. The seedlings become ready for transplanting after 35–45 days of sowing.

4.7.5 Cabbage and Cauliflower

The ideal time for raising cabbage and cauliflower seedlings in the nursery is October–November. About 1.25–1.50 kg of cabbage seeds are enough to raise seedlings for planting one hectare of area. However, seed requirement for cauliflower is high, that is 2.0–2.5 kg. To protect the seedlings from soil and seed-borne diseases, the seeds are treated with Captan or Thiram or Carbendazim @ 2–3 g/kg seeds before planting.

One to two months before bed preparation, well-rotten FYM (4–5 kg/m^2) is applied and plowed into the soil to an appropriate depth to ensure a good soil structure. For raising seedlings in the nursery, seeds can be planted either on flat beds or raised beds, the latter being more suitable especially during rains due to better drainage. After thorough soil preparation in the nursery, beds 3–5 m long, 1 m wide, and 15 cm high from ground level are prepared. The seeds are planted in lines on the raised beds. The lines are made by using a metal rod or wooden stick at a distance of 15 cm. The seeds are sown in these lines approximately 5 cm apart and 1 cm deep. The beds are irrigated immediately after sowing by using a sprinkler. Later, light irrigations are applied once or twice a day depending on soil moisture level and environmental conditions. When the seedlings are 40–45 days old, these become ready for transplanting.

4.7.6 Lettuce

In the nursery, lettuce seeds are grown during mid of September to mid of November for raising seedlings for outdoor planting. About 1 kg seed is sufficient to grow seedlings for planting on one hectare of land. The seeds are treated with a suitable fungicide such as Mancozeb or Captan @ 3 g/kg seeds before planting to prevent the seedlings from seed and soil-borne ailments.

The soil in the nursery is thoroughly prepared by adding 4–5 kg/m^2 well-decomposed FYM or other organic matter. Beds are prepared 3 m long, 1 m wide, and 15 cm high to provide better conditions for root growth and also to facilitate drainage during heavy rains. Seeds are sown on these beds in rows in the same way as described for cabbage and cauliflower. Seeds usually germinate within 3–4 days. When seedlings are 30–40 days old, these can be transplanted into the main field.

4.8 Conclusion

Establishing a meticulously managed vegetable nursery is paramount for achieving successful vegetable production outcomes. By providing optimal growth conditions, nurseries facilitate the development of robust, uniform seedlings, contributing to enhanced yields with improved produce quality. Implementation of effective disease and insect prevention measures, such as maintaining stringent hygiene protocols, utilizing high-quality materials, and adopting rigorous screening programs, is crucial for ensuring robust plant health. Despite the challenges posed by various factors such as transplant shock and additional costs associated with raising the nursery, the benefits of this practice, including efficient land utilization and precise harvest timing, outweigh the drawbacks. Looking ahead, advancements in nursery technology and practices hold significant promise for the future. Innovations in pest and disease management strategies, integrating automation and precision agriculture techniques are expected to enhance nursery efficiency and sustainability. By embracing and adopting these developments, growers can further optimize their production systems and contribute to establishing a more resilient agricultural sector.

References

Ahmad S, Ashraf I, Anjum MA (2017) Fruit and vegetable nurseries: establishment and management. In: Khan IA, Farooq M, Khan AS, Ziaf K (eds) Horticulture: science and technology. University of Agriculture, Faisalabad, pp 133–159

Arya PS (2003) Vegetable breeding, production and seed production. Kalyani Publishers, New Delhi

Hannah M (2002) Nursery stock production using the pot-in-pot technique. Department of Horticulture, The Ohio State University, Columbus

Indian Council of Forestry Research and Education (ICFRE) (2019) Modern Nursery Techniques. ICFRE Manual. Indian Council of Forestry Research and Education, Dehradun, India

Kumar R, Prajapati J, Upadhyay A, Yadav A (2023) Increase in production of different vegetable crops by using plastic mulch. Vigyan Varta 4(1):93–95

Palmer JW (1989) The effects of row orientation, tree height, time of year and latitude on light interception and distribution in model apple hedgerow canopies. J Hortic Sci 64:137–145

Pandey R, Behera SR, Khairiya P, Tewari V (2023) Planting techniques: sowing, spacing and soil considerations. In: Akash KS, Chandel SKS, Thakur P, Mukhia DJ (eds) Vegetable chronicles: vegetable science compendium, vol 1. Stella International Publication, Kurukshetra, pp 48–71

Pandiyaraj P, Kumar YR, Vijayakumar S (2017) Modern nursery raising systems in vegetables. Int J Agric Sci 9(52):4889–4892

Reddy KS (2019) Raising seedlings in nurseries. Bighaat. https://www.bighaat.com/blogs/kb/raising-seedlings-in-nurseries. Accessed 16 Jan 2023

Rymbai H, Reddy GSN (2011) Influence of open field and poly-house nursery on survival characters of rooted layers in guava (*Psidium guajava* L.). Life Sci Leafl 21:996–1002

Schoene G, Teager T, Haman D (2006) Survey of container nursery irrigation practices in West-Central Florida: an educational opportunity. HortTechnology 16(4):682–685

Sharmila BC, Mohan B, Alagudurai S (2014) Raising of hybrid vegetable seedlings under pro-trays. J Krishi Vigyan 2(2):64–68

Singh DK (1998) Regeneration of guava (*Psidium guajava* L.) cultivars by stooling with the aid of Paclobutrazol. Ann Agric Res 19(3):317–320

Singh J, Kumar A (2020) Plant tissue culture and its application in agriculture as biotechnological tool. Int J Curr Microbiol Appl Sci 11:274–284

Smart RE (1989) Solar radiation interception as a guide to the design of horticultural plantings. II. Twenty years of experience with grapevines. Acta Hortic 240:87–94

Thorpe TA (1981) Plant tissue culture: methods and applications in agriculture. Academic Press, New York

Thorpe TA (2007) History of plant tissue culture. J Mol Microbiol Biotechnol 37:169–180

Usman M, Shah MH, Zaman Q (2013) SOPs for raising containerized citrus and guava nursery. In: Usman M, Fatima B (eds) Citrus and guava nursery raising: a practical approach. University of Agriculture Faisalabad Press, Faisalabad, pp 23–47

Vishwakarma N, Sharma A, Sharma P, Kumar A (2018) Economic empowerment of farm women through management of the vegetable nursery. Int J Farm Sci 8(1):97–99

5 Vegetable Grafting: An Innovative Approach in Olericulture

Saddia Galani, Maria Babar, Muhammad Anees, and Muhammad Omar Kaseb

Abstract

Grafting is a horticultural technique widely practiced in vegetable crops with the main objectives of providing protection against soil-borne diseases, enhancing tolerance against biotic and abiotic factors, increasing harvest time, and ultimately improving fruit quality and yield. Compatibility of scion and rootstock is the first and the most important step in grafting to be successful. Successful grafting requires various physiological metabolites and a complex network of molecular mechanisms that are active at the graft junction and play a vital role in revascularization at graft union after grafting. They are significantly linked with the development of scion, which in turn enhances fruit quality and improves tolerance against both biotic and abiotic constraints. Although conventional breeding techniques and biotechnological tools can be utilized to evolve tolerant cultivars but they require considerable time. As an alternative, grafting has the ability to improve the performance of high-yielding cultivars for enhanced adaptability to various biotic and abiotic stresses. Creation of robust and resilient rootstocks can further help to expand the use of vegetable grafting and advance socio-economic-environmental sustainability in agriculture and food security.

S. Galani (✉) · M. Babar
Dr. A.Q. Khan Institute of Biotechnology and Genetic Engineering (KIBGE), University of Karachi, Karachi, Pakistan
e-mail: saddia.galani@kibge.edu.pk

M. Anees
Zhengzhou Fruit Research Institute, Chinese Academy of Agricultural Sciences, Zhengzhou, China

M. O. Kaseb
Horticulture Research Institute, Agriculture Research Centre, Giza, Egypt

Z. Iqbal et al. (eds.), *Sustainable and Innovative Vegetable Production in times of Climate Change*, https://doi.org/10.1007/978-981-96-6283-8_5

Keywords

Vegetable grafting · Scion · Rootstock · Compatibility · Biotic and abiotic factors

5.1 Introduction

Vegetable grafting is a horticultural technique in which two different parts of plants (scion and rootstock) are attached together which results in the development of a new plant after successful revascularization at graft junction. The scion develops into a canopy (trunk, branches, leaves), whereas the rootstock forms the root system. The point at which scion and rootstock are joined is called the union, where vascular bundles of scion and rootstock fuse together and develop into a plant. The scion is selected on the basis of its fruit quality, while the rootstock is chosen for its vigor and resistance against abiotic and biotic constraints.

Grafting of vegetables on a commercial scale was first begun in Japan and Korea after the 1930s to control soil-borne pathogens. This practice has now widely spread all over the world. In vegetables, intraspecific grafting (within the same species) and interspecific grafting (between different species) have extensively been reported. However, it has recently been demonstrated that plants, even those from distant families, can be effectively grafted (Notaguchi et al. 2020).

The grafted vegetables have extensively been planted throughout the world during the past few decades due to better resistance against pathogens, robust growth, and greater production. These advantages may lead to lesser inputs usage, such as fertilizers and pesticides, and prolonged harvest period with handsome revenue source for growers (Bie et al. 2017). Grafting also provides resistance against various abiotic factors like drought, flooding (Bhatt et al. 2015), extreme temperatures (Schwarz et al. 2010), soil salinity (Colla et al. 2010a), deficiency of nutrients (Huang et al. 2016), toxicity of heavy metals (Savvas et al. 2010), and soil organic pollutants (Schwarz et al. 2010). Grafting can be used as an effective tool to bridge yield gap and secure yield stability not only under single but also multiple stress conditions (Rouphael et al. 2018).

5.1.1 Applications of Vegetable Grafting

Over the past few years, grafted vegetable cultivation has gradually increased, become more widespread, and is now used to achieve a variety of goals. These include: (i) promoting plant growth and development, (ii) preventing wilt caused by pathogens, (iii) reducing fungal, bacterial, and viral infections, (iv) strengthening tolerance to thermal, water logging, and salinity stress, (v) increasing nutrient and mineral uptake to shoot, (vi) improving crop quality, and (vii) enhancing ornamental values for exposition and education (Ulas et al. 2019). Grafting also improves fruit quality as grafted watermelons have large-sized and heftier fruits compared to un-grafted ones, making preference for grafted type for high yield (Pessarakli

2016). Consumers have been paying more attention to organic and healthy products that avoid the use of pesticides and fertilizers. Global population has compelled producers to use grafting techniques for enhanced production to meet food requirements. In this regard, producers attempt to introduce appropriate scion and rootstock seeds with well-defined purpose (Davis et al. 2008a). In the twenty-first century, vegetable production and cultivation through grafting became an established practice having horticultural and biological uses to create new plant varieties with desirable characteristics.

5.2 The Process of Grafting

The grafting process is presented in Fig. 5.1.

5.2.1 Selection of Scion and Rootstock

Correct selection of scion and rootstock is the foremost important step for successful grafting. Scion is selected based on its fruit yield and quality attributes, and consumer demand. While rootstock is selected because of its vigorous nature and tolerance to various biotic and abiotic factors (Kyriacou et al. 2017). In order to produce the required trait (s) without introducing undesirable features, the rootstock must be compatible with the scion. The majority of rootstocks that are currently in use were developed by screening pre-existing germplasm (Fita et al. 2007). Few rootstocks were generated by performing crossings to create variation and choosing elite genotypes. Interspecific hybrids frequently produce superior rootstocks that significantly increase genetic richness but sometimes restrict graft compatibility to a few closely associated species (Bie et al. 2017).

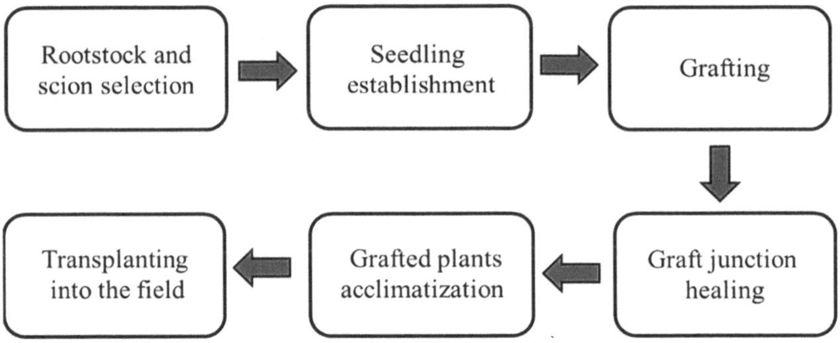

Fig. 5.1 The process of seedlings production through grafting (Modified from Bie et al. 2017)

5.2.2 Healing the Grafted Plants

Most of the cellular events that occur after grafting are common in both woody and vegetable plants. These events consist of attachment of scion and rootstock, division in callus cells, and formation of cross-bridge for revascularization at graft union (Pina et al. 2017). All these events help in the determination of graft compatibility and incompatibility. Tomato-grafted seedlings usually take 07 to 14 days to complete revascularization at graft union after grafting (Fan et al. 2015). According to Johnson and Miles (2011), vascular connection typically takes 05 to 08 days to establish and 14 days to completely develop graft union.

Commercial nurseries in Japan, Korea, China, Spain, and the United States use a variety of healing growth chambers with temperature and humidity control devices (Lee et al. 2010). Use of a healing chamber during grafting is gaining popularity in China and with this technique, 95% of success rate can be achieved on a commercial scale (Dumitrescu and Bogoescu 2022). The temperature should be between 27 and 28 °C for the first 48 hours, with 95% relative humidity. This 95% humidity is necessary to prohibit loss of water, since water loss during the first two days may cause the scion to wilt and ultimately result in failure of the grafting procedure. Commercial nurseries establish grafted transplants on benches in the greenhouse and cover trays with 0.01 mm polythene film for 05 to 07 days to increase humidity.

5.2.3 Transplanting into the Field

Though revascularization at the graft junction of grafted plants will be completed in 07 days, graft union does not entirely heal until at least 14 days after grafting (Johnson and Miles 2011). While transplanting, the graft union should be over the soil surface. The scion will root into the soil if the graft junction is placed beneath the soil, negating any benefits that the rootstock would have provided, such as resistance to soil-borne diseases.

5.3 Grafting Methods in Vegetables

There are several methods of grafting used in vegetable crops such as tongue grafting, splice grafting, and hole insertion grafting (Khankahdani and Zakeri 2012; El-Wanis et al. 2013).

5.3.1 Tongue Grafting

In tongue grafting (Fig. 5.2a), the rootstock and scion hypocotyls are split in such a way that the scion tongue is inserted into the rootstock, and then the graft union is protected with a grafting clip (Sinha 2018). Novice producers favor the tongue

Fig. 5.2 Grafting methods: (**a**) Tongue grafting, (**b**) Splice grafting, and (**c**) Hole insertion grafting (Kaseb et al. 2021)

grafting method because it is easy, simple, with a high success rate, and does not need healing chambers (Mohamed et al. 2014).

For practicing tongue grafting, both rootstock and scion should have at least one or two true leaves. A downward long narrow cut (having 45° angle) halfway through the rootstock hypocotyl is given below the cotyledons. A similar angled upward cut is given at the scion hypocotyl. The size and angle of the cuts must be identical. The rootstock and scion are brought together, joined in such a way they overlap. Then fixed with a grafting clip or securely wrapped with aluminum foil or parafilm. Five days after the grafting, upper rootstock portion and seven days later, lower scion portion is removed, and grafted plants become ready for planting after two days (Kaseb et al. 2021).

5.3.2 Splice Grafting

Splice grafting is a method of grafting, easier than the hole insertion method. The steps for splice grafting involve cutting the rootstock and scion both at an angle of 45° and then joining them together (Guan and Zhao 2016) (Fig. 5.2b). Diameters of rootstock and scion should be similar at the time of grafting. Commercial seedling nurseries have recently adopted it for vegetables. This technique was created for robotic cucurbit grafting (Sinha 2018). Despite being fairly straightforward, the splice method is not widely used because it results in a lower survival percentage if environmental conditions for healing are not ideal (Memmott and Hassell 2010). The splice method of grafting, however, produces a larger, vigorous, and faster graft junction than other techniques (Khankahdani and Zakeri 2012; El-Wanis et al. 2013).

After 7 to 10 days of sowing, rootstock plants are prepared for grafting when true leaves start to appear. Using a razor blade, one of the cotyledon leaves and apical meristem of rootstock and a matching portion of the scion stem are cut off. Grafting clips are used to hold the two cut surfaces together after being matched up. The grafted plants are then moved in a plastic tunnel or in a healing room, where temperature is maintained at 25 to 30 °C and relative humidity above 85%, and kept until the scion and rootstock become successfully attached (Kaseb et al. 2021).

5.3.3 Hole Insertion Grafting

In hole insertion vegetable grafting, scion is fitted in a hole drilled with a sharp needle into the rootstock. The rootstock stem diameter must be greater than that of the scion stem (Fig. 5.2c). This method is used for cucurbit crops such as watermelons. Squash and bottle gourd are most commonly used as rootstocks of watermelon. Watermelon (scion) seedling diameter should be smaller than rootstock seedlings. High-quality grafts can be produced by using the hole insertion method because it provides a large attachment area for rootstock and scion and also protects the graft union (Guan and Zhao 2016). The hole insertion method is preferred by seasoned watermelon growers because it requires less space and is quickly completed (Lee et al. 2010). However, it needs more time, skilled labor than other techniques and optimum environmental conditions (temperature, humidity, and light) after the grafting process (Hassell et al. 2008).

For germination, seedling trays are kept at a constant temperature of 30 °C. Rootstock plants are ready to be grafted 07 to 10 days after sowing when cotyledons and first true leaf are formed. Scion seeds are planted about 05 to 07 days in advance before grafting. Rootstock is prepared for grafting when the first true leaf has fully developed. With a razor blade, the apical meristem and true leaves of the rootstocks are cut off and a needle is used to make an outward-sloping hole between the cotyledon leaves. Scion hypocotyl is cut from both sides making it V-shaped and then inserted into the rootstock hole. Afterward, the grafted plants are shifted to a healing room or humidity chamber for a period of 12 to 15 days. Then the grafted plants are shifted in a greenhouse and kept at a temperature of 24 to 30 °C after the healing process is completed and until they fully survive (Guan and Zhao 2016; Kaseb et al. 2021).

5.4 Physiological, Anatomical, and Metabolic Changes During Graft Union Development

5.4.1 Physiological and Biochemical Alterations

The formation of graft union mainly affects physiological metabolites that are being regulated by the redox system. Various reactive oxygen species (ROS), i.e., superoxide (O_2^-), singlet oxygen (O_2), hydroxyl radical (OH), and hydrogen peroxide (H_2O_2) are involved in the regulation of the redox system. This redox system plays a crucial role during wound healing after grafting. In plants, grafting activates the redox system in both a positive and a negative way. It results in the elevation of ROS in grafted melon and pear seedlings (Irisarri et al. 2015). Increase in ROS also provides resistance against pathogenic bacteria and promotes lignification at the injury site. However, antioxidant enzymes naturally present in plants detoxify free radicals and inhibit their diffusion into plant cells of grafted region (Xu et al. 2015). Although grafting activates the antioxidant enzymes system, their catalytic activities along with the generation of free radicals may determine the success and failure of grafting.

5.4.2 Anatomical Studies for Examining Graft Union Development and Compatibility

Histological methods have been used in a variety of plants to characterize cellular events at the graft interface, from investigation of plasmodesmata development at the nanoscale level to callus tissue organization at the macroscopic level (Irisarri et al. 2015). The establishment of graft unions in tomato autografts and Arabidopsis has been extensively studied using light and electron microscopy (Flaishman et al. 2008). Callus cells ability to develop symplastic connections differed between compatible and incompatible grafts (Pina et al. 2012). At graft union, the attachment area is indicated by the continuity of rootstock and scion tissues at the junction of the graft. Tomato auto-grafts and intraspecific grafted seedlings showed comparatively large areas of attachment between rootstock and scion indicating a high degree of healing and compatibility when compared with tomato seedlings grafted on pepper plants and eggplant seedlings, which showed comparatively small attachment area (Babar et al. 2023).

5.4.3 Phenolic and Flavonoid Compounds as Conserved Markers for Graft Incompatibility

Graft compatibility and incompatibility have been investigated in many species, in relation to the accumulation of phenolic compounds (Canas et al. 2015). Higher plants naturally possess phenolic compounds (secondary metabolites) which play their distinct role in the metabolic pathways of plants. These inhibit lignification that plays a crucial role during scion and stock connection establishment at early stages. Generally, phenolic compounds are linked with small cells and do not form successful graft unions, thus resulting in incompatibility of grafting (Mng'omba et al. 2008). Several studies also reported that phenylpropanoid metabolites are greatly produced in incompatible graft combinations including grapevine (Canas et al. 2015), loquat (Mng'omba et al. 2008), pear (Musacchi et al. 2000), and plum (Zarrouk et al. 2010). Therefore, phenolic and flavonoid compounds are supposed to be the conserved graft incompatibility markers present at the graft interface (Pina et al. 2017).

5.5 Understanding Graft Incompatibility and Graft Union Development at Molecular Level

Molecular mechanisms at graft union also play a pivotal role in developing compatibility and incompatibility in various graft combinations (Irisarri et al. 2015). Successful grafting depends on cell division and differentiation and requires cell communication and hormones involvement. Phytohormones especially auxins are considered as key players during graft union development. The success of grafting in plants is usually influenced by the expression of auxin-signaling genes (Mo et al. 2018). At the initial stages of graft union development, auxin-related genes induced in the rootstock of Arabidopsis below graft union indicate that auxin is critical for revascularization at graft junction (Melnyk et al. 2018). A study on watermelon grafted on squash revealed that auxin-related genes changed their expression after grafting suggesting that these genes were crucial for auxin transport toward graft union. Graft success is also facilitated by the activity of cellulase-related genes, which are crucial for cell walls reconstruction after grafting (Notaguchi et al. 2020).

Successful grafting involves several complex molecular mechanisms which are active at the graft junction and linked to physiological and developmental events at the scion. In Cucurbitaceae hetero-graft (pumpkin × cucumber), the mRNAs, *CmNACP*, a member of the gene family of NAC (NAM Arabidopsis CUC2), necessary for development of apical meristem was directly transported from pumpkin (rootstock) and *CmNACP* expression upregulated in vascular tissues of cucumber scion (Xanthopoulou et al. 2018). Likewise, a molecule of mRNA (*CmPP16*) encoding the protein of phloem transported with its protein through graft junction and into the phloem of cucumber scion was also present in leaves, stems, and floral tissues of the scion (Harada 2010).

NGS (next-generation sequencing) provided new insight into the rootstock and scion interactions at the genomic level providing evidence that genetic transmission (RNA and DNA fragments) occurs at the grafting site between scion and rootstock (Wu et al. 2013). Additionally, scions of grafted seedlings exhibit crucial changes in the methylation of DNA, indicating that these mechanisms at epigenetic level may contribute to the effects of grafting (Jeynes-Cupper and Catoni 2023). In Cucurbits, intra-specific grafting resulted in alterations in secondary metabolic pathway, methylation pattern of global DNA, and expression patterns of miRNA in grafted scions (Xanthopoulou et al. 2018).

5.5.1 Differential Expression of microRNAs During Grafting

High-throughput sequencing has shown that microRNAs play a significant role in communication between rootstock and scion after grafting. Various miRNAs are differentially expressed in grafted watermelon, which help in stress alleviation (Liu et al. 2013). When cucumber was grafted on pumpkin or vice versa (hetero-grafts),

Li et al. (2013) identified grafted-responsive miRNAs that regulate the physiological processes of the hetero-graft. Expression profiles of miRNAs of tomato grafted on goji (hetero-grafts) and tomato self-grafts, had 163 and 43 miRNAs of differential expression in shoot and fruit, respectively (Khaldun et al. 2016).

5.6 Vegetable Grafting Using Automated Machines (Robots)

Due to the continuous demand of vegetable-grafted plants, more labor is required for grafting making it a costly process. An automated grafting robot can lessen the amount of labor required and increase production by performing planting, clamping, and cutting itself.

Grafting using automated machines can significantly enhance the efficiency of grafting and the survival percentage of grafted plants, which is an essential requirement for the development and adoption of vegetable grafting on a commercial scale (Guo-Ping et al. 2022). The first robot for vegetable grafting was developed in Japan, while the first robot for cucurbit grafting (GR800 series, Iseki and Co. Ltd., Matsuyama, Japan) came to be commercially accessible by 1993 (Sun et al. 2014). China has been making progress in vegetable grafting robots since 1993 and has invented various semi-automatic robots for vegetable grafting. These robots have now been converted to fully automatic grafting robots (Cheng et al. 2015). A single-arm automatic robot for grafting (2JSZ-600) of cucurbit vegetables was introduced in 1998 for picking, cutting, connecting, fixing, and rowing plant seedlings (Guo-Ping et al. 2022). Helper Robotech's GR-800CS is a fruit and vegetable grafting robot that is used for grafting vegetables of the Cucurbitaceae and Solanaceae families (Zhang et al. 2017).

Researchers have invented a variety of grafting robots that can perform splice, tube, insertion, approach, pinhole, and cleft grafting in vegetables. The grafting robot (2Jt-1600B) of melon scion on gourd rootstock was developed to resolve the existing problem of laborious manual grafting using the splice grafting method (Jiang et al. 2012).

Mitsubishi Group, a Japanese company, created the MGM600 automatic grafting robot using the tube grafting technique (Guo et al. 2015). T600 semi-automated robot for vegetable grafting was created allowing only one worker on the seedling establishing rack to automatically accomplish tasks of cutting, docking, fixing, casing, and other rootstock and scion activities (Song 2005). Another version, the Graft1000 automatic grafting robot (ISO Group, Gameren, The Netherlands) was made with a combined matching and grafting system for rootstocks and scions with a digital visual recognition system and a seedling carrying system. Later, they created Graft1200 automated grafting robot in 2010 with 100% success rate (Chen et al. 2010).

5.7 Intraspecific and Interspecific Grafting in Vegetables

There are three types of grafting such as intraspecific grafting in which rootstock and scion belonging to the same species are used; and interspecific and intergeneric grafting in which combinations from different species or genera, respectively are used (Louws et al. 2010).

5.7.1 Grafting in Solanaceous Crops

The most frequently grafted solanaceous crops are tomato and eggplant; however, pepper can also be grafted for particular uses. After the prohibition of methyl bromide as a soil fumigant (Rivard et al. 2010), the main aspect of grafting solanaceous crops was to prevent soil-borne diseases and nematodes attack (King et al. 2008). In solanaceous plants, grafting has been applied to confer protection against abiotic stresses like boron toxicity, salinity, extreme temperatures, and flooding in addition to fungal and bacterial diseases (Rivard et al. 2010).

For tomatoes, the most frequently used rootstocks are tomato hybrids and interspecific hybrids of tomato; however, eggplant rootstocks have also been advocated for particular situations like flooded or humid soil (Bhatt et al. 2015). The most commonly used commercially available tomato rootstock cultivars in the United States are "Maxifort" and "Beaufort", having resistance against tomato mosaic virus, Verticillium, Fusarium root rot, Fusarium crown rot, corky root, and nematodes (López-Pérez et al. 2006). Eggplant cultivar "Surya" is recommended as a tomato rootstock to prevent bacterial wilt in flooded or saturated soils (Black et al. 2003).

The main reason for eggplant grafting was to provide protection against soil-related diseases such as bacterial wilt, Fusarium wilt, Verticillium wilt, and root-knot nematodes. *S. integrifolium* is considered as the most compatible rootstock of eggplant and it enables increased harvesting (Sabatino et al. 2018). In Europe, the widely used eggplant rootstocks are tomato hybrids (Gisbert et al. 2011). This preference is possibly driven by the genetic progress made in tomato rootstocks relative to rootstocks of eggplant, or by gardeners' experience with tomato rootstocks. Grafting also modifies the resistance of pepper plants against Phytophthora blight and bacterial wilt without affecting fruit yield or quality (Kumbar et al. 2021).

5.7.2 Grafting in Cucurbit Crops

The most often grafted cucurbit crops are watermelon, muskmelon, and cucumber in China, South Korea, and Japan. In cucurbits, grafting is primarily used to improve responses of the entire plant to abiotic stresses as well as to give resistance against a variety of soil-borne diseases as well as fungal diseases such as Fusarium wilt, Verticillium wilt, vine decline, and gummy stem blight (Davis et al. 2008a).

Watermelon (*Citrullus lanatus*) grafting on bottle gourd (*Lagenaria siceraria*) rootstock was used to combat Fusarium wilt disease and flooding (Gaion et al. 2017). Bottle gourd, interspecific hybrids between *C. maxima* and *C. moschata,* and wild accessions of watermelon (*C. lanatus* var. *citroides*) are the most popular rootstocks (Davis et al. 2008b). Watermelon (cv. Charleston Gray) grafted on pumpkin (*Cucurbita pepo*) induced drought resistance with improved root length and number of leaves (Huang et al. 2016). Melon grafting on wax gourds (*Benincasa hispida*) and luffas *(Luffa cylindrica*) offers effective protection against a range of soil-related pathogens and also against numerous abiotic factors (Ulas et al. 2020). *Cucurbita* spp., *Cucumis* spp., bottle gourd, *Cucurbita* interspecific hybrids, wax gourd, luffa, and fig leaf gourd (*C. ficifolia*) all can successfully be grafted with cucumber (Bayoumi et al. 2021) for improving cold and Fusarium wilt tolerance (Davis et al. 2008b).

5.8 Grafting to Enhance Tolerance Against Biotic and Abiotic Factors

Grafting has been practiced to prevent soil-borne diseases and improve abiotic stress resistance in cucumber, pepper, tomato, and watermelon in several countries of Asia and Europe, (Lee et al. 2010). Eggplant rootstocks are recommended for grafting tomato scions due to improved tolerance against flooding, bacterial wilt and other soil-related pathogens, and root-knot nematodes (Keatinge et al. 2014). *Cucurbita maxima* grafted on *Cucurbita moschata* have demonstrated cold and copper toxicity tolerance (Rouphael et al. 2012). Cadmium titer in eggplant fruits was decreased when eggplant (*Solanum melongena*) was grafted on *S. torvum* rootstock (Arao et al. 2008). Under greenhouse production, salinity-tolerant rootstocks used for watermelon grafting improved yield by about 81% (Bie et al. 2010). When pepper scion grafted on Phytophthora blight and bacterial wilt-resistant breeding progenies (PR 920, PR 991, and PR 922), they had a higher rate of survival (Jang et al. 2012). Using *S. torvum* rootstocks for tomato grafting significantly increased flood tolerance and yield (Petran 2013).

5.8.1 Grafting-Induced Resistance Mechanisms

According to Guan et al. (2012), grafting induces the following resistance mechanisms:

1. Inherent resilience within rootstocks.
2. Alteration in microbiome diversity.
3. Contribution of vigorous root system.
4. Induction of systemic defense.

5.8.1.1 Inherent Resistance Within Rootstocks

As vegetable grafting is mainly utilized to combat soil-borne pathogens, the strategies of defense are typically linked to intrinsic resilience within rootstock (King et al. 2008). Non-host disease resistance is regarded as the most widespread and sustainable type of tolerance in plants because it provides tolerance to every member of plant species against all strains of specific pathogens. Interspecific rootstocks are mainly utilized in grafting plants of Solanaceae and Cucurbitaceae families. Utilizing non-host resilience in these rootstocks can be a practical strategy for combating certain infections that don't infect the rootstock species as compared to the cultivated species of scion. For instance, *F. oxysporum* f. sp. *lagenariae* commonly causes disease in bottle gourd while *F. oxysporum* f. sp. *niveum* primarily infects watermelon (Namiki et al. 1994). If watermelon cultivars are grafted on bottle gourd rootstocks, non-host resistance has shown as an effectual method for preventing Fusarium wilt because all genotypes of bottle gourd provide resistance to *F. oxysporum* f. sp. *niveum* races (Yetişir et al. 2007).

In host disease resistance, intraspecific grafting is primarily used because it reduces the chances of graft incompatibility. Host resistance is mainly categorized as horizontal and vertical resistance. Vertical resistance is governed by a single gene by cascade responses that are started by interaction between the plants *R* genes and pathogens avirulence (*avr*) genes. While horizontal resistance is governed by numerous genes having minor and non-divergent roles (Fry 1982). Many rootstocks of cucurbitaceous and solanaceous plants have *R* genes. The *Mi* gene gives resistance against *Meloidogyne arenaria*, *M. javanica,* and *M. incognita* in tomato by causing hypersensitivity reactions and reactive oxygen generation at the feeding site (Fuller et al. 2008).

5.8.1.2 Alteration in Microbiome Diversity

Rhizosphere microbes can be crucial in preventing the spread of soil-borne illnesses through a number of processes, including nutrient competition, antagonism, and parasitism (Broeckling et al. 2008). Soil-borne pathogen incidence and severity can also be understood by investigating the diversity of rhizosphere microbes in relation to plant species and genotypes (Yao and Wu 2010). *C. moschata* and cucumber grafting revealed the enhanced total number of beneficial actinomycetes and bacteria, while decreasing the population of fungi (Dong et al. 2010). Grafting of eggplant (susceptible scions) and *Solanum torvum* (rootstock) decreased Verticillium wilt incidence and enhanced the ratio of actinomycetes and bacteria to fungi (Yin et al. 2008).

5.8.1.3 Contribution of Vigorous Root System

The nutritional uptake and whole plant health status can be enhanced in grafted seedlings through vigorous and resistant root system, which can provide tolerance against soil-borne diseases. Grafting vegetables on robust rootstocks can increase the efficiency of nitrogen uptake and utilization (Colla et al. 2010b). Nitrate concentration in the xylem sap of muskmelon, which was grafted on *C. maxima* or *C. moschata* rootstock, was found higher compared to non-grafted melon seedlings (Salehi et al. 2010). Phosphorus concentrations analyzed in different grafted plant tissues such as eggplants, muskmelons, and watermelons have demonstrated that grafting with robust rootstocks can improve phosphorus uptake. Additionally, it was found that grafted seedlings had improved potassium, calcium, and magnesium absorption (Salehi et al. 2010). These nutrients besides playing crucial functions in the growth and development of plants can also play a direct part in defensive mechanisms.

5.8.1.4 Induction of Systemic Defense

The recognition of plant pathogens triggers a wide variety of defensive responses such as reactive oxygen species (ROS) accumulation, antimicrobial compound generation, pathogenesis-related gene expression, and hypersensitivity responses (Buchanan et al. 2015). Peroxidase, superoxide dismutase, and catalase are a few of those that are directly involved in the detoxification of ROS. When exposed to abiotic challenges including salt and low temperature, grafted plants showed stronger antioxidant enzyme activities in comparison to self-rooted plants (He et al. 2009). It was evident that cucumber grafted on *C. moschata* exhibited significantly high proteins expression in contrast to non-grafted cucumber. The proteomic analysis revealed that these proteins were involved in plant defense mechanism and photosynthesis (Li et al. 2009). Similarly, eggplant grafted on tomato rootstocks altered the genes involved in cell cycle regulation, transcription, translation, signal transduction, metabolism, and stress tolerance (Zhang et al. 2008).

The enzyme phenylalanine ammonia-lyase (PAL), which is essential to produce phenylpropanoids, is frequently used as an indicator of a plant's resistance to pathogen attack. Grafted eggplants exhibited higher PAL activity than non-grafted plants because of *V. dahliae* infection in both roots and shoots (Zhou et al. 1998). In addition, PAL activity observed in grafted pepper roots was also higher than in controls (non-grafted) both before and after *F. solani* inoculation (Jiang et al. 2010).

5.9 Effect of Grafting on Vegetable Fruit Quality

Vegetable grafting has the potential to be an essential strategy for improving fruit quality in addition to effectively managing soil-borne pathogens and the development of resistance against abiotic factors depending on the graft combination used (Rouphael et al. 2008). Different rootstock-scion combinations have variable effects on the quality of tomato fruits (Djidonou et al. 2016) by enhancing the efficiency of photosynthesis and absorption of nutrients (Milenković et al. 2019) which ultimately improve the texture of fruits in grafted vegetables (Cushman and Huan 2008). The texture of watermelon fruit was significantly enhanced when grafted on the rootstock of pumpkin (Huitrón-Ramírez et al. 2009).

Grafting can also affect carotenoids (lycopene and β-carotene) content in tomato fruit, but the combination (rootstock and scion) appears to be more critical. Recent research has shown that watermelon grafted on pumpkin, wild watermelon, and gourd rootstock resulted in an enhancement of lycopene content and concentration of trace elements in scion fruit. Additionally, by using this grafting approach, any potential grafting-related negative effects on fruit quality are reduced to a minimum (Nie and Wen 2023).

During the recent years, numerous studies have been conducted to better understand the process of grafting at the molecular level and to assess grafting effects on fruit quality. These studies further elucidated how different rootstock and scion combinations can influence taste, texture, and overall quality of different grafted vegetables. In a study investigating transcriptome changes in watermelon scions grafted on different cucurbit rootstocks, Garcia-Lozano et al. (2019) found that several genes regulating ripening and fruit quality had varied expression patterns in tissues of both scion and rootstock. According to Zhao et al. (2018), various pumpkin rootstocks can significantly modify the expression of genes regulating the production of sugar and aromatic compounds, which can affect fruit quality.

5.10 Effect of Grafting on Fruit Yield

Enhancing the efficiency of rootstocks to absorb nutrients and increase plant resistance are two major benefits of grafting. Therefore, grafted seedlings have improved nutrient absorption, which subsequently stimulate growth and development, and boost yield by regulating the synthesis of endogenous hormones (Zhang et al. 2021). In watermelon, grafts enhanced the availability of nutrients, water absorption, and assimilation of carbon dioxide regulated by rootstock leading to an improvement in fruit production (Rouphael et al. 2008). Melon grafted on pumpkin rootstock significantly enhanced fruit size and consequently improved total fruit yield (Bautista et al. 2011). Selection of rootstock is an important factor in predicting yields of grafted plants as it can increase the number of fruits produced per plant. Tomato plants grafted on vigorous rootstocks showed increased number of fruits per plant, average fruit weight, and yield compared to those plants grafted on less vigorous rootstocks (Schwarz et al. 2013; López-Pérez et al. 2006). In addition, tomato “Florida 47” grafted on “Multifort” or “Beaufort” rootstocks had higher fruit yield as compared to auto-grafted controls (Huitrón-Ramírez et al. 2009). Thus, the selection of appropriate rootstock is important for obtaining higher yields because it influences plant growth and ultimately total fruit yield.

5.11 Conclusions

Production of food in the future will be challenging due to the emergence of new pests and climate change. Plant grafting is one of the most effective techniques, which can be used in vegetable production to combat soil-borne diseases and other biotic and abiotic constraints. Resolving compatibility issues can help to expand grafting use in many other vegetables contributing to sustainable production and biodiversity preservation, which further can help in the expansion of grafting to distantly related plants. However, the production cost of grafted plants is significantly higher as this technique requires highly skilled workers to develop grafted plants. The cost of grafted plants can be greatly reduced by adopting automated grafting machines and robots. Moreover, the adoption of vegetable grafting on a commercial scale requires more investigation on molecular interactions between scion and rootstock. In the future, a deep understanding of grafting in plants will enable us to take all the advantages that plant grafting offers to manipulate agricultural productivity.

References

Arao T, Takeda H, Nishihara E (2008) Reduction of cadmium translocation from roots to shoots in eggplant (*Solanum melongena*) by grafting onto *Solanum torvum* rootstock. Soil Sci Plant Nutr 54(4):555–559

Babar M, Afzal N, Siddiqui K, Azhar A, Galani S (2023) Exploring graft incompatibility markers: Intraspecific and interspecific grafts of tomato (*Solanum lycopersicum* L.). Sci Hortic 310:111762

Bautista AS, Calatayud Á, Nebauer SG, Pascual B, Maroto J, López-Galarza S (2011) Effects of simple and double grafting melon plants on mineral absorption, photosynthesis, biomass and yield. Sci Hortic 130(3):575–580

Bayoumi Y, Abd-Alkarim E, El-Ramady H, El-Aidy F, Hamed E, Taha N, Prohens J, Rakha M (2021) Grafting improves fruit yield of cucumber plants grown under combined heat and soil salinity stresses. Horticulturae 7(3):61

Bhatt RM, Upreti KK, Divya M, Bhat S, Pavithra CB, Sadashiva AT (2015) Interspecific grafting to enhance physiological resilience to flooding stress in tomato (*Solanum lycopersicum* L.). Sci Hortic 182:8–17

Bie Z, Han X, Zhu J, Tang M, Huang Y (2010) Effect of nine squash rootstocks on the plant growth and fruit quality of melon. Acta Hortic 856:77–82

Bie Z, Nawaz MA, Huang Y, Lee JM, Colla G (2017) Introduction to vegetable grafting. In: Colla G, Pérez-Alfocea F, Schwarz D (eds) Vegetable Grafting: Principles and Practices. Croydon, London, UK, pp 1–21

Black LL, Wu DL, Wang JF, Kalb T, Abbass D, Chen JH (2003) Grafting tomatoes for production in the hot-wet season. In: International Cooperators' Guide. Asian Vegetable Research and Development Center, Shanhua, Taiwan. AVRDC Publication # 03-551, pp 1–12

Broeckling CD, Broz AK, Bergelson J, Manter DK, Vivanco JM (2008) Root exudates regulate soil fungal community composition and diversity. Appl Environ Microbiol 74(3):738–744

Buchanan BB, Gruissem W, Jones RL (eds) (2015) Biochemistry and Molecular Biology of Plants, 2nd edn. Wiley

Canas S, Assunção M, Brazão J, Zanol GC, Eiras-Dias JE (2015) Phenolic compounds involved in grafting incompatibility of Vitis spp.: Development and validation of an analytical method for their quantification. Phytochem Anal 26(1):1–7

Chen S, Chiu Y, Chang Y (2010) Development of a tubing-grafting robotic system for fruit-bearing vegetable seedlings. Appl Eng Agric 26(4):707–714

Cheng F, Bie Z, Qiusheng K, Yuan H (2015) The history, current status and future prospects of vegetable grafting in China. Acta Hortic 1086:31–39

Colla G, Rouphael Y, Leonardi C, Bie Z (2010a) Role of grafting in vegetable crops grown under saline conditions. Sci Hortic 127(2):147–155

Colla G, Suárez CC, Cardarelli M, Rouphael Y (2010b) Improving nitrogen use efficiency in melon by grafting. HortScience 45(4):559–565

Cushman K, Huan J (2008) Performance of four triploid watermelon cultivars grafted onto five rootstock genotypes: yield and fruit quality under commercial growing conditions. Acta Hortic 782:335–342

Davis AR, Perkins-Veazie P, Hassell RL, Levi A, King SR, Zhang X (2008a) Grafting effects on vegetable quality. HortScience 43(6):1670–1672

Davis AR, Perkins-Veazie P, Sakata Y, López-Galarza S, Maroto J, Lee SK, Huh Y, Sun Z, Miguel A, King SR, Cohen R, Lee JM (2008b) Cucurbit grafting. Crit Rev. Plant Sci 27(1):50–74

Djidonou D, Simonne A, Koch KE, Brecht JK, Zhao X (2016) Nutritional quality of field-grown tomato fruit as affected by grafting with interspecific hybrid rootstocks. Hort Sci 51(12):1618–1624

Dong L, Zuo Y, Li X, Wang Q (2010) Effects of grafting on cucumber soil biochemistry. J China Agric Univ 15(4):51–56

Dumitrescu IL, Bogoescu M (2022) Experimental research on the variation of microclimate factors in healing tunnels designed for grafted vegetable seedlings. Notulae Botanicae Horti Agrobotanici Cluj-Napoca 50(4):12623

El-Wanis MA, El-Eslamboly A, Salama MA (2013) Impact of different grafting methods on yield and quality of watermelon. Res J Agric Biol Sci 9(6):330–340

Fan J, Yang R, Li X, Zhao W, Zhao F, Wang S (2015) The processes of graft union formation in tomato. Hortic Environ Biotechnol 56(5):569–574

Fita A, Picó B, Roig C, Nuez F (2007) Performance of Cucumis melo ssp. agrestis as a rootstock for melon. J Hortic Sci Biotechnol 82(2):184–190

Flaishman MA, Loginovsky K, Golobowich S, Lev-Yadun S (2008) Arabidopsis thaliana as a model system for graft union development in homografts and heterografts. J Plant Growth Regul 27(3):231–239

Fry WE (1982) Principles of plant disease management. Elsevier eBooks. https://doi.org/10.1016/c2009-0-03045-1

Fuller VL, Lilley CJ, Urwin PE (2008) Nematode resistance. New Phytol 180(1):27–44

Gaion LA, Braz LT, Carvalho RF (2017) Grafting in vegetable crops: a great technique for agriculture. Int J Vegetable Sci 24(1):85–102

Garcia-Lozano M, Dutta S, Natarajan P, Tomason YR, López CMR, Katam R, Levi A, Nimmakayala P, Reddy UK (2019) Transcriptome changes in reciprocal grafts involving watermelon and bottle gourd reveal molecular mechanisms involved in increase of the fruit size, rind toughness and soluble solids. Plant Mol Biol 102(1–2):213–223

Gisbert C, Prohens J, Nuez F (2011) Performance of eggplant grafted onto cultivated, wild, and hybrid materials of eggplant and tomato. Int J Plant Prod 5(4):367–380

Guan W, Zhao X (2016) Techniques for melon grafting. Acta Hortic 1140:335–336

Guan W, Zhao X, Hassell RL, Thies JA (2012) Defense mechanisms involved in disease resistance of grafted vegetables. Hort Sci 47(2):164–170

Guo M, Yang M, Liu B, Niu P, Yang L (2015) Status and development trend of vegetable seedling industry in China. J Agric Mechanization Res 37(1):250–253

Guo-Ping Y, Feng M, Lin W, Huang Y, Tong R, Chen Y (2022) Review and prospect for vegetable grafting robot and relevant key technologies. Agriculture 12(10):1578

Harada T (2010) Grafting and RNA transport via phloem tissue in horticultural plants. Sci Hortic 125(4):545–550

Hassell RL, Memmott F, Liere DG (2008) Grafting methods for watermelon production. HortScience 43(6):1677–1679

He Y, Zhu Z, Yang J, Ni X, Bao-Cheng Z (2009) Grafting increases the salt tolerance of tomato by improvement of photosynthesis and enhancement of antioxidant enzymes activity. Environ Exp Bot 66(2):270–278

Huang Y, Jiao Y, Nawaz MA, Chen C, Liu L, Lu Z, Kong Q, Cheng F, Bie Z (2016) Improving magnesium uptake, photosynthesis and antioxidant enzyme activities of watermelon by grafting onto pumpkin rootstock under low magnesium. Plant Soil 409(1–2):229–246

Huitrón-Ramírez MV, Ricárdez-Salinas M, Camacho-Ferre F (2009) Influence of grafted watermelon plant density on yield and quality in soil infested with melon necrotic spot virus. Hort Science 44(7):1838–1841

Irisarri P, Binczycki P, Errea P, Martens HJ, Pina A (2015) Oxidative stress associated with rootstock–scion interactions in pear/quince combinations during early stages of graft development. J Plant Physiol 176:25–35
Jang Y, Yang E, Cho M, Um Y, Ko K, Chun C (2012) Effect of grafting on growth and incidence of phytophthora blight and bacterial wilt of pepper (*Capsicum annuum* L.). Hortic Environ Biotechnol 53(1):9–19
Jeynes-Cupper K, Catoni M (2023) Long distance signaling and epigenetic changes in crop grafting. Front Plant Sci 14:1121704
Jiang F, Liu XY, Liu W, Zheng N, Wang HT, Xi-Zhen A (2010) Relationship between root rot resistance and phenylaprapanoid metabolism in graft *Capsicum*. China Vegetables 08:46–52
Jiang K, Zheng W, Zhang Q, Guo R, Feng Q (2012) Development and experiment of vegetable grafting robot. Trans Chinese Soc Agric Eng 28(4):8–14
Johnson SJ, Miles CA (2011) Effect of healing chamber design on the survival of grafted eggplant, tomato, and watermelon. Hort Technology 21(6):752–758
Kaseb MO, Umer M, Mahmud E, Anees M, Diao W, Yuan P, Zhu H, Zhao S, Lu X, He N, El-Remaly E, Liu W (2021) Physio-anatomical study of polyploid watermelon grafted by different methods. Agronomy 11(5):913
Keatinge JDH, Lin L, Ebert AW, Chen W, Hughes J, Luther G, Wang J, Ravishankar M (2014) Overcoming biotic and abiotic stresses in the Solanaceae through grafting: current status and future perspectives. Biol Agric Hortic 30(4):272–287
Khaldun ABM, Huang W, Lv H, Liao S, Zeng S, Wang Y (2016) Comparative profiling of miRNAs and target gene identification in distant-grafting between tomato and Lycium (Goji berry). Front Plant Sci 7:1476
Khankahdani HH, Zakeri E (2012) Evaluation of different rootstocks and grafting techniques on graft union percent, yield and yield components of watermelon cv. "Crimson Sweet". World Appl Sci J 18:645–651
King SR, Davis AR, Liu W, Levi A (2008) Grafting for disease resistance. Hort. Science 43(6):1673–1676
Kumbar S, Narayanankutty C, Kurian PS, Sreelatha U, Barik S (2021) Evaluation of eggplant rootstocks for grafting eggplant to improve fruit yield and control bacterial wilt disease. Eur J Plant Pathol 161(1):73–90
Kyriacou MC, Rouphael Y, Colla G, Zrenner R, Schwarz D (2017) Vegetable grafting: The implications of a growing agronomic imperative for vegetable fruit quality and nutritive value. Front Plant Sci 8:741
Lee JM, Kubota C, Tsao S, Bie Z, Echevarría PH, Morra L, Oda M (2010) Current status of vegetable grafting: Diffusion, grafting techniques, automation. Sci Hortic 127(2):93–105
Li C, Li Y, Bai L, Zhang T, He C, Yan Y, Yu X (2013) Grafting-responsive miRNAs in cucumber and pumpkin seedlings identified by high-throughput sequencing at whole genome level. Physiol Plant 151(4):406–422
Li Y, Liang G, Liu X, Liu D, Fang C (2009) Proteomic study on grafted and non-grafted cucumber (Cucumis sativus L.). Acta Hortic Sinica 36(8):1147–1152
Liu N, Yang J, Guo S, Xu Y, Zhang M (2013) Genome-wide identification and comparative analysis of conserved and novel microRNAs in grafted watermelon by high-throughput sequencing. PLoS One 8(2):e57359
López-Pérez JA, Strange ML, Kaloshian I, Ploeg AT (2006) Differential response of Mi gene-resistant tomato rootstocks to root-knot nematodes (Meloidogyne incognita). Crop Prot 25(4):382–388
Louws FJ, Rivard CL, Kubota C (2010) Grafting fruiting vegetables to manage soilborne pathogens, foliar pathogens, arthropods and weeds. Sci Hortic 127(2):127–146
Melnyk CW, Gabel A, Hardcastle TJ, Robinson S, Miyashima S, Grosse I, Meyerowitz EM (2018) Transcriptome dynamics at Arabidopsis graft junctions reveal an intertissue recognition mechanism that activates vascular regeneration. Proc Natl Acad Sci USA 115(10):E2447–E2456
Memmott F, Hassell RL (2010) Watermelon (*Citrullus lanatus*) grafting method to reduce labor cost by eliminating rootstock side shoots. Acta Hortic 871:389–394

Milenković L, Mastilović J, Kevrešan Ž, Bajić A, Gledić A, Stanojević J, Cvetković D, Sunić L, Ilić ZS (2019) Effect of shading and grafting on yield and quality of tomato. J Sci Food Agric 100(2):623–633

Mng'omba SA, du Toit ES, Akinnifesi FK (2008) The relationship between graft incompatibility and phenols in Uapaca kirkiana Müell Arg. Sci Hortic 117(3):212–218

Mo Z, Feng G, Su W, Liu Z, Peng F (2018) Transcriptomic analysis provides insights into grafting union development in pecan (*Carya illinoinensis*). Genes 9(2):71

Mohamed F, El-Hamed KA, Elwan M, Hussien M (2014) Evaluation of different grafting methods and rootstocks in watermelon grown in Egypt. Scientia Hortic 168:145–150

Musacchi S, Pagliuca G, Kindt M, Piretti MV, Sansavini S (2000) Flavonoids as markers for pear-quince graft incompatibility. J Appl Bot 74:206–211

Namiki F, Shiomi T, Kayamura T, Tsuge T (1994) Characterization of the formae speciales of *Fusarium oxysporum* causing wilts of cucurbits by DNA fingerprinting with nuclear repetitive DNA sequences. Appl Environ Microbiol 60(8):2684–2691

Nie W, Wen D (2023) Study on the applications and regulatory mechanisms of grafting on vegetables. Plan Theory 12(15):2822

Notaguchi M, Kurotani K, Sato Y, Tabata R, Kawakatsu Y, Okayasu K, Sawai Y, Okada R, Asahina M, Ichihashi Y, Shirasu K, Suzuki T, Niwa M, Higashiyama T (2020) Cell-cell adhesion in plant grafting is facilitated by β-1,4-glucanases. Science 369(6504):698–702

Pessarakli M (2016) Handbook of cucurbits: growth, cultural practices, and physiology. CRC Press, Boca Raton

Petran AJ (2013) Interspecific grafting of tomato (*Solanum lycopersicum*) onto wild eggplant (*Solanum torvum*) for increased environmental tolerances. M.S. thesis. University of Minnesota, Minneapolis, USA

Pina A, Cookson SJ, Calatayud A, Trinchera A, Errea P (2017) Physiological and molecular mechanisms underlying graft compatibility. In: Colla G, Pérez-Alfocea F, Schwarz D (eds) Vegetable Grafting: Principles and Practices. Croydon, London, pp 132–154

Pina A, Errea P, Martens HJ (2012) Graft union formation and cell-to-cell communication via plasmodesmata in compatible and incompatible stem unions of *Prunus* spp. Sci Hortic 143:144–150

Rivard CL, O'Connell S, Peet MM, Louws FJ (2010) Grafting tomato with interspecific rootstock to manage diseases caused by *Sclerotium rolfsii* and southern root-knot nematode. Plant Dis 94(8):1015–1021

Rouphael Y, Cardarelli M, Colla G, Rea E (2008) Yield, mineral composition, water relations, and water use efficiency of grafted mini-watermelon plants under deficit irrigation. Hort Science 43(3):730–736

Rouphael Y, Cardarelli M, Rea E, Colla G (2012) Improving melon and cucumber photosynthetic activity, mineral composition, and growth performance under salinity stress by grafting onto Cucurbita hybrid rootstocks. Photosynthetica 50(2):180–188

Rouphael Y, Kyriacou MC, Colla G (2018) Vegetable grafting: a toolbox for securing yield stability under multiple stress conditions. Front Plant Sci 8:2255

Sabatino L, Iapichino G, D'Anna F, Palazzolo E, Mennella G, Rotino GL (2018) Hybrids and allied species as potential rootstocks for eggplant: Effect of grafting on vigour, yield and overall fruit quality traits. Sci Hortic 228:81–90

Salehi R, Kashi A, Lee JM, Babalar M, Delshad M, Lee S, Huh Y (2010) Leaf gas exchanges and mineral ion composition in xylem sap of Iranian melon affected by rootstocks and training methods. Hor tScience 45(5):766–770

Savvas D, Colla G, Rouphael Y, Schwarz D (2010) Amelioration of heavy metal and nutrient stress in fruit vegetables by grafting. Sci Hortic 127(2):156–161

Schwarz D, Öztekin GB, Tüzel Y, Brückner B, Krumbein A (2013) Rootstocks can enhance tomato growth and quality characteristics at low potassium supply. Sci Hortic 149:70–79

Schwarz D, Rouphael Y, Colla G, Venema JH (2010) Grafting as a tool to improve tolerance of vegetables to abiotic stresses: Thermal stress, water stress and organic pollutants. Sci Hortic 127(2):162–171

Sinha A (2018) Vegetable grafting for abiotic stress tolerance. J Pharmacogn Phytochem 7(4S):304–307

Song G (2005) An introduction on the development of industrial growing system of vegetable research grafting seedlings. J Agric Mechanization Res 27(6):52–54

Sun Q, Zhao D, Wang C, Zhao Y (2014) Design of a sapling branch grafting robot. J Robotics 2014:604746

Ulas A, Aydin A, Ulas F, Yetişir H, Miano TF (2020) Cucurbita rootstocks improve salt tolerance of melon scions by inducing physiological, biochemical and nutritional responses. Horticulturae 6(4):66

Ulas F, Aydin A, Ulas A, Yetişir H (2019) Grafting for sustainable growth performance of melon (*Cucumis melo*) under salt stressed hydroponic condition. Eur J Sustain Dev 8(1):201–210

Wu R, Wang X, Lin Y, Ma Y, Liu G, Yu X, Zhong S, B L (2013) Inter-species grafting caused extensive and heritable alterations of DNA methylation in Solanaceae plants. PLoS One 8(4): e61995

Xanthopoulou A, Tsaballa A, Ganopoulos I, Kapazoglou A, Avramidou EV, Aravanopoulos FA, Moysiadis T, Osathanunkul M, Tsaftaris A, Doulis AG, Kalivas A, Sarrou E, Martens S, Nianiou-Obeidat I, Madesis P (2018) Intra-species grafting induces epigenetic and metabolic changes accompanied by alterations in fruit size and shape of *Cucurbita pepo* L. Plant Growth Regul 87(1):93–108

Xu Q, Guo S, He L, Du N, Shu S, Sun J (2015) Physiological aspects of compatibility and incompatibility in grafted cucumber seedlings. J Am Soc Hortic Sci 140(4):299–307

Yao H, Wu F (2010) Soil microbial community structure in cucumber rhizosphere of different resistance cultivars to fusarium wilt. FEMS Microbiol Ecol 72(3):456–463

Yetişir H, Kurt Ş, Sari N, Tok FM (2007) Rootstock potential of Turkish *Lagenaria siceraria* germplasm for watermelon: plant growth, graft compatibility, and resistance to Fusarium. Turk J Agric For 31(6):381–388

Yin Y, Zhou B, Li Y, Fu Y (2008) Allelopathic effects of grafting on rhizosphere microorganisms' population of eggplants. Acta Hortic Sinica 35(8):1131–1136

Zarrouk O, Testillano PS, Risueño MC, Sánchez MÁM, Gogorcena Y (2010) Changes in cell/tissue organization and peroxidase activity as markers for early detection of graft incompatibility in peach/plum combinations. J Am Soc Hortic Sci 135(1):9–17

Zhang K, Chu J, Zhang T, Yin Q, Kong Y, Liu Z (2017) Development status and analysis of automatic grafting technology for vegetables. Nongye Jixie Xuebao/Trans Chinese Soc Agric Mach 48(3):1–13

Zhang ZH, Li MM, Cao BL, Chen ZJ, Xu K (2021) Grafting improves tomato yield under low nitrogen conditions by enhancing nitrogen metabolism in plants. Protoplasma 258(5):1077–1089

Zhang ZJ, Wang YM, Long LK, Lin Y, Pang JS, Liu B (2008) Tomato rootstock effects on gene expression patterns in eggplant scions. Russ J Plant Physiol 55(1):93–100

Zhao L, Liu AQ, Song T, Jin Y, Xu X, Gao Y, Ye X, Qi H (2018) Transcriptome analysis reveals the effects of grafting on sugar and α-linolenic acid metabolisms in fruits of cucumber with two different rootstocks. Plant Physiol Biochem 130:289–302

Zhou BL, Gao XY, Lin GR, Fu YW (1998) Relationship between disease resistance and electrolytic leakage, proline content and PAL activity in grafted eggplant. Acta Hortic Sinica 25(3):300–302

6 Methods of Open Field Cultivation of Vegetables

Muhammad Saqib, Aqsa Haram, Arslan Akbar, and Muhammad Akbar Anjum

Abstract

Vegetables are cultivated through various conventional and non-conventional methods in the world. Choosing the right cultivation method is the key for successful vegetable production. In Pakistan, vegetables are grown outdoors round the year as well as in polytunnels during winters. In open fields, vegetables are grown through various techniques including flatbeds, ridges, and raised beds depending upon the crop type, soil characteristics, and irrigation method to be used. In this chapter, different techniques of vegetable cultivation including broadcasting, dibbling, drilling, check row planting, ridge furrow planting, and raised bed planting under open field conditions are discussed. Each method is further described for its suitability and limitations for specific crop and growing conditions. In developed countries, experiments have also been conducted to grow vegetables in soilless media in outdoor environments due to the unavailability of suitable land as well as producing organic vegetables using organic growing media.

Keywords

Growing media · Irrigation practices · Line sowing · Outdoor cultivation · Ridge furrow planting

M. Saqib (✉)
Department of Horticulture, The University of Agriculture, Dera Ismail Khan, Pakistan
e-mail: muhammad.saqib@uad.edu.pk

A. Haram · A. Akbar · M. A. Anjum
Department of Horticulture, Bahauddin Zakariya University, Multan, Pakistan

Z. Iqbal et al. (eds.), *Sustainable and Innovative Vegetable Production in times of Climate Change*, https://doi.org/10.1007/978-981-96-6283-8_6

6.1 Introduction

Vegetables contribute a major share in global food production and play a significant role in food security. These are rich source of nutrients and are an essential part of our daily diet. These may serve as a primary food source as well as secondary additions to main vegetable dishes. In 2022, production of vegetables reached up to 1.17 billion metric tons in the world. China is the leading vegetable producer contributing 34% share in global production (Statista 2024). Pakistan produced approximately 1008 million tons of vegetables on an area of 500 thousand hectares during 2022–23 (GOP 2023). Vegetables are cultivated through various methods viz. conventional and non-conventional worldwide. Choosing the right cultivation method is essential for optimizing vegetable production. The sowing method is critical for large-scale productions. Cropping patterns and seed-planting procedures significantly impact seed germination, crop growth, and ultimately yield per hectare. Though most farmers in Pakistan cultivate vegetables through conventional methods (open field cultivation in traditional ways); however, with the awareness of new and more productive methods (indoor, organic, and innovative vegetable cultivation), farmers tend toward new modes of cultivation. Vegetables grown through different methods are discussed below.

6.2 Outdoor Cultivation of Vegetable Crops

Open field or outdoor cultivation methods for vegetables production are ancient. These traditional methods for planting have been practiced to date. Planting methods and planting distance of some commonly grown vegetables are given in Table 6.1. Different methods of planting the vegetable crops in open field are detailed below.

6.2.1 Cultivation on Flat Surface

In flat surface cultivation, vegetable crops are planted directly on the levelled field without altering the soil elevation. The soil is tilled and crops are planted in rows on flatbeds or broadcasted over the entire surface (Welbaum 2015).

6.2.1.1 Methods of Flat Surface Cultivation

On flat surfaces, vegetables can be cultivated in different ways as described below.

6.2.1.1.1 Broadcasting

In broadcasting, seeds are sown mechanically or manually on the seedbed. The sowing method is to sow the seeds evenly and cover them with planking. If there are many seeds, a mechanical spreader or broadcaster is used to do the job. This system has a very high seeding rate. Crops like spinach, fenugreek, purslane, cilantro,

Table 6.1 Planting methods and planting distances for vegetables commonly grown in Pakistan

Type of vegetable	Planting method	Row-row/ line-line distance	Plant-plant distance	References
Leafy vegetables				
Spinach	Broadcast, line sowing, ridge sowing for early crop	15–20 cm/ 30–45 cm	6–8 cm	Solangi et al. (2017)
Fenugreek	Flatbed sowing	20–25 cm	5–7 cm	Zaman et al. (2018)
Lettuce	Flatbed sowing, ridge sowing	45–60 cm	25–30 cm	Baloch (1994)
Root vegetables				
Radish	Double-side ridge planting	45 cm	6–8 cm	Jawad et al. (2015)
Carrot	Double-side ridge planting	60–75 cm	6–8 cm	El-Sayed (2021)
Turnip	Drilling on flatbeds	30–40 cm	10–12 cm	Abida et al. (2016)
Beetroot	Double-side ridge planting	40 cm	6–8 cm	Baloch (1994)
Vine vegetables				
Cucumber	Raised beds	2 m	30–45 cm	AARI (2024)
Pumpkin	Raised beds	3 m	45–60 cm	Thind and Mahal (2021)
Bitter gourd	Raised beds	2.5–3 m	20–30 cm	Rosin et al. (2017)
Sponge/ridge gourd	Raised beds	1–1.5 m	15–25 cm	Baloch (1994)
Musk melon	Raised beds	2.5 m	30–45 cm	AARI (2024)
Watermelon	Raised beds	3 m	50–60 cm	AARI (2024)
Solanaceous crops				
Tomato	Bed planting	1–1.5 m	60 cm	Baloch (1994)
Chilies	Single-side ridge planting	60 cm	30 cm	Zeb et al. (2018)
Brinjal	Single side ridge planting	60 cm	45 cm	Thind and Mahal (2021)
Potato	Ridges top	60–75 cm	30 cm	Gupta and Singh (1994)
Bulb crops				
Onion	Ridge planting on double side	60 cm	10–15 cm	Baloch (1994)
Garlic	Dibbling in lines	15–20 cm	7–8 cm	Thind and Mahal (2021)
Cole crops				

(continued)

Table 6.1 (continued)

Type of vegetable	Planting method	Row-row/ line-line distance	Plant-plant distance	References
Cauliflower	Single-side ridge planting	45–60 cm	45–60 cm	Baloch (1994)
Cabbage	Single-side ridge planting	45–60 cm	45–60 cm	Thind and Mahal (2021)
Broccoli	Ridges/flatbeds	45 cm	30–45 cm	Thind and Mahal (2021)
Others				
Peas	Dibbling on ridges/drill on flatbed	1.25 m / 45–60 cm	15–20 cm	Zaidi et al. (2019)
Okra	Ridge sowing on double sides/flatbed	60 cm / 45 cm	20–30 cm	Basit et al. (2021)

amaranth, cluster beans, turnip, carrot, etc. can be sown through broadcasting (Solangi et al. 2017; Zaman et al. 2018; El-Sayed 2021).

Usually, seed broadcasting is inadvisable because all planting procedures require equal seed placement and good seed-to-soil contact for better seed germination and seedling emergence, which cannot be obtained by broadcasting. When sowing by hand, achieving a uniform seed distribution is impossible due to uneven sowing between and within rows, creating clumps and gaps in the field. Control over sowing depth is limited. The plant population is adjusted using high seeding rates and thinning becomes necessary (Awan et al. 2016).

6.2.1.1.2 Dibbling

In the dibbling method, holes are made in the sowing area, the seeds are planted, and seedbeds are covered through the soil. The holes are made to a certain depth. A dibbler is used for this purpose. It is a cone-shaped tool that makes holes in the field. Bulb crops like garlic and onion and some other crops are sown through this method. Dibbling is also common for transplanting seedlings of different vegetables such as cole crops and solanaceous vegetables in the field.

Garlic (*Allium sativum* L.) is planted on flatbeds in lines. The detached individual cloves from the mother bulb are dibbled about 3 cm deep, with the distance of plant to plant 10 cm in lines and 20 to 25 cm between the lines in flatbeds. Although flood irrigation is usually practiced for garlic crops, proper drainage management is necessary in the rainy season to prevent root diseases (AARI 2024).

6.2.1.1.3 Drilling

In this method, furrows are created with a plow or using a bamboo or metal funnel attached to a plow for seed placement, and the seeds are dropped or drilled into the furrows or rows in a consistent stream. Afterward, the seeds are covered with soil, either manually or using machinery. This ensures that the right number of seeds is sown at the appropriate depth and spacing. Examples include spinach, fenugreek, lettuce, peas, grams, turnip, carrot, etc. (Solangi et al. 2017; El-Sayed 2021; Thind and Mahal 2021).

Solangi et al. (2017) while comparing the different sowing methods for spinach (*Spinacia oleracea* L.) concluded that sowing on flatbeds (drilling in rows) had better results in terms of plant height, number of leaves, time to first cutting, and yield ha^{-1}.

6.2.1.1.4 Check Row Planting

It is one of the traditional and ancient methods. The seeds are planted alongside straight, parallel furrows using a check row planter. The distance between rows and among the plants is kept uniform. The rows are typically in two perpendicular directions, which makes it easier to cultivate and harvest and helps to keep the fields free of weeds.

A study was conducted to design, manufacture, and test planters for selected vegetables, i.e., okra (*Abelmoschus esculentus* L.), cilantro (*Coriandrum sativum* L.), and cowpea (*Vigna unguiculata*). The aim was to sow the seeds in rows at the desired plant-to-plant distance and cover them with soil using a planter. The study's findings revealed that the field efficiency was 83.2% for okra, 83.03% for cowpeas, and 78% for cilantro. Additionally, the study demonstrated that this mechanical sowing method was more efficient in terms of field use and required less time compared to the traditional broadcast method (Shravan and Jogdand 2020).

Soil types, climatic conditions, irrigation methods, nutrients availability, and sowing methods affect the performance of different vegetable crops in different ways. Some vegetable crops like spinach, arum (taro), fenugreek, garlic, onion, cilantro, and cowpea can grow well on flatbeds and don't need a lot of drainage (Thind and Mahal 2021). Ali (2017) reported that managing the nursery seedlings of the autumn season onion (*Allium cepa* L.) crop was difficult because of high temperatures and monsoon precipitations, so skipping the step of raising the nursery and directly sowing onion sets in the field could be the wise decision for high yield. For cultivation of one hectare of land, around 1375 kg onion sets having diameter of 17–21 mm were required for flatbed planting with distance of 10 cm plant-plant and 25 cm between rows. Mid-August to first week of September was the recommended time for planting onions through sets.

6.2.1.2 Irrigation Practices for Flat Surface Cultivation

Traditionally, flood irrigation or irrigation in rows of flatbeds is done on flat surface cultivation. Irrigation intervals depend on the crop and soil type, growth stage of the crop, and weather conditions. Mal and Kaur (2019) reviewed the irrigation practices in vegetable crops and suggested that applying water at the perfect phenological phase could reduce disease and pathogen incidence. Time and intervals of irrigation could affect the phenological stages of plant. For example, onion as a shallow-rooted crop needed frequent irrigation. In onion crops, the most pivotal phases of water requirements were the period of bulb formation and the next 30 days of bulb development. Onions obtained most of their moisture needs (90%) from 30 cm of topsoil. The irrigation of onion crops was done by using the check basin method.

6.2.1.3 Advantages and Disadvantages of Flat Surface Cultivation

Flat surface cultivation has the following advantages:

- The flat surface method is simple, cost-effective, requires minimal soil preparation, and is economical for large-scale production.
- Flat fields facilitate the use of machinery, such as tractors and mechanical harvesters, which can reduce labor costs and increase efficiency.
- Suitable for crops that do not require significant drainage.

  However, it also has the following disadvantages:

- Flat surfaces may lead to waterlogging, especially in heavy rainfall areas or poorly draining soils.
- Crust formation in flat surfaces may occur in heavy soils which affects the germination of seeds.
- Poor drainage may lead to more incidence of root diseases like root rot.
- The uniform surface can encourage weed growth, which may require more intensive weed management strategies.

6.2.2 Cultivation on Ridges

Ridge cultivation involves making long, narrow, elevated ridges about 10–15 cm high with furrows between them, often used in areas with heavy rainfall or poorly draining soils. Crops are planted along the ridges, and furrows are maintained to manage water flow. Ridge size and height may vary according to crop type (Welbaum 2015). In the ridge cultivation method, water and fertilizers are used more efficiently. Ridge cultivation is a suitable growing method for root crops (turnip, radish, carrot, beetroot), tuber crops (potato, sweet potato), and some other vegetables (tomato, pea, peppers). It is a suitable method as it prevents water logging conditions and occurrence of root diseases, and also increases productivity.

Cultivation methods for potato (*Solanum tuberosum* L.) vary depending on the soil type and specific region of potato-growing areas. Gupta and Singh (1994) reported that potatoes grown through ridge and furrow method in sandy loam soils gave a high yield of tubers. In hilly areas, such as Kaghan valley of Pakistan, potato tubers are planted randomly without a certain distance between them. In the plains of KPK and Punjab, potatoes are grown on ridges made at 75 cm apart with plant-plant distance of 25 cm (Ali 2010). Zeb et al. (2018) evaluated the effect of ridge height on the performance of different cultivars of chili (*Capsicum annuum* L.) crop in agro-climatic conditions of Mansehra, Pakistan, and found that chilies planted at 40 cm high ridges produced higher yields and showed less incidence of root rot disease as compared with the same planted on smaller ridges of 20 and 30 cm height.

Basit et al. (2021) assessed different cultivating methods of okra (*Abelmoschus esculentus* L.) in a field study at Peshawar, Pakistan. Seeds of okra cultivar "Swat Green" were sown by two methods, i.e., on flatbeds and ridges. Different numbers of picking intervals (3, 4, and 5 days) were used in the study. The results revealed that okra plants grown on ridges with a 3-day picking interval of pods greatly affected plant growth and yield in terms of pod length (14.44 cm), pod weight

(22.41 g), fresh pod yield (0.165 tons ha^{-1}), seed weight (6.41 g), and seed yield (102.2 kg ha^{-1}) as compared to other treatments. Based on the results, okra seed plantation on ridges with approximately 3-day picking interval was recommended for the agro-climatic region of Peshawar, Pakistan.

Root crops grow best by ridge and furrow method. In the ridge planting method, an essential amount of water and nutrients can be applied to the crops. Jawad et al. (2015) analyzed the efficiency of nitrogen fertilizer with different sowing methods (ridges and flatbeds) in radish (*Raphanus sativus* L.) in the arid region of Dera Ghazi Khan, Pakistan. Results revealed that radish grown on ridges with the application of 150 kg ha^{-1} of nitrogen fertilizer gave the best results with maximum root yield (20.37 tons ha^{-1}).

Planting method has a significant impact on weed infestation. Generally, ridge planting faces less weeds infestation as compared with flat sowing as less area of land is irrigated in ridges. A study was conducted by Abida et al. (2016) to assess the impact of ridge and flat sowing on weed infestation in turnip crop (*Brassica rapa* L.) in agro-climatic areas of Peshawar, Pakistan. The results showed that the "Pahuja" cultivar of turnip planted by ridge method gave maximum yield (14.68 tons ha^{-1}) and minimum level of weed infestation. Thus, the ridge planting method was recommended for future sowing of "Pahuja" cultivar of turnip.

6.2.2.1 Irrigation Practices for Ridge Cultivation

In the ridge cultivation method, irrigation is applied through the furrow method. When watering crops, fields and ridges are never flooded, and water is supplied in furrows in an amount not exceeding 2/3 of the height of ridges. In areas where potatoes are grown on ridges, water can be applied in alternate furrows, saving 36% of irrigation water by losing 10.7% of yield. An irrigation depth of 05 cm is considered optimal for potatoes on sandy-loam soils. Sprinkler irrigation can also be applied to potato crops. Root vegetables like radish, turnip, and beetroot have shallow roots and primarily absorb water from the top 30 cm of the soil. The shallow furrow method is used to irrigate the root crops and Cole crops like cabbage and cauliflower planted on ridges (Mal and Kaur 2019).

6.2.2.2 Advantages and Disadvantages of Ridge Cultivation

There are the following advantages of ridge cultivation:

- The furrows between ridges allow efficient water management, preventing waterlogging and directing excess water away from crop roots.
- The ridge cultivation method improves root aeration and soil structure, promoting healthy plant growth.
- Weeding is easier to manage due to the ridge structure, which allows mechanical and manual methods to be used effectively.

This method has the following disadvantages:

- In sloped areas, ridges can be susceptible to erosion, especially during heavy rainfall.
- The ridge cultivation method requires more labor and management to form and maintain ridges contrasted with flat surface cultivation.

6.2.3 Cultivation on Raised Beds

It is a method of sowing vegetable crops where plants are grown on raised beds, with furrows or channels between them. These beds are typically elevated and may vary in width and height depending on the crop type, soil properties, and local environmental conditions. This system improves water management, aeration, and root development, making it a better choice for cultivation of many vegetable crops. Raised beds are usually about 15–30 cm high and 01–1.5-meter-wide (Mal and Kaur 2019; AARI 2024). Furrows between the raised beds are used for irrigating the field. For making bed and sowing seeds, bed planters are used. The raised beds warm up more quickly than flat surfaces, allowing for earlier planting and better growth in cool climates. Vegetable crops like tomato, aubergine (brinjal or eggplant), peppers, cucurbits, peas, etc., are grown on raised beds (Malik 1994; Ali 2017). Vegetables such as squash, aubergine, cauliflower, and cabbage produce good crops on raised beds on sandy loam soils during precipitating weather because the raised bed with furrow method provides a better drainage system and allows extra water to drain (Goswami 2005).

Bed size and planting distance are important factors in bed planting system. Pea crop was planted on raised beds keeping 05 cm high, 0.75 m wide with different plant spacings, i.e., 06–20 cm. The results revealed that in comparison to broadcast method, seed cost decreased by 291.9% through the pea planter. Thus, it was suggested that pea planter could improve vegetable mechanization in the country. By decreasing row-to-row distance in future, plant density and crop yield can also be increased (Zaidi et al. 2019).

In hilly areas of Pakistan like Kaghan Valley, where potato cultivation is practiced, potatoes grow well on plain wide beds with one side covered with soil. Distance between rows is usually kept 75 cm, while plant-to-plant distance is maintained at 20 cm, which gives higher yield (12.4 t/ha) as compared to other sowing methods like local haphazard planting, in furrows and on ridges. In the plain wide bed method, water and nutrient use efficiency increases thus more production of healthy tubers is obtained (Qasim et al. 2013).

Similarly, a comparison of different sowing methods was also made for spinach crop. Flatbed seed broadcasting, flatbed in rows through seed drilling, and ridge and raised bed planting were compared. The results highlighted that among all the four sowing methods, raised bed method performed better in terms of plant growth having 34.5 cm plant height with 21 leaves plant^{-1} and 2119 kg spinach yield ha^{-1}. The first cutting was taken 30 days after planting when grown on raised beds as compared to others. So, the raised bed sowing was recommended as an effective method for spinach (Solangi et al. 2017).

6.2.3.1 Hill Dropping

The hill-dropping planting method involves dropping selected seeds at pre-arranged intervals rather than in a continuous stream. The distance between plants in the same row is consistent. Hill dropping is the opposite of seed drilling, where the seeds are dropped in a steady stream, but the spacing from one plant to another in a row is not usually constant (Topakci et al. 2011). Usually, cucurbits like cucumber, muskmelon, watermelon, squash, pumpkin, etc. are sown through this method on raised beds (Thind and Mahal 2021; AARI 2024).

6.2.3.2 Irrigation Practices for Raised Bed Cultivation

Furrow irrigation is used when crops are sown on raised beds with one furrow between two raised beds. Tomatoes (*Lycopersicon esculentum*) have deep roots so they can draw groundwater from deeper layers. Crops are usually irrigated using bed furrow method, however, drip irrigation is also a suitable method. Peas grown on raised beds are usually irrigated by furrow irrigation. In cucurbits, for example, cucumber (*Cucumis sativus*), water stress can reduce yield of the crop and higher application of water increases fruit yield. So frequent irrigation is needed with furrow method while growing on raised beds (Mal and Kaur 2019; AARI 2024).

6.2.3.3 Advantages and Disadvantages of Raised Bed Cultivation

Cultivation on raised bed has the below given advantages:

- Raised beds enhance drainage, which reduces the risk of waterlogging and root diseases.
- Raised beds facilitate high-quality soil or compost, which enhances root development and promotes healthier plant growth.
- Raised beds help to reduce soil erosion and compaction, especially in areas with heavy rainfall or wind.
- Raised beds usually warm up rapidly during the spring, facilitating earlier plantings.
- The defined bed structure makes it easier to control weeds, manually or with herbicides.

Cultivation on raised bed has also some disadvantages, as given below:

- The construction of raised beds requires labor and machines, and can be expensive and time-consuming, especially for large areas.
- Raised beds may dry out more quickly and require more frequent watering.

6.3 Soilless Vegetable Production in Outdoors

Soil is the most abundant, cheap, easy to use, and naturally available growing medium for producing vegetables in fields. Globally, outdoor vegetable cultivation is done in the soil, though soil type varies from region to region. Soil type requirement also varies with the vegetable to be grown. For example, root and tuber vegetables including potato, carrot, and beetroot perform well under light loamy soil, while shallow-rooted vegetables viz. cabbage, lettuce, and beans give good production in heavy clayey soils. Although commercial vegetable production is economically suitable in field soils, however, soil degradation due to intensive cropping system, unavailability of suitable land, weed infestation, and less productive soils make successful vegetable production difficult in natural soils in open fields. In recent years, researchers from the University of Florida have evaluated different soilless media for growing vegetables outdoors in order to overcome the soil-related problems. Soilless culture as a system of vegetable crop production has been used for centuries around the globe and currently it is commonly used for vegetable production in greenhouses in Europe, the United States, Middle East, Japan, Canada, and other advanced countries. Initially, growers have selected some high-value vegetables including pepper, squash, cucumber, and onion, and adopted lay-flat perlite bag soilless system (Hochmuth and Hochmuth 2021).

6.3.1 Field Selection and Preparation

This system can be adopted at site in the field except flood-prone area and highly weed-infested areas. The soilless perlite bags are placed on raised beds covered with black polythene mulch. An alternative is to place the bags on nursery cloth. The perlite bags may be of the same size as used in greenhouses, i.e., 1 m in length and 20 cm in circumference. Bag can be even longer that is filled with soilless medium. Bags containing soilless medium are usually placed in a single row on the raised bed for growing vegetables such as tomato or melon, or these can be placed in twin rows for vegetables such as pepper or herbs. It is recommended to keep 30 cm distance between the bags while the plant-plant distance may be adopted as practiced in the normal field cultivation (Hochmuth and Hochmuth 2021).

6.3.2 Irrigation and Nutrient Management

Plants are fertilized and irrigated as in the case of glasshouses. Polyethylene tubing and micro emitters are used for individual plants. These polyethylene tubes are placed above the bags and water penetrates through the holes made in the bags near the plants. Tube is fixed with bag surface with some tape, hook, or wire to keep it at the right place. Drainage holes are also made at the bottom of the perlite bags after wetting the perlite. These perlite bags are useful for continuous cropping. The same bag can be used for several different crops grown in different seasons (Hochmuth and Hochmuth 2021; Gebreegziher 2023).

6.3.3 Crop Production

Crops raised in this system produce more yields per unit area. A higher fruit number and greater fruit size were recorded in tomato, pepper, cucumber, and zucchini. Moreover, fruit quality was improved in this soilless system probably due to better nutrition management and less weed and insect-pest infestation (Rumpel et al. 1996; Rumpel and Kaniszewski 1998; Hochmuth and Hochmuth 2021).

6.3.4 Limitations

This system is more expensive than the normal field production due to higher costs of soilless media, installation, irrigation system, labor, etc. The system is more suitable for high-value crops grown on smaller areas where higher produce quality matters. Technical knowledge of the grower is another important issue in this system. Moreover, soilless outdoor system is more adaptive for small farmers intended to grow a variety of crops preferably for direct marketing (Kalaivanan et al. 2022).

6.4 Conclusions

Vegetable cultivation is a versatile phenomenon and various vegetables can be grown in the field through different methods including flatbed, ridges, furrows, and raised bed cultivation. Some vegetables can be best grown using a single cultivation technique while others may be successfully grown by two or three methods. The suitability of a growing method for a specific vegetable depends on the vegetable type, soil conditions, water availability, and local topographical situation. Vegetable cultivation in soilless media in outdoor conditions is also under practice at experimental level in order to grow vegetables in areas meeting the climate requirements but with poor soils. Vegetables in soilless media produce higher yields of better quality with less weed and pest infestation.

References

AARI (Ayub Agricultural Research Institute, Faisalabad) (2024) Crop production plans / technologies. https://aari.punjab.gov.pk/infodesk_crop_plans

Abida A, Ayub G, Ilyas M, Baseerat B, Luqman L, Khan MI, Farooq S (2016) Effect of sowing techniques on weeds growth and yield of turnip cultivars. Pak J Weed Sci Res 22(3):417–423

Ali A (2010) Multitrophic interactions; potato, aphids, parasitoid system, influence of soil nutrition and nematodes herbivory. Ph.D. Thesis, Imperial College, London, p 221

Ali A (2017) Best practices of conservation agriculture in Pakistan. In: Pandey PR, Gurung TR (eds) Best Practices of Conservation Agriculture in South Asia. SAARC Agriculture Centre, Dhaka, pp 100–128

Awan DA, Ahmad F, Imdad M (2016) Influence of row spacing and frequency of cuttings on spinach (*Spinacia oleracea*) production. J Bioresour Mgmt 3(1):9. https://doi.org/10.35691/JBM.6102.0047

Baloch AF (1994) Vegetable crops. In: Malik MN, Bashir E, Bantel R (eds) Horticulture. National Book Foundation, Islamabad, pp 285–308

Basit A, Zahid A, Shah ST, Ullah I, Ullah S, Nawaz MK, Khalid MA, Ullah I, Ali F, Ali S (2021) Effect of sowing methods and picking intervals on fresh yield and seed production of okra (*Abelmoschus esculentus* L.). Pak J Weed Sci Res 27(3):407–418

El-Sayed SS (2021) Impact of some agronomical practices on carrot productivity and nutritional value. J Plant Prod 12(4):365–376

Gebreegziher WG (2023) Soilless culture technology to transform vegetable farming, reduce land pressure and degradation in drylands. Cogent Food Agric 9(2):2265106

GOP (2023) Pakistan Economic Survey 2022–2023. Economic Advisor's Wing, Finance Division, Government of Pakistan, Islamabad

Goswami SB (2005) Surface drainage practice for vegetables during high rainfall in lower Gangetic plains. Indian J Soil Conserv 53(1):31–34

Gupta CR, Singh PN (1994) Effect of method of planting and fertility levels on yield of potato: present and future. Proc Nat Symp Modipuram 1–3:113–114

Hochmuth GJ, Hochmuth RC (2021) Open-field soilless culture of vegetables. https://edis.ifas.ufl.edu/publication/SS504. Accessed 30 Sep 2024

Jawad R, Nawaz S, Hammad HM, Salik MR, Farhad W (2015) Nitrogen and sowing method affect radish growth and yield under arid environments of Pakistan. Sci Int 27(3):2245–2250

Kalaivanan D, Selvakumar G, Rathinakumari AC (2022) Soilless cultivation to secure the vegetable demand of urban and peri-urban population. In: Turan M, Argin S, Yildirim E, Güneş A (eds) Recent research and advances in soilless culture. Intechopen Publishers

Mal D, Kaur M (2019) Irrigation practices in vegetable crops: a review. Plant Arch 19(2):2177–2180

Malik MN (1994) Establishment of gardens. In: Malik MN, Bashir E, Bantel R (eds) Horticulture. National Book Foundation, Islamabad, pp 285–308

Qasim M, Khalid S, Naz A, Khan MZ, Khan SA (2013) Effects of different planting systems on yield of potato crop in Kaghan Valley: a mountainous region of Pakistan. Agri Sci 4(4):175–179

Rosin KG, Kaur R, Patel N, Rajput TBS, Kumar S (2017) Yield and irrigation water use efficiency of bottle gourd (*Lagenaria sicenaria* L.) in response to different irrigation methods and planting geometries. Int J Curr Microbiol App Sci 6(5):2475–2481

Rumpel J, Kaniszewski S (1998) Outdoor soilless culture of vegetables: status and prospects. J Veg Crop Prod 4(1):3–10

Rumpel J, Felczyniski K, Kaniszewski S (1996) Results of experiments with soilless open field tomato culture in Germany and Poland. In: Proceedings of the 9th International Congress on Soilless Culture, St. Helier, Jersey, 12–19 April 1996. International Society for Soilless Culture (ISOSC), pp 373–380

Shravan S, Jogdand SV (2020) Performance evaluation of manual single row vegetable planter for okra, coriander and cowpea crop. J Pharmacog Phytochem 9(3):1255–1258

Solangi M, Soomro A, Sheikh MJ, Baloch AW, Abro SI (2017) Effect of sowing methods on yield and growth of spinach. Pak J Sci 69(1):12–16

Statista (2024) Global production volume of vegetables from 2000 to 2022. https://www.statista.com/statistics/264059/production-volume-of-vegetables-and-melons-worldwide-since-1990/. Accessed 12 Oct 2024

Thind SK, Mahal JS (2021) Package of practices for cultivation of vegetables. Additional Director of Communication for Punjab Agricultural University, Ludhiana, p 39

Topakci M, Karayel D, Canakci M, Furat S, Uzun B (2011) Sesame hill dropping performance of a vacuum seeder for different tillage practices. Appl Eng Agric 27(2):203–209

Welbaum GE (2015) Vegetable production and practices. CABI, London

Zaidi MA, Amjad N, Mahmood HS, Shah SUS (2019) Performance evaluation of pneumatic planter for peas planting. Pak J Agri Sci 56(1):237–244

Zaman P, Md H, Sengupta K (2018) Fenugreek (Greek clover). In: Md H, Zaman P (eds) Forage crops of the world, volume II: minor forage crops. Apple Academic Press, New York, pp 141–150

Zeb B, Ali S, Ara N, Jan TA, Bacha SAS, Ahmad I, Shah SHA, Khan R, Khan IH, Ullah A, Khan NA (2018) Effect of ridges height on the growth, yield and root rot control of different chili cultivars in Mansehra. Intl J Biosci 13(5):464–472

Edible Landscaping: Planning and Designing

7

Zafar Iqbal, Muhammad Akbar Anjum, Shahid J. Butt, Alaiha Asif, and Rashid Mukhtar Balal

Abstract

Edible landscapes can be identified as sustainable components of urban planning which provide many ecological, economic, health, social, and cultural benefits. Edible landscaping also termed as foodscaping refers to the technique of incorporating edible plant species into ornamental landscapes in harmonious groupings. Advanced edible landscaping designs combine colorful vegetables, edible flowers, herbs, shrubs, fruit trees, medicinal plants, and berry bushes along with ornamental species. This chapter explains historical precedents of foodscaping, planning and designing of edible landscapes, water delivery system, use of smart and wise irrigation, choice of plant species for edible landscape, and attractive color schemes for edibles.

Z. Iqbal (✉)
Faculty of Agriculture, University of Sargodha, Sargodha, Pakistan
e-mail: zafarrohils@yahoo.com

M. A. Anjum
Department of Horticulture, Bahauddin Zakariya University, Multan, Pakistan

S. J. Butt
Department of Horticulture, Faculty of Agriculture, PMAS Arid Agriculture University, Rawalpindi, Pakistan

A. Asif
Horticultural Sciences Department, North Florida Research and Education Centre, University of Florida/IFAS, Quincy, Florida, USA

R. M. Balal
Department of Horticulture, College of Agriculture, University of Sargodha, Sargodha, Pakistan

Z. Iqbal et al. (eds.), *Sustainable and Innovative Vegetable Production in times of Climate Change*, https://doi.org/10.1007/978-981-96-6283-8_7

Keywords

Edible landscaping · Foodscaping · Biodiverse garden · Sun mapping · Color schemes for edibles

7.1 Introduction

Many institutions and non-governmental organizations (NGOs) are adopting regulations to support the growth and development of urban agriculture (Conway and Brannen 2014; Çelik 2017). Edible landscaping is a pretty, soft, and ecofriendly approach that combines conventional landscaping with micro-farming techniques. This practice ensures provision of fresh, residue-free, nutritious, and inexpensive food. Edible landscaping practices may be adopted by individuals, housewives, communities, and private/public organizations to create a friendly and sustainable ecosystem (Arthur 2017; Çelik 2017; Karaca 2019). Edible landscaping portrays green infrastructure as having diverse goals like aesthetics, functionality, health and wellness, and food sufficiency (Sanchez et al. 2021).

In traditional landscape, many fertile soil patches go wasted whereas in edible landscape, even few inches of land may be made productive with little effort of the gardener. Keeping in view sunlight and personal preferences, variety of ornamental and edible species may be interspersed, grouped, and arranged in a fascinating style. Edible landscapes can be designed in various forms at different scales to provide multifarious benefits to urban residents (Lovell 2010). They promote sustainable practices which enhance water and energy efficiency, support biodiversity, and reduce the use of hazardous chemicals in the landscape. These may be made an essential part of back/front yards, rooftop gardens, side lawns, house walls, public parks, community gardens, schoolyards, institutional campuses, urban forests, and walkways (Çelik 2017).

7.2 Historical Precedents of Edible Landscaping/Foodscaping

Although it appears to be a modern practice, an edible landscape is actually an ancient activity which dates back to medieval monks, ancient Persians, and Egyptians who grew a variety of vegetables, flowers, fruits, and herbs for edible, ornamental, and medicinal purposes (Geiger 2018). An old-fashioned type of edible landscaping is known as potager, or kitchen garden.

In the thirteenth century, the German friar Albertus Magnus designed the layout of an herb garden with the central lawn of the garden bordered by fragrant herbs like rue, basil, and sage. Magnus introduced in his writing about garden design, De Vegetabilibus. The medieval healing gardens introduced the idea of raised planter beds which are even used these days. Finding historical record in France in sixteenth and seventeenth centuries, small container gardens were designed to provide herbs for soups and later developed into big plots of diverse varieties. The monastic gardens in medieval times included flowers, medicinal herbs, and vegetables. Roman villa gardens in ancient Rome were both ornamental and productive. As diets of ancient Romans included fruits and vegetables, so kitchen gardens, orchards, and vineyards played an important role in their lives (Jashemski 2017). In Mesoamerican culture, horticultural and elaborate gardens were considered a pleasure of Aztec elites. Groves, orchards, and water gardens were also incorporated into the designs of more elaborate gardens (Evans 2000).

Robert Kourik, a landscape designer and environmentalist, used the term "Edible landscaping" in the 1980s for the type of gardening that combined pleasing appearance with crop production. In 1982, Rosalind Creasy introduced this theme through her famous publication titled, "The Complete Book of Edible Landscaping". The term "foodscape" first appeared in academic literature in 1995 (Yasmeen 1995). Being an old practice of English gardens, it was again started in 2009 by Queen Elizabeth initiating an organic edible landscape within Buckingham Palace Garden that included heirloom species of tomatoes, beans, lettuce, and some other comestibles (Geiger 2018).

Currently, edible landscaping has assumed some advanced and vertical trends. An increasing adoption of foodscaping or edible landscaping is due to rising global food prices, negative effects of urbanization, and growing concerns for environmental sustainability.

7.3 Benefits of Urban Edible Landscaping

7.3.1 Inter-Generational Equity and Safety

Effective urban edible landscape designing and strategies can meet 15–20% of the global food demand (Abdulkadir et al. 2012; Xie et al. 2019). Mixing flowers, vegetables, herbs, and other edibles surely display a unique, assorted, colorful, and fruitful landscape for citizens and other communities. Urban edible landscaping supports to achieve noble target of inter-generational equity, safety, security, and sustainability in food production. Food insecurity can occur if the diet is not nutritious and safe. The major impact of food insecurity is malnourishment which refers to deficiencies in a person's daily intake of energy and nutrients. Latent hunger and micronutrient deficiencies may be present in countries with high level of malnutrition (Mbow et al. 2019).

In the wake of COVID-19 pandemic, all the pillars of food security were threatened in rich and poor communities. Its consequences led to increased food costs, disrupted access to food markets, and provision of unsafe food (Béné et al. 2021). An edible landscape reduces unnecessary dependence on external food sources with unknown production systems. Turning home garden into edible landscape offers the best choice as it works outside the existing supply system and is more resilient to freakish and adverse weather conditions.

7.3.2 Provision of Nutrient Dense Food with Small Sodium Level

The prime purpose of growing edibles at home or other small spaces is to produce fresh crops of higher nutritive value and eating quality. Due to commercial packing, shipping, transit, storage, and supply chain problems, commercially grown vegetables and fruits are harvested before they attain best eating quality. Freshly picked fruits and vegetables have more vitamins and phytonutrients than weeks old produce of local or distant markets. Many non-conventional tasty varieties not available in food markets are available to home gardeners.

Plant foods contain fiber and nutrients with a small sodium level while frozen vegetables may contain sodium for preservation purpose. Unfortunately, over 90% people use an unhealthy amount of salt. The use of high amount of salt makes us prone to hypertension, high blood pressure, heart problems, obesity, and even cancer.

American Heart Association (AHA) recommends 2300 mg of sodium per day for an average healthy person while an ideal limit is 1500 mg per day for careful adults. It is wise to select vegetables for daily diet which are sodium free or have a very low level.

Some vegetables which have low sodium level include broccoli, bell pepper, carrot, sweet potato, radish, and artichoke. Vegetables which contain very low sodium include cauliflower, Brussels sprouts, cabbage, peppers, chickpeas, lettuce, collard greens, iceberg, okra, onion, scallions, parsley, peas, lima beans, endive, cantaloupe, rhubarb, and tomato. Vegetables which contain less than 05 mg sodium per reference amount are generally categorized as sodium free. These include cucumber, asparagus, eggplant, garlic, potato, green beans, lettuce, and squash (Earle et al. 2023; Tremblay et al. 2024).

7.3.3 Development of Sustainable Practices

Modern and big cities in the world face many ecological challenges including environmental degradation, unclean air, smog, water quality issues, insufficient green space, excess heat capture, and polluted and stormwater runoff. In contrast to conventional landscape, edible landscape offers manifold, long-lasting, and diversified benefits. It boosts sustainable practices that maximize water and energy efficiency, improves the structural integrity of soil, supports wildlife (Çelik 2017), and protects our planet from ecological damage caused by the residual effect of pesticides. Edibles raised in home gardens require no shipping, little refrigeration, and less or no energy for land preparation, cultivation, and other related agricultural operations.

Research trials have empirically proved that home gardeners use less quantity of water for the production of vegetables. Drip irrigation even saves more water as flooding wastes much water that otherwise may be used for raising of diverse crops. Edible gardens provide refreshing and cheap vegetables reducing daily food costs. The United Nations-ESCAP encourages the development of sustainable home gardening as a beneficial strategy for social protection, combined with health and nutrition education (UN-ESCAP 2009). It is an established fact that home or urban-grown vegetables minimize carbon emissions from the burning of fossil fuels. Growing own food at homes and left-over spaces needs much less inputs in terms of usable heat, power, and fertilizer starting from initial planting to harvesting. Maintenance requirements at home or nutrition gardens are also minimized by planting the right plant at the right place and matching the specific growing requirements of a plant with local conditions (Çelik 2017).

In the USA, "Food deserts" also pose a threat at some places as fresh and inexpensive food is not easily available. Reduced dependence on middlemen or market sources and increased reliance on homegrown edibles ensure eco-friendly provision of healthier and economical food. In the Indian state Gujarat, an effective model has been developed for backyard kitchen gardening, which aims to develop a healthier, prosperous, and sustainable life for women working at farmhouses (Tandel et al. 2015). A meta-analysis of 22 case studies on domestic/community gardens from different parts of the world has revealed improvement of physical, psychological, and social health having long-term impacts on well-being (Soga et al. 2017). Well-designed edible landscape improves air quality and protects residents from ill effects of pollutants like dust, smog, odor, and noise from year to year.

7.3.4 Educational Approach

An edible garden not only provides nutritious food but it is also a positive learning activity which imparts dynamic guidance and collaborative skills to curious gardeners. Edible landscape serves as an educational area and a learning academy for people of all ages. Children, youth, and adults living in urban areas can learn advanced and smart technological information about growing of fruits and vegetables in green areas of cities. Vega-phobic children and students develop a tendency to memorize names of vegetables and start picking and eating colorful veggies. Food gardening also provides physical education which includes a range of movements like stretching, bending, lifting, walking, digging, turning, twisting, pulling, dumping, and mingling.

7.3.5 Establishment of Biodiverse Gardens

Climate disaster and diminishing biodiversity have forced people to maintain natural equilibrium through promoting biodiverse gardens and sustainable habitats. Some important and precious insect species are on the verge of extinction from their place of habitation. It predicts a disastrous impact on the earth's ecosystems.

Urban biodiversity has now become the important segment of urban ecosystem as it promotes social, cultural, and ecological integrity. A biodiverse garden is a charming, pleasing, refreshing, and productive garden having a collection of different plant species grown in the available space and also in well-designed containers and pots.

Pro-environmental gardening practices aim to make gardens more desirable for wildlife. Growing native and naturally acclimatized species is a salubrious and sympathetic act for animals and other creatures. Butterflies, bees, mantis, beetles, avifauna, tachinid flies, and other beneficial species depend on a variety of plants for food and nectar. These insects, wildlife species, and birds love diverse landscapes with favorite choice. They are helpful in pollination and controlling the infestation of many pests or bugs.

Only a single bee "bumblebee" is very essential for our ecosystem which pollinates many plant species. A biodiverse garden proves to be a little heaven for pollinators as it provides attractive flora, pollens, and nectar-rich flowers. Its expanse of scenery offers fresh flora, colorful foliage, sensory pleasure, and pretty food for natural species.

Fragrant flowers, chirping birds, amazing flavors, whispering winds, communicating plants, variegated foliage, multiple plant textures, and unfurling fascination are some of the rewards of a biodiverse garden. Birds help in seed distribution, pollination, and pest control operations while insects assist in pollination in nature and also provide food for some wildlife species. Naturally existing soil organisms are also much beneficial as they enhance decomposition and some of them act as antagonists counteracting the harmful effects of other organisms. Soil microbes like fungi and bacteria inhabit rhizosphere and play an essential role in disease suppression and decomposition. Accumulation of decomposed material and organic matter facilitates long-term fertility and thriving of life below the soil.

In 1965, Singapore was considered a polluted territory. Due to the city's rapid growth, 90 percent of forests, 67 percent native bird species, and 40 percent mammals were lost there. Realizing loss of natural habitat and precious biodiversity, a plan was launched in 1967 to transform Singapore to a garden city. Gradually it transformed into a global green infrastructure and biodiversity model in the world. Now Singapore has attained a unique status as a green and biodiverse city in the world. Credit goes to Lee Kuan Yew who was known as "Chief Gardener" for his deep interest in biodiversity (UN 2018; Ferrer 2021).

7.4 Edible Landscaping: Planning and Designing

The present trend is to turn traditional and conventional lawns to edible delight and pleasure. Proper planning and designing are the essential components of urban edible landscaping. An edible garden can be started even in a small space or leftover/neglected area. Physical features (location, soil type, climatic conditions), availability of resources (water, energy, agricultural inputs), and type of species (local, exotic) must be considered. Before to start, yard can be mapped out with the help of suitable software.

The key considerations for planning and designing of a beautiful edible garden include:

7.4.1 Space Requirements

Various plant species have different space requirements to grow properly in edible landscape. An edible garden may be started at a little space, in containers or planters or an area of few square meters. Balconies, porches, courtyards, rooftops, side walls, walkways, and marginal areas are also feasible sites for edible gardening. Generally, a space of 3 × 3 meters to 6 × 3 meters (9–18 square meters) is suitable for a gardener who wishes to harvest edibles for his family on sustainable basis by just giving half an hour to an hour a week in the garden. If less space is available in the garden then the same may be maximized by vertical gardening techniques which ensure growing plants on vertical structures to vacate ground space for other edibles. Dwarf, semi-dwarf, miniature, and columnar varieties and shrubs that grow flat against a support or trellis are suitable for small spaces. Depending upon the availability of space, paths can also be designed to add beauty to landscape and facilitate access to adjacent areas.

7.4.2 Maintenance of Nutrient-Rich and Sustainable Soil

An edible garden needs nutrient-rich, well-aerated, and fine-quality soil. Soil must be free from contaminants, pollutants, weeds, undesired plants, and residues. Nutritional profile of different types of soils varies from region to region. For a new edible garden, soil sample should be analyzed to confirm the nutrients status.

Most vegetables thrive in rich, organic, and well-drained soil with pH ranging from 6.0 to 7.0. While preparing soil for vegetable cultivation, 02 to 05 centimeters layer of compost should be spread and mixed according to the type of soil in the beginning of each growing season. Other amendments including addition of chopped leaves, mulch, and well-rotten manure may also be done between successive and exhaustive crops. For raised beds, potting mix and compost should be added to the top 10 centimeters to revive the soil. Amendments retain moisture in sandy texture and aerate clay soil. Fertilization is also necessary to harness the full potential of different vegetables. The essential macronutrients include nitrogen, phosphorous, and potassium which are important for leafy green, root, and tuber vegetables and even all vegetables. Mulching with leaves, and clean straw may be done for tender vegetables.

7.4.3 Sun Mapping

Most of the plants obtain energy through photosynthesis phenomenon which requires sunlight. Due to insufficient sunlight, plants show lankiness. They produce yellow leaves and exhibit poor growth which doesn't conform to their true potential.

Some vegetables' cultivars don't produce flowers or fruits if proper sunlight is not received. Light initiates fruit ripening process by activating the production of sugars and pigments. This leads to pleasant taste and charming color of produce.

Sunlight is also essential for nutrient absorption in plants. It helps combat garden pests and diseases which otherwise flourish under humid and moist conditions.

The sun and shade map is required to explore the more and less productive areas of garden because sun-deprived plants give poor-quality produce. The containers may be placed at garden places where right amount of sunlight is received during the whole day. Many vegetables require morning sunlight with less intensity. Plants with the same or matching lighting requirements and life cycles may be placed in one group while herbs and annuals are combined in another section. Perennials need to be planted in garden areas where each plant gets a specific amount of light.

There are following fundamental sunlight conditions which explain the amount of sun during the growing season:

7.4.3.1 Full Sun

The areas with full sun receive direct sunlight for 06 to 08 hours per day. Peak sunlight hours are considered between 10 am and 02 pm.

7.4.3.2 Partial Shade

Partial sun areas receive 04 to 06 hours of direct sunlight and are shaded during the remaining part of the day.

7.4.3.3 Full Shade

Full shade areas receive 02 to 04 hours of direct sunlight or reflected light during the day. Shade helps in providing cool temperatures which prevent leafy greens from producing flowers too early or prematurely.

7.4.3.4 Deep Shade

Deep shaded areas get no sunlight at all, so vegetable cultivation is not recommended there. Vegetables can be classified into three groups on the basis of their light requirements:

1. Vegetables like cucumbers, winter squash, summer squash, okra, watermelon, cantaloupe, eggplant, tomatoes, peppers, bell peppers, zucchini, pumpkin, sweet potato, bitter gourd, peas and corn are full sun lovers. Herbs like rosemary, sage, caraway, dill, lavender, tarragon, anise, chamomile, fennel, calendula, valerian, Echinacea, cilantro, summer savory, chives, parsley, oregano, basil, chervil, and thyme also grow well in full sun.
2. Some vegetable crops have moderate light requirements as these can grow well at locations receiving 04 to 06 h of direct sun during a day length. Vegetables such as lettuce, beans, arugula, spinach, peas, beets, carrots, potatoes, broccoli, cabbage, onions, leek, radish, rutabaga, celery, cauliflower, asparagus, scallions, garlic, kohlrabi, rhubarb, Jerusalem artichokes, parsnip, New Zealand spinach, sorrel, watercress, endive, escarole, coriander, chervil, and turnips require partial sun.
3. Vegetables including endive, curly kale, mustard greens, chives, chard, mint, parsley, lettuce, spinach, onions green, amaranth, calabrese, kohlrabi and

Brussels sprouts need full shade. Asian greens like bok choy, tatsoi, and pak choi perform much better in full shaded areas.

7.4.4 Water Delivery System in Edible Landscape

Plant seeds need water for enzymes activation, germination, sprouting, and normal growth. Water is the primary input which whispers to seed to germinate at right time. It is essentially required for photosynthesis when plants use sun energy to manufacture food. Many important functions within plant tissue are accomplished in the presence of water. Water evaporates through a process known as transpiration which protects plants from overheating due to climatic conditions (Richmond 2021). Although edible plants need sufficient amount of water for their growth but excessive or less watering may cause stress, fruit drop and decreased vegetative, as well as reproductive growth. Excessive watering may cause suffocation, less aeration, and root rotting symptoms. Plants also become prone to attack of different pests and root diseases. Generally, vegetables need 2.5 to 5.0 centimeters of water per week depending upon the plant species, soil type, and prevailing climatic conditions. Increased water duration with less frequency is required for larger plants. Some vegetables may manifest splitting, yellowing, and wilting symptoms due to over or irregular watering. In addition to wastage of water, overwatering also disintegrates and erodes manure, fertilizer, and compost from fertile top 15 centimeters of soil. Underwatering can obstruct root system and hamper nutrients absorption and transportation to upper parts of the plant. Flower and fruit drop may occur during severe drought stress. Consistent working and frequent monitoring of edible landscape enable to optimize daily watering requirements.

7.4.4.1 Types of Irrigation

Water requirements of all vegetables, herbs, shrubs, and ornamentals don't match with each other so separate area or block can be allocated in the edible landscape for each group. Some species with almost similar water and light requirements can be raised as mixed crops. Generally, plants need to be watered during the first few weeks of development, after transplanting and during growth of edible plant parts. Wise and judicious irrigation methods are required to meet actual needs of various plants species growing in edible landscape which are detailed below:

7.4.4.1.1 Harvesting of Rainwater

Rain is natural water which falls from clouds when water vapors in the atmosphere are cooled down and condensed to water droplets. Rainwater is a gift of nature which provides nitrogen to the plants in the form of nitrates. Clean rainwater has a pH from 5.5 to 5.6 due to the presence of H + ions formed by the reaction of rainwater with carbon dioxide. Rainwater also contains more oxygen than tap water.

Collection of rainwater is an emerging trend these days as this is an important method of water conservation. It is the primary form of watering plants in the garden. Generally, 1.27 cm rainfall gives 1.13 liter of water per 01 square foot. Different types of containers may be used for collection and storage of rainwater for gardening purpose.

Rain barrels help capture rainwater from the roof of buildings or constructed structures to make it available for garden plantings. Most of the rain barrels can store up to 200 liters of water. Rain barrels available in the market have a water outlet, hose, overflow outlet, and a screen for trash prevention. Intermediate bulk containers also known as IBC tank or IBC tote or caged tote are industrial-grade containers commonly used for rainfall collection for the garden having capacity of 1040 to 1250 liters. Rainwater in the barrel or tote is naturally clean, soft, chlorine free, and less acidic with no dissolved minerals so it is ideal for gardening. Stored rainwater should be used for non-edibles or ornamental plants due to chemical, biological, and physical contaminations. However, if cleaned and filtered properly, then it can be used for edibles too. In areas of low rainfall, additional irrigation sources need to be adopted for precise watering of edible landscape.

7.4.4.1.2 Hand Watering

Hand watering is considered as the traditional and efficient method for watering of edibles, transplants, and new or mature crops. This method can be combined with micro-irrigation for larger gardens. Hand watering helps to adjust the watering needs of each specific crop. It ensures better growing conditions as every plant obtains water according to its specific requirements. In the hand watering method, soil is wetted around the plant base. Light and deep watering are provided for freshly seeded and established beds, respectively. Hand watering is also helpful in conserving water. Soil fungi proliferate in cool and moist conditions. Some fungi develop on leaves which remain wet for prolonged periods. So, vegetable crops which perform best with light watering or dry foliage like kale, cucumber, zucchini, Swiss chard, eggplant, spinach, pumpkin, tomato, thyme, fenugreek, and cabbage may be watered without unnecessary wetting of their foliage. Excessive flow to pathways and paved areas of garden may also be avoided. Hand watering provides ample time to monitor crops in the garden, observe their seasonal or varying requirements, and follow self-devised schedule. Hand watering is less affected by regional or global water restrictions. For small and disorganized spaces, container plantations, and vertical structures, there is no need to install expensive system rather hand watering is the best solution.

7.4.4.1.3 Hose-End Nozzle

A garden hose is an essential garden tool for watering of home gardens and edible landscape. It is actually a rubber pipe that determines the movement of water from water tank or source. The nozzle is attached to the end of hose which controls water pressure and adjusts spray patterns according to day-to-day gardening needs. This is a simple system which consists of different nozzles, attachments, and valves that may be connected to transform garden hose. The commonly used nozzles include pistol grip, dial wand, and fireman nozzle. If light pressure is required then a nozzle-producing mist is attached to the hose. For ornamentals and vegetables, a nozzle with flat or gentle spray type is quite suitable. Hose-end nozzle method is efficient and saves water.

7.4.4.1.4 Sprinkling System

The system simply consists of a long hose with sprinklers along the length or attached to a central system. Sprinkler system curtails manpower requirement for watering. Hose-end sprinklers or underground pipes with spray heads may be installed for broadcasting water to cover most of the garden area at one time. Attachment of an automatic timer makes both the systems more efficient, precise, and independent. Hose-end sprinkler is an economical and comparatively less expensive choice for watering edible landscapes. Spray patterns can be adjusted to avoid unnecessary runoff and wastage of water. Although installation of built-in sprinklers with underground pipes is expensive, however, these are feasible in the long run for permanent beds and well-designed green spaces. Spray heads like pop-ups, rotors, and bubblers are available that permit to select the pressure and direction of water delivery. With the help of automatic timer and moisture sensor, watering time may be easily programmed for the ease of absentee gardeners. Sprinkler system is best suited to irrigate large spaces requiring light irrigation. However, careful monitoring and good management are required to avoid losses due to evaporation and weather effects.

7.4.4.1.5 Micro–/Trickle/Drip Irrigation

Drip irrigation is a type of micro-irrigation system which has the capacity to conserve water and nutrients by permitting water to drip to root zone of plants. It is the most efficient, high-tech, and sophisticated method to supply water to each plant through tubing and emitters. In contrast to other irrigation systems, drip irrigation is almost 80 to 90% efficient. It performs better even in sandy soils and slopes and saves energy and input in larger gardens. It is being used effectively by commercial/semi-commercial vegetable growers, landscapers, and home gardeners. In drip or trickle irrigation system, water moves through a tube placed close to or below the soil surface and is supplied through holes or emitters. It consists of a short set up including pressure regulator, filter, and tube connector which makes its installation and water delivery much easy. Low pressure is required for drip irrigation system. Water is delivered drop by drop during the extended period of time enabling plant roots to absorb nutrients slowly but continuously. Drip irrigation is the best system between hand watering and sprinkling as it allows precise watering. Garden spaces arranged in long straight rows and vegetable and perennial gardens benefit much from this irrigation system. This system prevents plant diseases like downy and powdery mildew by minimizing unnecessary wetting of foliar parts. It saves time, inputs, labor, and water and reduces soil erosion and leaching of nutrients from rhizosphere (Yang et al. 2023).

The latest concept of co-application of water, fertilizer, and air makes drip irrigation technology the most efficient and useful method for growers as this ensures good water conservation, better soil condition, increased yields, and improved quality of produce (Zan et al. 2019; Zhu et al. 2019; Li et al. 2024).

There are five main types of drip irrigation systems:

- Soaker hose or porous soaker lines.
- Emitter hose.
- Drip tape.
- Micro-misting.
- Bubblers.

7.4.4.1.5.1 Soaker Hose or Porous Soaker Lines

It's a hose with equally spaced holes along its length and is placed alongside shrub lines, green spaces, and indoor vegetable plots. A soaker hose permits water to flow slowly to the ground avoiding runoff and ensuring direct root watering. Soaker hoses perform better for closely planted edibles and those in larger containers or garden beds. Soaker hose method saves water and does not allow the development of devastating foliar diseases. The only care is to arrange it on leveled ground instead of raised beds. If there are perennials and fruit trees in edible garden then a root feeder or root soaker can be used periodically for deep watering of plant roots. Soaker may be kept within the drip line of the plants and hose is attached for watering purpose. Soaker hose has the same advantages like drip irrigation system. However, it can be shifted to different raised beds to deliver water effectively and precisely to the soil surrounding the plants. Soaker hose is best suited for rows and beds of vegetables and annual plants, raised beds, newly planted trees, and perennials.

7.4.4.1.5.2 Emitter Hose

This system is made up of small hoses with evenly spaced nozzles also called as emitters. Emitter hose is designed to supply a precise amount of water, i.e., two or four liters per hour. Emitter irrigation system is the best choice for watering of landscape beds with perennial or permanent plants and also for vegetable gardens. It delivers water more efficiently than sprinklers as it avoids losses due to wind and runoff. It can be attached to timer, moisture sensor, filter, and pressure regulator for precise scheduling. This system delivers water directly to the soil proving it to be a good choice for watering.

7.4.4.1.5.3 Drip Tape

This is a very economical and easy-to-install system and works better when set up in straight lines. This is a preferred choice for annuals, cut flowers, and vegetables' patches. Drip tape can be fitted above ground, covered lightly with mulch, or buried underground (known as sub-surface irrigation). Drip tape has slits inside a thin-walled drip tube pipe. It is generally used in temporary or short-duration irrigation method.

7.4.4.1.5.4 Micro-misting System

Micro-misting, drip irrigation system is an assemblage of drip irrigation and sprinkler system and it distributes water as large droplets or a fan-like spray above the ground. This device called as micro-sprinkler is attached to irrigation hose system. Micro-sprinkler is feasible for use of orchardists and home gardeners and also for flower beds, hillsides, ground covers, and contours. Water is distributed slowly and accurately over root zones. This system may be used with nozzles having different distribution patterns.

7.4.4.1.5.5 Bubblers

Bubblers are devices which allow higher flow of water in multi- or circular or umbrella-like streams. They can deliver up to 98 liters of water per hour approximately. The flow from each nozzle is adjustable to determine the coverage area which ranges from 15 centimeters to 1.5 meters approximately. Timers and emitters may also be attached for water-specific plants. Instead of aerial spraying, they water plants in root area like drip irrigation system. Micro bubblers are suited for watering of container gardens and vegetables grown on raised beds. Bubblers with high flow rate can be much feasible to meet irrigation requirements of shrubs and fruit plants (Rivas et al. 2022). They are also known to enhance soil water content and encourage root growth (Dai et al. 2023). The growers and gardeners can easily adjust the watering schedule keeping in view their specific watering requirements. By combining bubbler system with essential accessories, they can watch their irrigation systems from any place using smartphones.

7.4.4.1.6 Furrow Irrigation

Furrow irrigation is a good watering system for row plantings. Most gardens can be watered with the furrow method depending upon the availability of space and suitable water source. Furrows do best when they are wide, shallow, and located on a plain ground. However, in sandy soils due to quick drainage of water, basins are required to be prepared to retain water around the plant roots for longer time. When crops are grown on beds, water is distributed through channels instead of extra soaking around the root system which may otherwise cause root problems. Furrow irrigation works well in loam soils. Furrow irrigation best utilizes rainwater. Presently, automated furrow irrigation system ensures precise amount of water for row crops. Auto furrow consists of main controller, field control nodes, moisture sensors, and flow sensors. Mobile app may be used for scheduling the watering according to the needs of cultivated crops.

7.4.4.2 Smart and Wise Irrigation for Different Vegetables

Water constitutes about 80 to 90% of vegetable and fruit weight. Most vegetables need 2.5 centimeters of water per week in summer season which soaks the root zone to a depth of 15 to 20 centimeters. This 2.5 centimeter of water per week equates to approximately 27.16 liters per square meter per week. Shallow-rooted vegetables need more water than deep-rooted ones because in hotter months, top soil may dry out quickly. It is wise to avoid over or under-irrigation in vegetable gardens. A smart irrigation schedule can be devised keeping in view the resources and actual varietal and seasonal needs.

Water requirements of different vegetable groups are detailed below:

7.4.4.2.1 Vegetables which Require 1.25–2.50 Centimeters of Water/Week

Peas, radish, cowpeas, pole beans, Jerusalem artichoke, amaranth, squash, pumpkin, and okra (Smith 2022; Hansen et al. 2023).

7.4.4.2.2 Vegetables which Require 2.50–3.75 Centimeters of Water/Week

Basil, beans, calendula, chives, cilantro, dill, garlic, horehound, leek, onion, Swiss chard, zucchini, spinach, cabbage, arugula, cauliflower, and sweet potato (Hansen et al. 2023; Spiridon 2024; Thornbro 2024).

7.4.4.2.3 Vegetables which Require 2.50–5.00 Centimeters of Water/Week

Artichoke, asparagus, beets, black salsify, cantaloupe, carrot, celery, chicory, cucumber, eggplant, French tarragon, honeydew, horseradish, hot pepper, lettuce, mint, mustard greens, New Zealand spinach, parsnip, potato, pumpkins, rutabaga, spinach, sweet pepper, tomato, turnip, cabbage, Brussels sprouts, broccoli, and Kohlrabi (Smith 2022; Hansen et al. 2023; Spiridon 2024; Thornbro 2024).

7.4.4.2.4 Vegetable which Requires 3.75–6.25 Centimeters of Water/Week

Popcorn (Brabham 2021).

7.4.4.2.5 Vegetables which Require 5.00–7.50 Centimeters of Water/Week

Water cress and garden cress (Grant 2021).

7.4.4.2.6 Vegetable which Requires 5.00–14.00 Centimeters of Water/Week

Sweet corn (Boeckmann 2021).

7.4.4.2.7 Vegetable which Requires 7.50–10.00 Centimeters of Water/Week

Wasabi (Cramer 2023).

7.4.5 Plant Species for Edible Landscape

Choice of plant species may vary depending upon aesthetics, flavor, texture, style, natural color scheme, climatic conditions, and specific requirements keeping in view local growing conditions. Various ornamentals, annuals, perennials, herbs, fruits, vegetables, mushrooms, and their combinations may be selected for this purpose. Herbs may be mixed into well-prepared beds and borders. Fragrant species like sage, rosemary, thyme, basil, lavender, oregano, Swiss chard, fennel, scallions, and Vietnamese herbs are combined with flowering perennials presenting a beautiful, fruitful, and beneficial landscape.

Fruiting shrubs like blueberry, currant, raspberry, mulberry, blackberry, and elderberry attract with beautiful canopy, flowers, fruits, and fragrance if grown at suitable sites in edible landscape. Cauliflower, cabbage, parsley, bell peppers, leafy greens, red radishes, and lettuce may be planted as borderline or edger plants in a bed designed for ornamentals. Eggplants, chili peppers, chard, okra, and kale display good look in well-designed groups. Pumpkin, bitter gourd, cucumber, and watermelon spread on ground after growing on beds. Spinach, cilantro, mint, and fenugreek grow on flat beds finding proper space in the garden. Aromatic vegetables like onion, carrot, celery, garlic, leek, parsnip, mushroom, shallot, scallion, bell peppers, thyme, coriander, and ginger are necessary for daily use particularly in Asian countries so these may be planted close to kitchen or cooking area. Species which support wildlife like butterflies, honeybees, hummingbirds, and ladybird beetle should also be interspersed in an edible garden to preserve precious biodiversity.

7.4.5.1 Selection of Multicolor Vegetables

A good edible landscape needs a pleasant and aesthetically consistent color scheme avoiding haphazard and disordered appearance. Different color schemes and themes may be mixed and matched according to own preferences. Color occupies a paramount position to present a unifying or contrasting theme. Purple color vegetables (cauliflower, basil, cabbage, lettuce, eggplant, bell pepper, chives), yellow and orange (squash, wax bean, golden beet, yellow chard, yellow bell peppers, yellow cherry tomatoes, yellow California poppies), white (radish, garlic chives, parsnip, oregano), red (radish, beet, onion, cherry tomatoes, cayenne peppers, red chard, Indian cress), green (cucumber, green beans, bok choy, cabbage, asparagus, coriander, fenugreek, spinach, pumpkin, okra, bitter gourd, basil, broccoli, kale, collards, parsley, sweet potato, zucchini), blue (blue potato, cabbage, beans, pepper, Queensland pumpkin, lettuce, mushroom, tomatoes, carrots, blue pumpkins, lettuce, curled kale) and multicolored (peppers, Swiss chard, carrots) add color diversity and beauty to edible landscape.

Color is the most attractive and charming factor in edible landscape. It is naturally present in foliage, stems, bark, floral parts, and fruits. Different colors may be used in garden fixtures, pavers, pebbles, stones, wood pieces, etc. A color scheme may be devised from 12 basic colors of color wheel which may give cool or warm effect. Green and blue are considered cool colors. Red, orange, and yellow are warm colors. Color schemes may be seasonal depending upon the types of varieties. An editing software can be used to explore different color combinations (Hansen and Alvarez 2019) for your edible landscape.

7.4.5.2 Color Schemes for Edibles

A wise selection of colors may transform entire place into a peaceful little paradise. Artistic and thematic designing of available spaces not only beautifies the garden but also reveals message of love and peace. Cool colors such as blue, green, and purple give a feeling of serenity and calmness in the surroundings. Warm colors, i.e., orange, red, and yellow inspire, excite, and energize both kids and adults. Different color schemes have their impact on gardeners, viewers, and visitors of edible landscape.

7.4.5.2.1 Monochromatic Color Scheme

Monochromatic color scheme uses one color or hue to present a harmonious design. Single color may have varying textures and light and dark shades or tints (Chappell et al. 2022; Balaj and Ahmeti 2022) which reflect attractive and appealing presentation.

7.4.5.2.2 Analogous Color Scheme

Analogous color scheme can be observed in nature and artistic creations as well. Although it is very close to monochromatic but it does not use single color, rather, it uses colors which adjoin it on the color wheel. It includes 03 to 05 adjacent colors on the color wheel, e.g., red, red-orange, orange, yellow-orange and yellow, and their light or dark shades like pink, pale yellow, and dark red (Hansen and Alvarez 2019; Chappell et al. 2022). Analogous is considered as low-contrast combination.

7.4.5.2.3 Complementary Color Schemes

Complementary color schemes combine opposite colors of color wheel having high contrast between them. These are violet and yellow, red and green, and blue and orange. Complementary colors are naturally present in flowers.

7.4.5.2.4 Analogous-Complementary Scheme

In analogous-complementary scheme, 04 to 05 colors are used with one as complement to give a stylish look to landscape.

7.4.5.2.5 Primary Color Scheme

In primary color scheme, three bright colors viz. red, yellow, and blue are used for beautification of gardens.

7.4.5.2.6 Secondary Color Scheme

Green, purple, and orange are secondary colors. They emerge from the combination of equal amount of two primary colors. Red and yellow are mixed to make orange color, blue and yellow give green, and red and blue produce purple shade.

7.4.5.2.7 Tertiary Colors

Tertiary colors are actually intermediate which are formed by equal mixing of primary and secondary colors.

7.4.5.2.8 Riotous Color Scheme

Riotous color scheme uses multiple colors with bright and vigorous combinations (Pennisi and Chappell 2022).

7.4.5.2.9 Pastel Color Scheme

In pastel color scheme, color tones are used to produce soft effect. Pastel colors are the softer variants of primary and secondary colors which appear by addition of white to original color (Velarde 2019; Chappell et al. 2022).

References

Abdulkadir A, Dossa LH, Lompo JP (2012) Characterization of urban and peri-urban agro-ecosystems in three west African cities. Int J Agric Sustain 10:289–314

Arthur B (2017) The foodscape revolution: finding a better way to make space for food and beauty in your garden. Lynn's Press, Pennsylvania, St

Balaj N, Ahmeti S (2022) Color schemes in landscape design: color theory. UBT international conference, p 5. Available at: https://knowledgecenter.ubt-uni.net/conference/2021UBTIC/all-events/5. Accessed 26 Nov 2024

Béné C, Bakker D, Chavarro MJ, Even B, Melo J, Sonneveld A (2021) Global assessment of the impacts of COVID-19 on food security. Glob Food Secur 31:100575

Boeckmann C (2021) Planting, growing, and harvesting sweet corn. Available at: https://www.almanac.com/plant/corn. Accessed 20 Aug 2023

Brabham B (2021) Growing popcorn. Available at: https://extension.wvu.edu/lawn-gardening-pests/gardening/wv-garden-guide/growing-popcorn-in-west-virginia. Accessed 20 Aug 2023

Çelik FD (2017) The importance of edible landscape in the cities. Turk J Agric Food Sci Technol 5(2):118–124. https://doi.org/10.24925/turjaf.v5i2.118-124.957

Chappell M, Davis B, Pennisi B, Sullivan M (2022) Landscape basics: color theory. University of Georgia. Available at: https://extension.uga.edu/publications/detail.html?number=B1396&title=landscape-basics-color-theory. Accessed 25 Jun 2023

Conway TM, Brannen K (2014) Who is tending their garden? Edible gardens as a residential landscaping choice. Cities and the Environment (CATE) 7(2):1–20

Cramer E (2023) How to grow wasabi: complex but rewarding roots. Available at: https://www.epicgardening.com/how-to-grow-wasabi/. Accessed 20 Aug 2023

Dai H, Wang R, Chen L, Wang L, Xiong C, Wang X, Zhang M (2023) Effects of different micro-irrigation methods on water use and the economic benefits of an apple–soybean intercropping system. Agronomy 13(4):1143. https://doi.org/10.3390/agronomy13041143

Earle WB, Ormseth G, Morales-Alvarez MC, Kaushik M, Juraschek SP (2023) Dietary sodium reduction is best for reducing blood pressure: controversies in hypertension. Hypertension 81(3). https://doi.org/10.1161/HYPERTENSIONAHA.123.20544

Evans ST (2000) Aztec royal pleasure parks: conspicuous consumption and elite status rivalry. Stud Hist Gardens Designed Landscapes 20(3):206–228. https://doi.org/10.1080/14601176.2000.10435621

Ferrer M (2021) Singapore: the city that learned how to blend nature with urban living. Available at: https://www.euronews.com/green/2021/11/11/how-has-singapore-learned-to-blend-nature-with-urban-living. Accessed 20 Jul 2023

Geiger M (2018) Edible landscaping—landscaping with taste. Iowa State University, Extension and Outreach. Available at: https://blogs.extension.iastate.edu/answerline/2018/05/03/edible-landscaping-landscaping-with-taste/. Accessed 28 Jan 2023

Grant A (2021) Care of watercress: growing watercress plants in gardens. Available at: https://www.gardeningknowhow.com/edible/herbs/watercress/growing-watercress-in-gardens.htm. Accessed 20 Aug 2023

Hansen G, Alvarez E (2019) Color in the landscape: finding inspiration for a color theme. University of Florida, IFAS Extension. Available at: https://edis.ifas.ufl.edu/publication/EP425. Accessed 15 Aug 2023

Hansen S, Beddes T, Barker B, Butler A (2023) Water recommendations for vegetables. Utah State University. Available at: https://extension.usu.edu/yardandgarden/research/water-recommendations-for-vegetables. Accessed 03 Aug 2023

Jashemski FW (2017) Gardens of the Roman empire. Cambridge University Press. https://doi.org/10.1017/9781139033022

Karaca E (2019) Edible landscapes as a solution to food security problem. In: Krystev V et al (eds) Theory and practice in social sciences. St. Kliment Ohridski University Press, pp 227–236

Li H, Ma Z, Zhang G, Chen J, Lu Y, Li P (2024) Performance of a drip irrigation system under the co-application of water, fertilizer, and air. Horticulturae 10(1):6. https://doi.org/10.3390/horticulturae10010006

Lovell ST (2010) Multifunctional urban agriculture for sustainable land use planning in the United States. Sustain For 2:2499–2522. https://doi.org/10.3390/su2082499

Mbow C, Rosenzweig C, Barioni LG, Benton TG, Herrero M, Krishnapillai M, Liwenga E, Pradhan P, Rivera-Ferre MG, Sapkota T, Tubiello FN, Xu Y (2019) Food security. In: Shukla PR et al (eds) Climate change and land: an IPCC special report on climate change, desertification, land degradation, sustainable land management, food security, and greenhouse gas fluxes in terrestrial ecosystems. Available at: https://doi.org/10.1017/9781009157988.007

Pennisi BV, Chappell M (2022) Landscape basics: color theory. https://extension.uga.edu/publications/detail.html?number=B1396&title=landscape-basics-color-theory. Accessed 28 Dec 2024

Richmond J (2021) How plants use water. Available at: https://extension.wvu.edu/lawn-gardening-pests/news/2021/03/01/how-plants-use-water#. Accessed 01 Aug 2023

Rivas J, Silvasy T, Harlow L, Barber L, Marvin JW, Momol E, McIntyre T, Freeman T (2022) Efficient irrigation for Florida-friendly edible landscapes. Available at: https://edis.ifas.ufl.edu/publication/EP617. Accessed 30 Jul 2023

Sanchez FC, Ilang-Ilang MC, Balladares MCE, Apacionado BV, Tayobong RRP, Medina NG, Torres RO (2021) Attaining food self-sufficiency through the edible landscaping (EL) technology: assessing the promotional activities of EL in The Philippines, amidst the COVID-19 pandemic. IOP Conf Ser: Earth Environ Sci. https://doi.org/10.1088/1755-1315/879/1/012014

Smith (2022) Nine drought tolerant vegetables. https://lawnlove.com/blog/best-drought-tolerant-vegetables/. Accessed 19 August 2023

Soga M, Gaston KJ, Yamaura Y (2017) Gardening is beneficial for health: a meta-analysis. Prev Med Rep 5:92–99

Spiridon S (2024) How often to water your vegetable garden (and more watering tips). Available at: https://lawnlove.com/blog/how-often-to-water-vegetable-garden/ Accessed 23 Nov 2024

Tandel BM, Nayaka P, Timbadia CK (2015) Cost-effective dietary backyard kitchen gardening: a success story. Rashtriya Krishi 10(1):113–114

Thornbro H (2024) Watering your vegetable garden: how much and how often. Available at: https://redemptionpermaculture.com/watering-your-vegetable-garden-how-much-and-how-often/. Accessed 24 Nov 2024

Tremblay A, Gagné M-P, Pérusse L, Fortier C, Provencher V, Corcuff R, Pomerleau S, Foti N, Drapeau V (2024) Sodium and human health: what can be done to improve sodium balance beyond food processing? Nutrients 16(8):1199. https://doi.org/10.3390/nu16081199

UN (2018) A city in a garden: Singapore's journey to becoming a biodiversity model. Available at: https://www.unep.org/news-and-stories/story/city-garden-singapores-journey-becoming-biodiversity-model. Accessed 21 Jul 2023

UN-ESCAP (2009) Sustainable agriculture and food security in Asia and the Pacific. UN ESCAP, Bangkok

Velarde O (2019) How to use pastel colors in your designs [+15 delicious pastel color schemes]. Available at: https://visme.co/blog/pastel-colors/. Accessed 16 Aug 2023

Xie Q, Yue Y, Hu D (2019) Residents' attention and awareness of urban edible landscapes: a case study of Wuhan. China Forests 10(12):1142. https://doi.org/10.3390/f10121142

Yang P, Wu L, Cheng M, Fan J, Li S, Wang H, Qian L (2023) Review on drip irrigation: impact on crop yield, quality, and water productivity in China. Water 15(9):1733. https://doi.org/10.3390/w15091733

Yasmeen G (1995) Exploring a foodscape: the case of Bangkok. Malays J Trop Geogr 26:1–11

Zan O, Tian J, Deng H, Yan X (2019) Effects of different aeration methods on dissolved oxygen in brackish water and medium water. J Drain Irrig Mach Eng 37:806–814

Zhu Y, Cai H, Song L, Chen H (2019) Aerated irrigation promotes soil respiration and microorganism abundance around tomato rhizosphere. Soil Sci Soc Am J 83:1343–1355

Urban and Indoor Vegetables' Production Techniques

8

Zafar Iqbal, Muhammad Akbar Anjum, Shahid J. Butt, Muhammad Adnan Shahid, and Muhammad Safeer Abbas Jafrey

Abstract

Vegetable production in urban areas and small home gardens is the best alternative approach to tackle food shortage, malnutrition, and inconsistent supplies. It has become a vital component of urban food production system as it has a yield potential of up to 50 kg per m^2. Urban vegetable production may be observed in a variety of ways viz. urban farming, urban homesteading, and indoor gardening. Indoor vegetable production methods not only provide fresh and safe food but also help to mitigate climate change effects and reduce carbon dioxide emissions. This chapter reveals significant and latest information on diverse vegetables' production techniques for urban home and small garden areas.

Keywords

Container gardening · Hügelkultur gardening · Inverted gardening · Keyhole gardening · Lasagna gardening · Square foot gardening · Vertical gardening

Z. Iqbal (✉)
Faculty of Agriculture, University of Sargodha, Sargodha, Pakistan
e-mail: zafarrohils@yahoo.com

M. A. Anjum
Department of Horticulture, Bahauddin Zakariya University, Multan, Pakistan

S. J. Butt
Department of Horticulture, Faculty of Agriculture, PMAS-Arid Agriculture University, Rawalpindi, Pakistan

M. A. Shahid
Horticultural Sciences Department, North Florida Research and Education Centre, University of Florida/IFAS, Quincy, Florida, USA

M. S. A. Jafrey
BinQain Agro Services, Lahore, Pakistan

Z. Iqbal et al. (eds.), *Sustainable and Innovative Vegetable Production in times of Climate Change*, https://doi.org/10.1007/978-981-96-6283-8_8

8.1 Introduction

Global agriculture faces great challenges due to shifting weather patterns, increasing urbanization, loss of arable land, and enhanced bioenergy use. Urban population is expected to increase from 50% to 70% by 2030. Overall world population will reach 9.6 billion by the year 2050. Conversion of agricultural land for urban infrastructure has resulted in food scarcity (Lovell 2010). The rising urban poverty is feared to be followed by hunger and food insecurity in the coming few years. The agricultural systems need to be transformed in the future to achieve the noble target of sustainability and self-sufficiency. To meet rising food demands, urban horticulture may greatly indemnify and contribute by tracing new potential spaces for the cultivation of food crops. A variety of techniques have been employed worldwide for successful indoor vegetable production of which significant ones are detailed here.

8.2 Indoor Vegetables' Production Techniques

8.2.1 Container Gardening

Container gardening is an ideal method for people who have little yard space or small garden area. Many vegetables and herbs can easily be grown in containers. Incredible mixing of colorful vegetables in containers adds elegance, charm, fascination, and diversity to residential gardens. The containers may be placed on ground, mounted on walls or vertical structures, and hanged from height in patio, porch, and adjacent areas in small garden or any large-sized farm (Figs. 8.1 and 8.2).

8.2.1.1 Requirements for Container Growing

8.2.1.1.1 Selection of Container

A container made from plastic, clay, wood, metal, fiberglass, or concrete may be used for vegetable growing. Nursery pots, plastic buckets, grow bags, plastic tubs, discarded drums, ice cream containers, bushel baskets, barrels, dishpans, trash cans, and grocery or cement bags may be used for growing of vegetables. Container materials may be classified into three groups viz. porous, non-porous, and semi-porous. Plastic, metal, glazed, varnished wood, and glass containers are non-porous while clay, unglazed ceramic, untreated wood, drapes, terracotta, and cardboard are porous. Semi-porous includes unsealed wood, pressed fiber, and unsealed natural stone. Plastic, cemented, and some wood containers may be placed safely outside in open space during the whole year (Mays et al. 2022). The size of container may vary for shallow and deep-rooted vegetables. Generally, the size of the container should be larger than the root ball of the plant to avoid suppressed and stunted growth.

Large containers (45 to 55 cm) are quite suitable for growing of multiple varieties of small vegetables. Vegetables like tomatoes, vine cucumbers, pole beans, carrots, peas, cauliflower, and okra are the best to be grown in big containers. Medium-sized containers (25 to 45 cm) best suit to bush or compact vegetables.

Fig. 8.1 Vegetables grown in paint buckets. (**a**) Spinach, (**b**) Turnip, (**c**) Coriander, (**d**) Tomato, (**e** and **f**) Carrot, (**g**) Mint, (**h**) Fenugreek, (**i**) Cauliflower

Fig. 8.2 Vegetables grown in cement bags. (**a**) Eggplant, (**b**) Tomato, (**c**) Mint

Vegetables like peppers, fresh beans, bush cucumber, lettuce, spinach, broccoli, carrot, eggplant, peas, chard, celery, orache, potato, endive, and sorrel are easily grown in containers of medium size. Small containers (15 to 25 cm) are ideally suitable for short-rooted vegetables and herbs like arugula, lettuce, spinach, cabbage, green onion, mint, coriander, radishes, fenugreek, chard, and kale.

8.2.1.1.2 Growing Medium

Soil collected from garden becomes compact and sometimes contains insect pests and soilborne plant pathogens. Instead, container plants perform best in light and well-drained potting mix containing sufficient organic matter, peat moss, compost, perlite, and rotted manure. Addition of organic mulch on top surface retains moisture to keep plants fresh and comfortable during hot summer days. Application of organic fertilizers ensures supply of essential macronutrients, micronutrients, minerals, and amino acids without any side effects.

8.2.1.1.3 Container Drainage

For proper drainage, holes are essential for containers made from any type of materials. Due to excess watering or water-logged conditions, pore spaces may be filled with water causing depletion of oxygen and mortality of plants. Self-watering containers may also be used which are useful for small plants requiring frequent watering. Few small or one large hole can facilitate desired drainage from bottom of container. For small containers, 03 to 06 drainage holes are required while large containers need 06 to 08 holes of the same size. Porous containers lose moisture rapidly but permit air circulation in the rhizosphere, whereas non-porous containers hold water for extended time but limit air movement necessitating drainage holes (Roll and Wilson 2020).

8.2.1.2 Types of Vegetables to be Grown in Containers

8.2.1.2.1 Green Beans (*Phaseolus vulgaris*)

All green beans, also known as string beans or snap beans, are tender annuals. More than 130 varieties of beans have been reported. They possess nitrogen-fixing properties and are highly productive even in containers or planters. Green beans beautify home garden with their attractive foliage and unique flowers. Beans are highly nutritive so added to daily diet. These can be sautéed, oven-roasted, and stir-fried to prepare vegetable mix.

8.2.1.2.1.1 Preferred Pot Size

For bush beans, 15 to 17 cm deep and 37.5 cm wide and for pole beans, 22 to 30 cm deep and 45 cm wide.

8.2.1.2.1.2 Method of Sowing

Sowing is done directly in a large-sized container, pot, or grow bag. Both bushy and climbing varieties of beans can be planted. Sowing of bush beans is done 2.5 cm deep and 5 cm apart while pole beans are sown 2.5 cm deep and 7.5 to 13 cm apart providing support for climbing. Bush beans may attain 60 cm height and occupy specified size of the container. Running a climbing variety utilizes vertical space so sparing garden place for other varieties.

8.2.1.2.1.3 Maturity/Harvesting

Bush beans mature in about 50 to 55 days while pole beans take 55 to 65 days till harvest.

8.2.1.2.2 Beets (*Beta vulgaris* L.)

Beet is a fast-growing root vegetable which grows well in small spaces and containers in most climates. It is a super nutrient-dense vegetable whose greens and root slices are added to salad. It is available in long, round, and globular shapes with varying colors including pink, purple, red, and yellow.

8.2.1.2.2.1 Preferred Pot Size

About 25 to 30 cm deep and 30 cm wide.

8.2.1.2.2.2 Method of Sowing

Seed coat of beet seed is tough so seeds are soaked in warm water for few hours before sowing to enhance germination process. Seed sowing is done in large-sized containers at a depth of 2.5 cm and at least 08 cm apart. Under moist conditions, germination starts within 05 to 08 days.

8.2.1.2.2.3 Maturity/Harvesting

Beet greens can be harvested after 15 to 21 days while beet roots mature in 40 to 60 days after planting.

8.2.1.2.3 Chard (*Beta vulgaris* L. var. *cicla*)

Chard, also known as Swiss chard, is a super food and good substitute for kale and spinach. The young attractive leaves of chard are eaten raw as salad while larger leaves and stalks are used in soups. It adds nutrition and color to pastas, sandwiches, pizzas, stews, casseroles, frittatas, and a variety of salads.

8.2.1.2.3.1 Preferred Pot Size

About 20 cm deep and 30 cm wide.

8.2.1.2.3.2 Method of Sowing

Seed soaking is done in water for 24 hours before planting to enhance germination. Seeds are sown 1.25 cm deep and 05 to 15 cm apart. Sprouting starts within a week after sowing. When plants produce 03 to 04 true leaves, thinning is done to ensure 10 to 15 cm distance (Boeckmann 2024).

8.2.1.2.3.3 Maturity/Harvesting

Harvesting of leaves starts after 60 days of sowing when the plants are 15 to 20 cm in height. Several pickings from individual plants may be done cutting old leaves and allowing young leaves to continue.

8.2.1.2.4 Chili Peppers (*Capsicum annuum*)

Hot chili peppers add beauty and bright color to home garden. As they possess the ability to cross-pollinate so they have different cultivars and hybrids for cultivation by growers. In addition to *Capsicum annuum*, there are five other *Capsicum* species viz. *C. baccatum*, *C. assamicum*, *C. chinense*, *C. frutescens,* and *C. pubescens*. Spicy flavor in hot peppers is produced by an alkaloid called as capsaicin (Xiang et al. 2021). Green and red chilies are used in stews, stir fries, marinades, curries, soups, rice dishes, and Asian cuisines.

8.2.1.2.4.1 Preferred Pot Size

About 30 cm deep and 30 to 35 cm wide.

8.2.1.2.4.2 Method of Sowing

Peppers can easily be grown in containers. They require less space as they grow upward. One transplant is suitable for growing in 20-liter container. Plants may be spaced 25 to 30 cm apart.

8.2.1.2.4.3 Maturity/Harvesting

Hot peppers mature between 90 and 150 days depending upon specific variety.

8.2.1.2.5 Sweet Peppers (*Capsicum annuum* L.)

Sweet or bell peppers are delicious, mild, and sweet flavored. Bell peppers display ivory, red, pink, green, purple, white, yellow, orange, and brown color. Their shapes include round, elongated, bell or block-shaped, and smooth cherry types. These can be used raw in salads, sandwiches, burgers, pizzas, and some Chinese meals and traditional cuisines.

8.2.1.2.5.1 Preferred Pot Size

About 30 cm deep with similar diameter.

8.2.1.2.5.2 Method of Sowing

Pepper seeds are small in size and are planted 0.6 cm deep in a pot. Seeds require about 14 days for germination. Transplants are spaced 30 cm apart in large-size containers. A single bell pepper plant performs better in a container with 30 cm diameter.

8.2.1.2.5.3 Maturity/Harvesting

Sweet peppers are harvested 60–100 days after transplanting depending on the type of variety.

8.2.1.2.6 Kale (*Brassica oleracea*)

Kale cultivars include:

- *B. oleracea sabellica* (Curly kale).
- *B. oleracea acephala* (Kale).
- *B. oleracea alboglabra* (Chinese kale).

Kale is a cruciferous leafy green biennial/perennial vegetable which belongs to cabbage family. This super food is best suited for growing in containers as it saves sufficient space and provides young leaves like spinach and lettuce. It can be added to salad, soups, pasta, and smoothies. Kale leaves are sweeter and nutritious.

8.2.1.2.6.1 Preferred Pot Size

About 25 cm deep and 45 cm wide.

8.2.1.2.6.2 Method of Sowing

Sowing of seeds may be done at 1.25 cm depth in a pot followed by transplanting to a large container. Kale seeds require almost a week for emergence. Thinning is done in large containers to maintain plant to plant distance of 40 cm.

8.2.1.2.6.3 Maturity/Harvesting

Its growing season is short as it matures after one to two months of planting. After cutting of young leaves, kale continues producing new leaves.

8.2.1.2.7 Lettuce (*Lactuca sativa* L.)

Lettuce is a leafy herbaceous annual or biennial plant raised for its leaves to be used as salad green. Lettuce is ideally suited for growing in containers due to its shallow root system, fast growing nature, and simple growing habit. It is a great source of antioxidants and becomes a part of a variety of recipes including burgers, sandwiches, soups, spring rolls, wraps, and tacos.

8.2.1.2.7.1 Preferred Pot Size

About 15 to 20 cm deep and 30 cm wide.

8.2.1.2.7.2 Method of Sowing

Lettuce can be grown from both seeds and transplants. Shallow seed sowing is done at a depth of less than 1.25 cm in a small pot or tray followed by transplanting in a large container. Plants may be thinned to 25 cm apart.

8.2.1.2.7.3 Maturity/Harvesting

Lettuce requires eight weeks from planting to harvesting. Harvesting of 10 to 15 cm tall leaves is quite suitable for use in kitchen.

8.2.1.2.8 Onions (*Allium cepa*)

Onion is a perennial vegetable which grows best in containers. It is used for cooking and salad preparation on daily basis in Asian countries. Green onion stalks are added to soups, salads, and other traditional dishes. They are also used in combination with most of the vegetables in global cuisines. All types of onions, i.e., red, yellow, or purple can be grown easily in suitable planters or pots placed in home gardens.

8.2.1.2.8.1 Preferred Pot Size
About 25 cm deep and 60 cm wide.

8.2.1.2.8.2 Method of Sowing
Onion seeds need to be planted 1.25 cm deep in a large container keeping green onions 05 cm and bulb onions 15 cm apart. In a large-sized container, 06 to 08 onion plants can be grown.

8.2.1.2.8.3 Maturity/Harvesting
Scallions can be picked after 30 to 35 days of seed sowing. Bulb onions are harvested after 90 to 120 days of planting but those planted from nursery sets become ready to eat after 40 to 80 days depending upon the type of cultivar.

8.2.1.2.9 Radish (*Raphanus sativus* L.)
Other scientific names of radish are:
Raphanus candidus
Raphanus raphanistrum sub sp. *sativus* (L.) Domin
Raphanus sativus var. *caudatus* L.H. Bailey
Raphanus sativus var. *niger* J. Kern
Raphanus sativus var. *oleiferus*
Raphanus sativus var. *oleiformis* Pers.
Raphanus sativus var. *sativus* (CABI 2019)

Radish is a most popular edible root vegetable globally. It is available in round or oblong shape and different colors including white, pink, red, and purple. Taste wise it is zesty, mild, crispy, peppery, sweet, and yummy. Salad radish is fast growing and grows in containers with harmony. Radish ideally fits into the theme of residential gardens or urban edible landscape.

8.2.1.2.9.1 Preferred Pot Size
About 15 to 30 cm deep and 30 cm wide

8.2.1.2.9.2 Method of Sowing
Seed sowing is done thinly just 1.25 cm deep in soil and later thinned to distance of 7.5 cm. Seedlings start sprouting within seven days of seed sowing. Six plants of small varieties while four of large ones can be grown in 30 cm wide container.

8.2.1.2.9.3 Maturity/Harvesting
Depending upon variety, radishes are harvested from 21 to 60 days of planting.

8.2.1.2.10 Spinach (*Spinacia oleracea*)
Spinach is quite suitable for growing in a container and even in a small partial shady place. It is considered as superstar of green leafy vegetables as it is highly rich in nutrients and water. It is added to food and snacks and baked items like brownies, chicken vegetable soup, and sausages.

8.2.1.2.10.1 Preferred Pot Size
About 15 to 20 cm deep and 35 cm wide.

8.2.1.2.10.2 Method of Sowing

Sowing of spinach seeds is done 1.25 cm deep and 2.5 to 5 cm apart in pots. Seeds take 05–14 days to germinate. Seedlings are later thinned to 7.5 to 12.5 cm apart.

8.2.1.2.10.3 Maturity/Harvesting

Spinach leaves can be harvested after 40 to 55 days of planting.

8.2.1.2.11 Tomatoes (*Solanum lycopersicum*)

Tomatoes add much beauty and delight to a garden with different shapes and colors. Green, orange, yellow, pink, and purple colored tomatoes give attractive look to the edible garden. These are used in salads, sandwiches, burgers, pastas, and for culinary purposes.

8.2.1.2.11.1 Preferred Pot Size

About 30 cm deep and 50 cm wide.

8.2.1.2.11.2 Method of Sowing

Tomato seeds are small in size so these are sown 0.5 cm deep in growing medium. When seedlings are 05 cm tall, they are transplanted to a large container. Containers are placed at sunny location receiving 06 to 08 hours of sunlight.

8.2.1.2.11.3 Maturity/Harvesting

Ripening is complete when tomatoes attain their natural color. They take 65 to 80 days to mature after transplanting.

8.2.1.2.12 Cabbage (*Brassica oleracea* var. *capitata*)

Cabbage is a fast-growing biennial leafy green vegetable. Mostly it is grown as an annual crop and harvested within the same growing season for good quality. It is closely related to vegetables like cauliflower, broccoli, Brussels sprouts, and kale. In raw form, it is added to salad and coleslaw. It is also used in soups, sandwiches, and Chinese foods. In Pakistan, Sri Lanka, and India it is liked to be cooked by adding potatoes.

8.2.1.2.12.1 Preferred Pot Size

About 25 to 30 cm deep and 45 cm wide.

8.2.1.2.12.2 Method of Sowing

Cabbage can be grown from seeds and transplants. Seeds should be planted 0.6 cm deep and 7.5 cm apart and later thinned to 30 to 50 cm apart when seedlings attain the height of 0.5 to 7.5 cm.

8.2.1.2.12.3 Maturity/Harvesting

Harvesting starts when heads become round, firm, and large. Generally, cabbage requires 60 to 80 days for full maturity after transplanting (Daveport et al. 2023) depending upon the type of variety.

8.2.1.2.13 Carrots (*Daucus carota* sub sp. *sativus*)

Carrots are among the best vegetables to be grown in containers. Fresh carrots picked from containers in own home garden are much nutritious and delicious. They can be used raw in salad or snack and cooked with other vegetables. Carrot roots can be used in various items, including cakes, puddings, jellies, pickles, and preserves. These are also processed to prepare chips, flakes, pickle, jam, and powder.

8.2.1.2.13.1 Preferred Pot Size

About 30 cm deep and 30 to 60 cm wide.

8.2.1.2.13.2 Method of Sowing

Seed sowing is done about 1.25 cm deep in a large-sized container and thinned to 7.5 cm apart. Carrot seeds require 10 days for germination so should be kept moist till sprouting. Thinning is an essential practice to provide little space between plants to facilitate the development of healthy roots. Finally, 10 plants may be planted in 20-liter container.

8.2.1.2.13.3 Maturity/Harvesting

Carrots require 60 to 80 days for full maturity.

8.2.1.2.14 Cucumber (*Cucumis sativus*)

Cucumber is the most favorite vegetable used in kitchens worldwide to make salads on a daily basis. It is related to pumpkins, watermelons, and zucchini. It gives very good taste when added to sauces, pickles, sandwiches, and burgers. Most of the bush and compact varieties perform well when grown in containers. Vining cucumber can be trained to be grown on trellis or a supporting structure.

8.2.1.2.14.1 Preferred Pot Size

About 30 cm deep and 50 cm wide.

8.2.1.2.14.2 Method of Sowing

Cucumbers require large-sized container to facilitate the development of root system for active growth. Cucumbers can simply be started from seed while transplanting is not feasible due to chances of root damage. Sowing is done by planting three seeds per pot about 1.25 cm deep. Soil is kept moist till the germination of seeds. Germination starts within 07 to 14 days. Four cucumber plants can be grown in a 50 cm wide container.

8.2.1.2.14.3 Maturity/Harvesting

Maturity time varies from 60 to 90 days depending upon the type of variety. Harvesting is done when cucumbers gain approximate length of 15 to 20 cm. Over-ripening should be avoided as it leads to bitter taste. Early harvest in the morning time ensures good texture and flavor.

8.2.1.2.15 Eggplant (*Solanum melongena*)

Eggplant is a perennial edible and ornamental vegetable which gains height of 60 to 90 cm on maturity. It is available in different shapes like small, round, long, and oblong. The color of peel varies from white to green or deep purple. Each plant produces 04 to 35 fruits depending upon variety, growing conditions, life cycle, and management practices. It can be added to soups, dips, fritters, and culinary dishes. Roasted, grilled, and steamed fruits add flavor to different dishes.

8.2.1.2.15.1 Preferred Pot Size

About 30 cm deep and 60 cm wide.

8.2.1.2.15.2 Method of Sowing

Seeds are sown 1.25 cm deep in a container and need to be kept moist and warm consistently. They start germination in 10 to 21 days. At least 25 to 30 cm space is required for one plant to grow to full potential. One plant can easily be grown in 60 cm container.

8.2.1.2.15.3 Maturity/Harvesting

Eggplant takes 70 to 120 days for maturity.

8.2.1.2.16 Peas (*Pisum sativum*)

Peas are herbaceous annuals that attract with their bright pink flowers in the gardens. Their three different types include snap, snow, and shelling peas. Garden peas, known as shelling or English peas, are grown for seeds enclosed in pods which are inedible. Snap peas are the result of a cross between garden peas and snow peas having edible pods. Snow peas are commonly used in Asian cuisines. Their pods are edible and added to soups, salads, and stir-fries. Some varieties are grown for ornamental purposes due to their unique growing style. Peas like cool conditions so early planting ensures good produce in the home gardens.

8.2.1.2.16.1 Preferred Pot Size

About 30 cm deep and 30 to 90 cm wide.

8.2.1.2.16.2 Method of Sowing

Seeds are sown 1.25 cm deep and 05 to 7.5 cm apart in a large-sized container. Tall varieties need to be supported by trellis or supporting structure. The containers are placed at a sunny location receiving 08 hours of sunlight.

8.2.1.2.16.3 Maturity/Harvesting

Peas take between 60 and 70 days to grow to maturity from seeds.

8.2.1.2.17 Summer Squash (*Cucurbita pepo*)

This is a warm season highly productive vegetable. It is available in different shapes, sizes, flavors, and colors. It has different types which include zucchini, crookneck, straight neck, and patty pan. It can grow 60 cm high and 90 to 120 cm wide. Compact and bush-type varieties are quite suitable for growing in containers. Squash can be used as raw, cooked, and shredded or grated in baked items. It can be roasted, grilled, fried, steamed, and sautéed. Squash flowers are also edible.

8.2.1.2.17.1 Preferred Pot Size

About 30 to 45 cm deep and 60 cm wide.

8.2.1.2.17.2 Method of Sowing

Two to three seeds are sown 2.5 cm deep in the soil. The plants are later thinned and spaced 60 cm apart. One plant per each large container performs better and gives good yield. The containers are placed at a sunny location which receives 06 to 07 hours of sunlight.

8.2.1.2.17.3 Maturity/Harvesting

Squashes can be harvested after 45 to 60 days of planting when they are 12 to 15 cm long.

8.2.1.2.18 Winter Squash (*Cucurbita maxima; C. argyrosperma; C. moschata*)

Winter squash or pumpkin is a good choice for growing in containers. It is a vining plant which requires more space and support for its growth. It is available in various shapes, sizes, and colorful patterns. Its common types include butternut, acorn, pumpkin, and spaghetti. It is rich in fiber and can be used in stews, grains, and greens-based salads, pizzas, quesadillas, baked pastas, casseroles, and gratins.

8.2.1.2.18.1 Preferred Pot Size

About 30 cm deep and 60 cm in diameter.

8.2.1.2.18.2 Method of Sowing

Direct seeding is the preferred method for starting squash, as the seeds germinate quickly within 07 to 10 days easily followed by deep root growth. Two to three seeds are sown 1.25 cm deep in a container and thinned later to viable seedlings after the formation of true leaves. Two plants should be grown in a 40-liter container.

8.2.1.2.18.3 Maturity/Harvesting

It takes 85 to 110 days for winter squash to mature.

8.2.1.2.19 Broccoli (*Brassica oleracea* var. *italica*)

Broccoli is a cool season crop and grows well under moderate climatic conditions. It is an annual high-fiber vegetable which can be planted in late summer or autumn. Compact varieties perform best in containers. Broccoli is used in many dishes worldwide including soups, salads, and protein-rich fajitas. Winter broccoli has sweeter flavor than one sown in warm season.

8.2.1.2.19.1 Preferred Pot Size

About 30 to 45 cm deep and 45 cm wide.

8.2.1.2.19.2 Method of Sowing

Seeds can be sown in trays or directly in containers. Seeds need to be planted 1.25 cm deep and 7.5 cm apart and later thinned to 30 to 50 cm apart. When seedlings are 03 to 04 weeks old, they are transplanted to larger container of desired size. Two to three plants can be grown in a 45 cm wide container.

8.2.1.2.19.3 Maturity/Harvesting

Harvesting is done from 50 to 65 days after sprouting.

8.2.1.2.20 Okra (*Abelmoschus esculentus*)

Okra, also known as lady's finger, is a warm season tropical vegetable which is much liked in Asian countries. It is also planted as ornamental plant due to its beautiful foliage and attractive blooms. Flowers and seeds of okra are also edible if used as cooked and raw. Dwarf varieties are quite suitable for growing in containers. Okra is steamed, fried, cooked, and added to soups or stews.

8.2.1.2.20.1 Preferred Pot Size

About 30 to 35 cm deep and 45 cm wide.

8.2.1.2.20.2 Method of Sowing

Transplantation of okra is not suitable due to its long tap root. Seed sowing is done 1.25 to 2.5 cm deep in the container. Soil should be kept moist till germination takes place. Germination occurs mostly within 05 to 10 days under warm conditions at a temperature of 24–32 °C. Plants need to be spaced 30 to 45 cm apart. The container should be placed in sunny area receiving 06 to 08 hours of sunlight.

8.2.1.2.20.3 Maturity/Harvesting

Okra matures in at least 60 to 65 days. Harvesting is done regularly as it is a cut and come again vegetable.

8.2.1.2.21 Potatoes (*Solanum tuberosum* L.)

Potato is considered as the fourth most important food crop in the world after rice, wheat, and corn. It is a versatile vegetable and dietary staple in many regions of the globe. Homegrown fresh potatoes can be boiled, baked, fried, and used in the number of recipes. They are desired for gravies, soups, sauces, stews, dumplings, pancakes, and salads. Potato starch is also used in many industries of the world. The value-added processed potato food products have high demand in global markets.

8.2.1.2.21.1 Preferred Pot Size

About 30 to 40 cm deep and 50 cm wide.

8.2.1.2.21.2 Method of Sowing

Sowing of potatoes is done by planting seed potatoes or potato tubers with sprouting eyes. Four seed potato pieces are placed on potting soil with the eye buds facing up in a 50 cm wide container and covered later with 10 cm of potting soil. Potatoes like moist but well-drained soil. In hot summer days, daily watering is required. Depending upon soil temperature, sprouts from seed potatoes emerge in 02 to 04 weeks. When potato plants attain the size of 15 cm, then hilling or earthing up is done by addition of few centimeters of soil around the potato plants covering the bottom of growing stems. Covering of stems enhances tuber production.

8.2.1.2.21.3 Maturity/Harvesting

Potatoes are ready to harvest between 90 and 120 days after emergence. Harvesting starts when potato plants show flowering. Few young potatoes can be picked for immediate use in kitchen. When plants become yellow, then all leftover potatoes can be harvested.

8.2.1.2.22 Oregano (*Origanum vulgare*)

Oregano is a culinary herb grown in most of the herb gardens. Compact leaves and ornamental flowers make oregano an ideal plant for containers and pots. It produces small white, pink, or blue flowers in summer. Oregano is a hardy perennial having two types viz. Mediterranean and Mexican. It is a versatile herb used in many recipes like pizzas, pasta sauces, Spanish stews, Greek salads, salsa, and meat mixtures. Dried or fresh leaves are also used in different recipes. Dried oregano is mostly preferred in Italian and Mediterranean foods.

8.2.1.2.22.1 Preferred Pot Size

About 20 to 30 cm deep and 45 cm wide.

8.2.1.2.22.2 Method of Sowing

Seeds are sown 0.6 cm deep and 25 to 30 cm apart in pots on the surface of well-drained seed compost. The pots are placed on a warm place as seeds need heat for germination. When emerged seedlings attain 15 cm height, they are transplanted 45 cm apart to larger containers placed at sunny location. Oregano prefers warm air and soil, and light watering throughout the growing season for its best growth. Its good companions include tomato, broccoli, cabbage, peppers, and many other vegetables.

8.2.1.2.22.3 Maturity/Harvesting

Oregano plants can be harvested when they attain height of 10 to 15 cm. Generally, they become ready for harvest after 80–90 days of sowing of seeds. However, leaves harvesting can be started between 42 and 56 days.

8.2.1.2.23 Sage (*Salvia officinalis*)

Sage is a perennial Mediterranean herb which is known for its beautiful velvety gray-green leaves. Its 900 species are grown globally. With rising temperature, it produces beautiful pink or purple flowers which attract pollinators. It is also named as culinary sage, common garden sage, and garden sage. It is an aromatic medicinal herb used for the treatment of stomach disorders, shortness of breath, sore throat, diabetes, high cholesterol levels, and memory loss. Sage possesses antimicrobial, anticancer, and anti-inflammatory properties. It is added to Mexican, Italian, and European cuisines. It can be used as fresh, dried, ground, or rubbed. Fresh leaves are eaten in salads and used for cooking in omelets, fritters, marinades, and soups.

8.2.1.2.23.1 Preferred Pot Size

About 20 cm deep and 30 cm wide.

8.2.1.2.23.2 Method of Sowing

Seed sowing is done in seed starting mix at a depth of 0.6 cm at a temperature of 15 °C or above. Germination starts within 02 to 03 weeks. As seed germination in sage is a slow process and takes much time so propagation through cuttings is preferred. A 7.5 to 10 cm cutting is clipped from the tip of a stem below the root node, treated with hormone, planted in sand or vermiculite, and placed in bright indirect light. Soil is kept moist to facilitate the formation of new roots. After the emergence of roots within 06 to 08 weeks, it is transferred to a small pot. When root ball is formed then again it is transferred to a large-sized pot. Combination of potting mix and compost, good drainage, air circulation, and sunlight (06 hrs.) favor good growth of sage.

8.2.1.2.23.3 Maturity/Harvesting

Harvesting may be done during the first year according to the requirement of the gardener. Sage requires 75 days after planting for the first harvest. Harvesting of leaves is done from properly established plants. Its life span as a perennial expands from 03 to 05 years.

8.2.1.2.24 Thyme (*Thymus vulgaris*)

Thyme is a fragrant perennial herb that is successfully grown in home gardens. It has shallow roots enabling it best suited for container growing. It flourishes well in hot conditions and doesn't require continuous watering or special care. Companion plants include cabbage, tomatoes, eggplants, broccoli, fenugreek, and Brussels sprouts. Common varieties include English, German, Lemon, and Creeping thyme. Being a culinary staple, its leaves, flowers, and oil are added to foods for aroma and flavor.

8.2.1.2.24.1 Preferred Pot Size

About 25 to 30 cm deep and 30 to 45 cm wide.

8.2.1.2.24.2 Method of Sowing

Thyme flourishes in a well-drained light soil at sunny and slightly protected location. For growing of thyme from seeds, sowing is done in small pots measuring 05 cm diameter in potting compost covered later with thin layer of compost. The pots are kept at warm position in a greenhouse. Seedlings emerge in a week time period which are thinned later to 02 to 03 seedlings. After hardening for a week, plants are transplanted to larger pots keeping 30 cm apart. Thyme is best propagated from cuttings or layering. For propagation from cutting, about 05 to 08 cm long cutting is clipped from tip of a stem, treated with rooting hormone and planted in sand or vermiculite. Roots are emerged after 06 weeks of planting. The seedlings are first transferred to a small pot. After formation of root ball, seedlings are shifted to larger pot.

8.2.1.2.24.3 Maturity/Harvesting

Thyme matures in 70 to 90 days. It is harvested when plant is 10 to 13 cm in diameter. Its harvesting is done periodically throughout the growing season.

8.2.1.2.25 Arugula (*Eruca sativa*)

Arugula is a fast-growing annual leafy vegetable and performs well in containers or window boxes. It's also named as rucola, Italian cress, colewort, and salad rocket. It is eaten raw in salads and also used as add-on topping for wraps, pizza, nachos, pastas, and sandwiches.

8.2.1.2.25.1 Preferred Pot Size

About 15 cm deep and 40 cm wide.

8.2.1.2.25.2 Method of Sowing

Arugula can be sown in two seasons viz. spring to early summer and later summer to fall. Seed is sown 0.6 cm deep and 2.5 cm apart in containers and covered with thin layer of soil. As germination requires moist environment so water is sprayed lightly to moisten the seeds and potting soil. Container needs to be placed in cool sunny environment at a well-ventilated place. Seedlings should be thinned 15 cm apart after proper emergence.

8.2.1.2.25.3 Maturity/Harvesting

Arugula is harvested after 20 to 50 days of seed sowing. Young leaves are sweet in taste so it's wise to harvest leaves when they attain 05 to 07 cm size.

8.2.1.2.26 Scallions (*Allium fistulosum* L.)

Scallions are quite suitable for growing in containers in home gardens because of shallow root system and fast-growing habit. They produce cylindrical stalks having hollow green tops. The whole scallion plant with white bulbous part growing underground and green stalks is edible. Scallions can be chopped and eaten as raw on top of soups, salads, and other foods.

8.2.1.2.26.1 Preferred Pot Size

About 15 to 20 cm deep and 30 cm wide.

8.2.1.2.26.2 Method of Sowing

8.2.1.2.26.2.1 Seeds Sowing

Tray is filled with potting compost. Rows of about 1.25 cm deep and 07 cm apart are made followed by sprinkling of seeds into the rows. Seeds are covered with compost and the trays are placed in a sunny position in a greenhouse. Emergence of shoots occurs within two weeks. After additional growth of 14 days, transplantation may be done in the container of choice.

8.2.1.2.26.2.2 Planting Green Scallions from Scrap

The leaves of tender green scallions are cut for use in kitchen leaving 07 to 10 cm of white bulbs and later buried 1.25 cm deep in fresh and fertile soil. Growth of green stems initiates within few days.

8.2.1.2.26.2.3 Planting Bulbs

Bulbs are placed with the root side down while top side up covering the lower portion with soil. Bulbs are spaced 05 cm apart in the container.

8.2.1.2.26.3 Maturity/Harvesting

Scallions are ready to harvest after 04 weeks of planting before bulb formation. These can be cut into slices to store in freezer bags for adding them to various recipes. Green scallions having small bulbs are harvested after 05 weeks. In both cases, only leaves can be clipped leaving white bulb in the soil to continue growing for next harvest.

8.2.1.2.27 Garlic (*Allium sativum*)

Garlic is a staple ingredient and can be grown easily in containers for use in kitchen. There are two types of garlic viz. soft and hard neck. They have similar flavors but prefer different climatic conditions. Soft neck garlic flourishes best in mild climate as it is less winter hardy while hard neck prefers colder climate due to its hardiness. Garlic is used to flavor many recipes in the kitchen like salad dressings, marinade, sauces, stews, and soups. Its extract possesses antifungal, antiviral, and antibacterial properties. When garlic is chopped or sliced, the thiosulfinate chemical in garlic is converted to allicin which is a natural bioactive antibiotic.

8.2.1.2.27.1 Preferred Pot Size

About 20 to 25 cm deep and 60 cm wide.

8.2.1.2.27.2 Method of Sowing

Cloves are split from a large-sized head. Then papery sheath or wrapping around the cloves is removed leaving the outer coat unbroken. Cloves are planted 08 cm apart keeping pointed side up in a container filled with fertile compost and manure. Four to 06 plants can be grown in a 60 cm wide container. Garlic needs chilling period under 4.5 °C for two months for proper development. Containers are placed in full sun after emergence of leaves and watered regularly.

8.2.1.2.27.3 Maturity/Harvesting

Long green shoots or scapes produced by hard neck varieties can be clipped when they are young, succulent, and twisting. These are added to flavor salads and some roasted vegetables. After harvesting of green shoots, bulbs show increase in size getting more nutrients for vigorous growth. Generally, garlic takes almost 06 to 09 months from planting to harvesting.

8.2.1.2.28 Cauliflower (*Brassica oleracea* var. *botrytis*)

Cauliflower is a commonly grown cool season vegetable. It has a shallow root system so it can be raised easily in containers. It is a good source of fiber and antioxidants. It is steamed, sautéed, stir-fried, and used raw in salads. It is also used in pizzas, tortillas, and many other recipes.

8.2.1.2.28.1 Preferred Pot Size

About 20 to 30 cm deep and 45 cm wide.

8.2.1.2.28.2 Method of Sowing

Container is filled with well-drained potting soil and seeds are sown 1.25 cm deep spaced 05 to 08 cm apart. Soil is kept moist but soggy or unnecessary wet condition must be avoided. Spray bottle may be used for watering after seed sowing. Seeds start germination within 08 to 14 days. Seedlings are later thinned 45 to 60 cm apart allowing one plant per pot. The containers are placed at sunny location with partial shade during hotter months.

8.2.1.2.28.3 Maturity/Harvesting

Cauliflower matures between 50 and 90 days after planting. Harvesting continues during the growing season depending upon the type of variety.

8.2.1.2.29 Celery (*Apium graveolens*)

Celery is an herbaceous aromatic vegetable. Its succulent leaf stalk, leaves, and hypocotyl are edible while seeds yield 02–03% essential oil and 17–18% fatty oil. Celery is added to dips and spreads and also used as salad. Its seed powder is added to recipes as a spice.

8.2.1.2.29.1 Preferred Pot Size

About 20 cm deep and 30 cm in diameter.

8.2.1.2.29.2 Method of Sowing

Seeds are sown 1.25 cm deep and 05 cm apart in nutrient-rich soil amended with organic compost. Soil is kept moist to promote proper growth. Germination starts within 02 weeks of sowing. The containers are placed at sunny location receiving sunlight for 06 hours. The seedlings are thinned when attain height of 07 to 08 cm and transplanted to large pots. One plant grows best in one pot.

8.2.1.2.29.3 Maturity/Harvesting

Celery takes about 90 to 120 days for maturity after planting.

8.2.1.2.30 Turnip (*Brassica rapa* subsp. *rapa*)

Turnip is a versatile cool season vegetable generally cultivated for its leafy greens and tender roots. It is considered as the hardiest and oldest vegetable. It is ideally suited for growing in containers. Turnip is used raw in salads or slaws and also as baked, boiled, or steamed. Turnip greens are rich source of calcium. They are used in omelets, sandwiches, casseroles, soups, and stews.

8.2.1.2.30.1 Preferred Pot Size

About 30 cm deep and 40 cm wide.

8.2.1.2.30.2 Method of Sowing

Seeds are sown 1.25 cm deep and 05 to 08 cm apart in the preferred pot containing fertile, well-drained, and porous soil. Good quality potting mix enhances root growth. When the seedlings become 10 cm tall, the weak ones are thinned out to maintain the desired space of 10 to 15 cm. The soil is kept moist but overwatering needs to be avoided to ensure proper development of roots. The container is placed at sunny location receiving full sunlight for better growth of the seedlings.

8.2.1.2.30.3 Maturity/Harvesting

Young green leaves of seedlings, which are removed during thinning, can be used in salads. Later, turnip greens can be harvested when the plants are 04 to 06 inches tall. Most of the turnip varieties mature within 35 to 60 days. Roots are harvested when they are 05 to 7.5 cm in diameter.

8.2.1.2.31 Parsley (*Petroselinum crispum*)

Parsley is a fast-growing biennial aromatic herb. There are 30 varieties of parsley which differ from each other on the basis of green shade and shape of foliage. The common types include curly-leaved and flat-leaved or Italian. Curly varieties with mild flavor are used for decoration purpose while flat-leaved ones possess strong flavor and are added to soups, sauces, vinaigrettes, and fillings. Parsley leaves are used in sauces, dressings, smoothies, stews, pasta, and stuffing.

8.2.1.2.31.1 Preferred Pot Size

About 30 cm in depth with similar diameter.

8.2.1.2.31.2 Method of Sowing

Seeds are soaked in water overnight prior to sowing to promote the germination process. Seeds are sown 1.25 cm deep in pots containing seed starting mix and covered with a thin layer of soil. The soil is kept moist avoiding waterlogged condition. The pots are placed at sunny location in the garden or residential area. The seeds show good germination within 14 to 28 days. Transplanting is done in large containers containing fertile soil after 35 to 42 days of sowing.

8.2.1.2.31.3 Maturity/Harvesting

Harvesting starts when plants become 15 cm tall. Most of the parsley varieties mature between 70 to 90 days after planting.

8.2.1.2.32 Basil (*Ocimum basilicum*)

This is an annual herb grown for its aromatic and peppery leaves. It can be grown easily in planters or containers. Its leaves show shades of green, red or purple color. It is used in pizzas, pastas, sauces, soups, sandwiches, salads, and even ice creams. Some common basil types include sweet basil (*Ocimum basilicum*), Thai basil (*Ocimum thyrsiflora*), lemon basil (*Ocimum citriodorum*), holy basil (*Ocimum tenuiflorum* or *sanctum*), hairy or American basil (*Ocimum americanum*), and Mexican or cinnamon basil (*Ocimum basilicum* 'Cinnamon').

8.2.1.2.32.1 Preferred Pot Size

About 20 to 30 cm deep and 40 cm in diameter.

8.2.1.2.32.2 Method of Sowing

Seed is sown 0.6 cm deep in pre-moistened potting mix and covered with a thin layer of soil. Watering is done regularly to avoid drying of soil. The container is placed at sunny position to receive direct sunlight for 06 to 08 hours. Seeds show germination within 05 to 10 days of sowing. After emergence, the seedlings are initially thinned to 08 to 10 cm and finally 30 cm apart. Transplanting is done to a larger container having well-drained fertile soil. Basil can also be grown from a cutting of 10 cm size placed in water. After the development of roots, plant is shifted to the desired container.

8.2.1.2.32.3 Maturity/Harvesting

Basil leaves become ready for harvest within 50 to 60 days. Outer leaves can be clipped leaving few on the plant for future growth.

8.2.2 Framed Raised Bed Gardening

A raised garden bed is an elevated framed plot-like structure containing soil on the surface of ground. Box-like frame provides structural strength and helps avoid soil erosion. Frames can be made from non-toxic materials like bricks, solid concrete blocks, corrugated metal and synthetic lumber or wood which can withstand rot and decay. The materials for bottom layer of raised bed include straw, mulch, leaves, grass clippings, and cardboard. Rich soil can be prepared for framed bed by mixing garden soil, manure, and compost keeping in view the requirements of the specific vegetable. Raised bed allows non-compaction of soil, free water movement, better air circulation, and controlled management. Better root proliferation and development lead to good crop stand. Raised bed in garden is a good choice for old or disabled persons as less work is required in bending position. Smart irrigation practices reduce extra exertion of the gardeners. Raised beds also solve problem of excess watering and facilitate good drainage. Raised beds work well in wet climate and on compact soils. In raised beds, all agricultural practices starting from sowing to harvesting are better performed and monitored. Fertilizers and other inputs are not wasted. Raised beds warm up in shorter time permitting prolonged season for growth. Framed raised beds can be prepared in different designs and sizes keeping in view the available space in patio, terrace, balconies, and home garden area. The beds should be 15 to 30 cm deep, 0.9 to 1.2 m wide, and 1.8 to 2.4 m long to permit the development of most of the vegetables and easy working. The design may be rectangular, hexagonal, round, triangle, table type, or L-shaped.

8.2.2.1 Vegetables for Framed Raised Bed

Vegetables like peppers, bush beans, pole beans, tomatoes, cucumbers, eggplant, carrots, sweet potatoes, leafy greens, beets, lettuce, radish, spinach, summer squash, mint, peas, fenugreek, celery, cabbage, potatoes, arugula, Swiss chard, zucchini, and onions can be easily and successfully grown in a framed raised bed.

8.2.3 Square Foot Gardening

The term square foot garden was introduced by Mel Bartholomew in the early 1980s. Square foot gardening or intensive gardening is a space-efficient method for maximization of yield. In square foot gardening, generally 4 × 4 feet (1.2 × 1.2 m) raised bed filled with mixture of potting mix and compost is prepared and then divided into 16 squares using canes, string, or wooden strips. In each square, different vegetables or herbs may be planted. Planting in any square depends upon the size of the plants. In case of larger size, one seedling has to be planted in one square. If size and canopy are small, then four plants may be grown in one square. When seedlings become 08 to 10 cm tall, then mulching may be done for moisture retention, soil enrichment, and weed suppression. All the squares are utilized during the growing season by planting another crop in the square when the first one is harvested. Square foot gardening can be started in the garden patch with plain surface receiving 06 to 08 hours of light daily. This method is comparatively inexpensive requiring minimum management practices. Vegetables of own choice can be grown with this method with less risk of weeds or undesired plants. Square foot gardening is ideally suited for growing of herbs and compact vegetables. This method is also suitable for gardening in patios and balconies. It is a very good method for companion and rotational planting.

8.2.3.1 Vegetables for Square Foot Design

- 16 plants per square foot: Red radishes, carrots, chives, beets, turnips, spinach, etc.
- 09 plants per square foot: Beans, Arugula, lettuce, spinach, scallions and garlic.
- 08 plants per square foot: Peas, pole beans, fenugreek, etc.
- 04 plants per square foot: Beets, leeks, garlic, kale, onions, lettuce, thyme, basil, Kohlrabi, parsley, chard, and Asian greens.
- 02 plants per square foot: Broccoli, cauliflower and lettuce.

01 plant per square foot: Cabbage, pepper, tomato, eggplant, celery, okra, Brussels sprouts, collard greens, potato, asparagus, oregano, zucchini and sage.

8.2.4 Vertical Gardening

It is an innovative and creative method for growing of diverse house plants, annuals, perennials, shrubs, vegetables, and herbs in a vertical design or fashion. The first vertical gardens are reported to be originated in 3000 BCE in the Mediterranean area (Leishman 2018). A vertical garden may be erected in patio, balcony, yard, small garden, and on porch wall and rooftop of buildings. It not only looks beautiful but also ensures provision of fresh and safe food to the residents. It may be designed and structured as indoor or outdoor activity in various ways by preparing living wall, moss wall, hanging planters/baskets, decorated drink bottles, hanging pots, and wall-mounted pots.

8.2.4.1 Vertical Gardening Structures

These structures include trellises, tuteurs, arbors, obelisks, teepees, tripods, arches, towers, pergolas, gazebos, wire cages, window box, netting, pallet, bamboo poles, posts, trampolines, ladders, pyramids, wooden stakes, tree trunks, and fences.

8.2.4.2 Vegetables for Vertical Gardens

Vegetables and herbs like pole beans, climbing peas, sweet potatoes, vining tomatoes, zucchini, cucumber, radish, lettuce, fenugreek, garlic, onion, pepper, spinach, kale, basil, thyme, parsley, sage, gourd vines, microgreens, oregano, thyme, coriander, luffa, and squash can be grown on vertical gardening structures.

8.2.5 Inverted/Upside-Down Gardening

This method was first tried by a gardener Kathi Lael Morris for growing peppers and tomatoes in 1998. A 05 to 07 cm hole is made in the bottom of a bucket or container to facilitate growth of a plant. The bucket is filled with fertile potting soil. After planting of desirable vegetable plant, the bucket is kept hanging from a suitable place with the help of a hook. Daily watering during hot summer and alternate day watering during mild days are required. An upside-down garden utilizes small place for the benefit of gardeners who have little or no growing area. This method saves space, prevents pests and fungal diseases, increases exposure to sunlight, and ensures good air circulation (Fig. 8.3).

8.2.5.1 Vegetables for Inverted Gardens

Many plant species can be grown by this method but the best ones include cherry tomatoes, peppers, eggplants, herbs, cucumber, beans, and squash. The top of upside-down container can also facilitate raising some vegetables including cress, lettuce, herbs, fenugreek, coriander, and mint.

Fig. 8.3 Inverted gardening: Tomatoes grown in paint bucket

8.2.6 Hanging Baskets

Hanging baskets provide a good opportunity for the gardeners who don't have sufficient space for domestic food production. This method helps beautifies interior and exterior of house with a variety of colorful and nutritious vegetables. The size of the basket should be 30 to 40 cm holding 3.5 to 4.0 kg of potting mix. These baskets can be hung in balcony, porch, patio, and on outside walls. Roots of young plants flourish well in fertile and well-drained potting mix in the hanging basket. During hot summer months, frequent watering is required to prevent wilting or withering. Vegetables and herbs grown in baskets comparatively remain safe from insect pests, diseases, and soil-borne problems (Fig. 8.4).

8.2.6.1 Edibles for Hanging Baskets

8.2.6.1.1 Vegetables

Tomato, eggplant, dwarf peppers, tomatillo, dwarf beans, zucchini, lettuce, spinach, chard, radish, mustard greens, Asian greens, spring onions, and cabbage.

Fig. 8.4 Vegetable growing in hanging baskets. A) Spinach, B) Fenugreek

8.2.6.1.2 Herbs

Thyme, oregano, basil, sage, mint, parsley, chives, hyssop, fenugreek, and coriander.

8.2.7 Window Box Gardening

Window box gardening is a feasible, pleasing, and attractive method for growing vegetables at home. It can be hung on or below the window sill, fence, or balcony of a home. Window of kitchen or living room is also the best site for the placement of a box containing short-duration vegetables. The size of window box should be 10 to 15 cm deep for shallow rooted and 25 to 30 cm for root vegetables or plants requiring more space. The length may be decided according to the availability of space. The box should be filled with good-quality potting soil mix. It should have drainage holes to avoid soggy conditions and rotting of roots.

8.2.7.1 Vegetables for Window Box Garden

Spinach, small red radish, lettuce, carrot, microgreens, tomato, kale, beets, arugula, leeks, and bush beans.

8.2.7.2 Herbs for Window Box Garden

Thyme, chives, parsley, sage, oregano, rosemary, basil, lavender, and fenugreek.

8.2.8 Keyhole Gardening

Keyhole gardens were started in the 1990s for the population of Lesotho in Southern Africa facing issues like drought and soil erosion. They were popularized after successful food production. In keyhole gardening, a circular garden bed is prepared with notch or wedge-shaped structure and a central composting well or basket. The bed may be prepared using stones, bricks, rocks, concrete blocks, or logs. Keyhole garden is generally 02 to 2.5 m wide and 0.6 to 0.9 m high. The well is filled with kitchen waste and leaf litter. This material breaks down and provides essential nutrients to the plants. The raised bed is filled with garden soil and preferably top layer with compost. Watering is done in the composting well wherefrom water seeps into the raised bed supplying necessary nutrients to the plants.

8.2.8.1 Vegetables for Keyhole Garden

Vegetables which grow best in keyhole garden include onions, garlic, carrots, radish, beets, turnip, spinach, lettuce, peas, tomatoes, and herbs.

8.2.9 Lasagna Gardening/Sheet Composting/Sheet Mulching

Lasagna gardening is an organic gardening method which doesn't require digging or tilling-like operations. It is an eco-friendly approach as it effectively utilizes kitchen scraps and garden waste for growing plants. Raised boundaries are erected using wood, stones, or related materials. Twigs and woody materials or cardboards are required to make a base layer. The second layer is approximately 05 cm thick and includes brown material like dry leaves, straw, hay, yard waste, mulch, etc. The third layer is 05 cm thick and contains green materials like kitchen waste, grass clippings, coffee grounds, and rotted manure. Following alternate layering, a 45 to 90 cm tall bed is finally formed. In hot and dry climates, the layers need to be kept moist. The organic materials are later decomposed to make fertile compost within 04 to 05 months. This method ensures best water retention, requires less labor and input, and improves poor soil.

8.2.9.1 Vegetables for Lasagna Garden

Most of the vegetables, herbs, and fruits can be grown in Lasagna garden. Tomatoes, peas, lettuce, peppers, green beans, squash, carrots, radish, and beets give good yield in Lasagna garden.

8.2.10 Straw Bale Gardening

A raised vegetable garden is started in straw bale without soil. These bales serve as good medium for the growth of vegetables and herbs. Straw is the best option for bale gardening as it is often seed or weed free after harvesting. The bales are placed at sunny locations receiving 06 to 08 hours of sunlight. Conditioning is done by application of nitrogen fertilizer on the bales followed by watering daily for few days. Straw is rich in carbon and addition of nitrogen enhances bacterial growth and composting. When decomposition of straw occurs then 05 to 08 cm layer of potting soil and compost is spread on bales. The seeds are sown directly in top layer or seedlings may be planted by making a suitable hole according to the type of vegetable. Watering is done regularly and almost daily in hot summer days. This is an easy, convenient, and less laborious method for growing of vegetables.

8.2.10.1 Vegetables and Herbs for Straw Bale Gardening

Vegetables like pumpkins, peppers, kale, spinach, eggplants, tomatoes, zucchini and cucumber while herbs like basil and parsley are best suited for growing in straw bales.

8.2.11 Core Gardening

Core gardening is an old gardening method used in dry places in Sub-Sahara Desert regions in Africa and Middle East. This method helps to feed the soil, maintain its fertility and conserve moisture. Compost acts as fertilizer and provides nutrients to the plants. In this method, raised garden bed of desired size (30 cm or more) is prepared to facilitate seepage of moisture to the soil through the core. A horizontal 20 to 25 cm deep trench is made in the center of raised garden bed. The core can supply moisture up to 60 cm on both sides. The organic materials like grass clippings, coffee grounds, plant cuttings, food scraps, kitchen waste, straw, twigs, and rotten manure are added to core to a depth of 10 to 12.5 cm. Composting soil is added on the top up to 12.5 cm on raised bed and core. Finally, plants of choice are planted on the bed followed by complete soaking of core.

8.2.12 No Till or No Dig Gardening

No till gardening, almost similar to lasagna gardening, ensures provision of essential nutrients on top of the soil. Unnecessary ploughing or tilling of soil is avoided. Cardboard or newspapers are spread as the first layer on the ground. Then a layer of organic materials, slow release fertilizers, rotted manure, or mulch is added on it. Finally, wood chips are added to make the third layer. These materials along with leftover plant roots decompose to form fertile compost for better growth of vegetable plants. No till gardening increases soil fertility, aeration, soil drainage, water conservation, microbial diversity, and ultimately crop yield. This method can easily be used in small urban areas and gardens by making small beds less than 1.2 m to avoid stepping and hardening of soil. Less labor, work, and time are required for growing of vegetables. Generally weeding operation is not required because the use of cardboard as weed suppressor in the first layer helps to suppress weeds.

8.2.12.1 Vegetables for No Till Gardening

Pepper, squash, tomato, cucumber, okra, eggplant, beans, spinach, kale, radish, Swiss chard, lettuce, beets, carrots, broccoli, leafy greens, arugula, onions, cabbage, garlic, peppers, and parsnip.

8.2.13 Hügelkultur/Hugel Gardening

In German, Hügelkultur means "hill culture". This method has already been practiced in Eastern Europe and Germany as a component of permaculture system. In this centuries-old method, a garden bed is prepared from rotten logs and plant residues. A patch measuring 2.4 × 1.2 m is selected at a sunny location in the garden. A 30 to 45 cm deep trench is made in this patch. Wood, logs, branches, and twigs-like materials are placed into this trench forming a base. Then layers of organic materials including grass clippings, straw, leaf litter, manure, etc. are spread also filling gaps and spaces of the foundational material. The top is covered with 05 to 07 cm of already removed garden soil carrying turf roots. This method is the best one to recycle garden waste including twigs, branches, logs, and leaf litter. The composted materials help to retain water in dry climates, facilitate good drainage, improve soil fertility, and also provide nutrients to vegetables.

8.2.13.1 Vegetables for Hügelkultur Gardening

Cabbage, cauliflower, eggplant, beets, lettuce, onion, parsnip, squash, pumpkin, beans, carrots, cucumbers, asparagus, leeks, shallots, and garlic grow best in Hügelkultur keeping in view the seasonal requirements.

8.2.14 Urban Roof Gardening

A rooftop garden is a small-scale representation of a beautifully designed agricultural farm. It displays pretty plants or vegetables raised in containers or small framed beds like structures placed on terrace or rooftop of residential and commercial buildings. Rooftop gardening is an eco-friendly approach which benefits gardeners with plenty of safe and healthy food, recreation, and even experience of learning. Lightweight special stands or structures can be prepared for growing of shallow-rooted vegetables. Vine-like varieties can be grown close to walls and railings. Rooftop gardening improves urban lifestyle by keeping the environment cool and calm for the residents. Air quality index in some cities of the world is reaching an alarming position. Rooftop gardening reduces heat absorbed by concrete materials, dust, pollution and stress levels and improves air quality by filtering gases and pollutants.

8.2.14.1 Vegetables for Roof Gardening

Lettuce, cucumber, beans, tomatoes, coriander, fenugreek, peppers, eggplant, squash, broccoli, cabbage, cauliflower, chard, spinach, turnip, artichoke, okra, garlic, and herbs can easily be grown on rooftop according to seasonal requirements.

8.2.15 Succession/Successive Planting

In succession planting, crops with short maturity dates are selected. By this method, a garden with little space can be made much productive giving non-stop and endless produce. This method makes harvesting of a specific vegetable possible in three seasons viz. spring, summer, and fall. Sowing of any desired vegetable can be done from 07 to 21 days to continue its harvesting at different times.

8.2.15.1 Vegetables for Succession Planting

Different vegetables can be planted in succession. After harvesting of one vegetable, the same space can be used for another suitable vegetable. Vegetables like basil, beans, arugula, beets, carrots, broccoli, endive, green onions, lettuce, kale, radish, turnip and spinach are best for succession planting. Intercropping or companion planting of different vegetables is also feasible to get enhanced yield from the same space. According to climatic conditions, cool and warm season vegetables can be grown in succession in the garden. Different varieties of same vegetable with different maturity time can be grown to get produce at desired intervals. Vegetables for this purpose include tomatoes, squash, kale, peas, summer squash, carrots, cabbage, Brussels sprout, collard and broccoli (Richmond 2023).

References

Boeckmann C (2024) How to grow swiss chard plants: the complete guide. Available at: https://www.almanac.com/plant/beans. Accessed Nov 28 2024

CABI (2019) *Raphanus sativus* (radish). Available at: https://www.cabidigitallibrary.org/doi/10.1079/cabicompendium.46796. Accessed 31 Aug 2023

Daveport M, Smith P, Doubrava N (2023) Cabbage and Chinese cabbage. Available at: https://hgic.clemson.edu/factsheet/cabbage-chinese-cabbage/. Accessed 02 Sep 2023

Leishman G (2018) The historic origins of vertical gardens. Available at: https://blog.growup.green/blogs/growupdates/the-historic-origins-of-vertical-gardens-gallery. Accessed 29 Nov 2023

Lovell ST (2010) Multifunctional urban agriculture for sustainable land use planning in the United States. Sustain For 2:2499–2522. https://doi.org/10.3390/su2082499

Mays D, Richter K, Bradely L, Sherk J, Kistler M, Neal J (2022) Plants grown in containers. NC state extension. Available at: https://content.ces.ncsu.edu/extension-gardener-handbook/18-plants-grown-in-containers. Accessed 27 Aug 2023

Richmond J (2023) Succession planting. https://extension.wvu.edu/lawn-gardening-pests/gardening/garden-management/succession-planting. Accessed 17 Dec 2023

Roll M, Wilson CR (2020) Container gardens. Colorado State University Extension. https://extension.colostate.edu/topic-areas/yard-garden/container-gardens-7-238/. Accessed 27 Aug 2023

Xiang Q, Guo W, Tang X, Cui S, Zhang F, Liu X, Zhao J, Zhang H, Mao B, Chen W (2021) Capsaicin—the spicy ingredient of chili peppers: a review of the gastrointestinal effects and mechanisms. Trends Food Sci Technol 116(7). https://doi.org/10.1016/j.tifs.2021.08.034

Protected Olericulture

9

Anam Noor, Sakeena Tul Ain Haider, Arslan Akbar, Khurram Ziaf, and Muhammad Akbar Anjum

Abstract

In recent decades, the demand for food has constantly increased many folds due to the rapid increase in global population. It has necessitated the development of rigorous agricultural practices that can enhance productivity with better produce quality in achieving sustainable food security. Protected cultivation is a specialized production technology, in which natural environmental factors, i.e., light, temperature, humidity, and CO_2 concentration, are modified with the aim to enhance growth, increase crop yield with better produce quality, and extend the growing season. Thus, it is a comprehensive system of controlled environment in which some or all natural environmental aspects are modified for optimum crop growth and maximum economic return. This chapter describes different practices and systems of protected production of vegetables and their advantages to extend the accessibility of vegetables round the year. It also demonstrates the factors which affect the successful production of off-season vegetables to get higher productivity and return.

Keywords

Greenhouses · Hydroponics · Microclimate · Off-season vegetable production · Plastic tunnels · Soilless culture · Vegetable crops · Vertical farming

A. Noor · S. T. A. Haider · A. Akbar · M. A. Anjum (✉)
Department of Horticulture, Bahauddin Zakariya University, Multan, Pakistan
e-mail: akbaranjum@bzu.edu.pk; anjumbzu@yahoo.com

K. Ziaf
Institute of Horticultural Sciences, University of Agriculture, Faisalabad, Pakistan

Z. Iqbal et al. (eds.), *Sustainable and Innovative Vegetable Production in times of Climate Change*, https://doi.org/10.1007/978-981-96-6283-8_9

9.1 Introduction

Agricultural production has been primarily outdoors, since early ages. Fast urban expansion and industrial development on fertile lands have decreased the cultivable land area every year. Therefore, off-season vegetable production can be a better alternative of land use systems ensuring efficient use of natural and other resources. Agricultural production is entirely climate and weather-dependent and in fact for vegetable crop production, the most important factor is environment because environmental conditions during production time affect the final yield and quality of produce (Trivedi and Singh 2015). Open field production of vegetables may get damaged by the natural calamity when they are exposed to natural catastrophes, e.g., intense rains, thunderstorms, hails, solar radiations, humidity, and temperature levels above or below plant growth optima (Max et al. 2009), fungal diseases (Mishra et al. 2010), and high infestation of insect-pests (Nguyen et al. 2009). Protected cultivation of vegetables under adverse climate is the sustainable approach and vegetables produced under protected culture result in higher yields with better quality in terms of size, shape, and color (Mishra et al. 2010). Protected cultivation, also called as comprehensive system of controlled environmental agriculture, is the alteration of the natural environment to attain optimal plant growth (Jensen 2002). Modification can be made to both the aerial and root environmental factors like temperature, light intensity and frequency, wind velocity, humidity, water, and mineral nutrients, which are manipulated with the aim to enhance crop growth, yield, better produce quality, and extension of the growing season (Layne et al. 2013).

Protected cultivation is mostly used for vegetables, small fruits, cut flowers, and ornamental foliage production, though it has a variety of agricultural applications. Protected agricultural technologies (windbreaks and soil mulches) or protective structures (greenhouse/net house, plastic tunnels, and row covers) have been developed to produce horticultural crops year-round and regulate the internal environmental conditions (Trivedi and Singh 2015). The need of protected cultivation has been dramatically increased since the last twenty years. The intent is to produce virus-free seedlings, grow crops with earlier maturity, extend the period of harvesting, get higher crop production with better quality, enhance the stability of production, conserve moisture efficiently, and ensure the availability of commodities frequently (Negi et al. 2013).

In Pakistan, indigenous protected horticulture for off-season vegetable production involves only poly-tunnel (plastic tunnel) farming; however, our protected horticulture potential has not yet been explored owing to energy crises and higher input costs. Vegetable crops having small size, short duration, and high value are considered most suitable under protected cultivation. The cultivation of vegetables like capsicum, tomato, and cucumbers under different protective structures particularly in naturally open ventilated polyhouses has proved more remunerative than open field cultivation (Negi et al. 2013). There is great potential for protected vegetables' production, which utilizes marginal lands and also extends the availability of different crops at affordable cost by controlling the environmental factors. Similarly, during spring and/or autumn, different vegetables like cauliflower, cabbage, broccoli, peas, and beans can be cultivated under controlled conditions in hilly areas of Khyber Pakhtunkhwa province, Gilgit Baltistan, and Azad Jammu & Kashmir.

9.2 History and Current Status

Production of high-quality vegetables and other crops under protected conditions dates back to the ancient times when these were grown in pots. At night or during cold periods, these pots were placed indoors to protect these valuable plants. Almost 2000 years ago, moveable beds were used by Romans to grow melons and cucumbers and during cold nights these beds were shifted to caves. Between 14 and 37 AD, the Romans used tale window covering to extend plant growth beyond its normal season. During the middle ages, around castles and monasteries, horticultural crops were grown where walls were utilized to protect these crops. However, growers in northern Italy protected their citrus trees from cold by using artificially heated glass enclosures. During the sixteenth century, botanical gardens were established and to protect young plants, glass cloches were used. Later on, the prototypes of the greenhouses appeared primarily in England, the Netherlands, France, and Japan. During the late 1600s and early 1700s, stove and heating fuels were used in glasshouses (Dalrymple 1973). In mid-eighteenth century, protective structures for several crops began to appear regularly. First lean-to greenhouses were constructed in the Netherlands and central Europe. The two-sided greenhouses emerged later in the 1800s leading to these early developments in protected cultivation. Commercial greenhouse crop production was well established by the end of nineteenth century. Nowadays, automated computer control is a common feature especially in commercial greenhouses and not only environmental factors but water and nutrient supplies are also controlled. Among vegetables, lettuce and radish were the first few crops grown under a protected environment. The cultivation of tomato in greenhouses became possible in the 1870s and 1880s. By the end of the nineteenth century, there were about 1000 greenhouses in the United States, which were built primarily to force winter vegetables. Since the 1960s, dramatic changes occurred in greenhouse production. Pensive tomato cultivation in greenhouses became the attractive and expensive venture in European countries. By 1980, there was an estimated area of 150,000 ha under greenhouses (glass, fiber, plastic made) globally for growing high-value food crops (Romanowski 1981).

Europeans are considered the pioneers in greenhouse technology that is more than about 200 years old. Later during World War II, a new development in technology took place with the start of the use of plastics. The worldwide protected crops extended in Asian countries like India, Korea, and especially China and have witnessed 30–40% growth since 1991 (Jiang et al. 2004). In China, the first protected cultivation started in the 1990s and nowadays total area under protected cultivation is more than 2.5 million hectares, of which 90 percent is under different vegetables. Further, in China, different low-cost structures and materials such as plastic mulches, plastic low tunnels, and walk-in tunnels are in use on 80 percent of the total area for protected cultivation and possibly this is the simple reason that China is today the largest producer of vegetables in the world (Jiang and Yu 2008).

In the Netherlands, for the growing of flowers/ornamentals and vegetables, protected cultivation under glass-shielded greenhouses was used earlier which increased rapidly. Globally according to an estimate, over 20 m ha area worldwide is now under protected cultivation and to meet the needs of the increasing population, the cultivation under protective structures is increasing (Yaling 2008). In Europe, Spain is the leading country for protected cropping as the area under economical poly-houses is 51,000 ha, under plastic mulches about 150,000 ha, and under low tunnels and floating covers, 20,000 ha. In Japan, the use of mesh, plastic tunnels, and glass-houses is increasing rapidly and there are about 54,000 ha under protective structures. Tomatoes, green pepper, brinjal, cucumbers, pumpkins, watermelons, lettuce, and strawberries are grown in greenhouses.

Pakistan's economy is largely dependent on the agriculture sector that plays a key role in the food security of the country, as the majority of the population is directly or indirectly involved in this sector. The contribution of the agriculture sector is almost 24% in the GDP (Gross Domestic Product) of Pakistan. The geographical location of the country allows cultivation of all horticultural crops that could be used for food purpose and also as a raw material in the processing industry. In Pakistan, vegetables are grown on about 0.46 million hectares of land with an annual production of about 8.37 million tons which is low in comparison to advanced countries.

To cater the minimum vegetables requirement of the growing population, there is tremendous pressure to increase vegetables production. Introduction and adaptation of protected horticulture in Pakistan is by far in comparison to developed or neighboring countries. Recently interest in protected cultivation of vegetable crops has been observed countrywide among progressive farmers and business enterprises that is a remarkable change in the area of protected cultivation with challenging opportunities.

In Pakistan, to sort out constraints faced by the agriculture sector, the use of plastic and glasshouses was introduced in the early 1990s and nowadays the concept of cultivation under protective structures has gained popularity and become a multimillion-dollar industry. With the increase in the country's population and dietary habits improvement, the consumption and ultimately production of off-season vegetables have increased. Production of vegetables under protected conditions is the best option for better utilization of land and other resources as this technology very efficiently and precisely uses land area, water, mineral nutrients, energy, and climatic resources such as sunlight, temperature, relative humidity, and atmospheric CO_2. By adopting protective structures such as greenhouses, cold frames, hot beds, poly-tunnels, floating row covers, etc., the high-value vegetable crops with good export potentiality can be produced round the year even under adverse weather and climatic conditions (Pervez et al. 2017). The protected horticulture industry enables to eliminate the big gap between present production and future targets in a sustainable way and also has potential to fetch higher net return per unit area. The major challenge for the protected cultivation adoption is the lack of awareness and the cost involved. The protected cultivation industry will showcase remarkable growth in the next few years if these aspects are taken care of by the policymakers and the industrial players.

9.3 Objectives

Protected cultivation of vegetable crops includes the following objectives:

- Environmental modification by structures and practices for maximum crop productivity with better produce quality.
- Introduction and utilization of advanced technology for continual production, advantage of market seasonality, and higher prices.
- Land utilization for commercial production of vegetables by providing favorable microclimatic conditions for the crops under protective structures.
- Continuous supply of disease and residue-free vegetables to the consumers.
- Raising of early/off-season nurseries.
- Hybrid seed production.

9.4 Protected Vegetable Production Advantages

Vegetable crops in open field conditions may experience short cropping season and adverse climatic conditions (high or low temperatures, dry, wet, or cloudy ambient) affecting their rates of photosynthesis, making them vulnerable to weeds, diseases, insects and other pests, depleted soil moisture and plant nutrients resulting in lower yields with poor produce quality. Protected culture ensures to control one or more of these factors or alter to desirable limits.

9.4.1 Protection from Frost and Low Temperature

Vegetable crops can be protected from unfavorable weather conditions when grown under cover. Greenhouse production system is very suitable in cold and wet climates. Due to the heating effect of high irradiation, there is a gradual increase in the internal temperature of the greenhouse. In fact, raising the indoor temperature in cold climate is responsible for the growing of vegetables in the greenhouse. If the mean night temperature drops and becomes unsuitable for optimum biological activity (12 °C), the night temperature in the greenhouse can be increased by closing the ventilators in the afternoon. This closed condition of greenhouse can retain the daytime high temperature for the night. Likewise, in greenhouses, use of wind machines, heating systems, or overhead sprinkler irrigation also prevents the damaging effects of frost and low temperature.

9.4.2 Protection from High Temperature

Each vegetable crop requires a specific range of temperature for its proper growth and development; although several vegetables can be produced in a wide range of temperature. During summer season, high temperature can be reduced to few degrees by some artificial forced ventilation (using fans) that can minimize the inside temperature during daytime. However, a cooling system is required for further lowering of inside temperature, e.g., general shading in very hot and humid condition and evaporative cooling mechanism in hot and dry conditions are used for cooling the greenhouse during daytime, as during this period temperature may rise beyond the critical limit. Further, vegetable crops may suffer from incidence of diseases that are produced under adverse climatic conditions, i.e., high temperature, high humidity, and flooding.

9.4.3 Control of Insect-Pests and Diseases

In protective structures, vegetable crops can be produced with less or free of insect-pests and diseases attack by maintaining a proper environment inside the structures. Insect-proof screens can be used to cover the openings of ventilation that keep the insect-pests away and prevent their entry inside the house. Further, it is easier to perform plant protection measures in the protective structures. Soil-borne pathogens can be killed by fumigation that can also be easily done inside the greenhouses and if the problem is severe, growing medium can be replaced or removed. Use of soilless substrates for protected vegetable production has proved to be very effective to reduce the chances of soil-borne pathogens.

9.4.4 Efficient Water Utilization

Water is the key factor for the production of vegetable crops because vegetables are tender and succulent in nature. It is not possible to grow vegetable crops during heavy rainfalls because it will badly affect the produce quality. Protected cultivation is the most suitable production system to reduce the high wind and heavy rainfall consequences. Water requirements for plants can be reduced by growing vegetables under covered structures and using water conservation techniques, like organic or plastic mulches, which save moisture and control weeds that in turn would enhance the productivity and crop profitability. Further, water can also be saved by the intervention of drip or sprinkler system in the greenhouses or poly-tunnels, which provides a better solution to water scarcity.

9.4.5 Manipulation of Production Time

Production times of vegetables can be manipulated through the protected production and it also helps to feed large growing population through continuous supply of vegetables. By providing suitable environmental conditions in the protective structures, production time can be adjusted depending upon market demands and profitability.

9.4.6 Expansion of Production Areas and Growing Season

Protected cultivation has made possible to grow crops indoors in those areas where these crops cannot be produced successfully in open fields (outdoors). Moreover, by providing optimum environmental conditions in protective structures, the growing season and availability time for vegetable crops can be extended noticeably. In this way, net returns and profitability can be increased by growing different vegetables at specific times of the season on large areas under controlled conditions. It also helps to fetch the high prices in market by avoiding the glut period when the produce available in the market is surplus.

9.4.7 Increased Crop Yields

Higher yields of vegetable crops are obtained due to the maintenance of optimum growing conditions and also using good quality planting materials and inputs under protective structures (polyhouses and shade-net houses). Moreover, when gourds, melons, and cucumber (cucurbits) are trained vertically as vine crops, their yield per unit area is increased many folds.

9.4.8 Improved Produce Quality

As indoor-grown vegetables are protected from natural calamities and weather extremities, so availability and quality of the produce are enhanced due to the provision of congenial environmental conditions to off-season cultivation. Vegetables grown under controlled conditions are safe to use due to less chemicals use. Therefore, for growing population increasing vegetables demand can be accomplished by using this technology.

9.4.9 Market Prices Stability

Protected vegetables' production helps in fulfilling the daily food requirements of consumers by ensuring a regular supply of quality produce that helps in stabilizing market prices. This also results in enhancing the productivity that promises high profits to the growers.

9.5 Important Considerations for Successful Protected Olericulture

Socio-economic factors such as availability of labor at specific time, cost of labor, and logistic conditions are important in addition to soil and climatic factors. These factors directly affect production and change with area, time, and crop being grown. The technology of growing crops under protective structures requires higher start-up investment for structure development, covering materials and operational costs. This greenhouse technology requires more precision and diligence as it is much more advanced than open-field growing and all these factors govern the competitiveness of protected cultivation of vegetable crops (Wittwer and Castilla 1995).

The feasibility of suitable structures for vegetable growing depends largely on climatic conditions of a specific site. In spite of this, the selection of these structures is also dependent upon several factors like economic status, type of soil, electricity, and access to the assured market.

9.5.1 Locality

Commercial production of vegetable crops depends primarily on the climatic conditions of specific region that also determine the selection and use of protective structures. The greenhouse horticulture in the temperate countries was developed to preserve heat for growing plants during cold periods. Thus, this attribute limits the use of the protected horticulture in the tropics, giving some markedly differing needs. Further, the protected cultivation is also affected by the market conditions and population in that region. While planning a new horticultural venture, the site should also be located away from the urban area at a place which is safe from the rapid expansion of the nearby city.

9.5.2 Environmental Conditions

For protected cultivation of vegetables, the selection of suitable site with appropriate climatic conditions is the basic requirement. Crop yield and quality can be managed by controlling the climatic conditions inside the protective structures. In favorable climatic conditions, protecting structures will be needed only for protection against rain and insect-pests, and it will reduce the cost of protected cultivation. If climatic conditions are enormously inappropriate for crop growth in a particular region, it may not be feasible for commercial protected cultivation because it will increase the cost of production.

9.5.3 Solar Radiation

For growing plants or raising crops, it is imperative to know the mean daily sum of solar radiation energy for each month. This information will help to devise proper protected technology that can use this solar radiation in favor of crop grown by controlling light intensity (irradiance), light duration, and light quality. A minimum daily radiation of 500–1000 W/m^2/day is the limit for sufficient growth of a plant. In tropical regions, cultivars suitable to shorter day length should be chosen for proper use of high mean daily radiation.

9.5.4 Temperature

Temperature is one of the most important climatic factors that should be considered in two different ways. First, the outside temperature on which the structural design of a greenhouse depends. Second is the inside temperature, which is modified in favor of the crop grown within the greenhouse. Temperature influences the final yield of crops grown in protective structures and also the partitioning of photo-assimilates between plant organs. Temperature requirements of most crops range within 10–25 °C, although it may differ depending on their production in normal season. Vegetable crops can be classified into warm season and cool season crops; and their requirements for temperature are generally on the basis of optimum growing temperature and night temperatures. Soil temperatures along with air temperatures are very important for vegetable crops. If the temperature goes higher than the required, it influences plant growth and due to rapid translocation of photosynthates, crop maturity is hastened. Night temperature affects root development, microbial growth, nutrient, and water absorption. For growing vegetable crops, production temperatures must be considered because temperature at freezing point or above 40 °C damages the crops. The course of temperature in a specific location is determined by factors like season, altitude, latitude, wind condition, and intensity of solar radiation. The greenhouses in the climatic zones around 20° to 30° north and south latitudes may require both the cooling and heating systems to grow crops throughout the year. It will avoid problems of extreme temperatures and danger of frost. However, the local variation in microclimate has to be examined properly to establish a greenhouse.

9.5.5 Light

Light is the most important factor required for photosynthesis that is being used by plants to manufacture their food. Light in protective structures greatly influences the production of horticultural crops. Protective structures should intercept maximum light to accomplish the plants' photosynthetic requirements. Light intensity, light quality, and photoperiod affect plant growth and play key role in quality crop production. Light intensity varies with elevation, latitude, clouds, fog, smoke, season, and time of day that also governs the photosynthesis rate. Vegetables such as cole crops, lettuce, carrot, spinach, and onions require less light intensity, while cucurbits, okra, and sweet potatoes with high light intensity can be grown successfully. Photoperiods differ at different latitudes and also in different seasons of the year so various crops require specific day lengths and are photoperiod sensitive. Specific day lengths are needed for reproductive growth of plants that determine crop production time. Plants which flower with longer day length are known as long day plants, while plants which flower when day length gets shorter are called as short-day plants. Vegetable crops, which remained unaffected by day length, are called day-neutral. Thus, desired photoperiods must be maintained in the protective structures, if flowering and fruiting are required in the grown vegetable crops.

9.5.6 Relative Humidity

The humidity of air outside and inside of a greenhouse needs to be meticulously evaluated and considered while planning a greenhouse (Von Zabeltitz 2011). Further, both absolute humidity and relative humidity have to be properly measured and considered with due importance. Plant growth and development slow down or stop when the relative humidity in air is lower than 30% or higher than 90%. Usually, high humidity within a protective structure results in rapid fungal growth or disease spread. Therefore, to ensure high yield of vegetables, maintenance of humidity level is very important that could be monitored regularly by installing hygrometers permanently in the protective structures.

9.5.7 Gases and Ventilation

In plastic tunnels and greenhouses, CO_2 levels may drop which affect plant growth and yield because CO_2 is a raw material for photosynthesis (Gruda 2005). In protective structures, inappropriate ventilation becomes a limiting factor for crop growth and quality due to low level of photosynthetic activity. Therefore, to sustain plant growth and quality, appropriate aeration is required. In greenhouses and tunnels for increasing vegetable yield significantly, the introduction of supplementary CO_2 applied at 1000–1500 ppm on days with poor ventilation has proved successful.

9.5.8 Wind

In the protective structures, air movement and ventilation take place by a gentle speed wind but the structures or even crop may get damaged due to heavy winds. Heavy windy areas generally require more expensive and sturdier structures, so the use of low-cost structures in regions of heavy cyclones or storms is the better option which can be altered rapidly.

9.5.9 Soil and Growing Substrate

Soil properties like texture and structure, availability of nutrients, organic matter content, water holding capacity, and topography are imperative for protected crop production. For the production of vegetables, the selection of suitable soil is very important; however, for good quality production and higher yields, the appropriate growing substrate is needed if the soil is not being used for growing the crops. In advanced countries, for protected horticultural crop production, several substrates are in use that include peat, perlite, vermiculite, rock wool, coco-coir, rice hulls, cotton waste, fruit and vegetable waste, corn cobs, sugarcane bagasse, and press mud. These growing substrates are very important to support plant growth, hold moisture and nutrients, and ensure proper aeration. These substrates can be used in various combinations with modification that is based on accessibility and cost in a specific area (Nelson 2012). However, in Pakistan, soil or soil mixes (containing well-rotten farmyard manure, leaf compost, etc.) are commonly used for protected vegetable production.

9.5.10 Fertilizer Application and Fertigation

The growing substrates act as a reservoir of nutrients for the growing plants, and these nutrients are absorbed by plant roots and then through vascular system are translocated to other plant parts. Proper and optimum application of nutrients in natural or chemical forms is the basic need of plants. Additional nutrients can be incorporated to the growing substrate in various ways and in protective structures the best method for nutrient application is through fertigation. Nutrients deficiency may cause poor crop growth and yield reduction with poor quality, while toxicity may occur due to surplus use of fertilizers. Generally, liquid fertilizers containing specific nutrients are applied to plants through drip irrigation system in protective structures.

9.5.11 Irrigation and Water Management

Water, being part of metabolic processes, is very important for ideal plant growth. In plants, water regulates opening and closing of stomata, and keeps balance of gases. Water should be of good quality having pH of 6.5 to 7.2 and free of unnecessary salts, heavy metals, and microbes. Vegetables are very sensitive to water deficit conditions and for successful production, availability of good quality water is very important during different critical periods of crops. For example, sweet pepper and tomato plants use 1.5–2 liters water on a daily basis, so for drought periods, storage facilities of sufficient volumes need to be constructed.

Concerning different methods of irrigation, drip irrigation is considered the best and efficient method for growing any kind of crop both in open field and protective structures. Proper monitoring during installation and running of a drip irrigation system are necessary to avoid clogging of emitters, which imposes one of the major problems of this system. So, it is imperative to determine water quality before planning for installation of a drip irrigation system. Analysis of water not only provides information about its applicability according to crop requirement but also gives an idea about the possible clogging problem in the system. There are three main factors associated with irrigation water that cause 90 percent of clogging in a drip irrigation system. Physical features (suspended solids), chemical reasons (precipitation reactions), and biological aspects (bacterial and algal growth) of irrigation water have been rated as 35, 22, and 37 percent responsible for clogging, respectively.

9.5.12 Resources

All essential resources such as availability of seed or plant material, synthetic fertilizers, pesticides, water, and energy should be accessible at reasonable prices which are required for profitable crop production. In Pakistan, the cultivation of vegetables under protective structures is hampered due to interrupted supply of energy and its unaffordable cost. Therefore, all resources that are required for quality production should be available at proper time to ensure optimum growth of plants with good quality yields.

9.5.13 Physical Infrastructure

Vegetable crops are perishable in nature, so access to the protected production sites, suppliers and the market is very crucial for the success of protected vegetable industry. It is very important that where needed, cool chain should be provided to maintain quality of produce. Moreover, mechanized systems for harvesting, washing, grading, etc., may lead to operations more efficiently.

9.5.14 Tools and Instruments

Use of specialized tools and instruments such as meters used to measure temperature, light, relative humidity, electrical conductivity, and pH will permit higher levels of efficacy. These tools will indicate the status of the protective structure environment and also provide guidelines about the management strategies that are better for plants. Computerization/automation may also permit the management of larger facilities more efficiently by using telecommunications or satellite technology.

9.5.15 Capital

In fact, the capital available for investment determines the size or area under protected cultivation. In Pakistan, it is difficult for small growers to manage the resources required for protected olericulture at individual level because successful operations for protected cultivation involve high capital investment. Therefore, producers should not be reliant on commission agents to obtain capital, rather they should make necessary capital arrangement before entering in the business.

9.5.16 Labor

In Pakistan, unavailability of trained labor is a major issue, but availability of unskilled labor is not a problem. Therefore, to perform the specialized operations timely, availability of cheap and trained labor should be ensured. More skills are needed to perform some specific operations in protected cultivation that differ greatly from traditional open field operations. Additional activities like intensive pruning, trellising, pollination, electrical conductivity (EC), and pH measurements require more technical skills to perform. Labor cost in poor Asian and African countries is not an issue because unskilled labor is easily available and wages are very low. In America, Australia, and Europe, the production operations have been automated due to higher wages and unavailability of skilled and unskilled labor, which enhances the laborer's efficiency.

9.5.17 Crop Protection

Crop health maintenance is the key factor for successful farming in terms of both yield and quality of produce. The vegetable crops grown in protected structures should be monitored frequently for any disease incidence. GAP (Good Agricultural Practices) should be adopted to protect the vegetable crops from the pests.

Planting material should be treated with suitable pesticides to avoid soil-borne diseases and attack of sucking insects at early stages of growth. Integrated Pest Management (IPM) strategies should be adopted and if needed, only registered pesticides should be used at the recommended dose. Indiscriminate use of chemical pesticides must be avoided. Pesticides should only be used when pest populations are expected to cause severe losses. Proper record of all pesticide applications should be maintained. Protective clothing, glasses, and gloves should be used while applying pesticides. To avoid continuous use of chemicals of the same group, rotation of different chemicals should be done so that the pathogens and insect-pests do not develop resistance.

9.5.18 Canopy Management

It is the practice performed on a plant that gives it a specific retainable structural framework by way of regular pruning and training. Actually, the direction of main stem and/or spacing of selected branches is managed through pruning in such a way that they provide a proper framework allowing more sunshine and air movement.

This increases the efficiency of biomass and reduces the land area covered by canopy of that plant to accommodate high-density planting.

9.5.19 Marketing

In the protected cultivation of vegetables, production cost is very high so their timely marketing is very important because high-quality produce with consistent supply is a critical need of most markets. A well-established domestic market will strengthen the protected industry, lower the cost of transportation, and ensure rapid marketing.

9.6 Protective Structures

For protected vegetable production, different types of structures are used that have particular function and purpose and these structures fit well into various sizes of farms. Similarly, low cost structures that can be replaced rapidly are preferred in regions with frequent summer storms. In such areas, for greenhouses roofs, plastic coverings are preferred on glass and walls are made with a wooden or metallic frame. In Asia, the protective structures being used vary from plastic tunnels, simple rain shelters, or shade houses to undeviating modern structures enclosed with plastic or glass. These structures may have totally automated environmental controls connected to soil-based or soilless production systems. The feasibility of relevant and appropriate structures for vegetable cultivation depends largely on the climatic conditions of a particular location, initial cost of installation of the protective structure, and its operational expenses. In spite of this, choice of these structures is also dependent upon various factors like economic status of a grower, electricity supplies, type of soil, and availability of assured market to fetch high return. Environmentally controlled glasshouses, automated greenhouses, cold frames, hot beds, plastic tunnels (low, medium or walk-in and high), and insect proof net houses are being used in different parts of the world for round the year and off-season cultivation of vegetables (Bhardwaj et al. 2011). Some commonly used structures for early, off-season, or year round production of vegetables are described below.

9.6.1 Floating Row Covers

Floating row cover is a plastic film fabric used without any mechanical support to protect crops from insect vectors, sucking pests, hoppers, and beetles. Floating row covers are also being used for early growth enhancement of vegetable crops (Wells and Loy 1985). It is viable production practice of vegetables when the temperature is below optimum and early production is desired. Single or paired row is covered with the fabric. Row covers protect the plants and soil from high light intensity, rainfall, hail, and insect-borne viruses. Covers improve the plant uniformity and fruit quality. They also extend the harvest season and increase the total yield. Red covers are generally more suitable when used in combination with black polythene-mulched beds or rows. Immediately after sowing or planting, cover is spread over the row. The edges of sheets are secured with soil through bricks or wire pins. These floating covers can be left over plants for 03 to 08 weeks, depending on crop growth stage and weather conditions.

Row covers are quite suitable for vine crops like cucurbits, and need to be removed when vines start flowering. This practice facilitates the pollinators for pollination purpose. The row cover for transplanted tomatoes and peppers should be removed after 04 to 07 weeks depending upon the season to avoid extremely hot condition which aborts flowers and fruits.

For self-pollinated crops and leafy vegetables, the floating covers can be had for the entire duration of the crops. The crops like musk melon, watermelon, cucumbers, squash, tomato, cabbage, leafy vegetables, lettuce, pea, green beans, carrot, and radish are grown under floating covers. Floating covers made of spun bonded or non-woven fabric of 10–50 g/m^2 density are used for frost protection. The porous covering material allows water and air movement. Either single rows or a number of rows at a time are covered. Heavier covers of densities higher than 30 g/m^2 are used primarily for frost and freeze protection. The edges are secured by burying in the soil.

9.6.2 Plastic Tunnels

Plastic tunnels or poly-tunnels are used to fulfill the needs of off-season vegetable crops and for enhancing the productivity of vegetable farming. Furthermore, these tunnels are suited especially for small landholding growers to take early production with higher yields (05–10 times) of better quality (Abbasi et al. 2008). Usually, three types of plastic tunnels are used based on their height, as described below.

9.6.2.1 Low Plastic Tunnels

Low plastic tunnels with iron, wire, or bamboo hoops are flexible transparent polyethylene coverings that are installed over rows of individual bed or several beds in open field during winters that usually provide protection against low temperature, frost, winds, and insect-pests (Fig. 9.1a, b). Low plastic tunnels have better effect in conjunction with plastic mulches and drip irrigation and also provide a passive control of plant microclimate. Low tunnels warm the plants' microclimate and can also warm soil that helps to increase the growth of plants and protect them from natural disasters like hails injury and cold winds. Low tunnels may enhance the harvest by about 30 days than their normal season. The polyethylene sheet coverings need to be removed during sunny periods to ensure adequate ventilation and prevent build up of high temperature and humidity inside for some sensitive crops. In such condition, the tunnels are closed before sunset to store sufficient heat inside to compensate for the night heat loss. Low tunnels are cheaper but as cultural practices (preparation of soil, spraying, picking) are difficult to perform, therefore, these are

Fig. 9.1 Low plastic tunnels (**a** and **b**); Medium or walk-in plastic tunnels (**c** and **d**); High plastic tunnels (**e** and **f**); and Conventional greenhouses (**g** and **h**)

used for the early production of summer vegetables. Low tunnels permit early crop with significantly higher yields of melons, cucumber, bitter gourds, summer squash, etc.

In vegetables, low tunnels are usually less than 1.0 m in height. Clear plastic films or polyethylene sheets are stretched over low (up to 1.0 m high) hoops made of steel wires, bamboo strips or cane to cover the plants. Low tunnel of convenient size (0.75–1.0 m wide) using UV-stabilized LDPE (low-density polyethylene) films or plastic sheets are made. These sheets are spread over the frames to provide support to sheet in tunnel shape. Polyethylene films of about 50-micron thickness with ventilating holes (04% of surface area) are usually used. The width of the tunnel depends upon the width of bed and height of hoops. The length of tunnels should not be more than 30 meters.

9.6.2.2 Medium or Walk-in Plastic Tunnels

Walk-in plastic tunnels are intermediate between high and low tunnels for height and also give high yields than low tunnels (Fig. 9.1c, d). These are protective structures made from pre-galvanized pipes and covered with UV-stabilized plastic films or polyethylene sheets. They have height about 02 meters from soil surface, while width may vary depending upon the number of beds covered and bed size. A man can easily walk inside these structures. Walk-in tunnels are used to provide favorable controlled environment like temperature, light intensity, humidity, ventilation, soil media, irrigation, fertigation, disease control, and other cultural practices throughout the growing period irrespective of the natural outdoor conditions. Tomatoes, cucumbers, sweet pepper, and hot pepper are best suited to grow in medium tunnels as these provide more vertical space and better crop management than low tunnels.

9.6.2.3 High Plastic Tunnels

A high plastic tunnel used for off-season protected production of vegetable crops offers better management practices and it is of the tallest height among all types of tunnels (Fig. 9.1e, f). These structures regulate the temperatures of air and soil that is very useful to extend the early or late season production of vegetables, especially the production of summer season vegetables during winter season. These structures lack the precision of environmental control conditions and are simple unheated greenhouses that are used to protect the crop from environmental menaces like low temperature and chilling winds. Tunnels shape is like a rounded Quonset and may be a single stand-alone house or made multi-bay tunnels. The size of the tunnel ranges from 10–30 m long and 05–10 m wide which also varies according to the region and production system needs. Frames are made of metallic pipe, wood, or PVC, and covered with two layers of plastic that offer better padding and also protect from freezing (Coneva 2021). In high tunnels, water is supplied through drip irrigation and it also helps to avoid the spread of certain insects and pests that reduces the use of pesticides. High plastic tunnels can be used with minimal maintenance to produce crops for several years. Crop growth, yield, and quality are better than field-grown vegetables and due to height and width, high tunnels also facilitate in cultural practices (soil preparation, spraying, harvesting). High tunnels are considered best to grow tomatoes, sweet peppers, cucumbers, and other cucurbits.

9.6.3 Greenhouses

Greenhouse is a specially built structure for growing plants under controlled environmental conditions. It is covered with a transparent material that allows the entry of natural sunlight. Greenhouse is also known as glasshouse or polyhouse depending upon the covering material used (Fig. 9.1g, h). Greenhouse is an artificially heated structure for overcoming climatic adversity while making use of solar radiation. The greenhouses are generally used for cultivation of high-value crops. Greenhouses protect the crops against cold, rain, hail and wind and provide plants with improved environments. In the modern era, in different parts of the world, use of greenhouses has increased enormously for profitable horticultural crop production. Further, in greenhouses, crops can be produced out of their normal season of outdoor production or year round, and their yields and quality are higher than those grown in open fields. However, several factors such as locality, input cost, quality production, good marketing information, and timely availability of inputs and skilled/trained labor should be necessarily considered for profitable greenhouse operation (Nelson 2012).

Greenhouse covering material (plastic, glass, or fiberglass) allows the entry of solar radiation through it but traps the thermal energy radiated by the plants inside which in turn increases the temperature of greenhouse. The role of glazing or covering material is transmittance of light. The greenhouse covering material must be durable, economical, and clean enough to provide optimum light transmissions. The covering materials decide the amount of light that comes from the sun and is received inside the greenhouse. Traditional greenhouses are made from glass, which allows maximum transmission of light. Glass greenhouses are durable lasting more than 25 years. Glass transmits about 90% of the sunlight that strikes its surface. It requires only regular cleaning and sealing of edges whereas plastic films need to be replaced every 2–3 years. However, glass greenhouses are more expensive to construct, maintain, and operate. Greenhouses made from polyethylene film contribute to higher humidity levels and dripping of condensation. Compared to glass, plastic panels are generally stronger and require fewer support frames. Using two layers of polyethylene sheet on the outside and inside as inner barrier helps to conserve the heat inside. The two layers of polyethylene transmit 75–87% of available light for the first 02–03 years and reduce heat loss up to 30–40%.

Greenhouses have infrared radiant heaters and solar radiation system, and thermostats are used to regulate the heat in greenhouse. Aspirated thermostat in commercial greenhouses provides effective control. Air movement and exchange are necessary to control temperature and humidity in greenhouses and natural ventilation, fan and pad cooling, fog or misting methods are used for greenhouse cooling. In greenhouses, high humidity is also a major concern for bacterial and fungal diseases infestation as it favors growth of fungal pathogens. Venting the humid air and replacing it with cooler drier outside air helps to decrease the relative humidity. Insect-pests and disease problems can be reduced by controlling the weeds under benches and around the greenhouse. In summer the temperature inside the greenhouse is increased because side walls are covered with glass or plastic film so it can be reduced by opening the side walls (Zhu 1999). Greenhouses have artificial light system to regulate light intensity and photoperiod. These are also equipped with an irrigation system for applying liquid fertilizers. Modern greenhouses are now getting popularity to fulfill the demand of local and export markets. Computer-based advanced technology is available to overcome some of the difficulties in greenhouses related to temperature, humidity, and irrigation control, and it helps to produce high-value vegetables for local consumption and long-distance supply (Papadopoulos et al. 1997).

9.6.3.1 Greenhouse Structures (Frame Types and Design)

Greenhouse frames are generally made of aluminum or galvanized steel and may be 30 m long and 6 m wide because they are lightweight and rust-resistant. The shape varies depending on the expected load, covering material (glass, fiberglass, or plastic), and the height of the crop to be grown. A-frame design is the most common structure of greenhouse. In A-frame design the weight of the greenhouse rests on the straight sidewall. The height of the gutter is 05 m; however, in most greenhouses, height is commonly kept as 2.5–3.5 m. High-side walls increase the greenhouse area relative to the growing space, thereby increasing heating costs. The height of curtain wall is usually kept at 0.75–0.90 m. Gothic arch-frame structures, which are peak at the top with curving sides, can provide adequate sidewall height without loss of strength. Another important consideration is the greenhouse orientation. Tall crops, such as tomatoes, grown in gutter-connected and multi-bay greenhouses should be oriented with the gutter (or ridges) toward North-South to improve light uniformity to all locations.

In all types of greenhouses, the most important consideration is to keep as little of the overhead area as possible. A free-standing greenhouse (lean-to greenhouse) generally allows entry of more light than a gutter-connected or multi-bay greenhouse, because of less overhead structure. In connected-type greenhouses, several structures of one style are joined together to create a large, undivided space for a large operation. This arrangement makes the structure more economical to heat on per unit area basis.

9.6.3.2 Types of Greenhouses

Greenhouses are constructed in several styles. There are three basic types of greenhouse; attached (i.e., lean-to), detached or freestanding (i.e., even span, uneven span, Quonset, The Arch, Gothic arch or Dome), and connected (ridge and furrow, gutter or saw-toothed). The lean-to greenhouse is built against a wall of a building usually south-facing, while the supporting wall is on the north side. The building structure provides support to the roof; therefore, it is less expensive but its sizes and uses are affected by the characteristics of the buildings to which it is attached. This style of greenhouse might look as half of a normal even span greenhouse (Fig. 9.2a).

The detached greenhouse also known as freestanding has none of the walls or the roofs attached to a pre-existing structure. The free standing even span greenhouse is a single structure that has roof with both even pitch and even width. It has symmetrical roof like a house with a double-sloped or A-shaped straight sides (Fig. 9.2b). Its framework is made from aluminum, steel, or wood. The covering is likely to be of permanent or semi-permanent material such as pure glass or fiberglass. This is the most common design for a glass greenhouse. The free standing can take maximum advantage of environmental factors such as light and wind. Uneven span greenhouse has unequal pitch and width. Use of this style of greenhouse is mostly limited to hillsides only (Fig. 9.2c). Quonset is greenhouse having a semicircular arch which starts usually from ground level (Fig. 9.2d). The Arch also has a semicircular arch which starts from top of a sidewall (Fig. 9.2e). In Gothic arch style, side walls are bent over the frame with pointed arch roof (Fig. 9.2f). Greenhouses are also built in dome-shaped design, which allow constant temperature, maximum sunlight, and effective air movement thus with reduced cooling and heating costs. Natural sunlight from every aspect enables to grow fresh vegetables all year round (Fig. 9.2g).

Connected greenhouses are usually ridge and furrow-connected, gutter-connected, or saw-toothed types. Furrow or gutter serves to carry rainwater and melted snow away. In case of ridge and furrow-type greenhouse, two or more even span greenhouses are connected to one another along the length (Fig. 9.3a). Walls between these greenhouses are removed to create more growing space resulting in a

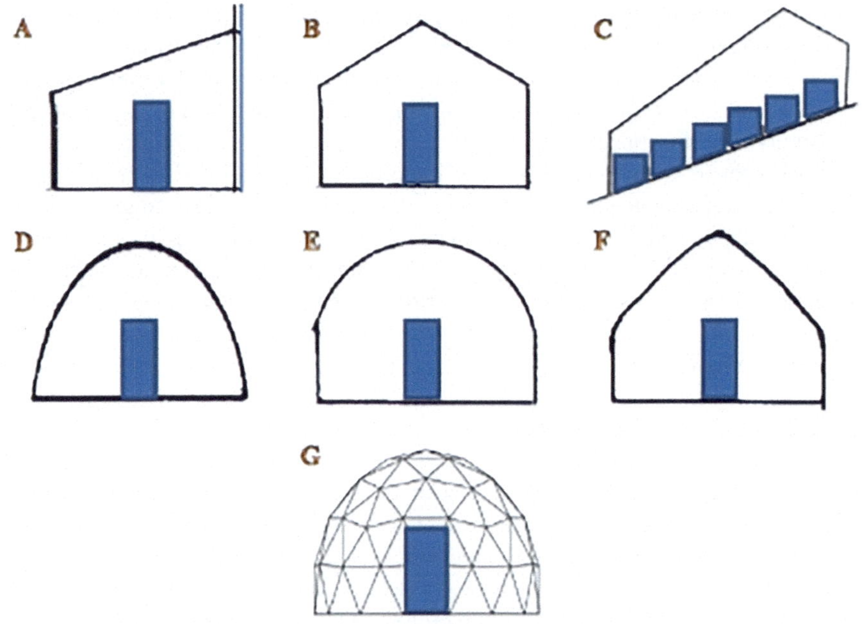

Fig. 9.2 Types of single greenhouses; (**a**) Lean-to, (**b**) Even span, (**c**) Uneven span, (**d**) Quonset, (**e**) The Arch, (**f**) Gothic arch, and (**g**) Dome type

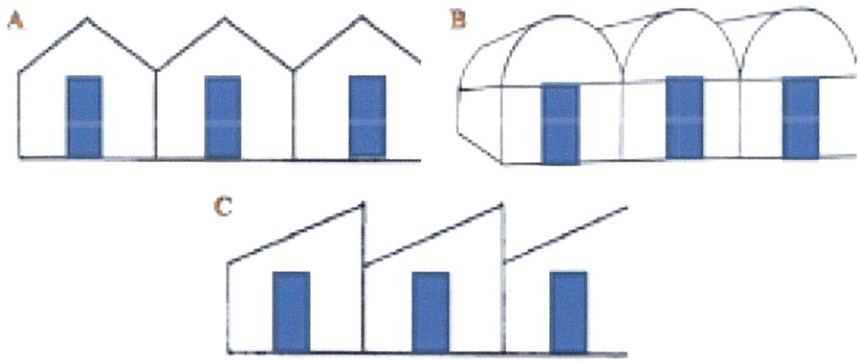

Fig. 9.3 Types of connected greenhouses: (**a**) Ridge and furrow, (**b**) Gutter connected, and (**c**) Saw-toothed

single structure with larger interior. In a gutter-connected greenhouse, two or more arch or gothic arch types are connected together at the gutter level. These are most economical when constructed on an area greater than 2000 m^2 (Fig. 9.3b). If natural ventilation is provided in ridge and furrow-connected greenhouse, it is called as saw-tooth type greenhouse due to the development of specific natural ventilation flow path (Fig. 9.3c).

9.6.4 Hot Beds

Hot beds are built for the same purpose as greenhouse but for small-scale operations. Hot beds are used for raising vegetable crops or their seedlings in the winter season to take the early season production. Seedlings of various vegetables like tomato, peppers, muskmelon, watermelon, and cucumber can be successfully grown in hot beds.

The frame of hot bed is usually made of wood or cement and brick stone. It is set above the ground level. The pit is filled with mixture of soil, well-rotten FYM, and ash. The hot beds are covered with glass or polythene lids during cold nights of winter and early springs, especially for tender herbaceous plants such as tomato, eggplant, peppers, and melons. Heat in beds is provided by making provision of electric heating wires, hot water, and steam pipes or by hot air flues. Decomposing manure or compost may also be used as potential heat source.

9.6.5 Cold Frames

Cold frames are small wooden or concrete structures used for early start of horticultural crops and to harden seedlings under low-temperature conditions. Structurally these are alike the hot beds but without provision of any artificial heating system.

Cold frames have plastic or glass cover, which traps solar heats, making the inside temperature several degrees higher than the outside air. It also raises the soil temperatures that in turn increase seed germination and plant growth. Straw or cloth covers may also be used to reduce heat loss during cold nights. Ventilation in such structure is made by raising the top slightly.

9.7 Protected Production Techniques

Demand for top-quality vegetables is increasing day by day in the global markets. Protected vegetable production manipulates the crop microenvironment for better performance of plants and early flowering. It expands production time, gives higher production with good quality and also keeps the crop protected from natural hazards (Vox et al. 2010; Giacomelli et al. 2012; Gruda and Tanny 2014). Advanced crop cultivation and protection techniques are employed to the production system that helps in improving yield through these practices. These technologies very efficiently use climatic resources, for example, sunlight, relative humidity and atmospheric CO_2 and also utilize land, water, mineral nutrients, and energy very well. So, high-value vegetable crops with good export potential may be produced round the year even under adverse weather and climatic conditions. This technology has the potential to fetch higher net return per unit area. The following are the major protected cultivation practices in vogue.

9.7.1 Mulching

It is a practice of covering soil surface around cultivated plants to make more favorable conditions for them by regulating soil temperature, conserving soil moisture, suppressing weed growth, enhancing activity of microbes, and keeping rhizosphere more friable permitting soil aeration conducive to soil microflora and root growth, etc. (Amani and Avagyan 2014). Mulch is a material used to modify the microclimate just the top or lower surface of soil and also to cover soil surface (Fig. 9.4a, b). Hard-baking and drying of soil are caused by evaporation of water when soil is directly exposed to high temperature and winds. Use of mulch ensures uniform growth of the plants by maintaining soil temperature. Effective covering materials (mulches) for crop production could be natural, like leaves, wheat straw, rice straw, hay, rice hulls, grass clippings, sawdust, peat moss, compost, etc. or synthetic, like polyethylene. Thickness and color of polyethylene (generally black) depend on ambient conditions that also enhance yield and crop quality. Organic mulches have several disadvantages, e.g., increased cost of labor, reduced soil warming if applied in early spring and during decomposition, and need more application of nitrogen to compensate for the nitrogen loss (Nelson 2012). Use of polyethylene mulch owing to better control over insects and diseases has gained universal popularity during the last two decades. Transparent film can be used for soil fumigation, to enhance the temperature of soil and to improve weed control that acts in a similar way to a

Fig. 9.4 Mulching (**a** and **b**); Vegetables production in hydroponics (**c** and **d**); and Vertical farming (**e** and **f**)

greenhouse (Decoteau et al. 1989). Conversely, polyethylene mulch is costly compared to organic mulches and is difficult to dispose of.

9.7.2 Soilless Culture

In soilless culture, soil is not being used as rooting medium to grow crops and nutrients are supplied through irrigation water. In Pakistan, soilless cultivation has significantly increased in recent decades. For protected cultivation of vegetable crops, hydroponics and aeroponics are relatively new technologies being used in soilless production systems (Gruda and Tanny 2015). Although in open fields and in greenhouses due to continuous cropping, various soil-borne diseases attack on crop plants; however, the prominent growing medium being used in Pakistan is still soil.

Soil-borne diseases can be controlled by several ways like crop rotation, fumigation, etc. Heavy investment in protective structures requires high-value crop production, so growing of crops for several years in the same soil is not possible. Thus, pathogen-free culture with higher yields of good quality may be ensured by shifting vegetable crop production gradually to soilless culture. Several types of substrates (peat moss, perlite, vermiculite, coconut fiber, rock wool, rice hulls, peanut hulls, and coco peat) are being used as growing media for vegetables production which keep the crops free from different soil-borne infections (Savvas et al. 2013).

Locally available materials like wheat straw, rice straw, rice hulls, corn cobs, fruit and vegetable waste, cotton waste, sawdust, sugarcane bagasse or press mud, etc. have been proved beneficial for vegetables production. Soilless cultivation is getting fame in simple polyhouses as well as fully equipped modern glasshouses because it is a valuable and effective method to control soil-borne diseases. Further, the costly, laborious, and time-taking tasks of soil pasteurization and soil amelioration or reclamation can be avoided with soilless culture (Nelson 2012). Although soilless culture has significant advantages, it requires technical expertise and higher initial capital for investment to establish successful crop production (Savvas et al. 2013). Several types of soilless systems such as hydroponics, aeroponics, bag culture, and container culture can be used depending upon the suitability of the crop to be grown.

9.7.3 Hydroponics

The word "hydroponics" is derived from two Greek words meaning water and labor (working). The term hydroponics refers to hydroculture means water culture or nutrient solution culture of plants without any supporting medium. However, growing plants in solid media to provide root anchorage and feeding with nutrient solution is also included in hydroponics and this technique is usually known as aggregate system. To provide mechanical support to the roots, inert substrates such as gravel, vermiculite, perlite, rock wool or coco coir can be used. All nutritional requirements of plants in hydroponic systems are accomplished through irrigation water (Fig. 9.4c, d). This is an age-old practice but uncommon in Pakistan. Hydroponics is a specialized technique that requires technical knowledge, good production awareness, skills, and financial investment than some other greenhouse systems. The most commonly grown vegetables hydroponically include tomato, lettuce, and cucumber. For vegetable production, a number of different hydroponic systems have been used successfully at commercial level (Singh and Singh 2014). The most common hydroponic techniques include nutrient film technique and deep floating technique, as described below.

Nutrient film technique (NFT) is a recirculating hydroponic system where plants are grown in a very shallow water stream and roots are directly exposed to solution containing all essential nutrients required for plant growth. This system consists of a reservoir or storage tank, a pump, and channels made up of flexible sheet. A thin film (0.5 mm) of nutrient solution continuously flows through channels. The seedlings are set in small openings on top of the enclosed channels. Their roots remain suspended in channels and when grown, form a mat in the channels. The success of the system largely depends on right channel length, flow rate, and channel slope (Graves 1983). The channel's maximum length should not exceed 10–15 m and for every 40 to 50 cm of horizontal length, one cm of drop (slope) is recommended. The nutrient solution in the channels is pumped from higher to the lower end that flows by gravity, feeding the root mat. The flow rate of the nutrient solution is adjusted to 01–2 liters per minute depending on the channel length and crop growth stage. For tall growing plants mechanical support is required. Inability to withstand unexpected power outage, low oxygen in nutrient solutions and fluctuations of temperature particularly during summer in nutrient solutions are the disadvantages of this system. Leafy vegetables for example lettuce can be grown successfully by this method.

In deep floating technique (DFT), containers are used filled with nutrient solution. Panels or floating rafts are placed over the containers to cover the surface of nutrient solution and insulate root zone from ambient air. Plant roots are continuously submerged in nutrient solution about 10–20 cm deep. In this system, there is no slope because nutrient solution is not pumped up. Air pumps are constantly used to increase oxygen flow or evenly aerate the solution. Among different hydroponic systems, this is more frequently used in Japan and Indonesia for growing plants. This system is better as compared to nutrient film technique (NFT) because it helps to maintain the temperature and oxygen and shows resistance to sudden failure of power. The system is more suitable for the vegetables with shallow root system rather than those with long deep roots. It is used for growing lettuce, cucumber, peppers, and tomatoes.

9.7.4 Aeroponics

The word "aeroponic" is derived from the Greek words *aero-* (air) and *ponos* (labor). It is a technique of raising plants without using soil or any other substrate. Plants are anchored in holes in Styrofoam panels and their roots are in growing chambers suspended in air. The growing chamber is a sealed box to avoid entry of light to facilitate the root growth and inhibit algal growth. Mist nozzles are set below the plants which spray nutrient solution onto the roots in the form of a fine mist continuously or in intervals of every 02–03 mins for few seconds. This is enough to keep roots moist and aerated. The advantages of this system include less and efficient use of water, maximum utilization of space, and production of plants free of soil particles to harness for export benefits. The aeroponic culture is usually practiced in greenhouses and suitable for the leafy vegetables like spinach and lettuce. This system is also best suited for growing of eggplant, bell pepper and tomato as well as for baby greens. However, potatoes and root crops like carrot, radish, and turnip cannot be grown.

9.7.5 Bag Culture

Bag culture is a system of growing vegetables in a soilless medium contained in polyethylene bags. These bags are opened at the top and a single plant is grown in each bag filled with growing medium. Alternately, the bags can be sealed around the growing medium, laid flat on the ground/floor and plants are grown in planting holes made at the top surface of the bags. Bags are made of 04 mil thick UV-stabilized polyethylene and usually filled with peat moss, vermiculite, perlite, coconut coir, composted saw dust, peanut hulls, rice hulls, or their mixes. The size of bag may vary depending on the crop and growing conditions. Drainage slits are made in the bottom edge of the bags so that excessive water and nutrient solution can be drained. Water or nutrient solution may be added manually to these bags or delivered through a drip irrigation system directly to the growing medium through drip tap or emitter. The amount of water applied varies with prevailing weather conditions and stage of crop growth. The bags may be kept in paired rows on the floor depending on the crop to grow keeping walking space in between (Patten 2007). Bag culture is quite simple and requires less substrate volume, however, to fill the bags needs more labor cost. Nursery crops, cucumber, tomatoes, and peppers are grown successfully using bag cultures.

9.7.6 Container Culture

Practice of growing plants, including vegetables, exclusively in containers instead of planting them in soil is termed as container culture or container gardening. A container is a small, enclosed, and usually portable object such as a pot, tub, tin, planter box, barrel, or basket filled with soil or some other growing medium used for raising plants. In Pakistan, clay pots (terracotta), plastic pots, and polyethylene bags are traditionally used to grow plants. Size of container varies depending upon the nature of vegetable crop and its growing duration. Containers may be filled with potting mix, coir dust, peat moss, vermiculite, or some other suitable medium. Fertilizers may be mixed with growing medium at the time of planting or these can be applied in the form of nutrient solution. Container-grown plants need frequent watering, however, excessive water must be drained through drainage holes provided at the bottom. Baby greens such as cilantro, mint, lettuce and spinach and vegetables such as tomato, peppers, eggplant are grown successfully and may need additional support to bear the fruits weight.

9.7.7 Vertical Farming

Vertical farming is an innovative approach to produce higher quantities of best quality leafy vegetables on soilless medium in vertically stacked layers (Fig. 9.4e and f).

Therefore, this technique is called as vertical grow technique and vertical farm as plant factory. The main aim of vertical farming is to get maximum crop yield from a limited space. Thus, crops are grown vertically in a tower-like structure by using natural and artificial lights in combination and adopting soilless culture. Usually, peat moss, coco peat, or some other non-soil medium is used to grow plants. The system is energy efficient and uses significantly less (70 to 95%) water. Through this system regular market supply is ensured and yield also increases many folds. Almost up to eight times more yield is obtained relative to other soilless systems. This system is suitable for raising leafy vegetables such as lettuce, kale, and spinach.

9.8 Off-season Vegetables Production

Direct sowing or nursery raising and transplanting periods for off-season vegetables with their "plant to plant" and "row to row" distances and optimum temperatures required for different off-season vegetables are given below in Tables 9.1 and 9.2.

9.9 Cultivation of Cucurbits in Low Plastic Tunnel

For early vegetable production and higher returns, cucurbits are usually grown in low-plastic tunnels. Earliness in fruit harvesting over normal season crop may vary from 25 to 60 days depending upon the crop (Table 9.3).

Table 9.1 Sowing and transplanting periods

Vegetables	Nursery sowing time	Nursery transplanting time	Direct sowing time	Plant to plant distance (cm)	Row to row or bed to bed distance (cm)
Tomato	1-15th Oct	1-15th Nov.	–	45	90
Hot pepper	1-15th Oct.	1-15th Nov.	After 16th Jan.	50	75
Sweet pepper	1-15th Oct.	1-15th Nov.	After 16th Jan.	30	75
Cucumber	–	–	16-31st Oct.	30	90
Bitter gourd	–	–	After 16th Jan.	60	120
Bottle gourd	–	–	After 16th Jan.	50	120
Sponge gourd	–	–	After 16th Jan.	50	120

Table 9.2 Optimum tunnel temperatures for different vegetables

Vegetables	Temperature (°C)
Tomato	21–29
Hot pepper	21–24
Sweet pepper	21–24
Cucumber	18–24
Bitter gourd	21–29
Pumpkin	18–24
Melons	18–24

Table 9.3 Time of transplanting of seedlings and expected earliness in fruit harvesting over normal season crop in cucurbits

Vegetables	Time of transplanting	Earliness in fruit harvesting (days)
Summer squash	1st week of December to last week of January	30–60
Cucumber	3rd week of January to first week of February	30–35
Bitter gourd	3rd week of January to first week of February	25–45
Bottle gourd	1st week of January to first week of February	30–40
Watermelon	2nd week of January to first week of February	30–40
Muskmelon	2nd week of January to first week of February	30–40
Long melon	1st week of January to first week of February	40–50

9.9.1 Nursery Raising

To grow early crop of cucurbits, their seedlings are raised in a soilless medium in plastic trays. Mostly perlite, peat moss, and vermiculite are used for filling of trays as root medium. Seeds are sown in cells (one seed in each cell). For raising healthy seedlings, optimum temperature of 24–25 °C is required in winter season for germination. After sowing of seeds, water is applied uniformly through sprinkler or boom in each tray. The seedlings of bottle gourd, summer squash, and pumpkin become ready for transplanting in about 25–28 days after sowing the seeds, whereas muskmelon, watermelon, and bitter gourd seedlings take about 30–35 days after sowing. The first week of December to the first week of February is a good time for transplanting (Saravaiya and Kumar 2013).

9.9.2 Transplanting of Seedlings

Before transplantation of seedlings on beds and to support plastic tunnels, 1.5–2.0 m apart galvanized iron arches are fixed manually. Width of 02 hoops (45–60 cm) with the same height (45–60 cm) are kept above the beds level for covering plastic and making a low tunnel over the plants depending upon crop and period of covering needed. Transparent non-perforated plastic of 30–50 microns thickness is used as covering material. It keeps the temperature higher in low tunnels than outside due to partially reflection of infrared radiation. Seedlings are transplanted in beds. After transplanting, plastic is covered on beds manually. Small vents on the eastern side of the plastic tunnel are made at 02 m spacing for proper aeration in the tunnels. The size of vents may be increased with increase in ambient temperature in field. Even whole plastic can be removed from beds depending upon the crop growth and ambient temperature.

9.9.3 Fertigation

Chemical fertilizers are applied through drip irrigation. Water is applied at 06–07 days interval along with fertilizer solution of NPK during first month of crop (December–January). During the second month, irrigation is applied at 04 days interval with fertilizer solution till the beginning of flowering. In March, fertigation can be done at 03–04 days interval.

9.9.4 Plant Protection

Aphids attack cucurbits at an early stage of crop. They can enter in tunnels through vents. Confidor, a systematic insecticide, can be applied through drip to control aphids. The plastic on tunnels can be completely removed from mid-February to the first week of March depending upon prevailing temperature and weather conditions.

9.9.5 Harvesting

First picking of fruits starts by the first week of February and continues till mid-March for summer squash. Muskmelon transplanted in the first week of February is ready by the end of April for first harvesting. Similarly, other crops can be harvested 30–50 days earlier than their normal season.

References

Abbasi NA, Habib U, Ahmad T (2008) Recent development in research and application in protected agriculture. In: Abbasi NA, Habib U (eds) Protected Horticulture. Department of Horticulture, Pir Mehr Ali Shah Arid Agriculture University, Rawalpindi, Pakistan, pp 35–44

Amani M, Avagyan G (2014) Effect of polyethylene bunch cover on fungal diseases control of banana (*Musa acuminate* L.) in Iran. Int J Farming Allied Sci 3:1054–1057

Bhardwaj ML, Bhardwaj RK et al (2011) Off season vegetable production technologies. Dept. of Vegetable Science, Dr. Y. S. Parmar University of Horticulture and Forestry, Nauni, Solan 173 230 (HP), pp 52–62

Coneva E (2021) Choosing a Fruit Crop for High Tunnel Production. The Alabama Cooperative Extension System. Alabama A&M University and Auburn University, Auburn

Dalrymple DG (1973) Controlled environment agriculture: a global review of greenhouse food production. In: Foreign Agriculture Economic Report No. 89. Economic Research Service, United States Department of Agriculture, Washington, D.C

Decoteau DR, Kasperbauer MJ, Hunt PG (1989) Mulch surface color affects yield of fresh market tomatoes. J Am Soc Hortic Sci 114:216–219

Giacomelli GA, Sase S et al (2012) Greenhouse production systems for people. Acta Hortic 927:23–38

Graves CJ (1983) The nutrient film technique. Hortic Rev 5:1–44

Gruda N (2005) Impact of environmental factors on product quality of greenhouse vegetables for fresh consumption. Crit Rev Plant Sci 24:227–247

Gruda N, Tanny J (2014) Protected crops. In: Dixon GR, Aldous DE (eds) Horticulture: Plants for People and Places. Springer, Netherlands, pp 327–405

Gruda N, Tanny J (2015) Protected crops – Recent advances, innovative technologies and future challenges. Acta Hortic 1107:271–278

Jensen MH (2002) Controlled environment agriculture in deserts, tropics and temperate regions – A world review. Acta Hortic 578:19–25

Jiang WJ, Yu HJ (2008) Present situation and future development for protected horticulture in Mainland China. Acta Hortic 770:29–35

Jiang W, Qu D, Mu D, Wang L (2004) Protected cultivation of horticultural crops in China. In: Janik J (ed) Horticultural Reviews. Wiley, pp 115–162

Layne DR, Wang Z, Niu L (2013) Protected cultivation of peach and nectarine in China – Industry observations and assessments. J Am Pomol Soc 67:18–28

Max JFJ, Horst WJ, Mutwiwa UN, Tantau HJ (2009) Effects of greenhouse cooling method on growth, fruit yield and quality of tomato (*Solanum lycopersicum* L.) in a tropical climate. Scientia Hortic 122:179–186

Mishra GP, Singh N, Kumar H, Singh SB (2010) Protected cultivation for food and nutritional security at Ladakh. Defence Sci J 61:219–225

Negi VS, Maikhuri RK, Rawat LS, Parshwan D (2013) Protected cultivation as an option of livelihood in mountain region of central Himalaya, India. Int J Sust Dev World 20:416–425

Nelson PV (2012) Greenhouse operation and management, 7th edn. Prentice Hall, Inc, pp 35–37

Nguyen THN, Borgemeister C, Max J, Poehling HM (2009) Manipulation of ultraviolet light affects immigration behaviour of *Ceratothripoides claratris* (Thysanoptera: Thripidae). J Econ Entomol 102:1559–1566

Papadopoulos AP et al (1997) Integrated management of greenhouse vegetable crops. In: Janick J (ed) Horticultural reviews. Wiley, pp 1–39

Patten GFV (2007) Gardening indoors with soils and hydroponics, 5th edn. Van Patten Publishing, USA, p 366

Pervez MA, Ahmad I et al (2017) Protected horticulture. In: Khan AS, Ziaf K (eds) Horticulture Science and Technology. University of Agriculture, Faisalabad, pp 194–222

Romanowski RR (1981) Environmental control structures. In: Plucknett DL, Beemer HL Jr (eds) Vegetable farming systems in China. Westview Press, Inc, Boulder, Colorado, pp 119–127

Saravaiya SN, Kumar S (2013) Protected cultivation of cucurbits in urban and peri-urban localities. In: National Workshop on Urban and Peri-urban Horticulture. Navsari (Gujarat), India, pp 112–123
Savvas D, Gianquinto G, Tuzel Y, Gruda N (2013) Soilless culture. In: Duffy R (ed) Good Agricultural Practices for Greenhouse Vegetable Crops. Food and Agriculture Organization of the United Nations, Rome, pp 303–354
Singh A, Singh SK (2014) Hydroponics or soilless culture. In: Singh KP, Bahadur A (eds) Olericulture: Fundamentals of Vegetable Production. Kalyani Publishers, New Delhi, India, pp 249–264
Trivedi AK, Singh VK (2015) Potential for improving quality production of temperate horticulture crops under protected cultivation. In: National Workshop cum Seminar on Emerging Prospects of Protected Cultivation in Horticultural Crops under Changing Climate. Precision Farming Development Center, Lucknow
Von Zabeltitz C (2011) Integrated Greenhouse Systems for Mild Climates, 1st edn. Springer, Berlin, Heidelberg, pp 59–135
Vox G, Teitel M, Pardossi A et al (2010) Sustainable greenhouse systems. In: Salazar A, Rios I (eds) Sustainable Agriculture: Technology, Planning and Management. Nova Science Publishers, Inc, New York, pp 1–79
Wells OS, Loy JB (1985) Intensive vegetable production with row covers. Hort Science 20:822–826
Wittwer SH, Castilla N (1995) Protected cultivation of horticultural crops worldwide. Hort Technology 5:6–23
Yaling L (2008) Present status of protected horticulture worldwide. In: Abbasi NA, Habib U (eds) Protected horticulture. Department of Horticulture, Pir Mehr Ali Shah Arid Agriculture University, Rawalpindi, Pakistan, pp 1–10
Zhu GR (1999) Protected peach culture (in Chinese). Chinese Agricultural Technology Press, Beijing

Hydroponic Vegetable Production

10

Shahid Javed Butt, Zafar Iqbal, Servet Varis,
and Kamran Shehzad

Abstract

In the present scenario of climate change and food insecurity, it has become imperative to provide nutritive food to the growing population to meet dietary requirements for maintaining better physical and mental health. Agronomic crops like wheat, rice, and other cereals are insufficient to meet the entire food demand and alleviate hunger and malnutrition. Hence, horticultural edible crops in general and vegetables in particular provide lucrative opportunity to combat the food risk factor in sustainable agriculture. Vegetables are excellent source of minerals and vitamins in addition to fundamental food components. Nevertheless, the conventional production of vegetables in soils poses several threats such as disease and pest infestation and requires several tedious cultural operations like soil preparation, irrigation, weeding, plant protection measures, etc. to secure the ultimate quality crop of required quantity. Hydroponically grown vegetables offer viable answer to resolve these problems. Hydroponics facilitates production of manifold yields with high quality regardless of climate change, which is a burning issue adversely affecting growth and development pattern of crop plants. Different systems of hydroponics either nutrient water culture (nutrient

S. J. Butt (✉)
Institute of Hydroponic Agriculture, PMAS-Arid Agriculture University,
Rawalpindi, Pakistan
e-mail: sbutt61@yahoo.com

Z. Iqbal
Faculty of Agriculture, University of Sargodha, Sargodha, Pakistan

S. Varis
Department of Horticulture, Namik Kemal University, Tekirdag, Türkiye

K. Shehzad
National Agricultural Research Center, Islamabad, Pakistan

Z. Iqbal et al. (eds.), *Sustainable and Innovative Vegetable Production in times of Climate Change*, https://doi.org/10.1007/978-981-96-6283-8_10

film technique (NFT), deep water culture (DWC) and aeroponics), or substrate culture (use of coco peat, perlite, rockwool, etc.) can effectively contribute to improve water use efficiency with positive impact on vegetables production and availability. The future of food security is in hydroponics where the guaranteed supply of nutritive diet for every individual lies with superior quality and greater quantity of vegetables.

Keywords

Food security · Hydroponics · Soilless vegetable production · Sustainable agriculture · Systems of hydroponics

10.1 Introduction

Growing plants without soil by using a water-based nutrient solution is called "Hydroponics". The term is commonly attributed to Dr. W.E. Gericke of California, who was active in the early 1900s. Originally, hydroponics means "working water". The term "hydroponics" is used to describe the practice of growing plants in water. However, hydroponics can also refer to the practice of growing plants without the use of soil in any other medium, such as peat moss, coconut fiber, straw bales, grape leftovers, and so on (Gericke 1937). Therefore, the term "soilless culture" is used interchangeably in the context of hydroponics. When the United States entered in World War II in 1941, vegetables produced on polluted soils were allegedly a major source of amoebic illness among the troops stationed on deserted islands in the Atlantic and Pacific oceans (Thomaier et al. 2015; Carolan 2016). As a result, it became imperative to devise a plan to eradicate the problem. Robert and Alice began using gravel as a rooting medium to grow fresh vegetables for the troops stationed there. Hydroponically grown food quickly became a global phenomenon within few years. Hydroponics is a youthful and active method of commercial plant cultivation, despite the fact that it has been used for decades (Manos and Xydis 2019). Later, when more and better experiments were conducted, it became clear that hydroponics yielded the best plant growth results. Further, advancements in the improvement of a wide range of horticultural crops made it possible to develop this innovative production system (Butt et al. 2015).

10.1.1 Hydroponics across the Globe

In 1973, the Danish Rockwool-Industry popularized its hydrophilic GROAN-rockwool as a horticultural substrate, ushering in the modern era of hydroponics. Substrate culture has emerged as a dominant method of cultivation, accounting for over 60% of the glasshouse land in the Netherlands and 50% in both Belgium and France. Hydroponics was used to grow crops on 800 hectares in Spain in 1992–1993. Standard methods were utilized, such as open system rockwool slabs (108 ha), coarse sand grow bags (490 ha), and perlite grow bags (205 ha). Greenhouse expansion in Türkiye averaged 06–08% growth each year between 1972 and 1995, and the country now possesses more than 12,000 ha of greenhouse space. The total area used for vegetable production is about 651,000 ha, including 37,000 ha dedicated to greenhouses (Kumar et al. 2023). Türkiye is a worldwide tomato producer with over 6,300,000 tons per year, mostly used for industrial processes. It is self-sufficient and exports 150 thousand tons per year, and ranks third with 8.5 million tons production per annum (Aboutalebi et al. 2013). The Netherlands has a strong position in the West European fresh vegetables market, with 92% of the area used for hydroponics under glasshouse conditions. New Zealand has also moved to hydroponics, and Sardinia island (Italy) has over 60,000 m^2 dedicated to the cultivation of tomatoes on grapes marc (grapes residue). China has developed Eco-organic soilless culture systems, which are low cost and more suitable for local conditions. In the year of 2000, there were over 25,000 ha with soilless culture worldwide (Gao et al. 2018).

10.2 Hydroponics and Vegetables' Cultivation

Time changes everything, including the way that man used for farming. Newer and more advanced farming methods have emerged in recent decades, which are more productive than their traditional predecessors. Farmers today use a broad variety of production methods, all of which can benefit from the application of modern farming technology. A single crop is grown using hybrid seeds of a carefully chosen type, high-tech machinery, and massive amounts of energy subsidies in the form of irrigation water, fertilizers, and pesticides. Increased productivity and higher-quality harvests have improved farmers' bottom lines as a result of agricultural modernization. To meet their business goals and increase their economic returns, more and more farmers are adopting contemporary farming of protected cultivation being practiced around the world. The production of vegetables in protected environment is enormously increasing in agriculture business. This is because there are two ways to overcome the food shortage, firstly using production facilities in a comprehensive and efficient manner, and secondly changing the consumer's habit to use more vegetables in place of cereals (Butt and Al-Haq 1993; Adak et al. 2018).

Protected horticulture uses greenhouses and other structures to grow horticultural crops. It involves growing plants in closed structures where they are shielded from the climatic elements and conditions can be carefully managed to meet the specific needs of each crop for maximum productivity (Ali et al. 2012). In modern greenhouse production technology, both the ventilation and heating systems are controlled by a specific climate control system. In the automated system of greenhouses, the climate control set points are programmed based on the climatic requirement of each stage of plant growth of a specific crop. The essentially required components of environment for hydroponically grown vegetables are:

- Temperature (heating and cooling source for plants' homeostasis).
- Humidity (controlled by air moisture).

- Light (natural or artificial lamps as per plant requirement).
- Air circulation (to maintain exchange of gases for plant growth).
- Ventilation (an aiding component to maintain temperature, humidity, and exchange of gases from and to the greenhouse).

Crop plants show their response toward growth and development under different set of environments. The undersoil environment of a plant is as important as aerial one. For providing most congenial facility of crop production, soilless culture offers diverse systems, from purely hydroponics based on water and nutrients only to those based on artificial mixes or purely organic or inorganic (inert) substrates used for hydroponic crop production (Butt 2002). Hydroponic methods have been refined and improved with the help of scientific research as time went on and are rapidly expanding in popularity all over the world. The conventional methods of production have a number of constraints to develop healthy and quality plants which affect their yield. In hydroponics, one of the major advantages is the elimination of different kinds of pathogens that cause drastic reduction in crop yield in terms of quality and quantity. The global market of hydroponically grown produce grew from $226.45 M in 2016 to $724.87 M in 2023, which indicates a tremendous contribution meeting market demand of food supply and sustaining food security (Devecchi et al. 2023).

Some of the benefits of hydroponic vegetable production over traditional field farming include:

- Hydroponics is ideal in situations like deserts and mountainous regions where cultivatable land is scarce.
- Crops yield nutritious food because they are provided with everything they need to thrive.
- Unlike traditional farming, both nutrients and water can be reused thus reducing the cost and water consumption.
- More oxygen reaches the plant roots, keeping them healthy.
- Nematodes, insects, and pathogens that harbor in the soil are less likely to cause any harm to the plants.
- The total span of crop production is significantly reduced in comparison to field grown. Hence the number of harvests is increased. For example, leaf lettuces are the quickest to harvest and usually take about 30–45 days after seed sowing depending on varieties, thus securing manifold yield in a year compared with field-grown.
- Costs associated with land preparation for planting, hoeing, and weeding are decreased.
- Crop rotation is not necessary.
- Regulated environmental conditions mitigate the problem of plants' biotic and abiotic stresses.
- However, there are also some drawbacks of hydroponics which include:
- Its initial investment is greater than that of a conventionally cultivated crop.
- A competent and technically trained workforce is essential for the successful operation of a hydroponic system.

- Hydroponics increases the risk of losses due to the rapid spread of certain diseases caused by various pathogens.

10.3 Requirements for Hydroponic Vegetable Production

The requirements for hydroponic cultivation of vegetables depend on the kind of growing system, i.e., Organic or Inorganic Substrate Culture, Nutrient Film Technique (NFT), or Deep-Water Culture (DWC). Similarly, the difference of requirements is significantly pronounced in view of horizontal, vertical, or layered style of hydroponic system. The main requirements for a hydroponic system are given in the following sections.

10.3.1 Water

River water, pond water, collected rainwater, or even water from deep underground can be well utilized to make fertilizer solutions. In hydroponics, a limited amount of water can serve the purpose to chanalize the crop production. The drainage water after proper treatment can also be reused for crop production in hydroponics (Butt 2007). The pH and mineral content of water can vary widely depending on where it comes from. In order for water to be used as a nutrient solution, its pH must be closer to 7. Therefore, water quality testing is required before it can be used in hydroponics. Filtered (reverse osmosis water purification) and disinfected water are required for use in nutrient solutions. Sterilization will eliminate any disease-causing microorganisms in the water supply. UV light or heat can be employed for this purpose.

10.3.2 Nutrient Solution

Hydroponics makes use of a nutrient solution, which consists of dissolved nutrients in water necessary for plant growth and development. Essential nutrients include nitrogen, phosphorus, and potassium, which must be present in a nutrient solution. The chemical reagents most commonly used in hydroponic formulations include calcium nitrate $Ca(NO_3)_2$, potassium nitrate (KNO_3), potassium dihydrogen phosphate (KH_2PO_4), magnesium sulfate ($MgSO_4$), etc. A number of additional chemicals added in small amounts to the nutrient solution include boric acid (H_3BO_3), copper sulfate ($CuSO_4$), manganese sulfate ($MnSO_4$), and zinc sulfate ($ZnSO_4$). The nutritional need of plants ranges very widely; hence the mixture is tailor-made for the crop at hand to add up the right proportion of nutrients. Both the solution's pH and EC are kept stable during crop growth periods. The nutritional solution can be poured into the vessels manually or by using dosers for more precision. Dosers allow addition of measured quantity of a solution into moving water. Importantly the time of fertilizer application depends when the plant is in need to keep it biologically active from sowing to harvesting to attain the required production. The yield is usually manifold higher compared with that obtained in conventional method of farming (Butt and Varis 1996).

10.3.3 Rooting Medium

Hydroponic rooting medium is the substance used to anchor the plant roots. There are a variety of growing substrates used as rooting materials in hydroponic methods. Some of the soil-free growing media (organic and inorganic) for hydroponic crop production (Butt et al. 2007a) are enlisted below:

- Peat compost (decomposed litter or sphagnum fiber from peatlands).
- Coir (processed fiber of coconut husk, trunk wood chips, and/or dried fronds).
- Clay pellets (clay aggregates).
- Pomace or marc (the solid remains of grapes, olives, or other fruits after pressing for juice or oil, containing skins, pulp, seeds, and stems of the fruits).
- Straw bales (an organic source of wheat straw).
- Perlite (a high-heated amorphous volcanic material of alumino-silicate).
- Vermiculite (high-heated expended hydrous phyllosilicate mineral particles).
- Rockwool (fibrous substance made from rocks like basalt, limestone, and coke).
- Coal-ash or Ash-ball (waste material produced after combustion of pulverized coal).
- Polyurethane ether foam/PUR (inorganic substrate).
- Polystyrene (used mostly in the form of compressed sheets/boxes).
- Sand (inorganic natural source of river beds).
- Gravels (aggregates of rock fragments).

10.4 Systems of Hydroponics for Vegetable Production

Unlike soil or soil-based growing media, hydroponics involves growing of plants in nutrient-enriched water; even the use of an organic or inert substrate requires supplements of nutrient solution for faster production and larger harvests. In a global meta-analysis, the results showed that the productivity of the growing systems varied between crop categories (Goh et al. 2023). Therefore, to start hydroponic vegetable production on a commercial basis, the most important is to decide the type of system to be installed in view of prevailing conditions, finance availability, facilitated protected structure, and the flow of marketing chain at local, national, and international prospects of selling the crop round the year (Butt et al. 2007b). Different types of systems used for hydroponic production technology of vegetables are described below.

10.4.1 Deep Water Culture (DWC)

Research proved that DWC is a highly suitable system concerning the crop performance resulting in higher yields of superior quality. It is based on water, oxygen, and nutrients and widely used for small to medium-capacity growers for production of vegetables because of comparatively cheaper setup and not requiring lot of components. In this system, many vegetables like lettuce and tomatoes can be grown faster and successfully as the roots are directly submerged in nutrient solution with continuous oxygenation through running of an air pump fixed in the reservoir. This type of system is now modified as recirculating deep-water culture (RDWC) in which a number of reservoirs are connected with the additional water pump but the nutrient solution is oxygenated through central air pump, thereby the system can be expanded for getting larger harvests of a grown vegetable by dint of less nutrient input and incurring low cost of larger system. However, it requires continuous but enough quantity of water, and the pH, water level, nutrient concentration, and nutrient temperature fluctuate frequently.

10.4.2 Hydroponic Bucket System

This system is a modification of DWC in which each bucket or container acts as reservoir or medium bed and also referred as Dutch bucket system. It is practiced for a crop having vine growth habit or bigger in size at fruiting stage. Tomatoes, beans, peppers, eggplant, and cucumber are few examples of commercial crops. Inside each bucket or in an alternate container, a mesh pot is placed to anchor the plant and avoid it to sink in the nutrient solution. The system is widely extended for bulk commercial production of vegetables where supply of nutrients is managed through emitters/drippers. The major advantage of separate buckets in the system is efficient pest and disease management as an infected bucket can be removed from the system without loss of whole row of buckets. The system can be designed based on the technical specifications to install irrigation and other necessary equipment. As a usual practice, the required nutrient solution is supplied to plants introduced in growing media through drippers connected with a pipe fixed with reservoir pump. The solution runs through the medium and drains out of the bucket. A timer will ease out to have required quantity of nutrients for optimum growth and development of the crop. The system offers simplicity, low cost, reliable results, easy monitoring, and most importantly the versatility to accommodate a variety of vegetables especially tomatoes, peppers, and cucumbers.

10.4.3 Ebb and Flow (Flood and Drain)

This system is the favorite of a large number of hydroponic growers. The system is based on the principle to wet the roots of plants for a short period of time and then nutrient solution is emptied for drying the roots to receive oxygen and aeration for optimum growth and subsequent manifold yield of the grown crop. In this way, the plants are not immersed continuously in liquid solution rather find a good time to breathe, without drying up completely. The system comprises two tanks: one grow tank and the other acts as reservoir (storage tank/sump). The prepared solution is poured in the reservoir which moves in two directions by reversible water pump. The nutrient solution goes out of the grow tank and subsequently sucked back to the reservoir. So, the drained (ebbed) solution will make the roots zone nutrient solution free that facilitates root aeration for absorption of oxygen getting better plant growth. A timer (optional) and air pump are connected with the reservoir for calculated time application of nutrient solution and oxygenation. Therefore, two phases are set in: one is irrigation phase and the other dry phase.

In ebb and flow system, the use of growing medium is decided by the system designer, depending on feasibility and turnkey basis. Similarly, irrigation time set is determined by the expert grower, which is of paramount importance for the success of this system. For two cycles viz. irrigation phase and dry phase, the timer is usually fixed for 10–15 minutes as irrigation phase after the interval of 02–03 hours during day time. The time of irrigation phase can be increased or reduced depending on the irrigation need of the vegetable crop. During day time, either natural sun energy or artificial lights are used to meet the required intensity of light. The total duration of light corresponding to the frequency of light is determined by the technical grower of greenhouse for pursuing the optimum mechanism of photosynthesis in relation to the night/dark hours for higher crop production. Usually about two parts of a complete day are scheduled for phototropism, therefore, during night time, the supply of nutrients is stopped to balance the plants' metabolism. Yet during hot and long night, little irrigation can be planned to sustain the growth of plants. The time and total nutrient quantity for vegetable production depend upon climatic factors, such as temperature, air humidity, and the nature of the crop grown. For this reason, it is not advisable to grow different crops at one time because growing

Fig. 10.1 Lettuce production in a designed system of Ebb and Flow

requirements of one crop may differ from the other. If such practice is inevitable, then a separate compartment of each crop is designed under the same roof of a greenhouse. An additional benefit of ebb and flow is that instead of only water nutrient solution, a growing substrate such as coco peat or coco coir can be used for hydroponic vegetables' production (Fig. 10.1). This will lower number of irrigations; hence cost of nutrient is significantly reduced.

The salient advantages of this system are given below:

- In a year, different vegetable crops can be grown in this system preferably growing of one crop at one time with the option of another crop within the same year. For instance, growing of tomatoes followed by lettuce production is practiced very successfully in one greenhouse as tomatoes bear heavily in summer and lettuce flourishes in winter.
- There is very less possibility of contamination in the form of bacteria, fungi, or other contaminants because oxygenated water continuously flows intermittently in root zone of growing plants.
- It facilitates to provide oxygen as aeration to plants' root zone.
- The nutrient feed can be changed as per climatic conditions and crop needs.
- The irrigation phase incorporated with dry phase saves the roots or a tuber vegetable from rot and decay.

On the other hand, the system is not a favorite of amateurs due to the following reasons:

- The system is not cost-effective.
- It is not suitable for small setup. This is because it requires a thorough knowledge of the crops as how to grow and meet their nutritional, watering, and humidity needs.
- The maintenance of wet and dry phases is difficult due to technicalities. If a pump gets clogged, the dry roots for a longer period will spoil the system as a whole.

10.4.4 Nutrient Film Technique (NFT)

NFT is a hydroponic system in which water-containing nutrients are directly provided to plant roots. The system is very much prevalent for growing commercial crops particularly the vegetables of premium quality with high demand in the market. Horizontal or vertical farming can be adopted in this system as per area or space available. The heavy population pressure in many countries has drifted attention to apply NFT either in a greenhouse or a chamber of container designed in multiple layers. NFT system is not passive but makes continuous use of a water pump to circulate nutrient solution throughout the system. The roots suspend in shallow water with nutrients in the form of a thin layer or film in channels or gullies (without any organic or inert substrate). A series of growing trays or pipes is fixed with a slight slope for continuous flow of water. The water pump forces the nutrient solution to reach the highest point in the system and the slope of pipes naturally drains the solution down to bring back in the reservoir from where it is again pumped to reach the growing points. In this way, nutrient solution flows in pipes/trays without any break, where plants grow rapidly taking nutrients from the solution. The system is perfect in balancing nutrients and oxygen supply to the roots. NFT is considered a highly recommended production system for leafy green vegetables and herbs. However, the fruiting vegetables like tomatoes can also be grown but with care in design to accommodate the efficient growth of larger root system.

In NFT system, the long sections of pipes are connected together either horizontally or vertically depending on the availability of space and resources. In the present scenario, the system is designed with vertical compact pattern for short of land. NFT has the following benefits of vegetable production:

- For larger crop harvests, NFT keeps the balance between nutrients and oxygen during its working.
- There are no or very less chances of chocking the pipes due to continuous flow of nutrient solution. This makes use of nutrient solution in the vegetable production technology very efficiently as it avoids nutrient build-up in the system.
- NFT, as an essential component of production, does not require any growing medium.

Concerning the restrictions, NFT system can only be kept operative by electrical or an alternate energy source like solar, to maintain regular flow of nutrient solution to feed the growing plants. Further, it does not support production of crops like tomatoes in a smaller growing area of NFT system unless space is not a problem for setting up the system.

10.4.5 Aeroponics

The use of advanced micro-clean technology of aeroponics for crop production is taking momentum steadily where water and nutrients are used to stimulate rapid plant growth by accelerating root development without depending on land use (Jagesh et al. 2020). The plants so produced are pathogen free and also curtail growth span of the crop. The experiments have shown that the intended crop can be harvested two to four times earlier than that from the standard plant-growing approach (Lakhiar et al. 2018). Aeroponics is a way forward for crop production technology which is based on competent understanding and realistic correctness.

This is an active system where high-efficiency air pump is used to spray the nutrient solution on roots of the plants. The mister and spraying heads are fixed in the system for this purpose. In aeroponics, the plants are kept suspended in air, yet their roots take abundant oxygen along with the required nutrients in the form of mist. Currently, this system is considered extremely efficient for highest yield and fastest production of vegetables in any hydroponic set-up due to economical use of water with fewer nutrient inputs. The plants are held by introducing small holding material like foam, rubber plant collars, or net cups filled with a small amount of suitable grow medium. The pressure and frequency of misting the nutrient solution are adjustable depending on climatic requirements and nature of plants in terms of their growth and development. Similarly, the set-up of horizontal or vertical mode of aeroponics is decided on the basis of availability of space and area. The maintenance and management of this system demand precision and close attention as a slight error or negligence can cause serious loss in crop yield.

Constructing aeroponics towers for vegetable production in vertical farming is also quite feasible where space scarcity is an excuse and this modified aeroponics is equally beneficial for domestic as well as commercial production of leafy and green vegetables. However, the overall set-up cost is notably higher and it is not easy to keep the system functional since the specialized air pumps are working which require consistent maintenance for smooth running of the system.

The aeroponics has the following merits:

- Highly efficient, giving greater yield of good quality.
- Significantly reduces number of days to complete the production cycle, thus can have more harvesting the year round.
- Cost-effective in terms of water and nutrient inputs.

The system has the following demerits:

- Higher initial cost.
- Frequent maintenance and troubleshooting to avoid the loss of crop.
- No compromise on the interruption of electricity. A standby arrangement of alternate energy source is a must.

10.5 Hydroponic Production of Commercial Vegetables

10.5.1 Tomato Production

As a first step, nursery is raised and ready seedlings are transplanted into the hydroponic system (Khan et al. 2023). For raising nursery seedlings, tomato seeds need to be soaked in water for at least 24 hours (seed priming) before sowing so that these can imbibe water to facilitate germination. Once the propagation trays are filled with a suitable medium (usually peat, perlite, vermiculite, or rock wool), the seeds are placed carefully in the medium with shallow depth. Trays are placed where these are receiving plenty of sunlight and air circulation. Seed germination usually takes place within 10–15 days. Once true leaves appear, the seedlings are then transplanted in bigger blocks of growing medium (coconut coir/coco slabs) (Khan et al. 2017; Qadeer et al. 2020). The seedlings and subsequent transplants are grown hydroponically by using different substrates of organic and inorganic nature (Butt et al. 2004). The same technique can be employed to several other horticultural crops of commercial importance.

Only that medium should be selected for tomato production which provides sufficient anchorage to plants, allows effective exchange of gases, and ensures good water holding capacity.

For tomato production, peat, perlite, rockwool, and coir are recommended for production system after getting successful seedlings on peat, perlite, or other medium. Tomatoes are conveniently grown on a hydroponic method like NFT, DWC, ebb and flow and drip system. Among these, ebb and flow method is mostly followed for tomato production. In this system, the required solution flows from one side and drains the other end. This system is easy to setup and manage. Nutrient solution used contains all the necessary macro- and micro-nutrients. Concentration of these nutrients in the solution increases with the progress in the stage of plant development. The seedlings are moved into the rockwool cubes, and then the rockwool cubes are placed into the plastic pipes that make up the system. Both the temperature and pH value of the solution are kept constant at around 26 °C and 5.6, respectively. A source of light is provided in the form of fluorescent light bulbs. To prevent the plants from toppling over, stakes made of strings are used to support them. At the initial stage of plant growth (Fig. 10.2a), the shoots that are extra,

Fig. 10.2 (**a**) The initial fruit development in tomato crop; (**b** and **c**) Tomato fruits from maturity to ripening; and (**d**) Marketable tomatoes with bright red color

unhealthy, or growing in between other shoots, are removed (Utomo et al. 2023). Tomato plants are susceptible to assault by a wide variety of bacterial, viral, and fungal diseases. *Verticillium* and *Fusarium* are the pathogens which may attack the plants. In order to protect plants from these pathogens, they are inspected frequently to look for any symptoms and sprayed frequently by using effective chemicals. Tomato crop is also susceptible to attacks by a variety of pests, including leaf minors, pinworms, and whiteflies, all of which have the potential to inflict significant damage (Siddiqui et al. 2023). Monitoring the crop and spraying it on a regular basis are two necessary preventative measures. The initial fruit development (Fig. 10.2a) continues faster parallel to vegetative growth of tomato plants. When fruits have reached their full maturity and ripeness, the vegetative development of leaves becomes slower, thus tending to change the fruit color from green to pale yellow and subsequently light to bright red, indicating ready to be picked (Fig. 10.2b, c). The marketable packed fruits of premium price are shown in Fig. 10.2d).

10.5.2 Pepper Production

Hydroponics allows precise environmental control, resulting in larger and healthier pepper crops. The step-by-step guide includes: selection of cultivar, growing medium, growing system, nutritional supplement, harvesting and packing protocols, etc. The choice of pepper production system depends on growers and plant needs that can work better in system-specific circumstances. Pepper plants can be grown hydroponically using NFT, ebb and flow and drip irrigation. To keep roots healthy, NFT recirculates nutrient-rich water. Water temperature and pH are maintained almost constant. It gets flushed and refilled every few minutes. One NFT-grown pepper plant can produce 0.9 kg of peppers. Similarly, drip-grown peppers are popular where water and nutrients to plant roots are steadily delivered by using small hoses and tubes. This method prevents over- and under-watering. The drip system can be simply customized for individual plants. The water movement to nourish plants in ebb and flow system is well practiced where plants grow in containers above trays. When water level in the trays rises, it flows into a central channel. A timer switches water from the reservoir to the trays and back again. The universal ebb and flow principle has several modifications (Sharma et al. 2018). In wick system, water is delivered from a reservoir to plant roots via wicks. This system makes growing peppers indoors easy without pests or soil-borne diseases. Hydroponic pepper cultivation requires 14–16 hours of light daily. Sunlight can be supplemented with artificial lights (Othman et al. 2019). Peppers growing in hydroponics are fertilized by adding essential nutrients in the water. Nitrogen in the nutrient solution promotes growth of leaves, roots, and fruits. Peppers also require phosphorus, potassium, and magnesium for their growth and quality fruits (Cho et al. 2018). Further, high or low temperatures also impede or degrade plant fruit production. Peppers thrive best at 22.5–25.5 °C temperature and 50–60% relative humidity (Singh et al. 2019). Maintaining pH of nutrient solution for hydroponic pepper is necessary for plant health and fruit yield. Pepper plants prefer pH 6.0–6.8.

Fig. 10.3 (**a**) Pepper seedlings; (**b**) Vegetative growth, flowering, and fruit set of well-developed pepper plants; (**c**) Fruits at maturity stage; and (**d**) The harvested crop of giant peppers ready for marketing

If the pH is outside this range, plants cannot absorb nutrients from the water, exhibiting nutrient deficiencies. Growth period of peppers in hydroponics varies depending on plant type. Jalapenos grow in 40 days, while bell peppers take 70 days. Tiny seedlings of pepper, when introduced on an organic substrate of hydroponic such as coco peat (coir) in an automated greenhouse, undergo various stages of growth and development for producing big attractive and high-quality fruits as shown in Fig. 10.3a–d.

10.5.3 Cucumber Production

It is convenient and faster to grow cucumbers hydroponically with high yield of quality fruits for getting premium return by following production procedures in a right way. The crop flourishes successfully in warm-season and very slow in growth and development in cool-season, whereas it shows damaging effect in frosty or very cold climatic condition. The recommended temperatures for cucumber production in hydroponics are 24–30 °C and 15.5–24 °C during day and night, respectively with 60–70% relative humidity. For better plant growth, enough light is required for 12–14 hours daily; either natural sunlight or artificial lamps light. Hydroponic cucumbers need suitable plant-to-plant space (30–45 cm) and row-to-row (60–90 cm) with continuous air circulation under warm and humid climate. Plant prefers pH 5.5–6.5 for optimum nutrients absorption. The crop gives promising yield when balanced but adequate nutrients mix of nitrogen, potassium, and phosphorus is used (Wdowikowska et al. 2023). Before beginning hydroponic cucumber cultivation, it is important to decide what kind of hydroponic system is to set up. Selection of a cultivar (either slicing or pickled cucumbers) is another important criterion. The type of hydroponic system to be opted depends on type of cucumber crop and total area available for protected cultivation. In ebb and flow system, plants are introduced for ultimate crop production in a medium that is frequently flooded and drained. Presently a great deal of work is being done to use waste materials causing pollution to the environment as growing media. The waste of grapes or grapes marc (a pressed material, organic in nature, containing seeds and outer skin of ripped grape fruits) proved a potential organic substrate for the production of various vegetables (Butt and Varis 2005; Butt and Habib 2009). To provide oxygen, moisture, and nutrients to the plants, the growing medium is periodically flooded and drained. The seedlings are raised by germinating seeds using peat compost, and later these seedlings are transferred to the growing system by introducing coco-peat as a hydroponic substrate with a drip nutrient system. By providing adequate space, light, and temperature, a bumper harvest of crunchy, flavorful cucumbers results in. Cucumber is a vine crop with fast-growing habit. To establish and control its growth in hydroponic system, vine training or trellis training technique is adopted. The trained vines will occupy very little space and keep the fruits away from the soil thus safeguarding from diseases and insect-pests attack. Different structures can be used for training cucumbers to trail vertically. A tall, sturdy support is usually selected that won't topple due to the weight of fruits. Cucumber plants need a height of 1.25 to 2.0 meters to reach the maximum potential of fruit bearing (Samba et al. 2023). Cucumbers can be harvested at varying sizes depending on the cultivar, use, and fruit characteristics. The skin of cucumber is a preferential indicator of when it is ready to be picked. If the skin is tough to scratch with your fingernail or thumbnail, it is ready to be picked. The harvesting of cucumbers is done by cutting fruits from the vine at an angle of 90 degrees with a sharp sterilized knife. Harvesting through pulling the fruits should be avoided to avoid damage to the fruits as well as the gowning vines.

10.5.4 Lettuce Production

Hydroponics simplifies lettuce cultivation to get ideal crop of supreme quality, saving a lot of space and avoiding risk of pests and diseases infestation. A variety of lettuce can successfully be grown on different types of hydroponic systems.

Usually, a hydroponic system that needs a little space for producing huge number of quality and marketable lettuce heads is preferred. Further, the hydroponic system adopted should be cost-effective for fetching maximum economic benefits through greater leafy yields. In other words, choosing the right hydroponic system is vital for lettuce production. In a deep-water system, lettuce is more susceptible to diseases and less efficient in growth. However, aeroponic system delivers greater benefits as it efficiently uses water and nutrient solution and also eliminates pathogens attack. For numerous reasons, lettuce producers prefer drip hydroponic system because it recycles nutrients, irrigates slowly and its low-pressure pump makes it quite fit for a smaller space. Ebb and flow system usually results in higher yields and is considered perfect for the manifold yield of lettuce crop. Majid et al. (2021) evaluated DWC and NFT in comparison to conventional soil-based cultivation system. DWC resulted in higher yields with superior quality heads, while NFT was the most water-efficient with about 64% water savings. To start lettuce growing, seedlings are raised by planting seeds in trays containing a suitable medium. Different media can be successfully used for the production of seedlings including peat moss, perlite, vermiculite, and rock wool. The seedling trays are placed in a sunny or well-lit area. Watering to nursery seedlings is essentially required for every other day. The optimum temperature for seed germination and seedling growth ranges from 18 to 27 °C. When the seedlings are 05 cm tall and have 04 true leaves, these can be transplanted in a hydroponic system. Transplanting of seedlings needs careful handling from cells to net pots as these are very fragile. Each net pot is placed in the reservoir of hydroponic system with the holes being bored into the lid or floating platform. Lettuce grows well with less light than other plants, i.e., 10–14 hours photoperiod per day (Tripathi et al. 2022). Lettuce flourishes well under lower temperatures. For better growth and development of plants, a temperature of 24 °C during the day and 13 °C at night is considered optimum. When the temperature gets too hot, the plants may bolt and taste bitter. The pH of the nutrient solution is maintained slightly acidic to practically neutral as it impacts nutrients absorption by plant roots. Maintaining a pH of 5.5–6.5 is considered optimum for getting higher yields (Dhal et al. 2023). Lettuce heads can be harvested after 04–08 weeks of transplanting depending on a variety of a specific group as well as market demand (Fig. 10.1). The proper air circulation in the greenhouse will keep the crop healthy, assist to get enough carbon dioxide, and avoid mildew and bacterial growth. The major marketable flaw in lettuce production is appearance of tip burn, a physiological disorder caused by environmental and nutritional imbalances. It increases with the increase in growth rate. Secondly, low Ca with high N uptake triggers tip burn especially when temperature drops below 15 °C (Butt and Varis 1998; Butt and Varis 2000). Thus, for getting high yields of supreme quality heads, lettuce plants should be provided with balanced nutrition and optimal environmental conditions.

10.6 Conclusion

Thickly populated countries need to improve their agricultural performance to address the challenge of food security. Hydroponics has the potential to increase vegetable yields with improved quality to feed local population as well as to export and earn foreign exchange. The misconception that a lot of soil is needed to grow vegetable crops has been debunked by hydroponics. Hydroponic production of tomatoes, lettuce, peppers, cucumbers, and many other vegetables has tremendous prospects to meet the food demand in view of global population growth especially in Asian countries. Hydroponics has the potential to be a game-changing technology by ensuring a steady and adequate supply of vegetables. It uses 70–90% less water and significantly less land for farming than traditional soil cultivation. The use of hydroponics has a great opportunity to become a market leader if the technology is implemented properly. The initial investment in a high-tech hydroponic facility is higher than that of traditional soil farming, but the low ongoing expenses and guaranteed high yields and quality mean that the initial expenditures are quickly recouped. The progressive agro-based industrialists now understand the importance of futuristic approach of investment, thus can share a substantial part in making the country's economy strong, and sustainable. Hydroponics' potential of manifold yield with high-quality vegetable production gives a ray of hope to ensure food security for the people of all communities.

Acknowledgments The authors are thankful to PMAS-Arid Agriculture University, Rawalpindi for providing few pics as courtesy taken from their hydroponic system.

References

Aboutalebi A, Jahromi MG, Farahi MH (2013) Evaluation of growth and yield of organically-grown basil (*Ocimum basilicum* L.) in soilless culture. J Food Agric Environ 11(1):299–301

Adak N, Gubbuk H, Tetik N (2018) Yield, quality and biochemical properties of various strawberry cultivars under water stress. J Sci Food Agric 98(1):304–311

Ali A, Ahmad A, Khaliq T, Akhtar J (2012) Phenology and yield of sunflower (*Helianthus annuus* L.) hybrids as affected by varying plant spacing and nitrogen levels under semi-arid conditions of Sargodha, Punjab. Pak J Sci 64(2):98

Butt SJ (2002) Soilless culture technology: a broad-based idea of hydroponics for vegetable and flower production – why not in Pakistan? COMSATS News Views 2(3&4):10–15

Butt SJ (2007) Farming without soil. The Daily Dawn (Sci-Tec World) April 14, 2007. www.dawn.com

Butt SJ, Al-Haq MI (1993) Vegetable cultivation, a secret of manifold prosperities. J Sci Tech Dev 12(3):46–51

Butt SJ, Habib U (2009) Advances of hydroponic crop production in protected horticulture. In: Proc Int Symp Protected Horticulture. Arid Agric Univ, Rwp, Pak, pp 125–134

Butt SJ, Varis S (1996) Production of tomato in different growing media in cold glasshouses. HASAD 12(137):17–23

Butt SJ, Varis S (1998) Production of lettuce in different growing media in cold glasshouses. HASAD 14(166):32–35

Butt SJ, Varis S (2000) Suitability of containerized rooting media using bag culture technique for quality vegetable crop production under unheated glasshouses. In: Proc 2nd Int Symp Agroenviron-2000, Tekirdağ, Türkiye, pp 179–186

Butt SJ, Varis S (2005) Beneficiary role of grape residue, an organic waste of agro-based industry causing environmental pollution – a new concept of crop production in hydroponics. Proc. In: 1st International conference environment sustain development, 712 June 2004, Islamabad, pp 1–3

Butt SJ, Varis S, Al-Haq MI (2004) Improvement of sensory attributes of tomatoes (*Lycopersicon esculentum* Mill) through hydroponics. Int J Agric Biol 6(2):388–392

Butt SJ, Abbasi NA, Mahmood T (2007a) Role of various growing substrates on the yield and quality of greenhouse crops. In: Proceedings International Symposium Prospects Hortic Ind Pak. Univ Agric, Faisalabad, Pakistan, pp 29–37

Butt SJ, Koral PS, Akhtar SA, Varis S, Kutuk MO (2007b) Development of hydroponics in Türkiye, Pakistan, and other countries. HASAD 22(260):66–72

Butt SJ, Varis S, Nasir IA, Sheraz S, Shahid A (2015) Micro propagation in advanced vegetable production: a review. Adv Life Sci 2(2):48–57

Carolan M (2016) Adventurous food futures: knowing about alternatives is not enough, we need to feel them. Agric Hum Values 33:141–152

Cho WJ, Kim HJ, Jung DH, Kim DW, Ahn TI, Son JE (2018) On-site ion monitoring system for precision hydroponic nutrient management. Comput Electron Agric 146:51–58

Devecchi M, Ghersi A, Pilo A, Nicola S (2023) Landscape and agriculture 4.0: a deep farm in Italy in the underground of a public historical garden. Horticulturae 9:417

Dhal SB, Mahanta S, Gumero J, O'Sullivan N, Soetan M, Louis J, Gadepally KC, Mahanta S, Lusher J, Kalafatis S (2023) An IoT-based data-driven real-time monitoring system for control of heavy metals to ensure optimal lettuce growth in hydroponic setups. Sensors 23:451

Gao Y, Yu H, Liu P, Ma C, Li Q, Jiang W (2018) Ending composting during the thermophilic phase improves cultivation substrate properties and increasing winter cucumber yield. Waste Manag 79:260–272

Gericke WF (1937) Hydroponics - crop production in liquid culture media. Am J Bot 85(2198):177–178

Goh YS, Hum YS, Chai LY, Lai YL, Khin WK, Yap WS, Tee YK (2023) A meta-analysis: food production and vegetable crop yields of hydroponics. Sci Hortic 321:112339

Jagesh KT, Devi S, Buckseth T, Ali N, Rajesh K, Singh RK, Zinta R, Dua VK, Chakrabarti SK (2020) Precision phenotyping of contrasting potato (*Solanum tuberosum* L.) varieties in a novel aeroponics system for improving nitrogen use efficiency: in search of key traits and genes. J Integr Agric 19(1):51–61

Khan MA, Butt SJ, Khan KA, Nadeem F, Yousaf B, Javed HU (2017) Morphological and physico-biochemical characterization of various tomato cultivars in a simplified soilless media. Ann Agric Sci 62:139–143

Khan QA, Butt SJ, Zia-Ul-Haq NI, Iqbal A, Sajid M (2023) Effect of various planting times on growth, yield and quality of hydroponically grown tomatoes (*Lycopersicon esculentum* Mill). Pure Appl Biol 12(2):206–211

Kumar V, Arun K, Singh J (2023) Trends in hydroponics practice/technology in horticultural crops: a review. Int J Plant Soil Sci 35(2):57–65

Lakhiar IA, Gao J, Syed TN, Chandio FA, Buttar NA (2018) Modern plant cultivation technologies in agriculture under controlled environment: a review on aeroponics. J Plant Interact 13(1):338–352

Majid MM, Khan JN, Ahmad Shah QM, Masoodi KZ, Afroza B, Parvaze S (2021) Evaluation of hydroponic systems for the cultivation of lettuce (*Lactuca sativa* L. var. *longifolia*) and comparison with protected soil-based cultivation. Agric Water Manag 245:106572

Manos DP, Xydis G (2019) Hydroponics: are we moving towards that direction only because of the environment? A discussion on forecasting and a system's review. Environ Sci Pollut Res 26:12662–12672

Othman Y, Bataineh K, Al-Ajlouni M, Alsmairat N, Hilaire RS (2019) Soilless culture: management of growing substrate, water, nutrient, salinity, microorganism and product quality. Fresenius Environ Bull 28:3249–3260

Qadeer A, Zia-Ul-Haq, Butt SJ, Asam HM, Mehmood T, Nawaz MK, Haidree SR (2020) Hydroponics as an innovative technique for lettuce production in greenhouse environment. Pure Appl Biol 9(1):20–26

Samba N, Nunomura O, Nakano A, Tsukagoshi S (2023) Effective training methods for cucumber production in newly developed nutrient film technique hydroponic system. Horticulturae 9:478

Sharma A, Acharya Kumar S, Singh J, Chaurasia A (2018) Hydroponics as an advanced technique for vegetable production: an overview. J Soil Water Conserv 17(4):364–371

Siddiqui Z, Hagare D, Liu MH, Panatta O, Hussain T, Memon S, Noorani A, Chen ZH (2023) A food waste-derived organic liquid fertilizer for sustainable hydroponic cultivation of lettuce, cucumber and cherry tomato. Food Secur 12:719

Singh H, Dunn BL, Payton M, Brandenberger L (2019) Selection of fertilizer and cultivar of sweet pepper and eggplant for hydroponic production. Agronomy 9:433

Thomaier S, Specht K, Henckel D, Dierich A, Siebert R, Freisinger U, Sawicka M (2015) Farming in and on urban buildings: present practice and specific novelties of zero-acreage farming (ZFarming). Renew Agric Food Syst 30(1):43–54

Tripathi V, Arora D, Ahlawat V, Lahari P, Kumar A, Kumari A (2022) Soilless cultivation in lettuce: a review. Eco Environ Conserv 28(4):1891–1898

Utomo PS, Al Wafa MAA, Helilusiatiningsih N, Irawati T (2023) Effect of seed age and variety types on the growth and production of tomato (*Solanum lycopersicum* L.) hydroponic systems. East Asian J Multidiscip Res 2(2):727–736

Wdowikowska A, Reda M, Kabała K, Chohura P, Jurga A, Janiak K, Janicka M (2023) Water and nutrient recovery for cucumber hydroponic cultivation in simultaneous biological treatment of urine and grey water. Plants 12:1286. https://doi.org/10.3390/plants12061286

Organic and Sustainable Vegetable Production

11

Safina Naz, Muhammad Akbar Anjum,
Hafiza Muniba Din Muhammad, Riaz Ahmad,
and Muhammad Ahsan Altaf

Abstract

Organic farming is a complete production management approach that contributes in sustainable productivity. It improves agroecosystem health, soil microbial activity, biological processes, produce quality, and economy of farmers. The growing of vegetables under organic farming is attaining much attention due to the increased demand of consumers. Organic vegetables have better organoleptic characteristics. These provide significantly higher levels of some nutrients and antioxidants and are also free from harmful chemical residues. Excessive and continuous use of chemical fertilizers and pesticides pose serious threats to both environment and human health. Organic farming is eco-friendly having a positive socio-economic impact on the environment. The main challenge to plant researchers is how to feed the increasing world population with organic vegetables. Several agricultural factors need to be taken into consideration for organic vegetable cultivation on a sustainable basis. Planting leguminous crops and fixation of biological nitrogen can be exploited to rapidly decompose organic manures, comprising an array of crop wastes and farm yard manure. Although organic farming is quite effective for quality production of vegetables at the same time, organic farmers may encounter some difficulties, i.e., non-provision of right inputs and technical expertise, yield challenges, mar-

S. Naz (✉) · M. A. Anjum · H. M. D. Muhammad
Department of Horticulture, Bahauddin Zakariya University, Multan, Pakistan
e-mail: safinanaz@bzu.edu.pk

R. Ahmad
Department of Horticulture, The University of Agriculture, Dera Ismail Khan, Pakistan

M. A. Altaf
College of Horticulture, Hainan University, Haikou, China

Z. Iqbal et al. (eds.), *Sustainable and Innovative Vegetable Production in times of Climate Change*, https://doi.org/10.1007/978-981-96-6283-8_11

keting uncertainty, and consumers' lack of trust in certified organic produce due to inconsistent regulations.

Keywords

Cover crops · Crop rotation · Organic fertilization · Organic manure · Vermicompost

11.1 Introduction

Consumers' growing desire for organic foods is mostly led by concerns about the health hazards caused by pesticide residues. Organically cultivated crops contain less contamination, even if there is no obvious correlation between organic growing practices and nutritional content of the produce (Wilkins and Hillers 1994). Organochlorine pollutants have, however, also been found on organic vegetables which were produced on soil that had been treated with pesticides 5–20 years earlier, or on traditional farms nearby that polluted the area with pesticides through combustion, wind distribution, rainfall, and leaching to surface and groundwater (Roitner-Schobesberger et al. 2008). According to extensive US data sets on pesticide residues detected in conventionally grown, organically grown, and fruits and vegetables cultivated using integrated pest management, organically produced foods have much fewer pesticide residues compared with all other growing systems (Baker et al. 2002). So, compared to foods cultivated using traditional pest management techniques, vegetables grown organically represent significantly lower pesticide hazards (Craddock et al. 2019). Moreover, if proper farm management techniques are employed, there is no information that organic vegetables may be more vulnerable to microbial contamination than traditional vegetable cultivation. However, more information should also be explored regarding organic cultivation and its crosstalk to pathogens and disease attacks (Barański et al. 2014).

According to available statistics, organic farming is practiced on more than 31 million hectares globally, earning over 26 billion US dollars yearly on a sustainable basis (Conway 2019). However, organic farming has been targeted for reliance on the soil's accumulation of phosphate and potassium from previous fertilization (Yiridoe et al. 2005). In developing nations like Bangladesh, India, and Pakistan, agriculture is essential for the socioeconomic growth of these countries. Organic farming does not deplete natural resources, so it is possible to ensure food security, reduce poverty, and save crucial natural resources with organic farming and some other methods that do not deplete natural resources (Yiridoe et al. 2005). The main issue that encourages plant researchers is how challenging it is to feed the world's expanding population with organic vegetables (Popa et al. 2019). For low-income or developing countries, there is a need to replenish their resources for safe production and a stable vegetables' supply. Thus, the idea of food security includes aspects of agricultural production, environmental management, occupational returns, marketing, nutrition and health, and government policies (Yiridoe et al. 2005; Sirieix et al. 2011).

Organic farming is a production approach that avoids or substantially reduces the use of chemical fertilizers, pesticides (herbicides, insecticides, fungicides, etc.), and artificial growth-regulating substances. Moreover, the foundation of organic farming is the pursuit of environmental, social, and economic sustainability (Fig. 11.1). Use of legume crops and biological nitrogen fixation can rapidly degrade organic manures, comprising an array of crop wastes and animal manure (Sirieix et al. 2011). The advantage of organic farming is that it minimizes the time between adding NPK to the soil and removing it from the soil by returning all residues to the soil, primarily through decomposition. According to previous studies, excessive and continuous use of synthetic fertilizers exert serious threats to both environment and human health (Gracia and De-Magistris 2008; Popa et al. 2019). Resultantly, farmers in industrialized countries are insisted to shift from conventional farming to organic farming. Consumer demand for organic food has increased as a result of better health awareness, and people are willing to pay high costs for this produce (Boccaletti and Nardella 2000). In contrast to contemporary agriculture, traditional farmers have their own concepts and standards to conserve soil health. Depending upon the nutrient source and its dose, the interaction between inorganic nutrients

Fig. 11.1 Organic production of vegetable crops

and organic matter may drop or improve soil nutrients status. It was revealed that the nutritional needs of the crop could not be fulfilled by using organic inputs alone, therefore, it is necessary to integrate the two forms to obtain superior crop yields (Chen 2007).

11.2 Challenges for Organic Vegetables Production

The growing of vegetables under organic farming system is challenging for farmers. However, organic cultivation is attaining much attention due to the increased demand of consumers for safer consumption (Saba and Messina 2003). For conventional growers, switching to organic entails significant adjustments that could cause stress regarding technical expertise and management practices. Seeds can be twice as expensive as regular seed or may be unavailable and the most significant food distributors need large quantities (Shepherd et al. 2005). Moreover, some food distributors seek less expensive organic vegetables. Other issues that organic farmers might face include lack of awareness, non-availability of inputs, limiting exchange between growers, big target of producing many vegetable crops at the same farm, lack of proper distribution channels, and consumers' mistrust in certified organic produce as a result of inconsistent policies within the country (Tsakiridou et al. 2008). Results are still contradictory based on prior research on the environmental effects of organic agriculture as opposed to conventional approaches. The conservation of water and soil, biodiversity, and other environmental benefits of organic agriculture have been mentioned in the literature. There is no conclusive proof that greenhouse gas emissions have decreased due to organic agriculture. Further, no definite evidence is available that organic production results in reduced nitrate leaching over traditional farming, despite some research reporting that less nitrogen is released into the groundwater from soil in organic production than conventional farming. For assessing the sustainability of vegetable production, some strategies have been developed including decision support models, life-cycle assessments, cost-benefit evaluations, monitoring of CO_2 emissions, energy analysis, biotic and abiotic integrity indices, and positioning analysis (Baker et al. 2002).

Researchers discovered a decrease in CO_2 emissions from a greenhouse tomato growing system to the environment that used recycled wastewater; however, the use of synthetic fertilizers was responsible for 30% of the air's acidity and 57% of the system's eutrophication impacts (Gruda 2019). Utilization of renewable energy sources must be kept into consideration to increase the sustainability of greenhouse organic production. When comparing nitrate leaching in conventional and organic mixed farms, the majority of fields showed similar losses except for conventional pasture which had fewer losses (Fracchiolla et al. 2020). Few of these studies, however, characterized environmental effects on the life cycle of vegetable cultivation systems because most of these researches were carried out in Europe. Similar to this, not much research has examined how organic farming practices affect postharvest aspects of vegetables. However, organic farming is necessary for the quality production of vegetables on a sustainable basis.

Several agricultural factors need to be taken into consideration for sustainable organic vegetable cultivation. The most crucial ones are: (i) soil's biological activities and properties, which form the basis of an organic monitoring program; (ii) application of suitable fertilizers and amendments at right stage of plant development to achieve optimum growth and causing minimum pollution to surface and ground waters; (iii) judicious irrigation management; and (iv) species and cultivars that satisfy consumers' requirement while also being well evolved to the growing conditions (Halbert-Howard et al. 2021). As synthetic fertilizers are prohibited in organic agriculture, farmers primarily depend on crop rotation, crop leftovers, composts, green manures, and farm yard manures. Mechanical methods, suitable cultural practices, and other techniques are used to manage insect-pests, diseases, weeds, and other pests (Bachinger and Zander 2007).

11.3 Germplasm Selection for Cultivation in Organic Farming

The selection of germplasm is important for cultivation under an organic farming system and is much imperative for sustainable productivity. A crucial management choice, selection of suitable vegetable crop and its cultivars, has a significant role in the profitable organic farming business. First, a decision must be taken based on the needs of the industry and the quality of the produce with appearance, flavor, nutritional content, and health-promoting ingredients, as well as shelf life and shipping tolerance. Species and cultivars should be chosen according to their productive capacity under certain growing circumstances (Annicchiarico and Pecetti 2010). Different cultivars should be chosen based on the type of soil and its health, extremities of temperature, humidity, light intensity, photoperiod, and production systems (field, tunnel, or greenhouse). Naturally, using cultivars that are tolerant or resistant to diseases will lower the prevalence of diseases (Reda et al. 2021). In organic farming, farmers should be vigilant regarding the origin of vegetable seeds because neither genetically modified (GM) seeds nor seeds of their successive progenies are permitted. Lack of access to certified seeds is a significant issue for organic farmers. Seed businessmen are well aware of this issue as demand for natural produce rises (Löschenberger et al. 2008). For the best production and quality, selecting the proper rootstocks for certain scion cultivars is also essential. To control soil-borne diseases, grafting has become a widespread practice for solanaceous as well as cucurbitaceous plants globally. Other advantages of grafting include resilience to cold temperatures, salt stress, water-logged soil, higher water and nutrient absorption and usage effectiveness, enhanced plant vigor, longer harvest times, and better fruit quality (Moreno et al. 2019).

11.4 Fertilization Management

The availability and balance of nutrients in the soil contribute to conserve soil fertility status. However, soil fertility goes low due to unavailable or imbalanced nutrients in the soil (Ullah et al. 2008). Fertilizer use in organic farming urges to adhere to some basic components: (i) increase soil nutrients, (ii) diminish utilization of non-renewable materials, and (iii) avoid polluting the agroecosystem. Primary method of introducing additional nitrogen or other nutrients is through organic manures, i.e., animal droppings or farmyard manure, compost and agricultural wastes, and crop rotation with legumes for biological nitrogen fixation (Aguilera et al. 2013). Moreover, organic farming regulations typically permit farmers to acquire mineral fertilizers, i.e., natural rock phosphate, crude potassium salts, natural calcium and magnesium carbonates, calcium chloride, calcium sulfate, magnesium sulfate, lime, elemental sulfur, stone meals, etc. (Carvalho et al. 2018). Dynamics of microbial communities and factors affecting health of beneficial soil microbes are also vital in organic farming (Gunapala and Scow 1998).

Synchronizing soil nutrient availability and distribution in the rhizosphere with crop requirements is a barrier to organic agriculture. The majority of nitrate leaching issues typically involve vegetables including lettuce, celery, spinach, and cabbage. Even when in the soil, mineral nitrogen is in smaller amounts at the time of absorption, vegetable residues can cause significant nitrogen losses by leaching during autumn or winter. Vegetable crops that grow quickly should not ever confront nutrient shortages since the organic sources mineralize too slowly to meet crop needs (Araújo and Melo 2010).

Legumes grown in succession can provide N to the subsequent vegetable crops. Moreover, compost is also an important source of N. At the time of transplantation, an animal feed and rock phosphate combination can be utilized to supply extra nutrients. Usually, its 50–66% is added in the soil prior to planting; while the remaining portion may be applied as per requirements at later growth stages. Rock powders can be used to supply comparatively high levels of P as well as K that tomatoes require. Growers frequently apply seaweed, fish emulsion, and compost as root and foliar fertilization in combination with soil-incorporated nutrients. Green manure, compost, animal wastes, and crop rotation with legumes are organic materials that provide plant nutrients, boost plant's ability to tolerate insect-pests and diseases, keep the soil moist, help in controlling weeds, and guarantee high yield of superior quality (Santos et al. 2012).

The use of rock minerals or foliar fertilization should be limited to enhance natural cycle of soil minerals. Liming increases soil fertility and plant growth with a pH range of 6.5 to 6.8. Due to their low C:N ratios, waste products like cow slurry, liquid cattle manure, and animal droppings are quickly utilized by soil microbes, which causes a rapid release of nutrients, particularly N, that are readily available to plants (Petersen and Sørensen 2008). Higher C:N ratios are typically found in organic fertilizers made from plant materials. When it comes to N mineralization or immobilization, the C:N ratio must be between 25 and 30 (Zubair et al. 2020). N, P, and K resources for organic fertilization of vegetables are presented in Table 11.1.

Table 11.1 Nitrogen, phosphorus, and potassium resources for organic fertilization of vegetables

Sr. #	Nitrogen source	Phosphorus source	Potassium source
1	Lettuce	Rock phosphate	Greensand
2	Sugarcane	Wood ash	Wood ash
3	Corn	Bone meal	Polyhalite
4	Cabbage	Poultry manure	Green manure
5	Soybean	Fish emulsion	Seaweed meal
6	Tomato	Compost	Compost
7	Coffee	Sewage sludge	Hardwood ashes
8	Cauliflower	Manure	Kelp meal
9	Soybean meal	Bat guano	Greensand
10	Cottonseed meal	Cover crops	Cover crops
11	Fish emulsion	Seafood	Banana peels
12	Cover crops	Legumes	Eggshells
13	Swine manure lagoon liquid	Nuts	Burned cucumber skins
14	Grass clippings	Crop residue	Illite clay

11.5 Soil Organic Matter

Organic matter has a significant impact on the physicochemical features of the soil, including its ability to hold more water and provide better aeration for seed germination and root growth. Majority of Pakistan's soils are deficient in organic matter. The amount of organic matter in the soil is declining over time because of continuous cropping resulting in reduced yield. Pakistan's soils have not only poor organic matter concentration but also show micronutrient deficiencies (Reeve et al. 2016). However, the application of compost improves nutrients' levels and physicochemical qualities of soil ultimately enhancing crop productivity (Lundquist et al. 1999). Moreover, addition of compost as well as other organic components lowers the alkaline soils' sodium adsorption ratio (SAR) and pH. Due to the production of organic acids and release of Ca, leaching of Na occurs, and therefore ECe of the soil is slightly increased. Moreover, soil organic matter content and all the key plant nutrients are also increased (Tu et al. 2006).

11.6 Cover Crops and Ridge Tillage in Organic Farming

Grasses and N-fixing legumes are examples of cover crops that are cultivated in conjunction with commercial crops. Contrary to green manure crops that grow quickly, cover crops are sometimes left on the soil surface for the whole growing season in between the commercial crops. To conserve N and manage weeds, rye (*Secale cereale L.*) is frequently utilized as a winter cover crop. In 3–6 months, a well-managed leguminous cover crop can help to fix up to 150–200 kg/ha of N. When a field is covered with a cover crop, N that would otherwise be lost through leaching or denitrification is retained in the soil (Gadermaier et al. 2012). This prevents nitrate pollution and keeps N in a natural form that is then made accessible to succeeding crops after the cover crop is mixed into the soil. Additionally, cover crops help to stop soil erosion. These advantages can also be obtained via intercrops, which can be harvested if they are profitable crops. For instance, intercropping with soybean or cowpea enhances corn production and N assimilation in tropical regions by 15–20%, a yield equivalent to that from providing 80 kg/ha of N fertilizer (Henneron et al. 2015).

The C:N of the substances, its level of absorption into the soil, the soil conditions, and the moisture levels all affect how quickly cover crops emit N into the environment. The C:N of the older crop materials is significantly higher because cover crop components are typically more developed during assimilation compared to those of green manure crops. Soil microbes may exhaust entire N available from decomposition of green manure crop in the soil by metabolic processes as they utilize the carbon-containing material for nourishment. However, in case of cover crops, just a little amount of soil N may be utilized by the microorganisms, and some N may be directly available to the grown crop. "N immobilization" describes the process of storing N in microbial biomass. The length of time needed for N outflow to surpass N immobilization varies depending on the aforementioned conditions. The farmer must determine whether additional N fertilizer is required and, if so, in what quantity, based on the timing of the greatest N demand by the succeeding crop (Berry et al. 2002).

In sustainable farming, ridge tillage is frequently used in conjunction with crop rotation and mixed cropping. At plantation time, this method just tills a zone of soil that is 15–20 cm wide as well as 5–10 cm deep. In addition to burying crop waste and weeds in the space between rows, the method produces an ideal seedbed condition. A few weeks after planting, between-row cultivation encourages mineralization of soil organic wastes, manages weeds, and reconstructs the ridge with a raised bed for subsequent crops (Naik et al. 2012). Additional animal manure, green manure, compost, and residues of a cover crop can enhance soil nutrients and water capacity as well as provide N and numerous other nutrients, although the timing of their impacts and processes of their actions may vary (Green et al. 2005).

11.7 Weeds Management

Weeds management is necessary for quality production of organic crops. Weeds control practices are very poor in organic farming of vegetables. In organic agriculture, the impacts of cultural techniques (i.e., fertilization and direct weed management) on crop-weed interactions frequently manifest themselves more slowly than in conventional agriculture. As a result, weed management should be approached over a long period of time and deeply integrated with other cultural practices, with goal of optimizing the whole cropping system and not only weed control. In this regard, cover crop management is vital due to its impact on soil, nutrients, pests, and weed control. It is emphasized that direct weed control (physical) can be effective only when preventive measures and cultural practices are used to diminish weed emergence (e.g., through suitable crop succession, cover crops, tillage) and enhance crop competitive ability (e.g., by sowing appropriate crop cultivar, and following proper planting pattern and fertilizer application strategy). There are several instances of system-oriented weed management methods created for organic farming, detailing issues and future directions (Bàrberi 2002). The methods of weed eradication for organic vegetable production are discussed below.

11.7.1 Manual Weed Control

Hand weeding is the manual way of weeds control during crop growth. Different hand tools, i.e., hand weeder, stirrup hoe, and weed puller are frequently being used for the destruction of weeds. Hand weeding is an efficient method to uproot the harmful annual weeds. However, perennial weeds are difficult to control through hand weeding due to their root separation from shoots during uprooting. So, their roots again develop new shoots. Hand hoeing is not effective to control perennial weeds (Sims et al. 2018). Further, hand weeding and hoeing take a lot of energy and time. Although manual way to control weeds is very effective for home gardening, it is not successful for commercial purposes.

11.7.2 Chemical Weed Control

The most efficient method to eradicate weeds is through application of herbicides, commonly called as chemical control. Herbicides can increase productivity, support low-till farming methods, and reduce labor costs. Herbicides can be categorized depending on their chemical nature, time of application (pre-plant, pre-emergence, and post-emergence), mode of action, formulation, site of absorption, and selectivity in function (Scavo and Mauromicale 2020). To minimize any negative environmental effects, herbicides use should be as low as possible. Further, in organic production, use of chemical herbicides is restricted. Farmers must, therefore, substitute herbicides with a mix of various practices, mainly preventive as well as other weed eradication methods.

11.7.3 Cultural Practices to Control Weeds

Mechanical weeding can manage weeds well even when other approaches are impractical. The crop type, soil form, implements, operating costs as well as personnel needs are some of the factors that influence choice of a mechanical weeding method. There are many types of tillage techniques used for mechanical weed management, but cultivation is the most important one for producing crops. It involves shallow tillage using a range of weed-controlling instruments (Monteiro and Santos 2022). Tillage is less labor-intensive and has some advantages, including the ability to break down soil organic matter, enhanced soil exposure to air, increased soil moisture level, and amended openness of organic wastes to decomposers.

Weed control is greatly impacted by mowing and cutting. In turf, vineyards, orchards, pastures, and forage crops, these operations are frequently used. Seldom are these methods effective enough to completely eradicate weed growth. Weeds' leaf area is decreased by cutting and mowing them, which also reduces their growth and inhibits or stops seed formation (Mia et al. 2020; Monteiro and Santos 2022).

11.7.4 Weeds Control with Mulching

Mulching materials (manure, straw, black plastic) can be helpful to suppress weeds, conserve soil moisture level, regulate soil temperature, and improve performance of highly valuable vegetable crops. Mulches are of several forms, such as natural mulches made from plant remains, weeds, straw, paper, sawdust, and synthetic mulches prepared from plastic (Mia et al. 2020). Black polyethylene has been utilized for weed control in a variety of agricultural production systems notably in many horticultural crops, e.g., eggplants, cucumbers, tomato, strawberry, and watermelon (Pannacci et al. 2017). Infrared light is utilized to warm the soil while photosynthetically active radiation is blocked by plastic mulches. These infrared-emitting mulches are successful for weeds elimination. Mulching is more suitable for suppression of annual weeds as compared to perennial ones (Mia et al. 2020).

11.8 Insect-Pests Management

11.8.1 Biological Control of Insects-Pests

Biological control is a method of managing insect-pests by using natural enemies (predators and parasitoids). Usually, beetles, predatory bugs, parasitic wasps, midges, mites, bacteria, fungi, entomo-pathogenic nematodes, and viruses are used for this purpose. It is advisable to use this sort of biological management in advance of pest outbreaks. There are more than 100 distinct species commercially available for supplemental biological control, most of which are used in greenhouse environments. These all-purpose predators consume insect-pests with soft bodies like aphids as well as their eggs (Gurr and You 2016). Parasitoid bugs hatch eggs that are laid in or on an opponent, and as the larvae evolve, they use the host's supplies until eventually perish. Insect parasitoids are skilled at locating their hosts and frequently have host preferences.

11.8.2 Chemical Control of Insect-Pests

Synthetic insecticides have largely been used to manage insects for the past 50 years (Gurr and You 2016). In organic farming, chemical insecticides are prohibited; however, the Organic Materials Review Institute has approved the use of a few organic insecticide alternatives. Plant-based extracts like neem oils, pyrethrums, ryania, rotenone, and sabadilla, compounds made from microbes like spinosad (made from naturally occurring soil bacterium *Saccharopolyspora spinosa*) are allowed. To control tiny, soft-bodied insects, detergents, mineral oils and potassium salts of fatty acids can also be applied. Diatomaceous earth and kaolin clay are examples of acceptable mineral products. Although some organic insecticides are rather durable, the majority have limited residual activity. For instance, when exposed to 22 days of sunshine, the two alkaloid components of the organic insecticide sabadilla, veratridine and cevidine, slowly broke down with 50% and 10% of degradation, respectively (Williams and Hammitt 2001).

11.8.3 Crop Rotation to Control Insect-Pests

The rotation of crops is also helpful to eliminate some of the insect-pests. In reality, rotations are not helpful to control migratory insects. Nevertheless, with correct rotation, less moveable insect-pests, like the Colorado potato beetle, underground pests, or pests with a single generation each year, can be significantly reduced. Further, potatoes must be kept 01 km away from prior potato crops in order to reduce insecticide sprays by 50% for Colorado potato beetle (Weisz et al. 1994). So, for the best results, rotations across susceptible species should last between 3 and 7 years.

11.8.4 Trap Cropping to Control Insect-Pests

Trap crops attract insect-pests into a specific site where they can be eliminated. Trap crops can be planted around or alongside the crop field, depending on the target pest. As a diamondback moth trap crop, it has proven beneficial to plant matched rows of mustard per 25 rows of cabbage. Mustard rows are planted twice 15 days before and 25 days after cabbage planting. Moreover, Indian mustard can be effectively used as diamondback moth trap crop as well as to attract leaf webber moths (Srinivasan and Veeresh 1986).

11.9 Disease Management

In organic agriculture, disease management mostly depends upon biological diversity and soil health management by following well-adjusted crop rotations that involve planting nitrogen-fixing legumes, cover crops and intercrops, adding manures and composts, and reducing soil tillage practices (Rawat et al. 2021). Most of the soil-borne diseases are normally repressed, although foliar diseases could be harmful at times. Only those pesticides permitted for organic farming are used when a significant disease outbreak is imminent. To overcome persistent disease problems, particularly in organic farming in tropical regions, a systematic strategy to manage diseases is essential, as is integrated approach (Bniggen and Termorskuizen 2003).

Changes in cultural practices, eradication of inoculum source, clean cultivation, application of organic amendments, development of organic pesticides, encouragement of biocontrol agents, planting cover and trap crops, using heat treatment, cold temperature, solar energy, and other methods can all be used to keep disease severity below the economic damage level in organic cultivation (Bruggen et al. 2016). Mechanical cultivation, proper cultural techniques, and other methods are used to manage insects, weeds, and other pests (Bachinger and Zander 2007).

11.9.1 Use of Tolerant Cultivars

The vegetables are prone to several diseases. To control these diseases, expensive chemicals are used which cause environmental pollution and health problems in humans due to their harmful residues. Resistant or tolerant cultivars of diverse vegetables have been successfully evolved against various soil-borne and foliar diseases (Espinosa et al. 1991; Maloy and Baudoin 2001).

The cultivation of tolerant cultivars is effective, safe, eco-friendly, easy, and cheap to reduce disease outbreaks. Several diseases, i.e., root rots, downy mildews, and powdery mildews can be controlled effectively by the cultivation of tolerant cultivars of vegetables to attain higher yields without leaving any toxic residues on crops (Abdel-Kader et al. 2013). Hence, viral, bacterial, fungal, and nematode diseases can be mitigated by the use of tolerant cultivars focusing on higher yields, superior quality, and safe consumption.

11.9.2 Crop Rotation and Cultural Practices to Reduce Diseases

Crop rotation for longer periods other than solanaceous crops can be helpful for the control of buckeye rot disease. Crop rotation of tomatoes with crucifers (turnip, cabbage, and broccoli) can reduce the incidence of viral diseases in tomato (Fadeev and Novozhilov 1987). Peppers and tomato are susceptible to buckeye rot disease which spreads by movement of water and rain splashes. The restriction in water movement between infected and healthy plants is effective for control of this disease. Transplantation date of vegetables crops also contributes in disease control by avoiding favorable conditions for pathogens. Green manuring of *Brassica* spp. can be supportive to avoid the disease incidence on plants (Jacobsen 2001). Late blight and buckeye rot can be effectively controlled by using mulches. Many fungal and bacterial diseases can also be managed through the use of mulching. Grass mulching is quite effective to reduce the occurrence of late blight in tomato (Fry 1982).

11.9.3 Disease Management with Biopesticides

Biopesticides are non-toxic and non-target organisms in nature (Ghananand et al. 2011). Therefore, their demand is increasing for sustainable production of organic vegetables. However, performance of biopesticides is lower as compared to chemical pesticides. Different diseases, i.e., brown spot, powdery mildew, and rust have been controlled by the application of bio-pesticide (bioagent: *Bacillus pumilus* QST 2808) in potatoes. The dollar spot and anthracnose diseases were controlled by using biopesticide (bioagent: *Bacillus subtilis* QST 708) in leafy vegetables and bulb crops. The occurrence of leaf blight, scab, rust, gray mold, powdery mildew, and bacterial spot diseases were controlled by the application of *Bacillus subtilis* strain QST 713 in vegetables (Cao et al. 2010).

11.10 Productivity of Vegetables Cultivated in Organic Farming

Since ancient times, organic matter has been applied to boost agricultural yields sustainably. Self-sufficiency in vegetables' production has been achieved due to high-input agricultural practices, but it has also brought some difficulties. Currently, there is a lot of debate about switching from mechanized agriculture to organic agriculture. In today's chemical-intensive farming, the cultivation of high-yielding cultivars with excessive fertilizer application, poor water management techniques, and indiscriminate use of pesticides has led to environmental pollution and land degradation (Berry et al. 2002). The production of most of the crops is declining. Therefore, transition from modern, chemically intensive agriculture to a more sustainable type, such as organic farming, is a dire need of the day. Thus, one of the possible options to sustain agricultural production at the desired level is to employ economical and locally accessible agro-inputs rather than using artificially prepared agrochemicals (Zaccardelli et al. 2018). It is generally considered that the productivity of several crops declines during the initial few years of organic cultivation, but the great market price of organic produce may be able to compensate for the reduction in yield.

It has been observed in various studies that vegetable production significantly increases with the use of well-managed organic farming ways. In contrast to chemical fertilizers alone, the use of chicken manure or mustard oil cake alone has successfully increased the fruit yield of brinjal. Vegetables respond well to organic plant nutrients ultimately benefitting farmers. According to a study, the addition of organic matter to the soil increased tomato yield by 2.5 times as opposed to using an inorganic nutrient source (Das et al. 2019; Fracchiolla et al. 2020). Vermicompost has also been reported to improve and enhance productivity in many vegetable crops like tomato, brinjal, and sweet pepper due to increased microbial activity.

The addition of vermicompost in the growing medium improves the fruit number in tomato. On a commercial level, it was discovered that applying vermicompost at a rate of 5 t/ha significantly improved tomato production in contrast to inorganic fertilizers (Ansari and Kumar 2010). In onion, plant size, leaf number, and root length were all noticeably improved by the addition of vermicompost in the soil. Similarly, the head weight of cabbage (700 g) was significantly enhanced, while earthworm compost usage @ 27 and 29 t/ha also dramatically improved yield (38 t/ha) (Kaur et al. 2015).

11.11 Quality of Vegetables Produced in Organic Farming

The basic step in organic production is to find out a substitute of inorganic fertilizers. Numerous materials of organic sources have been extensively utilized as fertilizers in organic cultivation (Ansari 2008). However, they should be used in modern agriculture with much care as animal waste and urban sewage waste could be enriched with pollutants, i.e., heavy metals, pesticide residues, and pathogens. To solve this issue, farmers have started employing a type of bio-fertilizer that is prepared by anaerobically fermenting organic wastes like fish byproducts, rice bran, oil seed sludge, etc. (Yadav et al. 2013). Such kind of organic fertilizer is odorless and can be kept for a prolonged period. Nevertheless, nutrient availability lasts longer compared to chemical fertilizers, but the problem is inadequate nutrient absorption at the initial growth stages of plants. The researchers are still striving to get a yield at par to that of traditional farming (Kumar et al. 2022). To elucidate the plant-restricting constraints and advantages of this organic fertilizer, two leafy vegetables were grown under greenhouse conditions, and the changing aspects of both the organic fertilizer and plant growth were deliberated (Bergstrand 2022). Organic fertilization increased the crop quality of both vegetables. The nutritional composition of leafy vegetables was studied by taking into consideration levels of ascorbic acid, sugars, and ionic-nitrate. During this assessment, a comparison was made between organically grown and conventionally produced vegetables (Mahmoud et al. 2009). According to the findings, organically fertilized vegetables had lower nitrate concentrations compared to chemically fertilized ones, but their accumulation of sugars and vitamin C was considerably higher. In case of chilies, carbohydrates and protein contents were substantially higher because they were managed to grow on organic waste containing vermicompost (Adebiyi et al. 2019). Organically produced potato tubers had much greater dry matter content (19%) that is considered appropriate for French fries making without altering the structure and flavor of the fries even when values exceed 23% (Haase et al. 2007).

11.12 Conclusion

Organic vegetable production has attained considerable popularity globally over the last few decades due to the increased demand of consumers for safe and nutritious food. Plant researchers have evaluated organic production of vegetable crops mainly based on some major components and factors, i.e., type of plant materials, soil management, crop fertilization, soil decontamination, and effective control of insect pests and diseases. Management of these major elements determines the quality of vegetable crops grown under organic conditions. A lot of work has been conducted on organic vegetable production systems and their effect on the yield and quality of the vegetables. However, it needs to be examined how organic farming practices affect postharvest life and the quality of the vegetables.

References

Abdel-Kader MM, El-Mougy NS, Lashin SM (2013) Biological and chemical resistance inducers approach for controlling foliar diseases of some vegetables under protected cultivation system. J Plant Pathol Microbiol 4:200

Adebiyi JA, Olabisi LS, Richardson R, Liverpool-Tasie LSO, Delate K (2019) Drivers and constraints to the adoption of organic leafy vegetable production in Nigeria: a livelihood approach. Sustainability 12:96

Aguilera E, Lassaletta L, Sanz-Cobena A, Garnier J, Vallejo A (2013) The potential of organic fertilizers and water management to reduce N_2O emissions in Mediterranean climate cropping systems: a review. Agric Ecosyst Environ 164:32–52

Annicchiarico P, Pecetti L (2010) Forage and seed yield response of lucerne cultivars to chemically weeded and non-weeded managements and implications for germplasm choice in organic farming. Eur J Agron 33:74–80

Ansari AA (2008) Effect of vermicompost on the productivity of potato (*Solanum tuberosum*), spinach (*Spinacia oleracea*) and turnip (*Brassica campestris*). World J Agric Sci 4:333–336

Ansari AA, Kumar S (2010) Effect of vermiwash and vermicompost on soil parameters and productivity of okra (*Abelmoschus esculentus*) in Guyana. Curr Adv Agric Sci 2:1–4

Araújo ASF, Melo WJD (2010) Soil microbial biomass in organic farming system. Ciênc Rural 40:2419–2426

Bachinger J, Zander P (2007) ROTOR, a tool for generating and evaluating crop rotations for organic farming systems. Eur J Agron 26:130–143

Baker BP, Benbrook CM, Groth E III, Benbrook KL (2002) Pesticide residues in conventional, integrated pest management (IPM)-grown and organic foods: insights from three US data sets. Food Addit Contam 19:427–446

Barański M, Średnicka-Tober D, Volakakis N, Seal C, Sanderson R, Stewart GB, Benbrook C, Biavati B, Markellou E, Giotis C, Gromadzka-Ostrowska J (2014) Higher antioxidant and lower cadmium concentrations and lower incidence of pesticide residues in organically grown crops: a systematic literature review and meta-analyses. Br J Nutr 112:794–811

Bàrberi P (2002) Weed management in organic agriculture: are we addressing the right issues? Weed Res 42:177–193

Bergstrand KJ (2022) Organic fertilizers in greenhouse production systems – a review. Sci Hortic 295:110855

Berry PM, Sylvester-Bradley R, Philipps L, Hatch DJ, Cuttle SP, Rayns FW, Gosling P (2002) Is the productivity of organic farms restricted by the supply of available nitrogen? Soil Use Manag 18:248–255

Bniggen VAH, Termorskuizen AJ (2003) Integrated approaches to root disease management in organic farming systems. Australas Plant Pathol 32:141–156

Boccaletti S, Nardella M (2000) Consumer willingness to pay for pesticide-free fresh fruit and vegetables in Italy. Int Food Agribus Manag Rev 3:297–310

Bruggen VAH, Gamliel A, Finckh MR (2016) Plant disease management in organic farming systems. Pest Manag Sci 72:30–44

Cao C, Park S, Gardener BBM (2010) Biopesticide controls of plant diseases: resources and products for organic farmers in Ohio. Ohio State University Extension, Columbus. https://ohioline.osu.edu/factsheet/SAG18

Carvalho DFD, Gomes DP, Oliveira Neto DHD, Guerra JG, Rouws JR, Oliveira FLD (2018) Carrot yield and water-use efficiency under different mulching, organic fertilization and irrigation levels. Rev Bras Eng Agric Ambient 22:445–450

Chen MF (2007) Consumer attitudes and purchase intentions in relation to organic foods in Taiwan: moderating effects of food-related personality traits. Food Qual Prefer 18:1008–1021

Conway G (2019) The doubly green revolution: food for all in the twenty-first century. Cornell University Press, Ithaca, pp 115–121

Craddock HA, Huang D, Turner PC, Quirós-Alcalá L, Payne-Sturges DC (2019) Trends in neonicotinoid pesticide residues in food and water in the United States, 1999–2015. Environ Health 18:1–16

Das A, Layek J, Ramkrushna GI, Babu S, Dey U, Suting D (2019) Integrated organic farming system: an innovative approach for enhancing productivity and income of farmers in the north eastern hill region of India. Indian J Agric Sci 89:1267–1272

Espinosa J, Despestre T, Camino V (1991) A new resistant sweet pepper variety. Capsicum-Newsletter No. 10, p 49

Fadeev IN, Novozhilov KV (1987) Integrated plant protection. Oxonian Press Pvt. Ltd, New Delhi, p 333

Fracchiolla M, Renna M, D'Imperio M, Lasorella C, Santamaria P, Cazzato E (2020) Living mulch and organic fertilization to improve weed management, yield and quality of broccoli raab in organic farming. Plan Theory 9:177

Fry WE (1982) Principles of plant disease management. Academic Press, New York

Gadermaier F, Berner A, Fließbach A, Friedel JK, Mäder P (2012) Impact of reduced tillage on soil organic carbon and nutrient budgets under organic farming. Renew Agric Food Syst 27:68–80

Ghananand T, Prasad CS, Lok N (2011) Effect of insecticides, biopesticides and botanicals on the population of natural enemies in brinjal ecosystem. Vegetos 24:40–44

Gracia A, De-Magistris T (2008) The demand for organic foods in the south of Italy: a discrete choice model. Food Policy 33:386–396

Green VS, Cavigelli MA, Dao TH, Flanagan DC (2005) Soil physical properties and aggregate-associated C, N, and P distributions in organic and conventional cropping systems. Soil Sci 170:822–831

Gruda NS (2019) Increasing sustainability of growing media constituents and stand-alone substrates in soilless culture systems. Agronomy 9:298

Gunapala N, Scow KM (1998) Dynamics of soil microbial biomass and activity in conventional and organic farming systems. Soil Biol Biochem 30:805–816

Gurr GM, You M (2016) Conservation biological control of pests in the molecular era: new opportunities to address old constraints. Front Plant Sci 6:1255

Haase T, Schüler C, Haase NU, Heß J (2007) Suitability of organic potatoes for industrial processing: effect of agronomical measures on selected quality parameters at harvest and after storage. Potato Res 50:115–141

Halbert-Howard A, Häfner F, Karlowsky S, Schwarz D, Krause A (2021) Evaluating recycling fertilizers for tomato cultivation in hydroponics, and their impact on greenhouse gas emissions. Environ Sci Pollut Res 28:59284–59303

Henneron L, Bernard L, Hedde M, Pelosi C, Villenave C, Chenu C, Bertrand M, Girardin C, Blanchart E (2015) Fourteen years of evidence for positive effects of conservation agriculture and organic farming on soil life. Agron Sustain Dev 35:169–181

Jacobsen B (2001) Disease management. In: Maloy OC, Murray TD (eds) Encyclopedia of plant pathology. Wiley, New York, pp 351–356

Kaur P, Bhardwaj M, Babbar I (2015) Effect of vermicompost and vermiwash on growth of vegetables. Res J Anim Vet Fish Sci 3:9–12

Kumar V, Eid EM, Al-Bakre DA, Abdallah SM, Širić I, Andabaka Ž, Kumar P, Goala M, Adelodun B, Singh J, Kumari S (2022) Combined use of sewage sludge and plant growth-promoting rhizobia improves germination, biochemical response and yield of ridge gourd (*Luffa acutangula* (L.) Roxb.) under field conditions. Agriculture 12:173

Löschenberger F, Fleck A, Grausgruber H, Hetzendorfer H, Hof G, Lafferty J, Marn M, Neumayer A, Pfaffinger G, Birschitzky J (2008) Breeding for organic agriculture: the example of winter wheat in Austria. Euphytica 163:469–480

Lundquist EJ, Scow KM, Jackson LE, Uesugi SL, Johnson CR (1999) Rapid response of soil microbial communities from conventional, low input, and organic farming systems to a wet/dry cycle. Soil Biol Biochem 31:1661–1675

Mahmoud E, Abd El-Kader N, Robin P, Akkal-Corfini N, Abd El-Rahman L (2009) Effects of different organic and inorganic fertilizers on cucumber yield and some soil properties. World J Agric Sci 5:408–414

Maloy OC, Baudoin A (2001) Disease control principles. In: Maloy OC, Murray TD (eds) Encyclopedia of plant pathology. Wiley, New York, pp 330–332

Mia MJ, Massetani F, Murri G, Neri D (2020) Sustainable alternatives to chemicals for weed control in the orchard – a review. Hortic Sci 47:1–12

Monteiro A, Santos S (2022) Sustainable approach to weed management: the role of precision weed management. Agronomy 12:118

Moreno MM, Villena J, González-Mora S, Moreno C (2019) Response of healthy local tomato (*Solanum lycopersicum* L.) populations to grafting in organic farming. Sci Rep 9:1–10

Naik VR, Kunnal LB, Patil SS, Guledgudda SS (2012) Organic and inorganic cultivation of chilli and its marketing—an economic analysis. Karnataka J Agric Sci 25:203–207

Pannacci E, Lattanzi B, Tei F (2017) Non-chemical weed management strategies in minor crops: a review. Crop Prot 96:44–58

Petersen J, Sørensen P (2008) Loss of nitrogen and carbon during storage of the fibrous fraction of separated pig slurry and influence on nitrogen availability. J Agric Sci 146:403–413

Popa ME, Mitelut AC, Popa EE, Stan A, Popa VI (2019) Organic foods contribution to nutritional quality and value. Trends Food Sci Technol 84:15–18

Rawat L, Bisht TS, Naithani DC (2021) Plant disease management in organic farming system: strategies and challenges. In: Singh KP, Jahagirdar S, Sarma BK (eds) Emerging trends in plant pathology. Springer, Singapore, pp 611–642

Reda T, Thavarajah P, Polomski R, Bridges W, Shipe E, Thavarajah D (2021) Reaching the highest shelf: a review of organic production, nutritional quality, and shelf life of kale (*Brassica oleracea* var. *acephala*). Plants People Planet 3:308–318

Reeve JR, Hoagland LA, Villalba JJ, Carr PM, Atucha A, Cambardella C, Davis DR, Delate K (2016) Organic farming, soil health, and food quality: considering possible links. Adv Agron 137:319–367

Roitner-Schobesberger B, Darnhofer I, Somsook S, Vogl CR (2008) Consumer perceptions of organic foods in Bangkok, Thailand. Food Policy 33:112–121

Saba A, Messina F (2003) Attitudes towards organic foods and risk/benefit perception associated with pesticides. Food Qual Prefer 14:637–645

Santos VB, Araújo AS, Leite LF, Nunes LA, Melo WJ (2012) Soil microbial biomass and organic matter fractions during transition from conventional to organic farming systems. Geoderma 170:227–231

Scavo A, Mauromicale G (2020) Integrated weed management in herbaceous field crops. Agronomy 10:466

Shepherd R, Magnusson M, Sjödén PO (2005) Determinants of consumer behavior related to organic foods. Ambio 34:352–359

Sims B, Corsi S, Gbehounou G, Kienzle J, Taguchi M, Friedrich T (2018) Sustainable weed management for conservation agriculture: options for smallholder farmers. Agriculture 8:118

Sirieix L, Kledal PR, Sulitang T (2011) Organic food consumers' trade-offs between local or imported, conventional or organic products: a qualitative study in Shanghai. Int J Consum Stud 35:670–678

Srinivasan K, Veeresh GK (1986) The development and comparison of visual damage threshold for chemical control of *Plutella xylostella* and *Crocidolomia binotalis* in cabbage in India. Insect Sci Appl 7:547–557

Tsakiridou E, Boutsouki C, Zotos Y, Mattas K (2008) Attitudes and behavior towards organic products: an exploratory study. Int J Retail Distrib Manag 2:158–175

Tu C, Louws FJ, Creamer NG, Mueller JP, Brownie C, Fager K, Bell M, Hu S (2006) Responses of soil microbial biomass and N availability to transition strategies from conventional to organic farming systems. Agric Ecosyst Environ 113:206–215

Ullah MS, Islam MS, Islam MA, Haque T (2008) Effects of organic manures and chemical fertilizers on the yield of brinjal and soil properties. J Bangladesh Agric Univ 6(2):6

Weisz R, Smilowitz Z, Christ B (1994) Distance, rotation, and border crops affect Colorado potato beetle (Coleoptera: Chrysomelidae) colonization and population density and early blight (*Alternaria solani*) severity in rotated potato fields. J Econ Entomol 87:723–729

Wilkins JL, Hillers VN (1994) Influences of pesticide residue and environmental concerns on organic food preference among food cooperative members and non-members in Washington State. J Nutr Educ 26:26–33

Williams PRD, Hammitt JK (2001) Perceived risks of conventional and organic produce: pesticides, pathogens, and natural toxins. Risk Anal 21:319–330

Yadav SK, Babu S, Yadav MK, Singh K, Yadav GS, Pal S (2013) A review of organic farming for sustainable agriculture in Northern India. Int J Agron 2013:718145

Yiridoe EK, Bonti-Ankomah S, Martin RC (2005) Comparison of consumer perceptions and preference toward organic versus conventionally produced foods: a review and update of the literature. Renew Agric Food Syst 20:193–205

Zaccardelli M, Pane C, Villecco D, Palese AM, Celano G (2018) Compost tea spraying increases yield performance of pepper (*Capsicum annuum* L.) grown in greenhouse under organic farming system. Ital J Agron 13:229–234

Zubair M, Wang S, Zhang P, Ye J, Liang J, Nabi M, Zhou Z, Tao X, Chen N, Sun K, Xiao J (2020) Biological nutrient removal and recovery from solid and liquid livestock manure: recent advance and perspective. Bioresour Technol 301:122823

Hybrid Seed Production in Vegetable Crops

12

Muhammad Abu Bakar Saddique,
Muhammad Ahmad Tariq, Tanveer Ahmad, Sadia Hakeem,
and Cai Linlin

Abstract

The vegetables demand is increasing at a greater pace because of the rising world population and changing food preferences. This ever-increasing demand can be fulfilled by increasing crop yield while land and other agricultural resources are already decreasing. In this scenario, heterosis breeding can help to increase crop yield from the same piece of land. The vegetables' gene banks keep potential genotypes which are sources of male sterility, self-incompatibility, and gynoecism. These mechanisms are critical for the production of hybrid seed on a profitable scale which is otherwise not possible by the conventional methods of hand emasculation and pollination. Using these mechanisms, different systems of hybrid seed production (three-line system, two-line system, and gynoecium line system) have been devised for vegetables. These systems are being used to produce hybrid seeds in tomato, onion, garlic, hot and sweet peppers, cucurbits, and cruciferous vegetables. The cultivation of these hybrid crops has boosted the

M. A. B. Saddique (✉) · M. A. Tariq
Institute of Plant Breeding and Biotechnology, MNS University of Agriculture, Multan, Pakistan
e-mail: abubakar.saddique@mnsuam.edu.pk

T. Ahmad
Department of Horticulture, MNS University of Agriculture, Multan, Pakistan

S. Hakeem
Institute of Plant Breeding and Biotechnology, MNS University of Agriculture, Multan, Pakistan

Vegetable Research Substation, Multan, Ayub Agricultural Research Institute, Faisalabad, Pakistan

C. Linlin
School of Life Science and Engineering, Southwest University of Science and Technology, Mianyang, China

Z. Iqbal et al. (eds.), *Sustainable and Innovative Vegetable Production in times of Climate Change*, https://doi.org/10.1007/978-981-96-6283-8_12

vegetable production manifolds. This chapter describes the mechanisms and systems of hybrid seed production in vegetable crops. It also explains the standards of seed production and seed certification system of Pakistan.

Keywords

Hybrid seed · Heterosis · Vegetables · Male sterility · Self-incompatibility · Gynoecious lines · Standards of hybrid seed production

12.1 Introduction

Heterosis, also known as hybrid vigor, brings forth superior F_1 plants having high yield, uniform crop stand, better quality, and fitness, when parents with different genetic makeup are crossed. In 1914, G.H. Shull, a botanist and geneticist, coined the term "Heterosis" during his work on maize at Cold Harbor in New York (Shull 1952; Kruse 1964). Since then, it has greatly benefited agriculture by bringing food security and economic gains by increasing yield and providing high-quality and nutritious food (Rai and Rai 2006). In current climate change and food security scenarios, it has immense potential to produce high-yielding and climate-resilient vegetable crops (Fujimoto et al. 2018). Since the first hybrid development (Sweet William × carnation) by Thomas Fairchild in 1717, heterosis has been exploited in almost all crops (Leapman 2012).

Various methods are used for the production of hybrid seed in vegetable crops including hand emasculation and pollination (e.g., in tomato, brinjal, pepper, and okra), staminate floral removal and hand pollination (in bitter gourd and other cucurbits), male sterility and hand pollination (in tomato and pepper), male sterility and natural pollination (in carrot, muskmelon, cauliflower, onion, cabbage, hot pepper, and radish), self-incompatibility and natural pollination (in cole vegetables), gynoecium and natural pollination (in cucumber and muskmelon), removal of staminate flowers and natural pollination (in cucurbits), and application of gametocides (Kumar and Singh 2004; Colombo and Galmarini 2017; Kumar et al. 2023). By using these mechanisms in the form of three lines, two lines, and gynoecious lines, a big list of hybrids has been developed in vegetables.

The accomplishment of hybrid seed production is associated with the understanding of flower structures, mode of pollination, and agronomic factors including weather, irrigation, fertilizer application, planting, and harvesting methods (Chakrabarty et al. 2023). Moreover, it also depends on the appropriate selection of parents that transfer their genes to the resultant F_1 progeny. The more diverse parents provide a broader genetic base and bring high heterosis for yield, quality, stress tolerance, root architecture, and early maturity (Hauck et al. 2015; Hochholdinger and Baldauf 2018; Farinati et al. 2023). The development of inbred lines with strong general and specific combining abilities is facilitated by genetic diversity among parents, and these lines are subsequently used to evolve promising hybrids (Kumar et al. 2023). Furthermore, there are opportunities to take advantage of heterosis and reduce input cost for producing hybrid seeds because of the genotypic potential of high seed production per cross and the prevalence of genetic mechanisms like male sterility (in tomato, onion, capsicums, cole crops, etc.), gynoecism (in cucurbits, mainly in bitter gourd and cucumber), and self-incompatibility (in cole crops) (Ram 1975; Chakrabarty et al. 2023).

This chapter highlights the value of heterosis breeding, significance of flower structure, and mechanisms (self-incompatibility, male sterility, gynoecism) involved in the production of hybrid seed. It also explains in detail the systems of hybrid seed production (conventional hybridization, three lines, and two lines systems) and standards for hybrid seed production (proper spacing, isolation distance, and elimination of off-type plants) in vegetables.

12.2 Heterosis in Vegetable Crops

Heterosis refers to the predominance in the characteristics of F_1 hybrids over two parental inbred lines. In the twentieth century, Hayes and Jones were the first to report hybrid vigor in cucumber (*Cucumis sativus*) (Hayes and Jones 1916). Moreover, the development of the first commercial hybrid of eggplant in Japan in 1925 marked an important milestone in hybridization efforts (Kakizaki 1931; Rai and Rai 2006). Due to their high production costs, marketable adoption of hybrids was initially inadequate. However, the growing accessibility of published research on heterosis in diverse vegetable crops has encouraged breeders to create new hybrids and investigate genetic mechanisms that can facilitate more effectual and financially feasible hybrid seed production systems. The understanding of floral structures and reproduction mechanisms in vegetable crops can impart a vital role in the production of hybrid seed at low cost.

12.3 Floral Characteristics of Vegetable Crops

It is imperative to understand the floral characteristics, anthesis timing, reproductive features, and mode of pollination to run an effective hybrid breeding program (Rai and Rai 2006; Kumar et al. 2014; Colombo and Galmarini 2017). Mostly in vegetables, flowers open in the morning, although exceptions exist among cucurbits (Jeffrey 1980). Dehiscence differs from transverse to longitudinal direction of anther, which plays a significant role in the determination of the mode of reproduction. Therefore, pollen fertility knowledge is necessary for appropriate seed setting (Pacini and Dolferus 2019). Moreover, in cases of unsynchronized flowering, information about pollen storage methods and their use in pollination process is valuable. This can significantly enhance the overall quality and efficiency of seed production (Hoekstra 1995).

12.3.1 Genes Controlling Floral Characteristics in Vegetable Crops

The transformation from a meristem to a reproductive organ is regulated by several genes. The quartet model identifies *AP1*, *AP3*, *PI*, *AG*, and *SEP* genes that are responsible for the development of floral organs including petal, sepal, stamen, and carpel (Fig. 12.1) (Smaczniak et al. 2012). Similarly, the ABCDE model also explains in detail the genes responsible for flower development in many plants (Coen and Meyerowitz 1991; Theissen 2001). This model summarizes that 5 homeotic genes (*A*, *B*, *C*, *D*, and *E*) regulate the development of flower organs (Fig. 12.2) (Rijpkema et al. 2010; Murai 2013). In line with the ABCDE model, six genes (*BraAP3*, *BraSHP*, *BraPI*, *BraAG*, *BraAP2*, and *BraSEP*) in Chinese cabbage, each representing a different class of ABCDE model, are actively involved in flower development (Meng et al. 2017). Moreover, the *PAP3* gene (gene of B-class) in pepper converts a protein into 226 amino acids which is quite similar to the MADS-box protein family. Viral-induced gene silencing (VIGS) knockdown of the *PAP3* gene causes pollen grains to shrink, resulting in male sterility, with no observed effect on the petal stage (Ning et al. 2016). In radish (*Raphanus sativus*), the *RsMADS* gene governs floral development (Li et al. 2016). Studies on expression profiles, evolutionary relationships, and dominant pathways related to flowering genes have also been conducted aiming to improve flowering and bolting in Brassicaceae (Nie et al. 2016), particularly in radish (Wang et al. 2017).

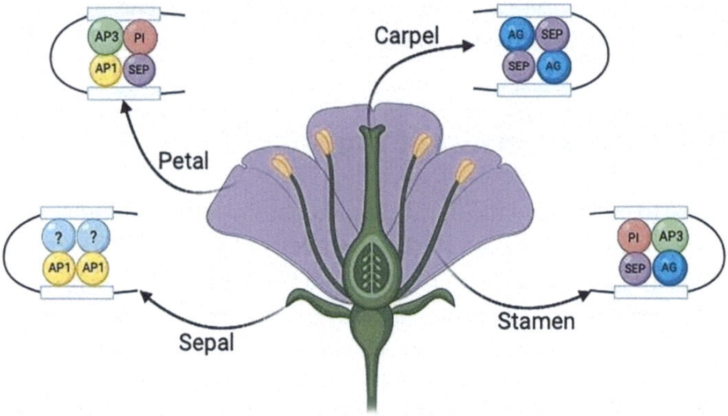

Fig. 12.1 Quartet model for floral organ specification. Floral homeotic genes (*AP1*, *AP3*, *PI*, *AG*, and *SEP*) provide overlapping floral homeotic functions

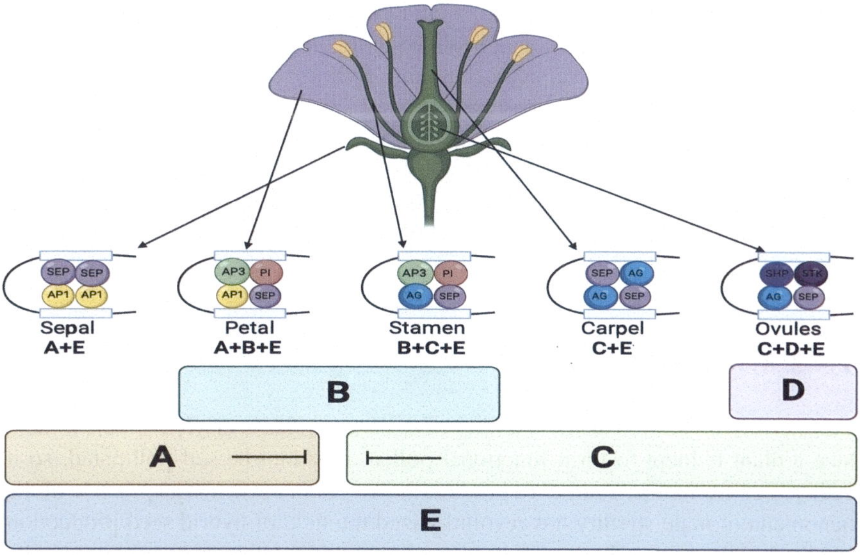

Fig. 12.2 ABCDE model for floral organ specification. Five sets of floral homeotic genes (*A*, *B*, *C*, *D*, and *E*) provide overlapping floral homeotic functions

12.4 Mechanisms for the Production of Hybrid Seed

Controlled pollination is a critical factor for hybrid seed production. It is being performed by the conventional method of hand emasculation and pollination or by exploiting genetic mechanisms, e.g., male sterility, self-incompatibility, and gynoecism (Pradeepkumar and Lekshmanan 2024). The use of genetic mechanisms is effective in terms of cost and seed purity and helps in reducing reliance on labor-intensive emasculation and manual pollination methods (Kumar and Singh 2004). Nowadays, these genetic mechanisms are being used in many vegetables and other crops to produce hybrid seeds.

12.4.1 Hand Emasculation and Pollination

Hybrid seeds of vegetable crops with hermaphrodite (brinjal, tomato, chili, sweet pepper, okra, pea, etc.) and andromonoecious flowers (muskmelon) are commonly produced through the method of pollination and hand emasculation. Typically, the emasculation is performed in the evening. For this purpose, unopened flower buds are selected that are expected to bloom the next morning. After emasculation, the pollination is performed in next morning from 08:00 am to 11:00 am. Pollen is collected by tapping the anthers, and pollination is achieved by immersing the protruding stigma into the pollen mass. According to Nishi (1967), the procedure involves five steps: emasculation, bagging of the emasculated flower, removal of the bag to pollinate the flower, pollination, and again bagging the pollinated flower.

In andromonoecious flowers (muskmelon), hybrid seed production employs a similar method (Pradeepkumar and Lekshmanan 2024). Hermaphrodite flowers in the female parent are emasculated 1 day before anthesis and protected from stray pollen by covering them with a bag. Male flowers from the desired male parent are gathered 1 day before anthesis and stored in a moist polythene bag (Shivanna and Tandon 2014). Pollination is performed the next day, followed by bagging and tagging.

12.4.2 Male Sterility

Male sterility is referred as the inability of a plant to release or yield viable pollen. Once a plant is inept to form functional pollen, it cannot be self-pollinated, so it needs pollen from other sources to produce seeds (Kumar 2013). The practice of the phenomena of male sterility has revolutionized the field of hybrid seed production by reducing the cost and time for seed production. However, male sterility can be either natural or induced by applying gametocytes. The naturally occurring male sterility is further subdivided into three classes, i.e., genetic male sterility (GMS), cytoplasmic genetic male sterility (CGMS), and cytoplasmic male sterility (CMS). These mechanisms have become favorites among breeders due to their effectiveness in producing hybrid seeds.

12.4.2.1 Genetic Male Sterility (GMS)

This sterility is caused by one or more genes, producing insufficient viable pollen, deformation of the pollen, and no opening of the anther (Fig. 12.3). In a recessive GMS system, a plant will be sterile if it has two copies of the recessive allele (homozygous) but it can produce pollen if it has at least one dominant allele (heterozygous) (Cheng et al. 2023). In the dominant system, male sterility is induced by just one dominant allele. GMS has been utilized in a limited number of vegetables, i.e., tomatoes, chili, okra, etc. (Singh et al. 2023).

Development of GMS Line To transfer a monogenic recessive gene that controls male sterility, a backcrossing program is required to introduce this trait into the desired genotype. The first step is to cross a recessive male sterile mutant plant with a plant of the same variety. In successive generations, starting from F_2, pollen from fertile segregants is collected to pollinate male sterile plants. Only the seeds from these male sterile plants are harvested for the next generation (Cheng et al. 2023). After four generations, homozygous male fertile (MsMs) plants are systematically excluded from the population, leaving only heterozygous male fertile (Msms) and male sterile (msms) plants. Subsequently, the genetically male sterile (msms) plants are maintained through backcrossing with a heterozygous isogenic line (Msms) to preserve the male sterility trait (Mishra and Kumari 2018).

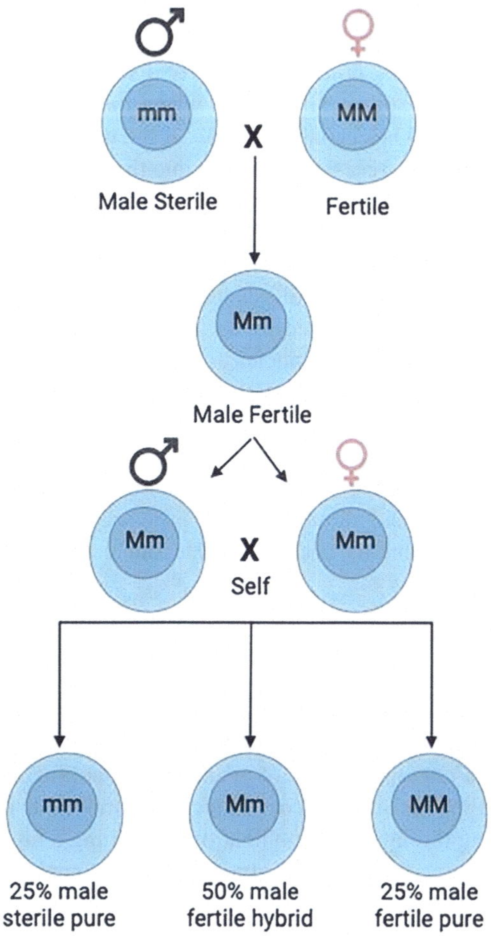

Fig. 12.3 Genetic male sterility. Male sterility is induced in the plant by the cross of recessive male sterile mutant (mm) plant with the fertile plant (MM) and then selfing of F_1 (Mm) is done until heterozygous isogenic line formed (during the selfing, only male sterile plants are selected)

12.4.2.2 Cytoplasmic Male Sterility (CMS)

The interplay of nuclear and cytoplasmic genes causes cytoplasmic male sterility. The sterility is because of the plant's cytoplasm, inherited from the egg cell rather than the pollen. The CMS results from a mutation in the mitochondrial DNA, which causes anthers failing to produce functional pollen (Xu et al. 2022). CMS is valuable in hybrid seed production since it is stable and transmitted through the female parent line, eliminating manual emasculation. CMS systems are critical in the Brassicaceae family, specifically cauliflower and cabbage.

Development of CMS Line

The formation of CMS line (designated as A-line) involves a backcrossing process. This process includes the crossing of a carefully selected maintainer line, known as the B-line, to an existing A-line for 6–7 successive generations. By doing so, A and B lines are established within the new genetic context. The restorer allele (Rf) can be introduced into a male parent through restorer breeding. However, if the homozygous restorer (Rf) gene is already present in the male parent, it can be directly used for the production of hybrid seed of the CMS line (Nahm et al. 2005). This technique is useful in cases where the vegetative part of crops (such as onion and cabbage) has economic value. If the maintainer gene (rf) is not present, then introducing the recessive gene (rf) into the desired line can help in establishing the maintainer line (Wan et al. 2007).

12.4.2.3 Cytoplasmic Genetic Male Sterility (CGMS)

CGMS is a blend of genetic and cytoplasmic male sterility. In this system, male sterility is generated by an interaction between a cytoplasmic component (sterile cytoplasm) and a nuclear component that can either restore fertility (fertile) or maintain sterility (sterile) (Guin et al. 2019). The CGMS facilitates the development of male-sterile lines and has the ability to restore fertility by crossing with a line carrying the therapeutic gene(s). This characteristic is especially useful in the production of hybrid seed since it eliminates the labor-intensive process of hand emasculation and allows for large-scale production. The CGMS systems are used in hot pepper, onions, and carrots (Hussain et al. 2018).

12.4.3 Self-incompatibility (SI)

SI is a crucial phenomenon in flowering plants, promoting genetic diversity and preventing self-fertilization. This physiological barrier hinders self-fertilization, even in the presence of abundant self-pollination. The SI was proposed by Pearson (1933) as a method of producing hybrid seeds in Brassica. The SI ensures seed setting through cross-pollination, sustaining a high degree of heterozygosity in a species by discouraging inbreeding. Two complementary and oppositional hypotheses (heteromorphic self-incompatibility and homomorphic self-incompatibility) have been advocated to describe the mechanisms behind self-incompatibility, revolving around the absence of essential substances or the production of inhibitors (Adamson 1965).

Pollen cytology, floral morphology, and genes play a role in self-incompatibility (Liu et al. 2023). The pollen behavior is governed by the pollen grain genotype, which can be S1, S2, S3, and so on. If a plant with the genotype S1S2 (female) is crossed with a plant with the genotype S3S4 (male), the cross will be compatible if the species displays sporophytic self-incompatibility (SSI). *Brassica oleracea* exhibits a sporophytic SI system. Sporophytic SI, reported in *Crepis foetida* and *Parthenium argentatum*, is determined by the sporophyte's genotype. However, if the species displays gametophytic self-incompatibility (GSI), the cross will be incompatible (Kao and Huang 1994) (Fig. 12.4). Gametophytic incompatibility, first described in *Nicotiana sanderae*, involves the pollen's genotype determining the incompatibility reaction.

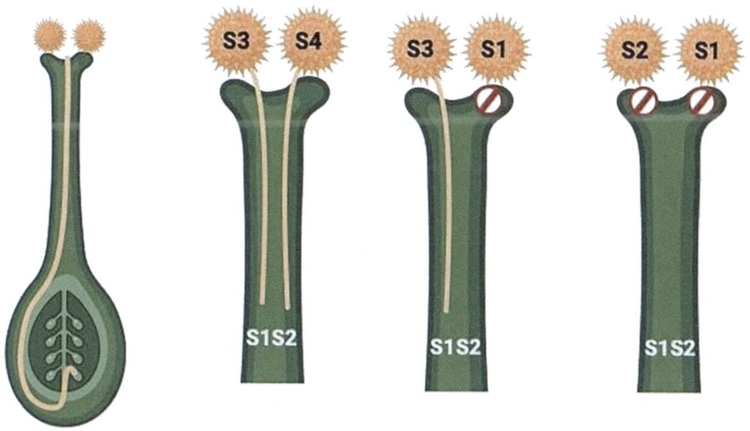

Fig. 12.4 Self-incompatibility. Pollen tube only initiates when the gene, that pollen carrying, is compatible with the one present in the female flower

12.4.3.1 Heteromorphic Self-incompatibility

In the heteromorphic system, morphological differences among flowers are controlled by a single gene, with the short style dominant over the long style. The SI is governed by a single locus S with multiple S-alleles (Gusakova and Gorodilov 1996). This SI system has been widely studied in the Solanaceae and Brassicaceae families, involving proteins at the S locus: SRK and SP11/SCR in Brassica.

12.4.3.2 Homomorphic Self-incompatibility

In the homomorphic system, there are no morphological dissimilarities among flowers. The incompatibility reaction is regulated by the diploid genome of the pollen-producing plant (sporophytic control) or the haploid genome of the pollen (gametophytic control).

The SI signaling, crucial for self/non-self-recognition between pollen and pistil, involves proteolytic events contributing to rejection of self-pollen and growth of compatible pollen. Recent studies highlight the involvement of specific proteins like ARM-repeat-containing1 (ARC1) in Brassicaceae-type SI and triggering a signaling cascade which results in programmed cell death in Papaveraceae-type SI. As SI continues to be researched and recognized, the use of simple, reliable, and affordable inbred line replication is beginning to emerge and will increase in the future (Ahmad et al. 2022).

12.4.4 System of Gynoecism

Gynoecism (femaleness) is primarily observed in vegetable crops, especially in the Cucurbitaceae family. Cucurbitaceae stands out for its diverse sex types, including monoecious, gynoecious, andromonoecious, androgynoecious, and gynomonoecious forms. It reflects the family's propensity for predominantly unisexual flower production as a result of evolutionary processes (Kaloo 1988; Robinson and Decker-Walters 1997).

The gynoecious flowering habit within cucurbits, especially in crops like cucumber and bitter gourd, has been used for hybrid breeding. This involves economically viable practices such as removing male flowers and controlling pollination (Ram et al. 2002; Behera et al. 2009).

12.4.4.1 Genetics of Gynoecium

The hereditary patterns of gynoecism are extensively studied in cucumbers (Kaloo 1988). The interaction of the female sex gene (In-F) and a mutant gene recessive for gynoecious expression controls pistillate characteristics (Robinson et al. 1976). In melon, the relationship between two genes *a* and *g* determines the stability of gynoecism (Robinson and Decker-Walters 1997). A single recessive gene *gy-1* governs the gynoecium in the bitter gourd (Ram et al. 2006) opposite to gynoecium in cucumber, which is controlled by a monogenic dominant gene (Pati et al. 2015).

12.5 Induction of Male and Female Flowers by Chemical Application

The study conducted by Laibach and Kribben (1950) reported the successful use of β-naphthalene acetic acid and β-indoleacetic acid to enhance the percentage of female flowers in cucumbers through chemically controlled induction of staminate and pistillate flowers in monoecious plants. This phenomenon is being extensively used for the development of hybrid seeds in cucurbits such as cucumber, melons, bottle gourd, and bitter gourd that have both male and female flowers (Robinson 1999).

Moreover, ethephon (ethylene-releasing compound) promotes the growth of pistillate and lags induction of staminate flowers. In many monoecious cucurbits, such as pumpkin (Mishra et al. 2015), summer squash (Shannon and Robinson 1979), cucumber (Robinson et al. 1976), and muskmelon (Behera et al. 2006), ethephon has been extensively used in regulating sex expression to produce economically feasible hybrid seeds. It has been demonstrated that ethephon can stop the formation of staminate flowers in muskmelon and watermelon for up to 40 days.

Plant growth regulators are also used to promote male flower development in gynoecious cucumber lines for large-scale hybrid breeding (Robinson 1999). Application of gibberellic acid induces male flowers but its effects are inconsistent. It has been substituted with more efficient silver nitrate (Peterson and Anhder 1960). The silver ions suppress ethylene activity (induction of female flowers in gynoecious lines) (Beyer 1976) and initiate the production of male flowers. It was suggested to use silver thiosulphate instead of silver ions due to their phytotoxic effects (Kohli and Vikram 2005). The staminate flower development can be stimulated by applying silver nitrate at concentrations of 50–100 mg/L or silver thiosulphate at concentrations of 25–50 mg/L to gynoecious plants at 2–3 leaf stage (Prajapati et al. 2015).

Further, it has been proposed that the growth of hermaphrodite flowers in bitter gourd can be induced by applying 6 mM silver nitrate solution (Mishra et al. 2015). Moreover, the performance of gynoecious lines is also influenced by environmental conditions. It has been found that temperatures above 30 °C and photoperiods of up to 12 h are ideal for gynoecious lines; at high temperatures, the photoperiod has no significant impact (Lai et al. 2018).

12.6 Hybrid Seed Production Systems

The above-mentioned genetic mechanisms (male sterility, self-incompatibility, and gynoecism) are used to develop systems of hybrid seed production in various crops. Currently, the systems of three lines, two lines, and gynoecious lines are extensively employed for hybrid seed production on a commercial scale (Rashid and Singh 2000).

12.6.1 Three-Line System

Hybrid seed production has revolutionized agriculture, offering increased yields, improved quality, and enhanced disease resistance in various crops. Among the techniques used for hybrid seed production, the three-line system is a widely used breeding approach in vegetable seed production (Colombo and Galmarini 2017). This system makes the use of three distinct lines: the male sterile line (A-line), maintainer line (B-line), and restorer line (R-line), each playing a critical part in the overall hybrid seed production process (Fig. 12.5). This system ensures high seed purity, bulk seeds production, and hybrid vigor. Moreover, it allows greater control over pollination and seed production. The main demerit of this system is that it requires the development and maintenance of three distinct lines which is a labor-intensive activity. This system is being successfully used in many vegetables including carrots (Morelock et al. 1996), onions (Pelletier et al. 1995), tomatoes (Petrova et al. 1999), peppers (Kalloo et al. 1998), and several other crops.

(i) *Male-sterile line (A-line):* The A-line, characterized by male sterility, is the foundation for hybrid seed production. This line lacks functional pollen, ensuring that no self-pollination occurs. The absence of viable seeds in the male-sterile lines is a crucial attribute for the success of the three-line system (Kumar et al. 2016).

(ii) *Maintainer (B-line):* The maintainer or B-line is responsible for maintaining the male-sterile condition in the A-line. It is genetically similar to the CMS line but possesses functional male fertility genes. The B-line plays a significant part in preventing the natural restoration of fertility in male-sterile plants. This is crucial for the continuous production of male sterile plants.

(iii) *Restorer (R-line):* The R-line is integral to the three-line system as it can restore fertility. The R-line carries specific nuclear genes that restore male fertility when crossed with the CMS line (A-line). Cross-pollination between the male-sterile (A-line) and the restorer (R-line) lines results in the formation of

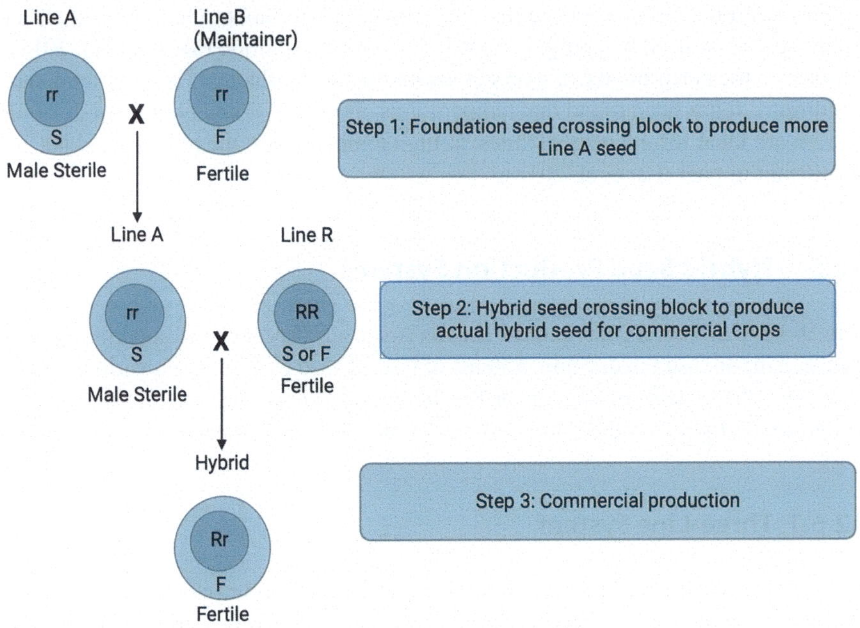

Fig. 12.5 Three-line system. The male sterile line crosses with the maintainer line to produce more male sterile plants so that it can cross with the restorer line to produce the hybrid seed

hybrid seeds. The resulting hybrid progeny exhibits average pollen production and fertility having desirable traits and vigor (Swamy et al. 2017).

12.6.2 Two-Line System

The two-line system is environment-dependent, as male sterility is controlled by environmental factors, e.g., temperature or photoperiod. Due to its environmental dependency, it is also referred as environment-sensitive genetic male sterility (EGMS) (Tomar et al. 2017). In EGMS, the male sterile (ms) gene is expressed within a specific array of temperature or photoperiod conditions. This allows the use of two lines, the male sterile line (A-line) and the restorer line (R-line), instead of three to develop hybrid seed (Fig. 12.6). There is no requirement of the maintainer (B-line) as the A-line becomes fertile under normal temperature or photoperiod. This system is economical as it reduces costs by eliminating the need for backcrossing. It can be used widely in various vegetable crops, including tomatoes, peppers, onions, etc. (Gupta and Singh 2016).

(i) *Thermosensitive genic male sterility (TGMS):* The sterility depends on specific temperature ranges. Plants become male sterile (A-line) once exposed to a particular temperature. This system is being used in pepper (Daskaloff 1972;

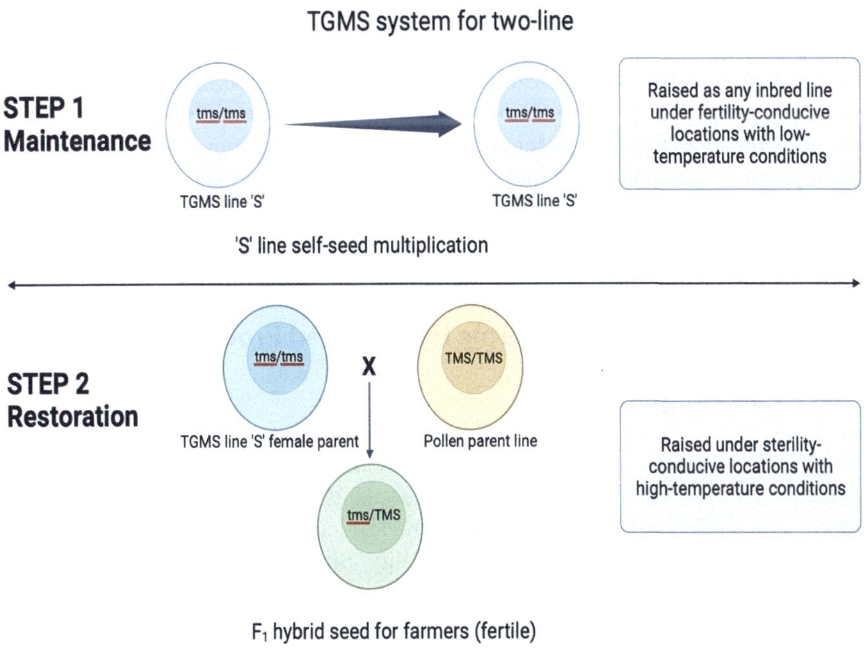

Fig. 12.6 Two-line system. Step 1 involves the maintenance of sterile lines by crossing TGMS at low temperatures. This helps to ensure that sterile lines remain free from unwanted pollination. Step 2 involves the restoration of fertility by crossing TGMS lines with the pollen parent to develop the hybrid. This process results in the desired traits of the hybrid

Shifriss 1997), cabbage (Rundfeldt 1961), carrot (Kaul 1988), Brussels sprout (Nieuwhof 1968), tomato (Sawhney 1983), and broccoli (Dickson 1970).

(ii) *Photoperiod-sensitive genic male sterility (PGMS):* The sterility depends on specific day-length conditions (Chakrabarty et al. 2023). Changing the duration of exposure to light induces male sterility and the plant behaves like an A-line while otherwise in normal conditions it is a fertile plant, e.g., cabbage (Rundfeldt 1961).

(iii) *Restorer line:* It contains genes that do not become sterile under environmental influence. This line is used to make a hybrid under specific conditions when the female line becomes sterile. Hence, it is used to develop hybrids commercially (Virmani et al. 2003).

12.6.3 Gynoecious Line System

This system is important for producing hybrid seeds in vegetables, particularly cucurbits (Pradeepkumar and Lekshmanan 2024) like cucumber (Chaudhary et al. 2001) and watermelon. In this system, the gynoecious plants are developed having predominantly female flowers. This system involves the use of three lines: the gynoecious, monoecious, and pollinizer. The gynoecious line bears female flowers, lacking functional male flowers (Colombo and Galmarini 2017). This allows controlled cross-pollination and ensures hybrid seed production. The pollinizer line comprises plants that produce abundant pollen. These plants play a critical role in providing pollen to fertilize the female flowers of the gynoecious line, enabling fruit development. The monoecious line contains plants with both male and female flowers, predominantly male flowers. The monoecious line is essential for the maintenance of germplasm for breeding programs (Kumar and Kumar 2017).

12.7 Standards for Vegetable Hybrid Seed Production

12.7.1 Land Requirement

For hybrid seed production, the field being used must be free from inoculum of soil-borne diseases and volunteer plants. Volunteer plants are plants of different species, varieties, or seed classes. The area should be leveled having satisfactory nutrients and organic matter with good drainage facility. Seed crop may require different plant population and row spacing from those used in commercial production (Singh et al. 2023). The proper plant spacing is crucial to promote flower growth, encourage pollinators, facilitate mechanical operations, and enable field assessments of the seed crop at distinct developmental stages. Some vegetable crops, such as melons, eggplant, pepper, and tomato, may require the same spacing for seed production as they do for commercial production (Kumar et al. 2016).

12.7.2 Isolation Requirement

The minimal separation required to prevent natural outcrossing between cross-compatible species or varieties is known as isolation distance (Kalloo and Chowdhury 2012). Satisfactory isolation of seed crops is crucial for maintaining purity in the following ways:

(a) By reducing possibility of cross-pollination between species or crops that are compatible with one another.
(b) By keeping different cultivars of the same crop from being mixed during harvest.
(c) By decreasing possibility of diseases and pests spread from alternate host crops.

There are three ways in which varieties can be isolated.

12.7.2.1 Temporal Isolation

Time isolation is used to accomplish various cultivars of the same crop at the similar location. To prevent outcrossing, it is recommended to sow seed crops either 15–20 days earlier or later than adjacent fields (Poehlman 2013). This practice is particularly beneficial for seed production of different maturity groups of cauliflower varieties.

12.7.2.2 Spatial Isolation

Distance isolation is required to keep different plant species separate from each other during pollination. The flowering behavior of the species, the method of pollination (vector and distance traveled) and the direction of insect flight (for insect-pollinated varieties) or the direction of winds (for wind-pollinated varieties) determine this distance (Ramsay 2005). The recommended isolation distances for some common vegetables are as follows: for beetroot, radish, turnip, spinach, cauliflower, and cabbage, the required distances are 1600 m for the foundation seed and 1000 m for certified seed. Similarly, for tomato the distances are 50 and 25 m; for chili, 400 and 200 m; and for onion, 1000 and 500 m for foundation and certified seeds, respectively (Pathak 1999; Trivedi and Gunasekaran 2013; Gabai-Hazera et al. 2018).

12.7.3 Isolation by Physical Barrier

To ensure purity of a seed crop, it is essential to prevent any unwanted foreign pollen from contaminating it. This is accomplished by creating a physical barrier around the seed crop, which can be accomplished by building structures such as buildings or net houses (Darmency et al. 2009). Alternatively, barrier crops that are densely planted around the seed crop can be used to protect against foreign pollen. Examples of such barrier crops include maize, pearl millet, sorghum, and other crops that grow taller than the seed crop (Creswell and Martin 1998). By isolating the seed crop, growers can prevent foreign pollen, producing high-quality vegetable seeds.

12.8 Genetic Purity and Seed Health Standards

Maintaining the genetic and physical purity of seed is essential for quality seed production. This involves the eradication of any spontaneous and odd plants from the male and female/restorer paternal lines throughout all growth stages. To accomplish this, crops are monitored at various phases of growth and those plants which do not meet the official variety description requirements are removed. The frequent task of rouging is essential to organized seed production as it guarantees the genetic integrity of a particular variety.

Rouging is a process of removing off-type plants that can negatively impact the quality of a seed crop (Bishaw et al. 2007). Off-types can appear in a seed crop for several reasons, including morphological variations within a vegetable crop. Vegetables that are self-pollinated usually show more uniformity and stability compared to those that are cross-pollinated. Identifying seeds produced by cross-pollination between a vegetable seed crop and other compatible types or wild plants can be challenging in the first generation but may become apparent later.

To maintain the integrity of hybrid seed, it is necessary to remove any plants that deviate from desired traits in male and female lines. Plants with seed-borne diseases should also be eliminated early to preserve optimal seed health in the parent population (Gupta and Kumar 2020). Variety descriptions based on morphological traits such as leaf form, flower color, fruit shape and color are often used to determine which plants to remove. To ensure seed purity rouging is performed at the following stages:

(a) Pre-flowering: Before flowers bloom, plants with varying characteristics, such as height, color, and foliage morphology, should be removed from fields to facilitate seed production.
(b) Flowering: Rouging is based on sex expression in cucurbit, flower initiation time in solanaceous crops, and curd maturity in cauliflower to prevent the mixing of different varieties of the same vegetable.
(c) At fruit development: The off-types are eliminated on the basis of true-to-type traits of developing fruits, such as fruit size, shape, and ripening color.
(d) At maturity: Late-maturing plants should be thrown away in the case of early maturing kinds (particularly in fruit and vegetable crops) and vice versa.

The health of the seed, which is defined by the absence of diseases mainly seed-borne, is a major determinant of the capacity to produce a healthy crop. The viability and capacity for storage of the seed are also influenced by its state of health. Selective diseases have potential to adversely impact reproductive efficiency, growth habits, and yield of seeds if they are passed down from generation to generation (Newton et al. 2011). The infestation level of anthracnose (*Colletotrichum capsici*) and leaf blight (*Phytophthora capsici*) in chili should not exceed 0.10% and 0.50% for foundation and certified seeds, respectively. Similarly, for cauliflower, the level of Black leg (*Leptosphaeria maculans*) should not exceed from 0.1% to 0.5% in the field for foundation and certified seed, respectively (Copeland et al. 1999; Trivedi and Gunasekaran 2013).

12.9 Seed Certification Standards in Pakistan

In Pakistan, seed certification is performed by the Federal Seed Certification and Registration Department (FSC&RD), which is associated with the Ministry of National Food Security and Research. The process of registration is based on the following steps:

1. To ensure genetic purity, FSC&RD examines the crop in the field grown for seed production. The field evaluation is the critical step performed by FSC&RD, used to evaluate a seed crop's genetic purity (Rana et al. 2022).
2. Following crop harvest, seed samples are taken, and examined in a laboratory to determine their quality using the specified analytical purity standards.

Moreover, pre- and post-control plots are set up to determine genetic purity of the seed and diseases transmitted by it. Only notified varieties are inspected, and their verification is based on physio-morphology traits that have been established through Distinctness, Uniformity and Stability (DUS) testing (Hariprasanna 2019). Before inspecting the crops, the following preliminary steps are confirmed:

(i) Checking source of the obtained seed through labels, seals, bags, or receipts.

(ii) Seed's field acreage and location.
(iii) Information about previous crops grown in the field used for growing the seed crop.
(iv) Isolation between seed crops and other crops.
(v) The crop's state and health to properly conduct crop inspection. The crop needs to be mostly intact and clean.
(vi) A producer or seed agency's request for a crop inspection application.
If each of the requirements listed above is met, a comprehensive examination of the crop is done to detect the following elements:
(a) Blending with different cultivars.
(b) The mixing of different species.
(c) The existence of weeds, especially obnoxious weed plants.
(d) Occurrence of seed-borne diseases.

12.10 Conclusion

The conventional setup of vegetable seed production is unable to fulfill the increasing vegetable demand. The exploitation of heterosis breeding is one of the environment-friendly ways that can help to meet the target of increasing demand for vegetables. It has immense potential to achieve the target of zero hunger. This chapter explains in detail the techniques of hybrid seed production in vegetables. It explains how male-sterility, self-incompatibility, and gynoecious plants are being used for the production of hybrid seed in vegetables. It also explains the standard protocols needed for the production of hybrid seed and system of seed certification in Pakistan. Thus, this chapter serves as a complete guidebook for vegetable seed production from hybrid development to variety approval.

References

Adamson RM (1965) Self and cross-incompatibility in early round-headed cabbage. Can J Plant Sci 45:493–497

Ahmad MH, Rao MJ, Hu J, Xu Q, Liu C, Cao Z, Larkin RM, Deng X, Bosch M, Chai L (2022) Systems and breakdown of self-incompatibility. Crit Rev Plant Sci 41(3):209–239

Behera TK, Dey SS, Sirohi PS (2006) DBGy-201 and DBGy-202: two gynoecious lines in bitter gourd (*Momordica charantia* L.) isolated from indigenous source. Indian J Genet Plant Breed 66(1):61–62

Behera TK, Dey SS, Munshi AD, Gaikwad AB, Pal A, Singh I (2009) Sex inheritance and development of gynoecious hybrids in bitter gourd (*Momordica charantia* L.). Sci Hortic 120:130–133

Beyer EJ (1976) Silver ion: a potent anti-ethylene agent in cucumber and tomato. Hort Sci 11(3):195–196

Bishaw Z, Niane AA, Gan Y (2007) Quality seed production. In: Yadav SS, McNeil D, Stevenson PC (eds) Lentil: an ancient crop for modern times. Springer, Dordrecht, pp 349–383

Chakrabarty SK, Basu S, Schipprach W (2023) Hybrid seed production technology. In: Dadlani M, Yadava DK (eds) Seed science and technology: biology, production, quality. Springer Nature, Singapore, pp 173–212

Chaudhary BN, Piluek K, Taychasinpitak T, Sagwansupyakorn C (2001) Development and maintenance of gynoecious lines of cucumber (*Cucumis sativus* L.). Agric Nat Resour 35(3):242–250
Cheng Z, Song W, Zhang X (2023) Genic male and female sterility in vegetable crops. Hortic Res 10(1):uhac232
Coen ES, Meyerowitz EM (1991) The war of the whorls: genetic interactions controlling flower development. Nature 353:31–37
Colombo N, Galmarini CR (2017) The use of genetic, manual and chemical methods to control pollination in vegetable hybrid seed production: a review. Plant Breed 136(3):287–299
Copeland LO, McDonald MB, Copeland LO, McDonald MB (1999) Seed certification. In: Principles of seed science and technology. Springer, Boston, pp 277–295
Creswell R, Martin FW (1998) Dryland farming: crops and techniques for arid regions, ECHO technical note. ECHO, North Fort Myers
Darmency H, Klein EK, De Garanbé TG, Gouyon PH, Richard-Molard M, Muchembled C (2009) Pollen dispersal in sugar beet production fields. Theor Appl Genet 118:1083–1092
Daskaloff S (1972) Male sterile pepper (*Capsicum annuum* L.) mutants and their utilization in heterosis breeding. Proceedings of the Eucarpia meeting on capsicum, 16–18 September, 1971 Torino, Italy, vol 7, pp 205–210
Dickson MH (1970) A temperature-sensitive male sterile gene in broccoli (*Brassica oleracea* L. var. *Italica*). J Am Soc Hortic Sci 95:13–14
Farinati S, Scariolo F, Palumbo F, Vannozzi A, Barcaccia G, Lucchin M (2023) Heterosis in horticultural crop breeding: combining old theoretical bases with modern genomic views. Front Hortic 2:1250875
Fujimoto R, Uezono K, Ishikura S, Osabe K, Peacock WJ, Dennis ES (2018) Recent research on the mechanism of heterosis is important for crop and vegetable breeding systems. Breed Sci 68(2):145–158
Gabai-Hazera A, Vaissière BE, Blacquière T, Freitas BM, Allsopp M, Chabert S, Dag A (2018) Protocol for using pollinators in hybrid vegetable seed production. ISF Working Group Vegetable Seed Production
Guin K, Sethi SK, Arya RK (2019) Genetic studies on *Triticum timopheevi* based cytoplasmic genetic male sterility (CGMS) system in relation to hybrid seed production in wheat (*T. aestivum* L.). Ekin J Crop Breed Genet 5(2):103–110
Gupta A, Kumar R (2020) Management of seed-borne diseases: an integrated approach. In: Kumar R, Gupta A (eds) Seed-borne diseases of agricultural crops: detection, diagnosis & management. Springer, Singapore, pp 717–745
Gupta A, Singh BK (2016) Development of hybrids and hybrid seed production of onion. In: Singh B, Pandey S, Singh N, Manimurugan C, Devi J, Singh PM (eds) Principles and production techniques of hybrid seeds in vegetables. ICAR-IIVR, Varanasi, pp 101–111
Gusakova GA, Gorodilov NA (1996) Selection of white head cabbage [*Brassica oleracea capitata*] on the basis of self-incompatibility. [Russian] Food and Agriculture Organization, Rome
Hariprasanna K (2019) Distinctness, uniformity, and stability (DUS) testing in sorghum. In: Aruna C, Visarada KBR, Bhat BV, Tonapi VA (eds) Breeding sorghum for diverse end uses. Woodhead Publishing, Duxford, pp 341–365
Hauck AL, Novais J, Grift TE, Bohn MO (2015) Characterization of mature maize (*Zea mays* L.) root system architecture and complexity in a diverse set of Ex-PVP inbreds and hybrids. Springer Plus 4:424
Hayes HK, Jones DF (1916) First generation crosses in cucumbers. Conn Agric Exp Stat Ann Rep 40:319–322
Hochholdinger F, Baldauf JA (2018) Heterosis in plants. Curr Biol 28(18):R1092
Hoekstra FA (1995) Collecting pollen for genetic resources conservation. In: Guarino L, Ramanatha Rao V, Reid R (eds) Collecting plant genetic diversity: technical guidelines. CAB International, Wallingford, pp 527–550
Hussain SM, Hussain K, Farwah S, Rizvi S, Rashid M, Saleem S, Andrabi N, Rashid H (2018) Male sterility in vegetable crops. J Pharmacogn Phytochem 7(3):3390–3393
Jeffrey C (1980) A review of the cucurbitaceae. Bot J Linn Soc 81(3):233–247

Kakizaki Y (1931) Hybrid vigor in egg-plants and its practical utilization. Genetics 16(1):1–25
Kalloo G, Chowdhury JB (eds) (2012) Distant hybridization of crop plants, vol 16. Springer, Berlin/Heidelberg
Kalloo G, Banerjee MK, Kumar S, Parkash C (1998) Hybrid vegetable technology in India – an overview. In: Souvenir, national symposium on emerging scenario in vegetable research and development. Project Directorate of Vegetable Research, Varanasi, pp 42–52
Kaloo G (1988) Vegetable breeding. CRC Press Inc., Boca Raton
Kao TH, Huang S (1994) Gametophytic self-incompatibility: a mechanism for self/non-self discrimination during sexual reproduction. Plant Physiol 105(2):461
Kaul MLH (1988) Male sterility in higher plants, Monographs on theoretical and applied genetics. Springer, Berlin
Kohli UK, Vikram A (2005) Hybrid cucumber. J New Seeds 6:375–380
Kruse R (1964) Genetic basis of heterosis. NACTA 8:6–13
Kumar S (2013) Male sterility in vegetables. In: Singh KP, Bahadur A (eds) Olericulture-1: fundamentals of vegetable production. Kalyani Publishers, New Delhi, pp 431–439
Kumar A, Kumar R (2017) Techniques of hybrid seed production in cucurbitaceous vegetables. Int J Farm Sci 7(4):98–101
Kumar S, Singh PK (2004) Mechanisms for hybrid development in vegetables. J New Seeds 6(4):381–407
Kumar AR, Lakshmanan V, Thondaiman V, Balamohan TN (2014) Botany of vegetable crops: a text book. Narendra Publishing House, New Delhi
Kumar R, Singh PM, Bhardwaj DR, Prasanna HC, Chaubey T, Pandey S, Tiwari SK, Singh BK, Gautam KK, Reddy YS, Mishra GP (2016) Innovative tools for hybrid seed production of vegetables. Indian Hortic (Magazine) 6(1):29–31
Kumar R, Saini M, Taku M, Debbarma P, Sharma D, Pandey R, Gaikwad K, Lal SK, Talukdar A (2023) Genetic studies and mapping of quantitative trait loci (QTL) for yield and yield-related traits in vegetable and seed soybean (*Glycine max* L.)-derived populations. Indian J Genet Plant Breed 83(3):301–309
Lai YS, Zhang W, Zhang X, Shen D, Wang H, Qiu Y, Song J, Li X (2018) Integrative analysis of transcriptomic and methylomic data in photoperiod-dependent regulation of cucumber sex expression. G3 8(12):3981–3991
Laibach F, Kribben FJ (1950) The effect of growth regulators on the flowering of cucumber. Naturwissenschaften 37:114–115
Leapman M (2012) The ingenious Mr Fairchild: the forgotten father of the flower garden. Faber & Faber, London
Li C, Wang Y, Xu L, Nie S, Chen Y, Liang D, Sun X, Karanja BK, Luo X, Liu L (2016) Genome-wide characterization of the MADS-box gene family in radish (*Raphanus sativus* L.) and assessment of its roles in flowering and floral organogenesis. Front Plant Sci 7:211403
Liu S, Guo R, Zhao J, Xia E, Tao Y, Zhou Q, Chen Z, Xie H, Zhu J, Wei C (2023) Cytological and transcriptional insights of late-acting self-incompatibility in tea plants (*Camellia sinensis*). Beverage Plant Res 3(1):19
Meng C, Gu A, Zhao J, Wang Y, Chen X, Shen S (2017) Expression analyses of ABCDE model genes and changes in levels of endogenous hormones in Chinese cabbage exhibiting petal loss. Hortic Plant J 3(4):133–140
Mishra S, Kumari V (2018) A review on male sterility-concepts and utilization in vegetable crops. Int J Curr Microbiol App Sci 7(2):3016–3034
Mishra S, Behera TK, Munshi A (2015) Induction and morphological characterization of hermaphrodite flowers in a gynoecious line of bitter gourd by silver nitrate, gibberellic acid, and silver thiosulfate. Int J Veg Sci 21:204–211
Morelock TE, Simon PW, Peterson CE (1996) Wisconsin wild: another petaloid male-sterile cytoplasm for carrot. Hort Sci 31:887–888
Murai K (2013) Homeotic genes and the ABCDE model for floral organ formation in wheat. Plan Theory 2(3):379–395

Nahm SH, Lee HJ, Lee SW, Joo GY, Harn CH, Yang SG, Min BW (2005) Development of a molecular marker specific to a novel CMS line in radish (*Raphanus sativus* L.). Theor Appl Genet 111(6):1191–1200
Newton AC, Johnson SN, Gregory PJ (2011) Implications of climate change for diseases, crop yields and food security. Euphytica 179:3–18
Nie S, Li C, Xu L, Wang Y, Huang D, Muleke EM, Sun X, Xie Y, Liu L (2016) De novo transcriptome analysis in radish (*Raphanus sativus* L.) and identification of critical genes involved in bolting and flowering. BMC Genomics 17:389
Nieuwhof M (1968) Effect of temperature on the expression of male sterility in Brussels sprouts. Euphytica 17:265–273
Ning M, Chen L, Wencai Y, Huolin S (2016) PAP3 regulates stamen but not petal development in *Capsicum annuum* L. Hortic Plant J 2(2):91–96
Nishi S (1967) F1 seed production in Japan. Proceedings of XVIII international horticulture congress vol 3, pp 231–257
Pacini E, Dolferus R (2019) Pollen developmental arrest: maintaining pollen fertility in a world with a changing climate. Front Plant Sci 10:429534
Pathak CS (1999) Hybrid seed production in onion. J New Seeds 1(3–4):89–108
Pati K, Munshi AD, Behera TK (2015) Inheritance of gynoecism in cucumber (*Cucumis sativus* L.) using genotype GBS-1 as gynoecious parent. Genetika 47(1):349–356
Pearson OH (1933) Breeding plants of the cabbage group. Calif Agric Exp Stati Bull 532:3–22
Pelletier G, Ferault M, Lancelin D, Boulidard L, Dore C, Bonhomme S, Grelon M, Budar F (1995) Engineering of cytoplasmic male sterility in vegetables by protoplast fusion. Acta Hortic 392:11–17
Peterson CE, Anhder LD (1960) Induction of staminate flowers on gynoecious cucumbers with gibberellin A3. Science 131:1673–1674
Petrova M, Yulkova Z, Gorinova N, Izhar S, Firon N, Jacquemin JM, Atanassov A, Stoeva P (1999) Characterization of a cytoplasmic male sterile hybrid between *Lycopersicon peruvianum* Mill. × *Lycopersicon pennellii* Corr. and its crosses with tomato. Theor Appl Genet 98:560–566
Poehlman JM (2013) Breeding field crops. Springer, New York
Pradeepkumar T, Lekshmanan DK (2024) Mechanism of hybrid seed production in cucurbitaceous vegetables. Euphytica 220(2):17
Prajapati S, Jamkar T, Singh OP, Raypuriya N, Mandloi R, Jain PK (2015) Plant growth regulators in vegetable production: an overview. Plant Arch 15(2):619–626
Rai N, Rai M (2006) Heterosis breeding in vegetable crops. New India Publishing Agency, New Delhi
Ram HH (1975) Level of self-incompatibility in cauliflower (*Brassica oleracea* var Botrytis L.). South Indian Hortic 23:64–66
Ram D, Kumar S, Banerjee MK, Singh B, Singh S (2002) Developing bitter gourd (*Momordica charantia* L.) populations with a very high proportion of pistillate flowers. Cucurbit Genet Coop Rep 25:65–66
Ram D, Kumar S, Singh M, Rai M, Kalloo G (2006) Inheritance of gynoecism in bitter gourd (*Momordica charantia* L.). J Hered 97:294–295
Ramsay G (2005) Pollen dispersal vectored by wind or insects. In: Poppy GM, Wilkinson MJ (eds) Gene flow from GM plants. Blackwell, New Delhi, pp 43–77
Rana AW, Gill S, Akram I (2022) Seed system of Pakistan: policy challenges and prospects. International Food Policy Research Institute, Washington, DC
Rashid MA, Singh DP (2000) A manual on vegetable seed production in Bangladesh, AVRDC-USAID-Bangladesh Project. Horticulture Research Centre, Bangladesh Agricultural Research Institute, Gazipur
Rijpkema AS, Vandenbussche M, Koes R, Heijmans K, Gerats T (2010) Variations on a theme: changes in the floral ABCs in angiosperms. Semin Cell Dev Biol 21(1):100–107
Robinson RW (1999) Rationale and methods for producing hybrid cucurbit seed. J New Seeds 1(3–4):1–47
Robinson RW, Decker-Walters DS (1997) Cucurbits. CAB International, Wallingford, pp 103–108

Robinson RW, Munger HM, Whitaker TW, Bohn GW (1976) Genes of the cucurbitaceae. HortSci 11(6):554–568
Rundfeldt H (1961) Untersuchungen zur zuchtung des kopfkohs (*B. Oleracea* var. *capitata*). Z Pflanzenzuchtung/J Plant Breed 44:30–62
Sawhney VK (1983) Temperature control of male sterility in a tomato mutant. J Hered 74:51–54
Shannon S, Robinson RW (1979) The use of ethephon to regulate sex expression of summer squash for hybrid seed production. J Am Soc Hortic Sci 104(5):674–677
Shifriss C (1997) Male sterility in pepper (*Capsicum annuum* L). Euphytica 93:83–88
Shivanna KR, Tandon R (2014) Reproductive ecology of flowering plants: a manual. Springer, New Delhi
Shull GH (1952) Beginnings of the heterosis concept. In: Gowen JW et al (eds) Heterosis: a record of researches directed toward explaining and utilizing the vigor of hybrids. Iowa State College Press, Ames, pp 14–48
Singh H, Sekhon BS, Kumar P, Dhall RK, Devi R, Dhillon TS, Sharma S, Khar A, Yadav RK, Tomar BS, Ntanasi T, Sabatino L, Ntatsi G (2023) Genetic mechanisms for hybrid breeding in vegetable crops. Plan Theory 12(12):2294
Smaczniak C, Immink RG, Angenent GC, Kaufmann K (2012) Developmental and evolutionary diversity of plant MADS-domain factors: insights from recent studies. Development 139(17):3081–3098
Swamy BN, Hedau NK, Chaudhari GV, Kant L, Pattanayak A (2017) CMS system and its stimulation in hybrid seed production of *Capsicum annuum* L. Sci Hortic 222:175–179
Theissen G (2001) Development of floral organ identity: stories from the MADS house. Curr Opin Plant Biol 4(1):75–85
Tomar BS, Jat GS, Singh J (2017) Advances in hybrid seed production of vegetable crops. In: Kanwar HS et al (eds) Advances in quality seed production of vegetable crops. Department of Vegetable Science, Dr. Y.S. Parmar University of Horticulture and Forestry, Solan, pp 19–37
Trivedi RK, Gunasekaran M (2013) Indian minimum seed certification standards. The Central Seed Certification Board, Department of Agriculture and Cooperation, Ministry of Agriculture, Government of India, New Delhi, p 569
Virmani SS, Sun ZX, Mou TM, Ali AJ, Mao CX (2003) Two-line hybrid rice breeding manual. International Rice Research Institute, Manila
Wan C, Li S, Wen L, Kong J, Wang K, Zhu Y (2007) Damage of oxidative stress on mitochondria during microspore development in Honglian CMS line of rice. Plant Cell Rep 26:373–382
Wang J, Qiu Y, Cheng F, Chen X, Zhang X, Wang H, Song J, Duan M, Yang H, Li X (2017) Genome-wide identification, characterization, and evolutionary analysis of flowering genes in radish (*Raphanus sativus* L.). BMC Genomics 18:981
Xu F, Yang X, Zhao N, Hu Z, Mackenzie SA, Zhang M, Yang J (2022) Exploiting sterility and fertility variation in cytoplasmic male sterile vegetable crops. Hortic Res 9:uhab039

Part III

Management and Care

Mineral Nutrition of Vegetable Crops

13

Zahra Batool, Muhammad Baqir Hussain, Niaz Ahmad, and Waseem Hassan

Abstract

Minerals are the ingredients of biomass structure in all life forms on earth. These essential minerals are acquired from air, water, and soil to meet the nutritional requirements of plants. Nutrients including carbon and oxygen (acquired from the air), hydrogen (acquired from water), nitrogen, phosphorus, potassium, sulfur, calcium, magnesium, chlorine, boron, nickel, zinc, molybdenum, copper, manganese, and iron (acquired from the soil) are needed to complete the life cycle of a plant. The world population increases continuously, intensifying the pressure to produce more from the available land resources. Where in the 1960s "The Green Revolution" ensured the food supply for the time's high population. But now the world population has increased twofold from the 1960s. So, we need to revisit the nutrient resources to make the growing population's food secure. Therefore, the present chapter reviews the significance, essentiality, management, and integration of mineral and organic nutrient resources to meet the nutritional security of vegetable crops. The health of the soil determines the production of healthy food which leads to the health of consumers. The supply of nutritious vegetables after maintaining healthy soil conditions through integrated nutrient management would contribute to food security.

Z. Batool · M. B. Hussain · W. Hassan
Department of Soil and Environmental Sciences, MNS University of Agriculture, Multan, Pakistan

N. Ahmad (✉)
Department of Soil Science, Bahauddin Zakariya University, Multan, Pakistan
e-mail: niaz.ahmad@bzu.edu.pk

Z. Iqbal et al. (eds.), *Sustainable and Innovative Vegetable Production in times of Climate Change*, https://doi.org/10.1007/978-981-96-6283-8_13

Keywords

Biofertilizers · Fertilizer requirements · Integrated nutrient management · Mineral nutrients · Nutrient use efficiency · Organic amendments · Vegetable production

13.1 Introduction

13.1.1 Importance of Mineral Nutrition in Vegetable Crops

Vegetable crops are negatively affected by low mineral nutrition in terms of growth, development, and productivity (Herrera et al. 2015). Vegetables, however, require a balanced supply of minerals to perform various physiological functions, including photosynthesis, cell division, and enzyme activation (Vatansever et al. 2016). Obtaining maximum yields and maintaining good crop health depend on mineral nutrition for vegetable crops (Nawaz et al. 2020). A variety of minerals play key role in biosynthesis of chlorophyll, the green pigment that absorbs light energy and converts it into chemical energy (Tränkner et al. 2018). An important function of minerals is to activate and regulate enzymes. For example, phosphorus (P) is a constituent of ATP (adenosine triphosphate) which is the energy currency for plants, while potassium (K) is involved in the activation of numerous enzymes that are important for metabolic processes (Pandey 2018). Lack of these minerals in sufficient amounts will stop the growth process, affect flowering, and reduce the yield of crop plants (Pantoja-Benavides et al. 2021). Minerals are involved in the structural strength of vegetable crops. Calcium (Ca) is of great importance because it is involved in the formation of cell walls. It helps to maintain the cell structure and rigidity. The proper level of Ca prevents blossom end rot in tomatoes and peppers, which can deteriorate crop quality and marketability (Reitz et al. 2021). Micro-nutrients are crucial for the good health of vegetable crops. In addition to macro-nutrients such as nitrogen (N), P, and K, micro-nutrients such as iron (Fe), zinc (Zn), manganese (Mn), copper (Cu), boron (B), and molybdenum (Mo) have lower needs but are nevertheless essential for plant growth and development. They play a role in photosynthesis, nitrogen fixation, and hormone control by acting as cofactors for various enzymes (Dimkpa and Bindraban 2016). Plants require well-balanced mineral nutrition to survive the effects of environmental change and illnesses. K levels, for example, determine how well a plant can withstand droughts by controlling water intake and osmosis (Cabot et al. 2019). Then, micro-nutrient deficits may weaken plant immune systems, making them more susceptible to diseases and pests. Finally, mineral nutrition plays a critical function in vegetable crop production (Aftab and Hakeem 2020). Minerals not only facilitate photosynthesis and enzyme activation, but also play an important role ensuring structural integrity and stress tolerance, making them essential for obtaining maximum yields, high quality, and sustainability in vegetable production systems. Farmers can protect health and productivity of their vegetable crops by appropriately applying mineral nutrient elements, hence promoting food security and sustainability of agriculture (Tripathi et al. 2022).

13.1.2 Role of Essential Minerals in Plant Growth

Essential minerals are vital components for the growth and development of plants, categorized into two main groups: macro-nutrients and micro-nutrients (Sahu et al. 2020). Among macro-nutrients which are required in relatively large amounts are N, P, K, Ca, magnesium (Mg), and sulfur (S) (Monib et al. 2023). N is a principal component of amino acids, proteins, and chlorophyll, which are used for vegetative growth and general plant activity. P is required for the transfer of energy, DNA synthesis, and root development, whereas K controls water uptake, enzyme activation, and strengthening of stress tolerance mechanisms (Ma et al. 2022). Ca functions as structural material and maintains cell wall stability for root growth and fruit quality, while Mg is an essential element for chlorophyll formation and activation of enzymes (Zheng et al. 2015). S is involved in protein formation, biosynthesis of chlorophyll, and whole metabolism in plants (Dawar et al. 2023). Micro-nutrients, such as Fe, Zn, Mn, Cu, B, and Mo, are needed in smaller quantities (Chakrabarty and Chakrabarty 2019). Fe is essential for chlorophyll and enzyme formation and Zn is necessary for enzyme activation and hormone regulation (Kobayashi et al. 2019). Mn is involved in photosynthesis, enzyme activation, and N-metabolism, and Cu takes part in various metabolic processes and lignin formation (Rai et al. 2021). B helps in cell wall formation, pollen germination, and carbohydrate transportation, while Mo is important for N-fixation and enzymatic processes. The lack of these important minerals could lead to chlorosis, stunted growth, poor fruit set, and low yields. Hence, maintaining a ratio of necessary minerals is the most vital task for optimum plant growth, productivity, and produce quality (Long and Peng 2023). Soil management practices, such as applications of fertilizers and soil amendments, help to provide mineral nutrition for plants which in turn supports sustainable crop yield and food security (Chen et al. 2021). The functions and deficiency symptoms of these mineral nutrients in vegetable crops are given in Table 13.1.

13.1.3 Role of Balanced Nutrition in Crop Production

In today's agricultural production systems, balanced nutrition is essential for achieving optimal crop output and quality. This term refers to the act of providing vital nutrients to plants at the appropriate times during their growth cycle (Nawaz et al. 2020). Balanced nutrition provides plants with an abundance of macro-nutrients like N, P, K, Ca, Mg, and S, as well as micro-nutrients like Fe, Zn, Mn, Cu, B, and Mo. The importance of balanced nutrition in improving crop yield and quality, and maintaining soil fertility and environmental sustainability is explained below.

13.1.3.1 Optimal Growth and Development

Appropriate feeding and nourishment containing all the macro- and micro-nutrients ensure the proper growth and development of the crops. Macro-nutrients in addition to micro-nutrients are especially important for attaining energy transfer and structural development, while micro-nutrients are also essential for ensuring the

Table 13.1 Mineral nutrients, functions, and deficiency symptoms in vegetable crops

Mineral nutrient	Function	Deficiency symptom
Nitrogen (N)	Vital component of amino acids, proteins, nucleic acids, and chlorophyll (Ravier et al. 2017)	Stunted growth, yellowing of leaves (chlorosis), and reduced yields (Tian et al. 2020)
Phosphorus (P)	Critical role in photosynthesis, root development, and flower and fruit formation (Malhotra et al. 2018)	Poor root growth, delayed maturity, and decreased fruit or seed production (Muindi 2019)
Potassium (K)	Regulates water uptake, osmotic balance, and enzyme activation in plants (Hasanuzzaman et al. 2018)	Weak stems, leaf scorching, and reduced resistance to stress (Xu et al. 2021)
Calcium (Ca)	Structural component of cell walls, contributing to their strength and integrity. It is involved in cell division, membrane permeability, and nutrient uptake (Thor 2019)	Disorders like blossom end rot in tomatoes and peppers, also leading to distorted growth and weakened cell walls (Hocking et al. 2016)
Magnesium (Mg)	Central component of chlorophyll, essential for photosynthesis and carbohydrate metabolism (Tanoi and Kobayashi 2015)	Chlorosis, particularly in older leaves, and impaired growth (Farhat et al. 2016)
Sulfur (S)	Essential for nitrogen fixation and chlorophyll formation (Li et al. 2020)	Yellowing of leaves similar to nitrogen deficiency and reduced plant growth and vigor (Becana et al. 2018)
Iron (Fe)	Essential for chlorophyll synthesis, electron transfer reactions, and enzyme activation, also involved in photosynthesis, respiration, and nitrogen fixation (Harish et al. 2023)	Interveinal chlorosis, reduced growth, and impaired nutrient uptake (Connorton and Balk 2019)
Manganese (Mn)	Essential for photosynthesis, enzyme activation, and defense against oxidative stress (Martins et al. 2020)	Interveinal chlorosis and affects overall plant growth (Li and Yang 2018)
Zinc (Zn)	Crucial for root development and stress tolerance (Hassan et al. 2020)	Stunted growth, distorted leaves, and reduced yields (Helfenstein et al. 2015)
Copper (Cu)	A cofactor for enzymes involved in lignin synthesis, photosynthesis, and respiration (Rai et al. 2021)	Wilting, reduced growth, and abnormalities in leaf development (Shabbir et al. 2020)
Boron (B)	Essential for cell wall formation, pollen germination, and carbohydrate transport (Long and Peng 2023)	Brittle, distorted leaves, and reduced fruit set (Yang et al. 2021)
Chlorine (Cl)	Aids in photosynthesis and osmotic regulation in plants (Bulman et al. 2019)	Stunted growth, leaf chlorosis, necrosis, water imbalance, and increased susceptibility to stresses (Wang and Grimm 2021)
Molybdenum (Mo)	A component of enzymes involved in nitrogen metabolism and nitrate reduction, crucial for nitrogen fixation in legumes (Banerjee and Nath 2021)	Chlorosis and stunted growth (Banerjee et al. 2021)

activation of enzymes and numerous other metabolic processes (Monib et al. 2023). When all the nutrients are supplied in sufficient amounts, plants can develop themselves genetically which in turn, results in vigorous growth, greater biomass, and high yield.

13.1.3.2 Improved Nutrient Use Efficiency

Nutrients can be absorbed and utilized more efficiently when the supply is balanced, hence avoiding wastage, runoff, leaching, and immobilization (Gong et al. 2020). Thus, carefully planned, optimized utilization of the input fertilizers or soil amendments guarantees farmers savings on cost while improving the benefits to crop yield.

13.1.3.3 Enhanced Stress Tolerance

Plants that are adequately fed are also better able to cope with stressors of the environment like drought, heat, cold, salinity, pests, and diseases. Balanced nutrition reinforces the plant tissues and accordingly, it aids in the development of roots and a wide spectrum of physiological processes that contribute to stress tolerance (Kumari et al. 2022). For example, adequate K levels can ensure water uptake efficiency and turgor pressure, so drought can have a smaller impact on crop yields.

13.1.3.4 Enhanced Crop Quality

Balanced nutrition not only drives crop yield but also plays a significant role in crop quality. Defects such as blossom end rot, tip burn, or fruit cracking, as a consequence of inadequate or imbalanced nutrients, result in poor quality produce, which leads to lower market value (Reitz et al. 2021). With adequate nutrients supply, crops may achieve more desirable characteristics, including color, flavor, texture, nutrient content, and shelf life, leading to more appealing products to consumers and higher prices in the markets.

13.1.3.5 Environmental Sustainability

With the help of a balanced nutrition approach, environmental sustainability can be maintained by minimizing the quantity of lost nutrients to the environment. Inappropriate use of some nutrients may result in water pollution, eutrophication of water bodies, and degradation of soils (Springmann et al. 2018). The use of nutrients by farmers only as much as crops need can be helpful in avoiding environmental problems and maintaining farm productivity to a better level.

13.1.3.6 Resistance to Pests and Diseases

If the nutrients are balanced in plants, this helps them build up natural defenses against pests and diseases. Plants that have certain nutrients lacking are commonly prone to attacks by insects, pathogens, and other stressors. Likewise, calcium represents an essential component for the integrity of cell walls, making plants less prone to pathogen penetration. In contrast, plants display weakened immune responses due to zinc deficiency, exposing them to the attack of disease agents (Tripathi et al. 2022). Through balanced nutrition, growers can build plants' resistance to pests and diseases, which significantly lowers the need for chemical interventions as environment-friendly pest management practices are promoted.

13.1.3.7 Uniformity and Consistency in Crop Production

Optimum nutrition ensures evenness and regularity in crop production. Nutrient deficiencies or excesses can trigger uneven growth patterns, therefore, the produce may have different maturity levels and sizes, both field to field and within individual plants. There should be a sufficient supply of nutrients throughout the growing season for uniform growth and development and consequently, there will be more predictable yields and higher-grade products for consumption (Cabot et al. 2019). This homogeneity has not only become a market requirement but is also a prerequisite for the ease of harvesting, processing, and distribution of crops. This has, therefore, led to a reduction of postharvest losses and enhanced economic returns for the farmers.

13.1.3.8 Long-Term Soil Health and Fertility

A well-balanced nutrients' application improves long-term soil health and fertility. Constant cropping without extracting residual minerals from the soil can deplete the land's fertility, resulting in reduced agricultural yields and productivity over time. Farmers can replenish lost nutrients while improving soil structure and fertility by adding fertilizers or organic amendments based on soil nutrient analysis and crop needs. A balanced nutrition plan also encourages beneficial microbial activity in the soil, which increases nutrient cycling and organic matter decomposition. This in turn improves soil health and resilience.

13.1.3.9 Adaptability to Changing Environmental Conditions

The climate change and severe weather conditions necessitate the stable nutrition for crops to adapt with new habitat circumstances. Nutrient availability and nutrient uptake are linked to different factors such as temperature extremities, drought, flooding, and soil pH (Anderson et al. 2020). If growers maintain optimal nutrient level in soils and plants, plants become tough enough to withstand environmental stresses and thrive in adverse conditions. It is the adaptability that is essential for stability and ability of the system to tackle the challenges of climate change.

A balanced nutrition is a key factor in achieving successful agriculture that provides high yields of superior quality. By giving plants the right levels of nutrient components, the growers will facilitate their strength as well as resistance to drought and pests, enhance crop quality, and contribute to a healthy ecosystem. Implementing good nutrient management principles leads to higher productivity of food systems by ensuring balanced nutrition and ecological sustainability.

Plants require minerals for growth and development because they participate in several key physiological processes (Pandey 2018). However, plants mostly require macro-nutrients with trace levels of micro-nutrients (Vatansever et al. 2016). Understanding the roles and importance of these important minerals is vital for enhancing plant health and productivity (Briat et al. 2015). Finally, crucial minerals are essential for plant growth and development because they play critical roles in a number of physiological processes. Maintaining optimum mineral nutrition is crucial for raising yields, enhancing crop quality, and ensuring overall plant health and vigor. Growers can better regulate mineral nutrition if they understand the various roles as well as requirements of macro- and micro-nutrients for each vegetable crop.

13.2 Factors Influencing Nutrients' Availability in Soil

Factors influencing nutrients' availability in soil are diverse and multifaceted, encompassing both natural processes and human activities. An explanation of some key factors is given in the following sub-sections.

13.2.1 Soil pH

The pH of soil determines whether it is acidic or alkaline, which has a major impact on nutrients' availability. Nutrients' solubility and microbial activity are pH-dependent (Malik et al. 2018). Acidic soil (low pH) has an abundance of aluminum, manganese, and iron but minimal calcium and phosphorus. In contrast, alkaline soils (high pH) may limit the availability of micro-nutrients like iron, zinc, and manganese. Achieving the suitable pH range for different crops is essential for nutrients' absorption (Riaz et al. 2020).

13.2.2 Soil Texture

Soil texture is a proportionate mixture of three particle size soil separates including sand, silt, and clay. Soils with dominantly fine particles (clay) tend to hold more water and nutrients but may suffocate plant roots due to low drainage conditions. However, few crops can be an exception for preference to fine textured soils where need for low drainage is essential. On the contrary, course (sandy) textured soils have wide pore spaces and least water or nutrient holding capacity, tend to starve the plants owing to high drainage. However, few exceptions are there for preference to sandy soils where loose and drainage conditions are required. Therefore, loam (display of equal properties of sand, slit and clay) textured soils are considered suitable for most of the crops for providing better anchorage, nutrient and water supply. Soil texture affects the nutrients supply with its cation exchange capacity property which is higher in clay, intermediate in silt and least in sandy particles. Most of the nutrients in soil solution are in cations form which can be held by the soil particles for maintaining optimal nutrient supply to plants. Meanwhile, the soil microbial activity is likely influenced by the soil texture which contributes to the plant nutrient supply through mineralization and microbial diversity in the rhizosphere.

13.2.3 Soil Organic Matter

Providing nutrients to plants in the soil and maintaining soil fertility depends greatly on soil organic matter content. Mineralization of organic matter by soil microorganisms serves the plants with nutrients (Yan et al. 2015). Additionally, organic matter helps to improve soil structure, water retention, and cation exchange capacity (CEC), which boosts nutrient retention and decreases leaching losses (Kane et al. 2021).

13.2.4 Soil Micro-organisms

Soil microbes are an integral part of the process of recycling and making nutrients available to plants. The development of symbiotic associations among plants and micro-organisms, such as bacteria, fungi, and actinomycetes, contributes to the release of minerals such as N, P, and K (Yadav et al. 2021). Pathogens and harmful microbes can likewise impact nutrient availability indirectly through crop diseases and directly by competing for nutrients with plants. Balancing the soil microbial community with crop rotation, cover crops, and organic amendments can increase nutrient solubility ultimately strengthening plant health (Nazli et al. 2020).

13.2.5 Soil Compaction

Human activities and natural forces like the use of equipment and machinery, and foot traffic can result in soil compaction. This limits the growth of roots and hence uptake of nutrients. Compacted soils pose a major problem of small pore space that does not allow water infiltration, air exchange, and root penetration (Wang et al. 2019). Thus, due to nutrient suppression, plants may stop growing properly and develop nutrient deficiencies. However, soils can be managed through tillage and aeration or cover cropping to reduce compaction and enhance the availability of nutrients (Qiao et al. 2017).

13.2.6 Nutrient Interactions

Complex interactions among nutrients in soil determine their availability for plants which could be synergistic, antagonistic, or neutral (Fageria 2001). Antagonistic interactions that check the uptake or use of one nutrient by another, consequently lead to nutrient imbalances and deficiencies (Kumar et al. 2021). Plants produce low molecular weight chelators to sequester Zn or Fe and keep them bioavailable for themselves while unavailable to other competitors (Sinclair and Krämer 2012). However, additive interactions, where one nutrient makes up for the deficiency of another nutrient, often lead to suboptimal plant growth. Bouain et al. (2014)

observed that increasing Zn concentration in lettuce decreased the P concentration. The awareness of the plant interactions among nutrients is critical in designing balanced fertilizers and in controlling nutrient management operations (Ma et al. 2016).

13.2.7 Soil Parent Material

The properties and features of the parent rock, from which soil originates, greatly affect the availability of nutrients. The content of essential nutrients in different rocks and minerals differs and, therefore, they influence the initial nutrient situation in the soil. For instance, limestone-derived soils have a higher calcium level, while granite-derived soils are rich in potassium and silicon (Sağlıker et al. 2018). Over the time, weathering processes transfer nutrients from parent materials to the soil, thus making the soil fertile. Learning the geological history of soil might give information about its nutrients and accessibility (Sarkar and Naidu 2015).

13.2.8 Soil Drainage and Water Management

Sufficient soil drainage is necessary for proper nutrient uptake and better plant health. Bad water drainage is attributed to many factors, for example, compacted soil layers, high water table, incorrect irrigation practices, etc., which can be the cause of waterlogging and anaerobic conditions. Consequently, roots cannot grow to their full potential (Carstensen et al. 2019). Moreover, plants are not able to absorb nutrients as they should. On the other hand, well-drained soils support oxygen dispersion throughout the soil profile as well as transport of nutrients within the soil profile. Management practices which include contouring, subsurface drainage system, and proper irrigation method and timing ensure adequate water retention, optimal soil moisture level, and nutrient availability (Reichardt and Timm 2019).

13.2.9 Land Use and Management Practices

Soil nutrient dynamics are greatly altered by human activities and land management interventions directed at agricultural soils. Intensive farming techniques like tillage, monoculture crops, and overuse of fertilizers can lead to a decline in soil organic matter, disruption of the soil structure, and irregularity of nutrient cycling (Wills et al. 2017). Some sustainable land management practices include conservation tillage, crop rotation, organic amendments as well as agroforestry which promotes soil health and nutrient cycling consequently increasing the nutrient availability over the long run. Attainment of both agricultural production and environmental guardianship is a critical element in building and sustaining soil fertility and productivity.

13.2.10 Soil Amendments and Fertilization

The addition of soil amendments and fertilizers has a direct effect on the mobility and availability of nutrients in soil. Organic amendments like compost, manure, and biochar contribute nutrients and soil organic matter for better soil structure and nutrient-retaining capacity. Inorganic fertilizers are the source of all the required specific nutrients that can be directly absorbed by plants (Pastuszczak et al. 2022). Although artificial fertilizers are highly productive, excessive dependence on them can result in nutrient imbalances, soil acidification, and environmental pollution. Setting management programs appropriately, based on soil testing and nutrient balance rules, delivers nutrients at optimal levels and minimizes environmental effects (Tripathi et al. 2020).

13.2.11 Microbial Activity and Soil Biota

The micro-organisms are the central participants in nutrient cycling and availability. Effective soil microbes additionally contribute to N-fixation, mineralization of nutrients, and decomposition of organic matter, which eventually releases plant-available nutrients. Mycorrhizal fungi are symbiotically connected to the roots of the plants and they offer many nutrients, especially P (Azcón-Aguilar and Barea 2015). The existing soil fauna, specifically earthworms and arthropods, are the driving force of nutrient cycling only through their movements. To ensure rich and healthier soil microbiota, practices like reduced tillage, cover cropping, and least pesticide use create favorable conditions for nutrient availability and ecosystem stability (Sapkota et al. 2020).

13.2.12 Urbanization and Soil Degradation

Soil degradation and nutrient loss may occur due to urbanization and land development activities. The conversion of native habitats to urban areas loses soil organic matter, breaks down the nutrient cycles, and further promotes soil compaction and pollution. Urban soils commonly present reduced fertility, excessive salinity, and pollutants like heavy metals and wastes which interfere with the nutrient supply and plant growth (van Delden et al. 2016). The use of green infrastructures, including urban horticulture and rain gardens, can limit soil degradation and help to enrich urban soil quality and increase nutrient availability for plant growth (Deeb et al. 2018).

All these factors described above lead to understand a complex soil system and various processes that determine soil nutrient availability. Integrated soil management approaches that address these factors individually and jointly foster sustainable agriculture, improve soil fertility, and guarantee longevity of agricultural ecosystems (Urmi et al. 2022).

13.3 Nutrient Management in Vegetable Crops

Appropriate nutrient management is at the core of successful vegetable growing practice, empowering plants to get the required nutrients that enhance intensive production as well as reduce negative impacts such as nutrient runoff and leaching. Soil test and nutrient analysis are powerful tools for nutrient management. These provide important data about the fertility of the soil and allow for making refined fertilizer recommendations (Congreves and Eerd 2015). Sources of various mineral fertilizers, their nutrient percentages, and uptake forms for plants are given in Table 13.2.

13.3.1 Soil Testing

Field soil sampling is the process of drilling or digging small samples of soil from the field to find out their chemical properties such as pH value, organic matter content, and texture using laboratory analyses (Voltr et al. 2021). The soil test supplies the farmer with knowledge about field's existing nutrient availability so that adequate fertilizers can be supplied based on this information.

Table 13.2 Sources of mineral fertilizers, nutrient percentages, and uptake forms for plants

Mineral fertilizer	Nutrient percentage	Nutrient uptake forms
Urea	N = 46	NO_3 and NH_4
Triple superphosphate	P_2O_5 = 46	H_3PO_4 and HPO_4
Muriate of potash	K_2O = 60	K^+
Diammonium phosphate	N = 18 P_2O_5 = 46	NO_3 and NH_4 H_3PO_4 and HPO_4
Calcium nitrate	N = 15 Ca = 19	NO_3 and NH_4 Ca^{2+}
Ammonium sulfate	N = 21 S = 24	NO_3 and NH_4 SO_4^{-2}
Magnesium sulfate	Mg = 10 S = 13	Mg^{2+} NO_3 and NH_4 SO_4^{-2}
Ammonium nitrate	N = 34	NO_3 and NH_4
Monoammonium phosphate	N = 11 P_2O_5 = 52	NO_3 and NH_4 H_3PO_4 and HPO_4
Potassium nitrate	N = 13 K_2O = 44	NO_3 and NH_4 K^+
Potassium sulfate	K_2O = 50 S = 17	NO_3 and NH_4 NO_3 and NH_4 SO_4^{-2}
Ammonium phosphate	N = 16 P_2O_5 = 20	NO_3 and NH_4 H_3PO_4 and HPO_4
Calcium ammonium nitrate	N = 26 Ca = 08	NO_3 and NH_4 Ca^{2+}

13.3.1.1 Sampling

Samples of soil are taken from the location of the field which is intended to represent different types of soil, slopes, and history in soil management practices. The depth and frequency of the sampling may differ with the types of crops, variety as well as management purposes.

13.3.1.2 Sample Preparation

Once collected, soil samples are air-dried, ground, and sieved to remove the larger particles and rubbish. Proper sample preparation helps to achieve valid and consistent findings.

13.3.1.3 Analysis

The soil samples are subject to laboratory determination of the major soil properties such as nutrient level, pH, and organic matter content. Different analytical approaches, like chemical extraction and spectroscopy, are typically used to get an objective and accurate figure for soil nutrient estimation.

13.3.1.4 Interpretation

The results of soil testing are interpreted using the crop nutrition standards and baselines for soil fertility. Farmers receive precise guidelines on fertilizer rate of application and time to align with nutrient needs.

13.3.2 Nutrient Analysis

Nutrient analysis is a technique for determining the amount of key nutrients in soil samples and whether they are freely available to plants. N, P, K, Ca, Mg, S, and other critical elements are routinely examined in soil. An overview of the nutritional analysis procedure is given below.

13.3.2.1 Nutrient Extraction

For the release of nutrients from the soil matrix and extraction, chemical extraction techniques are applied to soil samples. The methods of extraction separate only the specific nutrients for accuracy and reliability.

13.3.2.2 Measurement

Nutrients in soil samples are assessed using analytical methods such as spectrophotometry via flame, absorption or atomic absorption. Each nutrient component is independently analyzed to determine its concentration in the soil sample.

13.3.2.3 Data Interpretation

The nutrient analysis information is interpreted in the context of the crop nutrient requirements and soil fertility recommendations. Nutrient levels are compared to accepted sufficiency ranges or critical levels to identify deficiencies or excesses that need to be addressed.

13.3.2.4 Fertilizer Recommendations

Based on nutrient analysis results, farmers will be provided with customized fertilizer recommendations to increase nutrient supply and meet crop needs. Recommendations take into account variables such as type of crop, growing stage, yield aim, and environmental factors. Fertilizer recommendation for different vegetable crops in Punjab (Pakistan) are given in Table 13.3, while different methods of fertilizer application in these vegetable crops are presented in Table 13.4.

13.3.3 Benefits of Soil Testing and Nutrient Analysis

Soil testing and nutrient analysis offer several benefits for vegetable crop production, which are given below.

Table 13.3 Nitrogen, phosphorus, and potassium (N:P:K) recommendation for different vegetable crops in Punjab (Pakistan)

Vegetable	N:P:K		References
	kg acre^{-1}	kg ha^{-1}	
Cucumber	36:36:25	90:90:62	Niaz (2012a)
Musk melon	32:23:25	80:57:62	
Watermelon	64:46:25	160:115:62	
Vegetable marrow	60:35:25	150:87:62	
Bottle gourd	58:36:25	145:90:62	
Pumpkin	13:27:25	32:67:62	
Sponge gourd	38:36:25	95:90:62	
Bitter gourd	40:40:25	100:100:62	
Okra	25:35:25	62:87:62	
Brinjal	36:36:25	90:90:62	
Hot pepper	66:46;25	165:115:62	
Bell pepper	66:46:25	165:115:62	
Potato	100:80:50	250:200:125	Baloch (1994)
Tomato	90:45:75	225:112:187	Niaz (2012b)
Peas	35:35:25	87:87:62	
Cabbage	45:35:25	112: 87:62	
Cauliflower	48:32:16	120:80:40	Baloch (1994)
Turnip	20:23:25	50:57:62	Niaz (2012b)
Radish	20:23:25	50:57:62	
Carrot	45:45:25	112:112:62	
Spinach	46:27:0	115:67:0	
Fenugreek	46:23:0	115:57:0	
Lettuce	32:18:0	80:45:0	
Onion	48:34:25	120:85:62	
Garlic	55:25:50	137:62:125	
Sweet potato	24:34:75	60:85:187	Niaz (2012a)

Table 13.4 Methods of fertilizer application for different vegetable crops

Vegetable	Methods of fertilizer application	Nutrient use efficiency	References
Tomatoes	Broadcast application and side-dressing	Side dressing is more effective with approximately 80% nutrient use efficiency	Borisov et al. (2022)
Lettuce	Foliar feeding, top-dressing, and drip irrigation	Foliar feeding has a high nutrient use efficiency of around 80–90%	Khani et al. (2020)
Bell peppers	Band placement, fertigation, and mulch application	Fertigation is the most effective method with nutrient use efficiency of 80–90%	Junior et al. (2017)
Carrots	Pre-planting incorporation, side-dressing, and foliar feeding	Side dressing is typically the most effective method with nutrient use efficiency of around 80–85%	Colombari et al. (2018)
Cucumbers	Hilling, drip irrigation, and top-dressing	Drip irrigation systems have high use efficiency approximately 80–90%	Feng et al. (2019)
Broccoli	Broadcast application, side-dressing, and fertigation	Fertigation is generally more effective with a nutrient use efficiency of around 85–95%	Debbarma and Bhatt (2021)
Spinach	Broadcast application, side-dressing, and foliar feeding	Side-dressing is generally more effective with a nutrient use efficiency of around 80–90%	Mola et al. (2020)
Potatoes	Hilling, side-dressing, and fertigation	Fertigation is generally more effective with a nutrient use efficiency of around 85–95%	Kim and Rho (2023)
Onions	Pre-planting incorporation, side-dressing, and foliar feeding	Side-dressing is generally more effective with a nutrient use efficiency of around 80–90%	Abo-Dahab et al. (2016)
Peas	Pre-planting incorporation, side-dressing, and foliar feeding	Side-dressing is generally more effective with a nutrient use efficiency of around 80–85%	Lobon et al. (2020)
Cauliflower	Broadcast application, side-dressing, and fertigation	Fertigation is generally more effective with a nutrient use efficiency of around 85–95%	Verma et al. (2022)
Beans	Pre-planting incorporation, side-dressing, and foliar feeding	Side-dressing is generally more effective with a nutrient use efficiency of around 80–90%	Garcia et al. (2020)
Cabbage	Broadcast application, side-dressing, and fertigation	Fertigation is generally more effective with a nutrient use efficiency of around 85–95%	Wu et al. (2020)

(continued)

Table 13.4 (continued)

Vegetable	Methods of fertilizer application	Nutrient use efficiency	References
Radishes	Pre-planting incorporation, side-dressing, and foliar feeding	Side-dressing is generally more effective with a nutrient use efficiency of around 80–90%	Shan et al. (2020)
Eggplant	Hilling, side-dressing, and fertigation	Fertigation is generally more effective with a nutrient use efficiency of around 85–95%	Incrocci et al. (2017)
Pumpkins	Broadcast application, side-dressing, and fertigation	Fertigation is generally more effective with a nutrient use efficiency of around 85–95%	Yavuz et al. (2015)
Sweet potatoes	Hilling, side-dressing, and fertigation	Fertigation is generally more effective with a nutrient use efficiency of around 85–95%	Wang et al. (2022)

13.3.3.1 Precision

Soil testing helps in determining the soil nutrient status, which is essential in assisting the growers to apply fertilizers where they are most needed to ensure nutrients are used efficiently and waste is minimized.

13.3.3.2 Cost-Effectiveness

Optimized fertilizers use based on soil tests helps to reduce input costs and still produce good crop yield and quality.

13.3.3.3 Environmental Sustainability

Through a detailed soil testing method, we can avert the over-application of chemical fertilizers, thus minimizing nutrient runoff, leaching, and environmental pollution, which in turn enhances sustainable agricultural practices.

13.3.3.4 Long-Term Soil Health

Soil is tested for nutrient levels regularly and changes in soil fertility are monitored so that proactive soil management strategies are adopted to conserve soil health and productivity for future crop cycles.

Soil testing and nutrient analysis are inseparable components in nutrient management for vegetable crop production as they help growers to understand soil fertility better and guide them in making informed decisions to boost nutrient supply and uptake for sustainable agricultural practices. Likewise, regular soil testing along with appropriate nutrient management practices is an indispensable aspect of the achievement of healthy and productive vegetable crops with negligible effects on the environment.

13.4 Integrated Nutrient Management

Integrated Nutrient Management (INM) involves a mixture of organic amendments, inorganic fertilizers, microbial inoculants, and other nutrient sources to improve nutrient availability and harmonize with plant needs.

13.4.1 Balanced Nutrition

The INM allows crops to take advantage of the balanced supply of the basic nutrient elements (N, P, K, and micro-nutrients) necessary to stimulate their optimal growth, production, and quality.

13.4.2 Soil Health

The application of organic amendments, by way of improving soil health and fertility, augments soil structure, microbial activity, nutrient cycling, and water storage capacity. Healthy soils provide the foundation for healthy plants due to their ability to foster vigorous root growth and a normal nutrient uptake process.

13.4.3 Nutrient Cycling

INM ensures that nutrients are recycled efficiently within agro-ecosystems and their use is improved, hence, losses to the environment through leaching, runoff, or volatilization are minimized. Organic amendments build up long-term soil fertility and boost the soil to depend less on inorganic fertilizers.

13.4.4 Environmental Sustainability

Among others, implementing INM techniques such as organic amendments, will reduce the reliance on synthetic fertilizers, limit pollution of nutrients, and ensure that farms are sustainable. Sustainable nutrient management systems including the practices mentioned promote healthy ecosystems and resilience.

13.5 The Role of Organic Amendments in INM

INM comprises careful and balanced use of multiple fertilizer sources to achieve desired soil fertility, increase crop yields, and assure long-term viability of agricultural activities. Organic amendments originating from plants or animals provide most of the organic matter and vital nutrients during the INM process.

13.5.1 Types of Organic Amendments

13.5.1.1 Compost

Compost is a nutritious soil amendment resulting from the breakdown of organic materials like crop residues, farm manure, yard waste, and food waste. Composting is a process where microorganisms break down organic matter into humus, a stable matter with organic carbon, N, P, K, and several micro-nutrients. Compost improves soil structure, water retention, nutrient cycling, and microbial activity leading to enhanced soil health and fertility.

13.5.1.2 Manure

Animal manure, like cow, poultry, and horse manure, is a valuable source of organic nutrients for increased soil fertility. Manure is comprised of diverse quantities of N, P, K, and micro-nutrients if the animal type, feeding regime, and management techniques are taken into consideration. Employing manure in the soil offers organic matter, mineral nutrients, and beneficial micro-organisms that contribute to soil fertility, microbial activity, and plant nutrition.

13.5.1.3 Green Manure/Cover Crops

Green manure crops are grown for improving soil structure and enhancing soil nutrition using biomass production and nutrient cycling. Apart from leguminous cover crops (clover, vetch, and peas), there is also one more benefit of biological N-fixation. Through that symbiotic association with N-fixing bacteria, green manure crops enrich the soil with N. Non-leguminous cover crops, including rye, oats, and buckwheat, serve as a source of organic matter and also help to suppress the growth of weeds, minimize erosion, and enhance soil tilth. Crop rotation is accomplished by growing green manure crops which leads to healthier soils, higher nutrient availability, and sustainable ecosystem.

13.5.1.4 Biochar

Biocarbon is a carbon-rich soil enhancer generated via pyrolysis of organic wastes in a duration devoid of oxygen. Biochar's extreme stability and resistance to the breakdown by micro-organisms contribute to the long-term holding of carbon in the soil. Besides this, biochar improves soil structure, water retention, nutrient retention, and cation exchange capacity (CEC), which results in the betterment of soil fertility and nutrient availability. Biochar can be added to the soil directly or mingled with organic fertilizers to better utilize its potential.

13.5.2 Benefits of Organic Amendments in INM

13.5.2.1 Soil Fertility

Productivity of soil can be improved by providing the soil with organic substances. These organic amendments improve soil structure and fertility level.

13.5.2.2 Soil Structure

Organic amendments improve soil structure, aggregation, and porosity, and, therefore, can allow rhizosphere to expand, allow water to flow in, and supply air to the plant all of which are essential for growth.

13.5.2.3 Nutrient Retention

Organic matter accumulation can serve as a source of nutrients. This significantly reduces nutrient leaching and runoff, and it increases nutrient retention in the soil and ensures availability to the plants to uptake.

13.5.2.4 Carbon Sequestration

Application of organic fertilizers, like compost and biochar, helps to secure carbon in the soil, which in turn lowers greenhouse gas emissions and assists in combating global warming.

13.5.2.5 Resilience

By using INM techniques, good soil quality and different microbial species can alleviate agroecological stresses caused by drought, pests, diseases, and other elements. In short, organic fertilizers are considered an important element of INM approaches. These improve soil fertility, ensure yield quality, and guarantee sustainability of the environment. Through the use of organic amendments in nutrient management practices, the growers can improve health of the soil, maximize nutrient utilization, and increase resilience of the agricultural systems by ensuring sustainability.

13.6 Potential of Biofertilizers for INM

INM means the harmonious use of different sources of nutrients, including biofertilizers, with the objectives of improving soil richness and boosting crop production. Biofertilizers are biological agents with beneficial micro-organisms, which make nutrients available, stimulate development of the plants, and enhance soil quality. The application of biofertilizers to vegetable crops is a vital part of INM strategies because these improve the traditional sources of nutrients and make soil fertility sustainable. How biofertilizers are integrated into nutrient management is described below.

13.6.1 Supplementing Nutrient Inputs

Biofertilizers supplement synthetic fertilizers by providing bioavailable nutrients including N, P, and K to vegetable crops. Incorporating biofertilizers minimizes reliance on chemical fertilizers, corrects nutrient imbalances, and encourages ecologically responsible nutrient management techniques.

13.6.2 Improving Soil Health

Biofertilizers enhance soil biological activity, microbial diversity, and nutrient cycling processes, leading to improved soil structure, fertility, and resilience. Beneficial micro-organisms in biofertilizers contribute to organic matter decomposition, nutrient mineralization, and disease suppression, supporting sustainable soil health management.

13.6.3 Enhancing Nutrient Use Efficiency

Biofertilizers facilitate nutrient uptake efficacy due to stimulation of root growth, nutrient solubilization, and symbiotic nutrient acquisition systems in vegetable crops. For instance, biofertilizers with N-fixing traits form symbiotic associations with the roots of crops and then fix N from the atmosphere.

13.6.4 Mitigating Environmental Impacts

Biofertilizers become a solution to lessen environmental pollution resulting from nutrient losses through prevention of nutrient runoff, leaching, and greenhouse gas emissions which are associated with synthetic fertilizers. The use of biofertilizers in sustainable nutrient management is part of environmental conservation, water quality protection, and climate change mitigation.

13.6.5 Enhanced Crop Performance

Biofertilizers enhance plant growth, yield, and quality by improving nutrient availability, root development, and stress tolerance in vegetable crops.

13.6.6 Cost-Effectiveness

Biofertilizers are low-cost alternatives to synthetic fertilizers that maintain or increase crop productivity and profitability. Finally, biofertilizers are significant components of INM systems for vegetable crops, since they improve soil fertility, crop yield, and environmental sustainability. Combining biofertilizers with other nutrient sources and management practices can help the growers to enhance soil health, boost nutrient use efficiency, and create resilient and sustainable agricultural systems.

13.7 Conclusions and Future Perspectives

Mineral nutrition knowledge serves as a base for sustainable vegetable production. Nutrients are the building blocks for crop growth, which determine the yield, quality, and tolerance of crops to biotic and abiotic stress factors. Upon carrying out this exploration, it is evident that a balanced and holistic approach to nutrient management is necessary for successful crop growth and ensuring no negative impacts on the environment. By addressing soil integrity, nutrient provision, and crop preference, farmers can in turn implement methods to maximize nutrient uptake efficiency and sustainable agricultural practices. This involves soil testing, nutrient analysis, and incorporation of multi-nutrient sources such as organic amendments, biofertilizers, and inorganic fertilizers.

References

Abo-Dahab AM, Marey RA, Geries LSM (2016) Influence of some nitrogen fertilization rates and application methods on growth, yield and yield components of onion. J Agric Res 42:630–647

Aftab T, Hakeem KR (2020) Plant micronutrients: deficiency and toxicity management. Plant Micronutr 11:56–66

Anderson R, Bayer P, Edwards D (2020) Climate change and the need for agricultural adaptation. Curr Opin Plant Biol 12:34–45

Azcón-Aguilar C, Barea J (2015) Nutrient cycling in the mycorrhizosphere. J Soil Sci Plant Nutr 15:372–396

Baloch AF (1994) Vegetable crops. In: Malik MN, Bashir E, Bantel R (eds) Horticulture. National Book Foundation, Islamabad

Banerjee P, Nath R (2021) Prospects of molybdenum fertilization in grain legumes—a review. J Plant Nutr 45:1425–1440

Banerjee P, Das P, Sinha S (2021) Importance of molybdenum for the production of pulse crops in India. J Plant Nutr 45:300–310

Becana M, Wienkoop S, Matamoros MA (2018) Sulfur transport and metabolism in legume root nodules. Front Plant Sci 11:89–103

Borisov VA, Sosnov VS, Menshikh AM, Vasyuchkov IY (2022) Efficiency of the main application of fertilizers and top dressing of tomato under drip irrigation on ordinary chernozems of the Rostov region. Veg Crops Russ 4:60–64

Bouain N, Kisko M, Rouached A et al (2014) Phosphate/zinc interaction analysis in two lettuce varieties reveals contrasting effects on biomass, photosynthesis, and dynamics of Pi transport. Biomed Res Int 2014:548254

Briat J, Dubos C, Gaymard F (2015) Iron nutrition, biomass production, and plant product quality. Trends Plant Sci 20:33–40

Bulman DM, Mezyk S, Remucal CK (2019) The impact of pH and irradiation wavelength on the production of reactive oxidants during chlorine photolysis. Environ Sci Technol 53:4450–4459

Cabot C, Martos S, Llugany M et al (2019) A role for zinc in plant defense against pathogens and herbivores. Front Plant Sci 10:12–18

Carstensen MV, Børgesen CD, Ovesen N et al (2019) Controlled drainage as a targeted mitigation measure for nitrogen and phosphorus. J Environ Qual 48:677–685

Chakrabarty K, Chakrabarty AS (2019) Textbook of nutrition and health diseases. Springer, Singapore, pp 21–63

Chen L, Xie H, Wang G et al (2021) Reducing environmental risk by improving crop management practices at high crop yield levels. Field Crop Res 265:293–298

Colombari L, Lanna N, Guimarães L, Cardoso A (2018) Production and quality of carrot in function of split application of nitrogen doses in top dressing. Hortic Bras 36:15–19
Congreves KA, Eerd LV (2015) Nitrogen cycling and management in intensive horticultural systems. Nutr Cycl Agroecosyst 102:299–318
Connorton JM, Balk J (2019) Iron biofortification of staple crops: lessons and challenges in plant genetics. Plant Cell Physiol 60:1447–1456
Dawar R, Karan S, Bhardwaj S et al (2023) Role of sulphur fertilization in legume crops: a comprehensive review. Int J Plant Soil Sci 23:34–38
Debbarma S, Bhatt L (2021) Nutrient uptake, soil nutrient status and water use efficiency as influenced by drip fertigation in broccoli (*Brassica oleracea* var. *italica*). J Plant Nutr 45:674–685
Deeb M, Groffman P, Joyner J et al (2018) Soil and microbial properties of green infrastructure stormwater management systems. Ecol Eng 2:44–64
Dimkpa CO, Bindraban PS (2016) Fortification of micronutrients for efficient agronomic production: a review. Agron Sustain Dev 36:254–260
Fageria VD (2001) Nutrient interactions in crop plants. J Plant Nutr 24:1269–1290
Farhat N, Elkhouni A, Zorrig W et al (2016) Effects of magnesium deficiency on photosynthesis and carbohydrate partitioning. Acta Physiol Plant 3:21–38
Feng Q, Jingjing J, JiWen X et al (2019) Drip irrigation and fertilization improve yield, uptake of nitrogen, and water-nitrogen use efficiency in cucumbers grown in substrate bags. Plant Soil Environ 65:328–335
Garcia PL, Sermarini RA, Trivelin PCO (2020) Nitrogen fertilization management with blends of controlled-release and conventional urea affects common bean growth and yield during mild winters in Brazil. Agronomy 10:19–35
Gong Z, Xiong L, Shi H et al (2020) Plant abiotic stress response and nutrient use efficiency. Sci China Life Sci 63:635–674
Harish V, Aslam S, Chouhan S et al (2023) Iron toxicity in plants: a review. Int J Environ Clim Change 2:13–28
Hasanuzzaman M, Bhuyan M, Nahar K et al (2018) Potassium: a vital regulator of plant responses and tolerance to abiotic stresses. Agronomy 8(3):31
Hassan MU, Aamer M, Chattha MU et al (2020) The critical role of zinc in plants facing the drought stress. Agriculture 1:9–39
Helfenstein J, Pawlowski M, Hill CB et al (2015) Zinc deficiency alters soybean susceptibility to pathogens and pests. J Plant Nutr Soil Sci 178:896–903
Herrera LFR, Shane MW, López-Bucio J (2015) Nutritional regulation of root development. Wiley Interdiscip Rev Dev Biol 4:431–443
Hocking B, Tyerman S, Burton R, Gilliham M (2016) Fruit calcium: transport and physiology. Front Plant Sci 1:7–23
Incrocci L, Massa D, Pardossi A (2017) New trends in the fertigation management of irrigated vegetable crops. Horticulturae 1:3–37
Junior E, Medeiros J, Oliveira F et al (2017) Nitrogen and potassium fertigation in bell pepper cultivated in greenhouse using fertigation managements. Rev Bras Eng Agrícola Ambient 21:186–190
Kane DA, Bradford MA, Fuller E et al (2021) Soil organic matter protects US maize yields and lowers crop insurance payouts under drought. Environ Res Lett 16(4):044018
Khani A, Barzegar T, Nikbakht J, Ghahremani Z (2020) Effect of foliar spray of calcium lactate on the growth, yield and biochemical attributes of lettuce (*Lactuca sativa* L.) under water deficit stress. Adv Hortic Sci 34:11–24
Kim J, Rho I (2023) Effects of fertigation treatment using a subsurface drip system on spring potato (*Solanum tuberosum* L.) growth and yield. Open Agric J 17:119–132
Kobayashi T, Nozoye T, Nishizawa N (2019) Iron transport and its regulation in plants. Free Radic Biol Med 133:11–20
Kumar S, Kumar S, Mohapatra T (2021) Interaction between macro- and micro-nutrients in plants. Front Plant Sci 1:12–32

Kumari VV, Banerjee P, Verma VC et al (2022) Plant nutrition: an effective way to alleviate abiotic stress in agricultural crops. Int J Mol Sci 23:97–114

Li L, Yang X (2018) The essential element manganese, oxidative stress, and metabolic diseases: links and interactions. Oxidative Med Cell Longev 5:7580707

Li Q, Gao Y, Yang A (2020) Sulfur homeostasis in plants. Int J Mol Sci 1:12–21

Lobon S, Joy M, Casasus I et al (2020) Field pea can be included in fattening concentrate without deleterious effects on the digestibility and performance of lambs. Animals 10(2):243

Long Y, Peng J (2023) Interaction between boron and other elements in plants. Genes 14:23–27

Ma Y, Oliveira R, Freitas H, Zhang C (2016) Biochemical and molecular mechanisms of plant-microbe-metal interactions: relevance for phytoremediation. Front Plant Sci 1:17–29

Ma J, Chen T, Lin J et al (2022) Nitrogen, phosphorus, and potassium influence rice growth and development by regulating the efficiency of energy production. Rice Sci 29:166–178

Malhotra H, Vandana, Sharma S, Pandey R (2018) Phosphorus nutrition: plant growth in response to deficiency and excess. In: Hasanuzzaman M, Fujita M, Oku H, Nahar K, Hawrylak-Nowak B (eds) Plant nutrients and abiotic stress tolerance. Springer, Singapore, pp 71–190

Malik A, Puissant J, Buckeridge KM et al (2018) Land use driven change in soil pH affects microbial carbon cycling processes. Nat Commun 19:88–106

Martins A, Krum B, Queirós L et al (2020) Manganese in diet: bioaccessibility, adequate intake and neurotoxicological effects. J Agric Food Chem 10:102–641

Mola I, Cozzolino E, Ottaiano L et al (2020) Nitrogen use and uptake efficiency and crop performance of baby spinach (*Spinacia oleracea* L.) and lamb's lettuce (*Valerianella locusta* L.) grown under variable sub-optimal N regimes combined with plant-based biostimulant application. Agronomy 1:1–15

Monib AW, Alimyar O, Mohammad MU et al (2023) Macronutrients for plant growth and human health. J Res Appl Sci Biotechnol 2:268–279

Muindi E (2019) Understanding soil phosphorus. Int J Plant Soil Sci 1:1–18

Nawaz F, Shehzad M, Majeed S et al (2020) Role of mineral nutrition in improving drought and salinity tolerance in field crops. In: Hasanuzzaman M (ed) Agronomic crops. Springer, Singapore, pp 129–147

Nazli F, Najm-ul-Seher, Khan MY et al (2020) Soil microbes and plant health. In: Ul Haq I, Ijaz S (eds) Plant disease management strategies for sustainable agriculture through traditional and modern approaches: sustainability in plant and crop protection, vol 13. Springer, Cham, pp 111–135

Niaz S (2012a) Mosam garma ki sabzian. Directorate of Agricultural Information, Agriculture Department, Government of the Punjab, Lahore, pp 1–49

Niaz S (2012b) Mosam sarma ki sabzian. Directorate of Agricultural Information, Agriculture Department, Government of the Punjab, Lahore, pp 1–45

Pandey N (2018) Role of plant nutrients in plant growth and physiology. In: Hasanuzzaman M, Fujita M, Oku H, Nahar K, Hawrylak-Nowak B (eds) Plant nutrients and abiotic stress tolerance. Springer, Singapore, pp 51–93

Pantoja-Benavides AD, Garces-varon G, Restrepo-Díaz H (2021) Foliar growth regulator sprays induced tolerance to combined heat stress by enhancing physiological and biochemical responses in rice. Front Plant Sci 12:34–49

Pastuszczak M, Stanek-Tarkowska J, Kačániová M (2022) Impact of soil fertilized with biomass ash on depth-related variability of culturable bacterial diversity and selected physicochemical parameters in spring barley cultivation. Int J Environ Res Public Health 1:19–42

Qiao Y, Miao S, Han X et al (2017) Improving soil nutrient availability increases carbon rhizodeposition under maize and soybean in Mollisols. Sci Total Environ 60:416–424

Rai S, Singh P, Mankotia S et al (2021) Iron homeostasis in plants and its crosstalk with copper, zinc, and manganese. Plant Stress 1:100008

Ravier C, Meynard J, Cohan J et al (2017) Early nitrogen deficiencies favor high yield, grain protein content and N use efficiency in wheat. Eur J Agron 89:16–24

Reichardt K, Timm LC (2019) How plants absorb nutrients from the soil. Soil Plant Atmos 16:193–225

Reitz NF, Shackel K, Mitcham E (2021) Differential effects of excess calcium applied to whole plants vs. excised fruit tissue on blossom-end rot in tomato. Sci Hortic 290:34–47

Riaz MU, Ayub MA, Khalid H et al (2020) Fate of micronutrients in alkaline soils. In: Kumar S, Meena RS, Jhariya MK (eds) Resources use efficiency in agriculture. Springer, Singapore, pp 577–613

Sağlıker HA, Cenkseven Ş, Kızıldağ N et al (2018) Is parent material an important factor in soil carbon and nitrogen mineralization? Eur J Soil Biol 10:10–16

Sahu M, Gupta A, Cp MN (2020) Role of micronutrients in vegetable production: a review. Int J Chem Stud 8:1984–1986

Sapkota R, Santos S, Farias P et al (2020) Insights into the earthworm gut multi-kingdom microbial communities. Sci Total Environ 1:7–27

Sarkar B, Naidu R (2015) Nutrient and water use efficiency in soil: the influence of geological mineral amendments. In: Rakshit A, Singh HB, Sen A (eds) Nutrient use efficiency: from basics to advances. Springer, New Delhi, pp 29–44

Shabbir Z, Sardar A, Shabbir A et al (2020) Copper uptake, essentiality, toxicity, detoxification and risk assessment in soil-plant environment. Chemosphere 259:127436

Shan N, Chuan L, Liu J et al (2020) Application effects of fertilizer recommendation by nutrient expert system on radish. J Appl Ecol 31:3719–3728

Sinclair SA, Krämer U (2012) The zinc homeostasis network of land plants. Biochim Biophys Acta 1823:1553–1567

Springmann M, Wiebe K, Mason-D'Croz D et al (2018) Health and nutritional aspects of sustainable diet strategies and their association with environmental impacts: a global modelling analysis with country-level detail. Lancet Planet Health 2:451–461

Tanoi K, Kobayashi N (2015) Leaf senescence by magnesium deficiency. Plants 4:756–772

Thor K (2019) Calcium-nutrient and messenger. Front Plant Sci 10:440

Tian G, Qi D, Zhu J, Xu Y (2020) Effects of nitrogen fertilizer rates and waterlogging on leaf physiological characteristics and grain yield of maize. Arch Agron Soil Sci 67:863–875

Tränkner M, Tavakol E, Jákli B (2018) Functioning of potassium and magnesium in photosynthesis, photosynthate translocation and photoprotection. Physiol Plant 89:23–36

Tripathi S, Srivastava P, Devi RS, Bhadouria R (2020) Influence of synthetic fertilizers and pesticides on soil health and soil microbiology. In: Prasad MNV (ed) Agrochemicals detection, treatment and remediation—pesticides and chemical fertilizers. Butterworth-Heinemann, Oxford, pp 25–54

Tripathi R, Tewari R, Singh KP et al (2022) Plant mineral nutrition and disease resistance: a significant linkage for sustainable crop protection. Front Plant Sci 13:56–69

Urmi T, Rahman M, Islam MM et al (2022) Integrated nutrient management for rice yield, soil fertility, and carbon sequestration. Plan Theory 7:11–23

van Delden L, Rowlings DW, Scheer C, Grace PR (2016) Urbanisation-related land use change from forest and pasture into turf grass modifies soil nitrogen cycling and increases N_2O emissions. Biogeosciences 13:6095–6106

Vatansever R, Ozyigit II, Filiz E (2016) Essential and beneficial trace elements in plants, and their transport in roots: a review. Appl Biochem Biotechnol 181:464–482

Verma A, Tripathi D, Spehia RS et al (2022) Effect of drip fertigation levels on soil fertility and productivity of cauliflower cv. Pusa Snowball K-1. Int J Plant Soil Sci 34:1684–1692

Voltr V, Mensik L, Hlisnikovsky L et al (2021) The soil organic matter in connection with soil properties and soil inputs. Agronomy 11(4):779

Wang P, Grimm B (2021) Connecting chlorophyll metabolism with accumulation of the photosynthetic apparatus. Trends Plant Sci 10:1016–1028

Wang M, He D, Shen F et al (2019) Effects of soil compaction on plant growth, nutrient absorption, and root respiration in soybean seedlings. Environ Sci Pollut Res 26:22835–22845

Wang D, Chen X, Tang Z et al (2022) Application of humic acid compound fertilizer for increasing sweet potato yield and improving the soil fertility. J Plant Nutr 45:1933–1941

Wills S, Williams CO, Duniway M et al (2017) Human land-use and soil change. In: West L, Singer M, Hartemink A (eds) The soils of the USA, World soils book series. Springer, Cham, pp 351–371

Wu X, Bai M, Li Y et al (2020) The effect of fertigation on cabbage (*Brassica oleracea* L. var. *capitata*) grown in a greenhouse. Water 12:10–76

Xu Q, Fu H, Zhu B et al (2021) Potassium improves drought stress tolerance in plants by affecting root morphology, root exudates, and microbial diversity. Metabolites 11(3):131

Yadav AN, Kour D, Kaur T et al (2021) Biodiversity, and biotechnological contribution of beneficial soil microbiomes for nutrient cycling, plant growth improvement and nutrient uptake. Biocatal Agric Biotechnol 33:102009

Yan N, Marschner P, Cao W et al (2015) Influence of salinity and water content on soil microorganisms. Int Soil Water Conserv Res 3:316–323

Yang L, Pan J, Hu N et al (2021) Citrus physiological and molecular response to boron stresses. Plan Theory 11(1):40

Yavuz D, Yavuz N, Seymen M, Turkmen O (2015) Evapotranspiration, crop coefficient and seed yield of drip irrigated pumpkin under semi-arid conditions. Sci Hortic 7:33–40

Zheng C, Xiang L, Tan Q et al (2015) Soil application of calcium and magnesium fertilizer influences the fruit pulp mastication characteristics of Nanfeng tangerine (*Citrus reticulata* Blanco cv. Kinokuni). Sci Hortic 191:121–126

Irrigation Management in Vegetable Crops

14

Atique-ur-Rehman, Rafi Qamar, Axel Garcia y Garcia, Muhammad Shoaib, and Shakeel Ahmad

Abstract

Vegetables are high-value crops which require careful water management to achieve optimal growth, quality, and yield. Irrigation is a key input in vegetable production as water deficits or surpluses can directly affect both yield and quality. Vegetable crops are typically shallow-rooted with high transpiration rates, thus cannot withstand water stress. Proper irrigation ensures that plants receive adequate water during critical growth stages such as germination, flowering, and fruit setting. Water deficits can lead to stunted growth, reduced leaf size, flower drop, and poor fruit and/or edible part development. Conversely, excessive irrigation can cause reduced yields, poor produce quality, and increased vulnerability to insect-pests and diseases. Further, excess of water can result in root decay, nutrient leaching, and increased pest and disease pressure. Vegetable crops such as cucumbers, tomatoes, peppers, okra, and leafy vegetables have high water requirements, particularly during hot and dry growing seasons. This chapter explores irrigation management principles and practices for vegetable crops, focusing on methods to maximize water use efficiency while maintaining healthy crop growth. It also highlights different systems and methods of irrigating vegetable crops with their pros and cons and importance of irrigation scheduling.

Atique-ur-Rehman · S. Ahmad (✉)
Institute of Agronomy, Bahauddin Zakariya University, Multan, Pakistan
e-mail: shakeelahmad@bzu.edu.pk

R. Qamar
Department of Agronomy, College of Agriculture, University of Sargodha, Sargodha, Pakistan

A. Garcia y Garcia
Department of Plant Sciences, University of Wyoming, Powell, WY, USA

M. Shoaib
Department of Agricultural Engineering, Bahauddin Zakariya University, Multan, Pakistan

Z. Iqbal et al. (eds.), *Sustainable and Innovative Vegetable Production in times of Climate Change*, https://doi.org/10.1007/978-981-96-6283-8_14

Keywords

Capillary irrigation · Drip irrigation · Ebb and flow irrigation · Fertigation · Irrigation scheduling · Sprinkler systems · Subsurface irrigation · Surface irrigation

14.1 Introduction

Vegetables contain ample carbohydrates, proteins, dietary fibers, minerals, vitamins, and phytochemicals making them an excellent source of nutrients. Adequate water supply to vegetable crops during their growing period is critical to achieve better quality produce and higher returns (Vijayakumar et al. 2010). Conversely, improper application of irrigation results in poor growth and development of vegetable plants with lower yields and poor quality of produce. It also results in lower water use efficiency and significant loss of nutrients due to their fixation and leaching; hence lower nutrient use efficiency as well (Harisha et al. 2017). Surface irrigation remains an important method of water delivery for vegetable crops, particularly in regions where water is plentiful and flat terrains are common. While simple and cost-effective, it requires careful management to optimize water use efficiency and reduce potential drawbacks such as soil erosion and water wastage. Modern irrigation techniques are generally costlier compared to conventional methodologies and even depend heavily on outside professional support from vendors and distributors. Nevertheless, such technologies benefit a large number of farmers. Drip and sprinkler irrigation are looking to replace more conventional surface irrigation techniques of water distribution. These are capable of assisting farmers for better water management and waste reduction, which decrease the amount of water that has to be lifted. Planners and engineers frequently advocate adoption of these technologies so that water saved by these techniques could be used by other farmers. Drip irrigation delivers water periodically or continuously drop after drop in root zone of crop plants (Patil and Das 2015). Drip irrigation not only offers ideal growing conditions for root system, but it also supports good nutritional achievement and higher plant growth.

Drip irrigation also facilitates conservation of water wealth which might not be possible if over-utilized because of ineffective surface irrigation techniques (Meena et al. 2017). Smith et al. (2016) defined fertigation as drip irrigation with improved fertilizer usage. Fertigation improves nutrient usage through application of nutrients at crucial phases of the crop leading to better nutrient uptake and nutrient use efficiency, along with higher crop yield of good quality (Brewer et al. 2018; Mustafa et al. 2021; Zakir et al. 2023, 2024). Fertigation through the use of drip irrigation further lowers energy consumption, ensures disease as well as pest control, reduces labor costs, and is useful for rippling sloping terrain (Michael 2008). Whereas overhead sprinklers continue to be one of the most cost-effective mechanisms in regard to capital outlay. For improved water productivity, in addition to these high-efficiency irrigation systems, various resource conservation technologies like raised bed planting can be utilized. Not only method of irrigation and amount of water to be applied are imperative but frequency and time of irrigation are also equally important as vegetable crops have some critical periods; if irrigation is not applied at these stages, it will adversely affect yield and quality of the produce (Mubarak et al. 2022; Ali et al. 2013; Ahmad and Hasanuzzaman 2012; Zakir et al. 2025).

14.2 Importance of Irrigation for Vegetable Crops

Irrigation is crucial for vegetable crops because it provides a consistent and reliable water supply, which is essential for healthy growth, high yields, and good-quality produce. Vegetables typically have shallow root systems and high-water demands, making them particularly sensitive to moisture fluctuations. Irrigation ensures that plants receive necessary water for critical processes like photosynthesis, nutrient uptake, and cell growth. It supports root development, boosts vegetative growth, and enhances formation of fruits or other edible parts. Without sufficient irrigation, crops may suffer water stress, leading to wilting, stunted growth, and poor-quality produce. Water shortages during flowering or fruiting can significantly reduce yields and marketability.

Irrigation supply allows vegetable crops to be cultivated in regions with irregular rainfall or during dry seasons, making farming viable even in arid or semi-arid regions. This not only extends the growing season but also ensures food security by enabling year-round cultivation. Modern irrigation systems, such as drip or subsurface irrigation, enhance water use efficiency by supplying water directly to the plant root zone, minimizing evaporation and water runoff. Efficient irrigation also promotes better nutrient uptake and helps reduce the risk of nutrient leaching. By preventing water stress and ensuring optimal moisture levels, irrigation helps to maximize crop yield, improve quality, and increase profitability of vegetable farming. Additionally, irrigation plays a key role in precision agriculture, where automated systems and sensors ensure precise water application, conserving resources and reducing costs. In summary, irrigation is indispensable for vegetable crop production, especially in areas with limited and/or irregular rainfall, ensuring both high productivity and sustainability.

14.3 Types of Irrigation Systems for Vegetable Crops

The selection of an irrigation system plays a critical role in water use efficiency and crop productivity. There are several irrigation methods available for irrigating vegetable crops as given below, each with its pros and cons.

14.3.1 Surface Irrigation

Surface irrigation is one of the primitive and most widely practiced methods of irrigation in agriculture. It involves application of water across surface of the field, allowing it to infiltrate the soil and reach the plant roots. For producing vegetable crops, surface irrigation is commonly used in areas where water resources are abundant and the land is relatively flat. This method relies on gravity to distribute water over the field, making it simple and cost-effective compared to more advanced systems like drip or sprinkler irrigation. The primary forms of surface irrigation used in vegetable farming include furrow irrigation, border irrigation, and basin irrigation. Each method has its own specific applications depending on crop type, soil properties, and field layout.

14.3.1.1 Furrow Irrigation

Furrow irrigation is the most commonly used method of surface irrigation for row-planted vegetable crops. In this system, small channels or furrows are created between the rows of crops and water is introduced at the head of each furrow, flowing down the channel and infiltrating into the soil. Furrows are typically spaced 0.75 to 1.0 m apart, depending on the crop and row spacing. The furrow length can range from 100 to 300 m, although shorter furrows are preferred for better water distribution uniformity.

Furrow irrigation methods are low cost and easy to implement, requiring minimal infrastructure and energy inputs. It is suitable for a variety of vegetable crops like tomatoes, peppers, beans, and cucumbers. It can be adapted to different soil types with adjustments to furrow slope and length. However, there are certain disadvantages associated with furrow irrigation method which include uneven distribution of water along the furrow length, leading to over-irrigation at the top and under-irrigation at the bottom. Water losses can occur in this method due to deep percolation at the head and runoff at the tail end of the furrows. This method requires good field leveling for uniform water distribution. Furrow irrigation is ideal for the crops planted in rows such as tomatoes, beans, squash, and cucumbers, where the plants are spaced evenly in rows. It's widely used in regions with clay or loam soil that can hold water long enough absorbed by the root zone.

14.3.1.2 Basin Irrigation

In basin irrigation, the field is prepared, divided into small, flat basins or "checks" and water is applied to flood each basin. The water is contained within the basin by surrounding embankments, allowing it to infiltrate evenly across the soil surface. Basins can vary in size from small (for vegetable crops) to large (for cereal crops or pastures). The basin size may vary depending on the water supply, field slope, and soil type. Vegetable crops are often grown in smaller, more controlled basins, which allow for better water management. Basin method of irrigation is highly efficient in fields with fine-textured soils, such as clay, which hold water well. It minimizes water losses from runoff when basins are well designed. This method is simple to implement in flat terrains and does not require expensive infrastructure. Basin method of irrigation is not suitable for sloped fields or soils with low water-holding capacity, such as sandy soils. This method may also result in waterlogging if basins are overfilled, especially in heavy soils. Precise control of water inflow is required in this method to prevent uneven water distribution. Basin irrigation is often used for high-value vegetable crops that require uniform water distribution. Leafy vegetables like lettuce, spinach, and fenugreek benefit from basin irrigation because they provide even moisture distribution, which is crucial for consistent growth.

14.3.1.3 Border Irrigation

In the border irrigation method, the field is divided into long, narrow strips called "borders" and water is applied to flow down the length of the field between the borders. The borders help to confine the water to specific areas, allowing it to infiltrate into the soil as it flows downward. Borders can range from 03 to 10 m wide and are separated by low ridges or embankments. The length of the border is typically shorter for vegetable crops, usually around 50–150 m, to ensure even water distribution. Border irrigation is well-suited for vegetables planted in blocks rather than rows. It provides uniform distribution of water along the length of the border, reducing the risk of waterlogging or drought stress. It is suitable for both medium and heavy soils, especially when the field has a slight slope. This irrigation method, however, requires careful field leveling to ensure uniform water distribution. Border irrigation is effective for growing crops like carrots, onions, and leafy greens, where a uniform water supply across the block is essential. It is particularly useful on flat or gently sloping fields with medium-textured soil.

14.3.2 Sprinkler Irrigation

Sprinkler irrigation is a method of pressurized irrigation that is similar to natural rainfall in which water is applied through a network of pipes, pumps, and nozzles. It is a versatile and efficient irrigation method suitable for raising a wide range of vegetable crops under various soil and field conditions. In this method, water is delivered to the crops through sprinklers, which spray it over the plants in small and controlled amounts. This system can cover large areas with minimal labor and can be adapted for both small and large-scale vegetable farming.

Sprinkler irrigation is especially useful in areas with uneven terrain or where water conservation is essential. The flexibility and ease of operation make it one of the most popular irrigation methods for growing vegetables like tomatoes, peppers, carrots, and leafy greens. Sprinkler systems can vary widely in design, size, and complexity. The most common types of sprinkler irrigation systems used in vegetable farming include permanent (solid-set), portable, center pivot, and traveling sprinklers.

14.3.2.1 Solid Set or Permanent Sprinkler System

In a solid set system, a network of pipes and sprinklers is permanently installed in the field. The sprinklers are spaced evenly to ensure uniform water distribution across the entire area. Pipes are usually buried underground or placed on the soil surface, with risers that connect the sprinklers to the pipe network. The sprinklers are spaced based on their spray radius, typically 12–18 m apart, depending on the crop and the system's design.

Minimal labor is required in the system after installation, as the system is fixed in place. It is considered ideal for areas where frequent irrigation is needed, such as sandy soils or hot climates. Uniform water distribution is the key characteristics of the system, which is crucial for sensitive vegetable crops. However, high initial investment due to the cost of pipes, sprinklers, and installation is one of the disadvantages of the system. Limited flexibility, as the system is fixed and cannot be easily moved to different fields, is also another issue related with the system. Solid-set systems are commonly used in large vegetable fields, especially for crops like lettuce, spinach, carrots, onions, and tomatoes, where consistent moisture is critical.

14.3.2.2 Portable Sprinkler System

A portable sprinkler system consists of lightweight pipes and sprinklers that can be easily moved around the field. This system provides flexibility for farmers who grow vegetables in different plots and need to irrigate some areas at different times. Portable systems typically consist of aluminum or plastic pipes that can be manually moved to different sections of the field. Sprinklers can be detached and repositioned as needed. This system is often used in small to medium-sized fields. It is low cost and flexible, making it suitable for small-scale vegetable growers. This system is also easy to assemble, disassemble, and transport to different fields or locations. Nevertheless, this system is labor-intensive, as the system needs to be manually moved from one part of the field to another and water distribution in this system may not be as uniform as in permanent systems, especially on uneven terrain. Portable systems are ideal for mixed vegetable plots, where different crops with varying water requirements are grown. They are often used for crops like beans, cucumbers, leafy greens, and root vegetables.

14.3.2.3 Center Pivot Sprinkler System

A center pivot system consists of a long arm (or boom) mounted on a central tower that rotates around the field, applying water as it moves. Water is sprayed through sprinklers attached to the boom. This method is commonly used for commercial vegetable farming on a large scale. The system is designed in a circular layout, with the pivot point installed at the center of the field. The arm is supported by wheels that move the sprinklers in a circular motion, allowing water to be distributed evenly across the entire field. This system ensures high efficiency in water distribution and coverage of large area. Reduced labor costs, as the system is automated and covers large fields with minimal intervention, are another advantage of the system. However, this system is expensive to install, making it more suited for large-scale operations. This system is only applicable to circular fields, which may not be ideal for all types of vegetable farming. Center pivot systems are widely used at large commercial vegetable farms, particularly for crops like potatoes, onions, carrots, and cabbage.

14.3.2.4 Traveling Sprinkler System

Traveling sprinkler system consists of a sprinkler mounted on a wheeled cart that moves across the field. The cart is pulled along a path and water is sprayed as it travels. Traveling sprinklers can be designed to move on tracks or follow a cable laid across the field. Water is supplied through a flexible hose connected to the sprinkler unit. The system can irrigate rectangular or irregularly shaped fields. High flexibility and ability to cover irregularly shaped fields is the advantage associated with this system. This system is suitable for large fields and areas where fixed systems may not be feasible. High initial cost and complex maintenance requirements are the cons associated with the system. This system also requires flat or gently sloping fields for optimal performance. Traveling sprinklers are suitable for a wide range of vegetable crops, especially for medium to large fields with irregular shapes. Crops like corn, beans, and leafy vegetables benefit from this system.

14.3.3 Subsurface Irrigation

Subsurface irrigation is an efficient and precise irrigation method in which water is applied to the root zone of plants directly through a network of buried pipes or tubes. Unlike surface or sprinkler irrigation, water is delivered below the soil surface, minimizing evaporation and runoff. In this method, water is applied directly where it is needed, resulting in higher efficiency and improved crop productivity. Subsurface irrigation is especially suitable for vegetable crops that require consistent moisture levels and can benefit from reduced soil surface wetting. This method is predominantly useful in areas where water conservation is essential and commonly used in high-value crops such as tomatoes, peppers, cucumbers, and other vegetables grown in both open fields and greenhouses.

Subsurface irrigation can be classified into two main types: Subsurface drip irrigation and Sub-irrigation. Each method has its unique design and applications depending on the soil characteristics, crop type, and availability of water.

14.3.3.1 Subsurface Drip Irrigation

Subsurface drip irrigation system comprises drip lines (usually made of plastic materials) installed below the soil surface, usually at a depth of 15–30 cm, depending on the crop's root zone. Water is delivered to the root zone of crop plants directly through a network of small emitters in the drip lines, ensuring precise control of the water to be applied. The systems consist of mainlines, sub-mainlines, and laterals, with emitters spaced along the laterals to deliver water uniformly. The drip lines are buried beneath the rows of vegetable crops and water is applied at low pressure, minimizing energy use. Emitters are typically spaced 20–40 cm apart, based on soil characteristics, crop root depth, and crop water requirements. The subsurface drip irrigation is one of the most efficient methods of irrigation, with minimal water losses due to evaporation or runoff as up to 95% of the water applied is used by crops. Since water is delivered below the soil surface in this system, the soil surface remains dry, reducing weed growth and need for herbicides. Subsurface irrigation minimizes the risk of diseases caused by wet foliage, such as fungal infections, since the plant leaves remain dry. Once installed, the systems require minimal labor for operation and maintenance compared to surface or sprinkler systems. Subsurface drip irrigation supplies and maintains consistent moisture levels in the root zone, promoting uniform crop growth and reducing moisture stress. The subsurface irrigation system is particularly suited for high-value vegetable crops like tomatoes, peppers, eggplants, cucumbers, and carrots. It is used in regions where water resources are limited and efficient irrigation is critical. It is widely used in the regions where water conservation is a priority, i.e., arid and semi-arid areas.

The installation of the systems can be expensive due to the cost of materials, equipment, and installation, especially for large-scale operations. Emitters in the systems can become clogged due to sediments or poor water quality. Filtration systems are essential to prevent this problem. Since the system is buried underground, it can be challenging to detect leaks or malfunctions and repairs require digging up the pipes.

14.3.3.2 Sub-irrigation (Controlled or Water Table Irrigation)

Sub-irrigation involves maintaining a water table just below the rhizosphere of crops. Water is supplied through a network of ditches, pipes, or channels placed below the soil surface, allowing it to rise into the rhizosphere by capillary action. The water table is controlled by adjusting the water levels in the system, ensuring that the crops receive adequate moisture without over-saturating the soil. In sub-irrigation systems, water is stored in a network of underground pipes or trenches and its level is controlled by adjusting the inflow and outflow of water. These systems are commonly used in areas with shallow groundwater or where artificial water tables can be created. Sub-irrigation is highly efficient in reducing water losses by minimizing evaporation and runoff, as water is stored and distributed underground. This system allows precise control of the root zone moisture, ensuring that crops receive a consistent water supply without over-irrigation. Since water is not applied to the soil surface, sub-irrigation eliminates problems associated with soil erosion, making it suitable for sandy or fragile soils. However, sub-irrigation systems can be complex to install and manage. These require extensive field preparation, including creating ditches or laying underground pipes and maintaining the water table. When water table is too high, there is a risk of waterlogging, which can damage crops, so precise water table management is essential. Sub-irrigation is used in areas with high water tables or regions where creating a controlled water table is possible, such as low-lying lands or peatlands. It is commonly used for leafy vegetables, such as lettuce, spinach, and cabbage, which require consistent moisture levels for uniform growth.

14.3.4 Seepage Irrigation

Seepage irrigation is a method of irrigating crops that relies on the natural movement of water through the soil, where water is delivered indirectly to the crops' root zone through underground water tables or canals. This method is typically used in areas with high water tables and soils having good water-holding capacity, like clayey or loamy soils. It is commonly practiced in regions with shallow groundwater or where surface water can be channeled through ditches or canals to allow water to percolate into the rhizosphere. In seepage irrigation, water seeps laterally through the soil profile from the water source (such as ditches, canals, or underground water tables) and rises by capillary action to the root zones of the plants. Seepage irrigation has been widely used in certain regions, especially in low-lying agricultural areas, for crops like vegetables, rice, and sugarcane. Seepage irrigation is particularly effective for vegetables that can tolerate slow, steady moisture delivery and thrive in well-drained soils. Common vegetables that benefit from seepage irrigation include tomatoes, peppers, cucumbers, eggplant, cabbage, lettuce, beans, squash, melons, and broccoli. These crops, with relatively shallow root systems, can easily access the water that rises by capillary action through the soil.

Seepage irrigation works by utilizing natural subsurface water movement to irrigate crops indirectly. The primary components of this system are ditches or canals that are dug near the crop fields. Water is introduced into these ditches and it seeps laterally and upward through the soil to reach the plant roots. The basic principle involves creating a network of canals or furrows that allows water to infiltrate the soil. Over time, water spreads beneath the surface, moistening the soil and eventually reaching the root zone of vegetable crops through capillary action.

14.3.5 Capillary Irrigation

Capillary irrigation, also called as capillary wick irrigation or capillary mat irrigation, is a passive and water-efficient irrigation system that uses capillary action to deliver water to the root zone of vegetable crops. This method relies on the natural upward movement of water through small pores in the soil or growing medium, ensuring consistent moisture at the root zone without over-watering. It is particularly useful in controlled environments such as greenhouses, raised beds, and container gardens.

Capillary irrigation functions by exploiting the physical property of capillary action, where water naturally moves upward through narrow pores in the soil or wicking material. The system keeps the water level low enough to allow the soil to draw only as much moisture as needed by the plant roots, ensuring the optimal amount of water is delivered to the crop plants. The water from the reservoir is absorbed by the capillary mat or wick. This water is then transferred to the growing medium through capillary action. The plant roots absorb the moisture that rises into the growing medium. As the plant takes up water from the soil, more water is drawn from the reservoir to replace it, maintaining consistent moisture levels. Since the water is supplied gradually and directly to the rhizosphere, the soil never becomes waterlogged, and excess water is not wasted. This method is beneficial especially for those vegetable crops which are sensitive to overwatering like tomatoes, peppers, and leafy greens. Vegetable crops suitable for capillary irrigation include tomatoes, cucumbers, peppers, lettuce, spinach, herbs, and leafy greens.

Capillary irrigation is an efficient and low-maintenance system that delivers consistent water to vegetable crops by leveraging capillary action. It is ideal for small-scale, container-based, or greenhouse vegetable production, offering numerous advantages in terms of water efficiency, plant health and ease of use. While it is not as scalable as other irrigation methods like drip or sprinkler irrigation; capillary irrigation excels in water conservation, making it a valuable option for vegetable growers in controlled environments or regions with limited water availability.

Capillary irrigation systems come in many forms, all of which rely on the natural process of capillary action to deliver water to the root zone. Each method has its advantages and limitations depending on the scale of operation and the kind of vegetable crops to be grown. These systems are particularly useful in conserving water, providing consistent moisture, and reducing labor, making them ideal for small-scale farming, container gardens, or greenhouses. There are several types of capillary irrigation systems, as described below, each utilizing slightly different method or material to achieve efficient and consistent water delivery.

14.3.5.1 Capillary Wick Irrigation System

Wicking bed system is certainly not a novel innovation, rather a new irrigation method which relies on the evapotranspiration rate and water uptake of plants. This system uses wicks made of porous materials, such as cotton, synthetic fibers, or ropes, to draw water from a reservoir to the plants' root zone. Wicks are placed with one end in the water reservoir and the other in the growing medium where the plants are located. Water is drawn upward through the wick to keep the growing medium moist (Fig. 14.1). This system is best for container-grown vegetables, small garden beds, or potted plants. It is also suitable for vegetables like tomatoes, lettuce, herbs, and leafy greens. This system is simple to set up and maintain. It is also very water-efficient and works well for container gardens. This system requires regular monitoring of the wicks to ensure these don't clog or dry out. Although several scientists have revealed that capillary systems like capillary wicks and capillary mats require less water as compared to traditional techniques, all these systems have drawbacks which make them inadequate regarding urban agriculture.

Fig. 14.1 Wicking bed system

14.3.5.2 Capillary Mat Irrigation System

Capillary mats are commonly utilized for the cultivation of nursery plants in containers. Capillary mats are synthetic fibrous absorbent materials like geotextile fabric or carpet underlay on which containers are placed. This type of system uses a porous mat (usually made from felt, polyester, or other synthetic materials) placed under pots, trays, or directly on the bed of plants. The capillary mat is laid over a water reservoir or in direct contact with water. The plants or containers with plants are placed on top of the mat, allowing water to move upward through the mat and into the plants' root zone. This system is best suited to greenhouse plants, indoor container crops, and small-scale vegetable operations. It is used for growing herbs, lettuce, and smaller vegetables. This system provides even water distribution to many pots or trays simultaneously and helps in maintaining consistent moisture levels. Mats may need periodic cleaning or replacement due to clogging or algae growth. This system is limited to container systems or closely spaced plants.

14.3.5.3 Sub-irrigation Planters

Sub-irrigation planters are self-watering containers that use a built-in water reservoir at the bottom of the container and allow water to move upward through the soil by capillary action. A reservoir is placed at the bottom of a pot or planter. The soil or growing medium above the reservoir acts as the wick, pulling water up into the root zone as needed. Some designs also have a wicking system built into the planter. Urban gardening, container gardening, or small home vegetable gardens are ideal for the sub-irrigation planters. This system is ideal for tomatoes, peppers, and other deep-rooted crops. Minimal maintenance is required in this system once set up, reduces over-watering risk, and allows for long periods between refilling the reservoir. Initial setup in this system can be more expensive than other capillary systems. If not maintained periodically, the reservoir can dry out or get clogged with algae.

14.3.5.4 Capillary Bed Irrigation

This system involves placing a porous layer or substrate (such as sand, perlite, or special capillary soil) underneath the root zone of the plants to provide constant moisture. The capillary bed sits over a water reservoir or wicking system and water moves upwards through the bed into the plant roots by capillary action. These are often found in greenhouse vegetable production. This system is best suited to greenhouse crops or those grown on small raised beds. It is also suitable for vegetables like lettuce, spinach, or seedlings. This system provides uniform water distribution across the growing bed. Ideal for plants that need constant moisture. The capillary substrate needs careful selection and management to ensure consistent water delivery avoiding salt buildup. When compared to traditional surface irrigation, wicking beds can save approximately 90% of the water.

14.3.5.5 Capillary Rise Bed System

This system involves planting directly in soil or growing medium that is in contact with a water table or reservoir below. The growing bed is raised slightly above the water source, allowing water to move upward through the soil naturally. This creates a balance between moisture and air availability in the root zone. Raised beds, in-ground gardens, or fields with shallow water tables are best suited for this system. This system is often used for plants that need consistent moisture like cucumbers, lettuce, and melons. The capillary rise bed system can be more sustainable and natural, requiring minimal intervention once established. It is also scalable for larger vegetable crops; however, careful soil management is required to prevent waterlogging. This system may not be suitable for those crops which are sensitive to over-watering and for the areas with fluctuating water levels.

14.3.5.6 Self-watering Trays or Platforms

Self-watering tray uses a reservoir beneath it to store water, which is absorbed by the soil or growing medium in the tray through capillary action. These self-watering trays are like capillary mats but these are more structured. The trays with water reservoir are used for raising seedlings, herbs, or small vegetable crops in indoor or outdoor gardens. These can also be used in hydroponic setups. This system is easy to use for small crops and the trays can often be moved around easily. The system is excellent for starting vegetables like cucumbers, tomatoes, and peppers. The disadvantage of this system is that it is limited by the size of the trays, so it's not suitable for large-scale vegetable production.

14.3.5.7 Capillary Tube Irrigation

This method involves using small tubes, like wicks, that draw water through capillary action and deliver it to the plants. Tiny tubes or narrow pipes are placed from a water source to the growing medium and water moves through the tubes due to capillary pressure. These tubes are connected to the reservoir and placed near the plants. This method is most suitable for controlled environments like greenhouses or where individual plants need specific water delivery. This method allows more control over the water distribution and prevents over-saturation of the soil. The system can be more complex to install and maintain than simpler wick or mat systems.

14.3.5.8 Hydroponic Capillary Systems

In some hydroponic setups, a capillary action system can be integrated to move nutrient-rich water from a reservoir into a growing medium without the need for pumps. Water is delivered from a reservoir into the soilless growing medium via wicks or capillary mats. Nutrient solutions can be fed into the reservoir to provide essential nutrients to the plants. This method is most appropriate for small-scale indoor or greenhouse hydroponic vegetable production. This method delivers both water and nutrients to the root zone, reduces need for energy-consuming pumps, and is very water-efficient. However, hydroponic capillary systems are usually limited to smaller setups and may not be much efficient for large-scale commercial operations.

14.3.6 Ebb and Flow Irrigation

Ebb and flow irrigation (also called flood and drain irrigation) is a hydroponic or container-based system that efficiently delivers water and nutrients to vegetable crops by periodically flooding the plant roots and then draining the solution back into a reservoir. It is particularly well-suited for hydroponic vegetable production but can also be used in greenhouses or indoor environments where controlled irrigation is necessary. Ebb and flow system operates through cycles of flooding and draining. The system delivers water and nutrient solutions to the plant root zone for a set period and then allows the water to drain away, leaving the roots exposed to air. This cycle is repeated multiple times a day depending on the crop's water needs. In the flooding cycle, a pump activates and floods the growing tray or bed with water, submerging the plant roots. The water contains essential nutrients required for plant growth. The water is pumped from a reservoir into the grow bed or container where the vegetable crops are planted. The roots absorb both water and nutrients during the flooding period. This cycle usually lasts for 10–30 min depending on the crop and environmental conditions. The drain cycle starts after the flood cycle, the pump turns off and the water drains back into the reservoir through the same inlet or through a designated drainage outlet. As the water drains, oxygen is drawn into the root zone, allowing the roots to breathe. This oxygenation prevents root decay and promotes healthy plant growth. The drained nutrient solution is collected in the reservoir, where it can be reused for subsequent cycles, making the system water-efficient. The flood and drain process is repeated throughout the day at regular intervals, which can be automated using timers. The frequency of the cycles is based on several factors like crop type, plant maturity, environmental conditions, and growing medium. Typically, for vegetable crops, the system floods and drains 03–06 times a day. This system of irrigation is best suited to various vegetable crops like green leafy vegetables (spinach, lettuce, kale, etc.), herbs (parsley, cilantro, basil, mint, and other herbs), fruit-bearing vegetables (tomatoes, cucumbers, peppers, and eggplants), and root vegetables (radishes and carrots). Ebb and flow is widely used in hydroponic setups, where plants are grown without soil.

14.3.7 Drip Irrigation

Drip irrigation is considered as a precise and highly efficient method as it delivers water directly to the root zones of vegetable crops. It conserves water, reduces weed infestation, and improves plant growth by providing continuous moisture where it is needed most. This method is especially advantageous in those areas where water is scarce, but it is also increasingly popular in regions where water conservation and crop quality are priorities.

An important parameter for laying out the drip irrigation system for vegetable crops is emitter spacing along the lateral lines. Emitter spacing should match the crop spacing and soil conditions. Closely spaced crops (e.g., carrots, onions) may need emitters spaced 15–20 cm apart. Widely spaced crops (e.g., tomatoes, peppers) may need emitters spaced 30–45 cm apart (Fig. 14.2). Proper pressure management is essential for ensuring that water is delivered evenly across the field in drip irrigation system. The use of pressure-compensating emitters is particularly useful in large or uneven fields to ensure uniform water delivery.

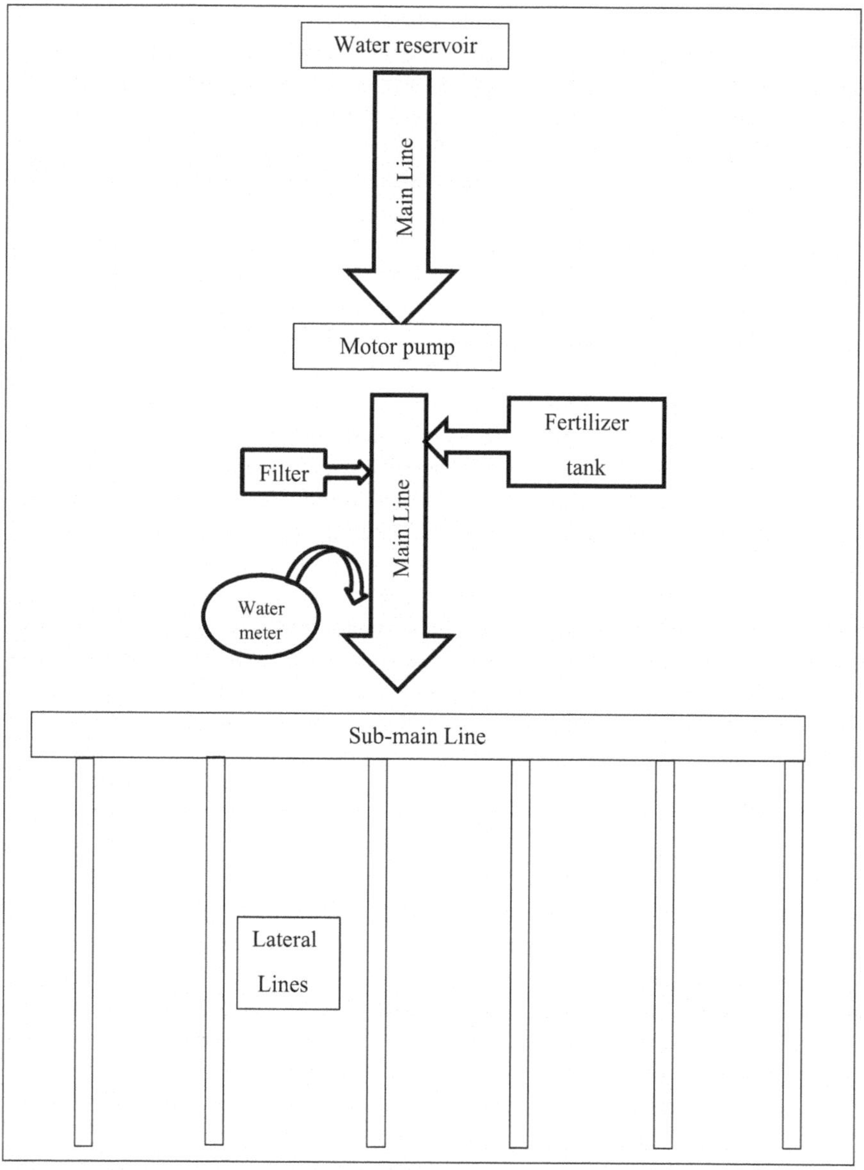

Fig. 14.2 Drip irrigation system layout for vegetables

Drip irrigation is particularly beneficial for high-value and water-sensitive vegetables. Common vegetable crops that thrive with drip irrigation include tomatoes, peppers, cucumbers, lettuce, carrots, onions, broccoli, cauliflower, eggplant, and spinach. Each crop has different water needs and the drip system can be customized to deliver water as per the specific needs of the crop.

Drip irrigation is a highly efficient system for growing vegetable crops, offering significant benefits in terms of water conservation, yield improvement, and nutrient management. It provides precise water delivery, reduces wastage, and promotes healthy crop growth, making it a vital tool for modern and sustainable vegetable farming. When designed, installed, and maintained properly, drip irrigation can lead to substantial improvements in the productivity of vegetable crops. Drip irrigation facilitates fertigation which improves fertilizer usage and hence nutrient use efficiency with higher yields of good quality. Fertigation also lowers energy consumption and reduces labor costs.

14.3.8 Drip Kit System

Drip kit irrigation is a simplified version of drip irrigation system, specifically designed to provide an affordable, easy-to-use solution for small-scale vegetable farmers, home gardeners, or those in regions with limited water resources. It combines the efficiency of traditional drip irrigation with a more user-friendly and portable setup. These kits are often pre-packaged and contain all the necessary components to establish a functioning drip irrigation system with minimal technical skills required. So, a drip kit irrigation system is a ready-to-use, pre-assembled drip irrigation package designed for small-scale farmers or home gardeners. The system includes everything needed to install a fully functional drip irrigation system in a relatively small area, typically ranging from 100 square meters to few hectares. These kits are designed for ease of setup and operation, allowing users to start irrigating their vegetable crops with minimal expertise. The drip kit delivers water slowly and directly to the root zone of the plants, just like a conventional drip irrigation system. However, it is typically smaller in scale, portable, and more affordable (Fig. 14.3).

Drip kit irrigation is an accessible, cost-effective, and highly efficient method for irrigating vegetable crops. Drip kits are widely used for vegetable crops due to their simplicity, affordability, and ability to optimize water use in small plots. It conserves water, improves crop yields, reduces labor, and requires minimal maintenance, making it an ideal solution for sustainable vegetable farming. Its ease of use and affordability make it a valuable tool for those seeking to optimize water use in small plots, while still providing the benefits of a traditional drip irrigation system.

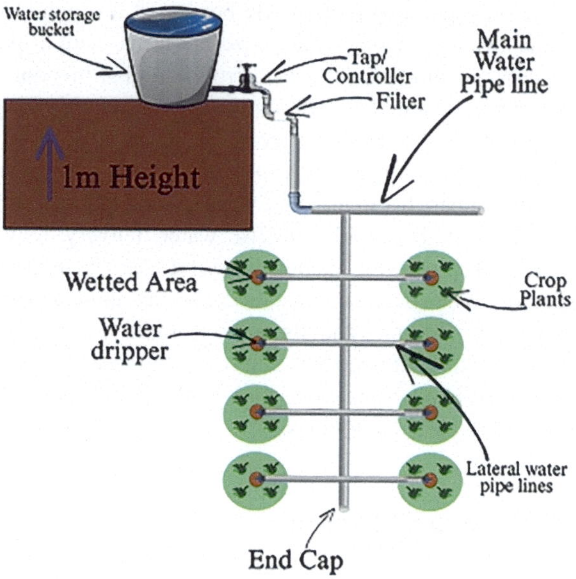

Fig. 14.3 Drip kit irrigation system for vegetables

14.4 Irrigation Scheduling for Vegetable Crops

Irrigation scheduling for vegetable crops is a systematic approach to determining the optimal times and amounts of water to apply to the crops, ensuring efficient water use, and promoting healthy crop growth. The objective is to align irrigation with the crop's water needs at different growth stages while minimizing water waste and avoiding the problems associated with over- or under-watering, such as reduced yields, poor-quality produce, and environmental degradation. The following are the key factors that need to be considered for successful irrigation scheduling.

14.4.1 Key Factors in Irrigation Scheduling

14.4.1.1 Crop Water Requirements

Different vegetable crops have specific water needs based on their physiology, root depth, and growth stage. For instance, leafy vegetables like lettuce, spinach, fenugreek, and cabbage have shallow root systems and require frequent irrigations with smaller amounts of water to maintain consistent soil moisture levels. On the other hand, deep-rooted vegetables like okra and cucurbits require less frequent but deeper irrigation. Critical stages like seed germination, flowering, fruit setting, and fruit development are particularly sensitive to water stress, so careful irrigation at these stages is essential. The water requirements of crops also vary across their life cycle. Generally, more water is needed during periods of rapid growth and less water as crops approach maturity.

14.4.1.2 Soil Characteristics

Soil type plays a significant role in determining irrigation frequency and duration. Soils with different textures (sand, loam, clay) have varying water-holding capacities. Sandy soils are highly permeable with poor water-holding capacity, meaning they lose water quickly to deep percolation and need frequent irrigation but in smaller amounts. Clayey soils have high water-holding capacity but low permeability, meaning they retain water longer but require more careful management to avoid waterlogging. Loamy soils, which are a mix of sand, silt, and clay, offer a balanced water-holding capacity and are ideal for most vegetable crops. Thus, understanding soil characteristics is critical for deciding the frequency of irrigation and the amount of water to be applied; as different soil types can impact the rate of water absorption and its retention in the root zone.

14.4.1.3 Climate and Weather Conditions

Local climatic conditions, i.e., temperature, relative humidity, solar radiation, and wind speed directly affect the crops' water demand and evaporation rates. In hot, dry, and windy climates, evapotranspiration (ET) occurs at a higher rate, requiring more frequent irrigation. Conversely, in cooler or more humid climates, irrigation needs are reduced. Seasonal variations in rainfall also influence irrigation schedules, with more frequent irrigation required during dry seasons or periods of low rainfall.

14.4.1.4 Growth Stage of the Crop

The water needs of vegetable crops vary significantly depending on their stage of development. Seeds require consistent moisture to ensure even germination and establishment. Irrigation during this stage is critical and frequent and light applications are often needed. During the early growth stages, when plants are developing leaves, stems, and roots, water requirements are moderate. Regular irrigation ensures the development of a healthy root system, which is necessary for nutrient uptake and resilience against drought. The reproductive stage is the most water-sensitive stage, as any water stress during flowering or fruit set can lead to fruit drop, reduced yields, and poor quality. At this stage, irrigation should be timed to ensure that soil moisture remains optimal. As crops reach near maturity, water requirements often decrease slightly. However, excessive irrigation during this phase, particularly for root crops (e.g., carrot, radish, beetroot, and onion) can result in waterlogged conditions, poor storage quality, or cracking the vegetables.

14.4.2 Irrigation Scheduling Techniques

14.4.2.1 Soil Moisture Monitoring

One of the most direct methods of irrigation scheduling is monitoring level of soil moisture in the root zone. This can be done manually by taking soil samples or using soil moisture sensors, which provide real-time data on soil water content.

These sensors allow growers to apply irrigation only when soil moisture levels fall below a certain threshold, ensuring that water is only applied when needed.

14.4.2.2 Evapotranspiration (ET)-Based Scheduling

ET is the quantity of water lost through evaporation from the soil surface and transpiration from the plants. ET-based irrigation scheduling uses weather data (temperature, humidity, solar radiation, wind) to estimate how much water the crop loses daily and then replenishes that amount through precipitation and irrigation. This method is particularly useful in regions with reliable weather data and is often combined with crop-specific ET values to fine-tune irrigation needs.

14.4.2.3 Tensiometer and Other Soil Water Measurement Devices

Tensiometer is a measuring instrument used to determine the tension or suction that plant roots must exert to extract water from soil pores. This provides a direct indication of how easily water is available to the crop and when the tension exceeds a certain threshold, irrigation is triggered. Similarly, devices like gypsum blocks or TDR (time-domain reflectometry) probes measure soil moisture levels and help in making irrigation decisions.

14.4.2.4 Crop Coefficients (Kc)

Crop coefficients are used to adjust ET values for specific crops and at different growth stages. Each crop has a specific Kc value that accounts for its water-use characteristics. This allows for more accurate irrigation scheduling by applying the precise quantity of water at each growth stage.

14.4.2.5 Computer Models and Decision Support System

Advanced irrigation scheduling often incorporates computer models that simulate soil water balance and crop water requirements. These systems use real-time data from meteorological observatories (weather stations), soil sensors, and crop growth models to provide precise commendations regarding when and how much to irrigate. Farmers can access these systems through mobile apps or web-based platforms, allowing for data-driven decisions (Ahmad et al. 2013; Ahmed et al. 2021).

14.4.2.6 Visual Indicators

While not as precise as other methods, visual indicators such as plant wilting or changes in leaf color can signal the need for irrigation. However, relying solely on visual signs of water stress can lead to late or excessive irrigation, so it is often used as a supplementary method in combination with more reliable tools.

14.4.2.7 Advances in Irrigation Scheduling

Recent advancements in irrigation technologies have revolutionized scheduling. Automated irrigation systems integrated with sensors and weather data allow for precise water applications, improving water-use efficiency and crop performance. Smart irrigation systems can adjust irrigation frequency and duration based on soil moisture status, meteorological conditions, and plant growth stage. This reduces water waste, lowers labor costs, and ensures that the crop receives the precise amount of water at the proper time.

14.5 Conclusion

Different irrigation techniques have been used for growing vegetables to decrease input costs and increase returns. Because the ultimate objective is to figure out the best irrigation method for growing vegetables with fewer inputs, each irrigation technique has distinct benefits and drawbacks under various soil and environmental conditions. Keeping in mind the advantages of drip fertigation especially soil buffering properties, moisture conservation, and soil temperature under various agro-climatic conditions, the use of such systems for vegetable production has been recommended in order to get high-quality produce, maximize profit, and reduce cultivation costs. Adopting these technologies is cost-effective for high-value crops like capsicum, tomato, cucumber, and greenhouse vegetables. Therefore, it is recommended that farmers should adopt drip fertigation to increase vegetable crop production.

References

Ahmad S, Hasanuzzaman M (2012) Integrated effect of plant density, N rates and irrigation regimes on the biomass production, N content, PAR use efficiencies and water productivity of rice under irrigated semiarid environment. Not Bot Horti Agrobot Cluj-Napoca 40(1):201–211

Ahmad S, Ahmad A, Ali H, Hussain A, Garcia y Garcia A, Khan MA, Zia-Ul-Haq M, Hasanuzzaman M, Hoogenboom G (2013) Application of the CSM-CERES-Rice model for evaluation of plant density and irrigation management of transplanted rice for an irrigated semiarid environment. Irrig Sci 31(3):491–506

Ahmed M, Ahmad S, Fahad S, Fayyaz-Ul-Hassan (2021) Potential Applications of DSSAT, AquaCrop, APSIM Models for Crop Water Productivity and Irrigation Scheduling. In: Fertigation Technologies For Micro Irrigated Crops, Goyal MR, and Ray LIP (eds.). CRC Press, Taylor & Franis Group

Ali H, Iqbal N, Ahmad S, Shahzad AN, Sarwar N (2013) Performance of late sown wheat crop under different planting geometries and irrigation regimes in arid climate. Soil Till Res 130:109–119

Brewer MT, Morgan KT, Zotarelli L, Stanley CD, Kadyampaken D (2018) Effect of drip irrigation and nitrogen, phosphorus and potassium application rates on tomato biomass accumulation, nutrient content, yield, and soil nutrient status. J Hortic 5(1):227

Harisha CB, Asangi H, Singh R (2017) Impact of drip irrigation and fertigation on residual soil nutrient status, nutrient uptake and nutrient use efficiency of fenugreek under semi-arid conditions. Int J Pure Appl Biosci 5(2):661–667

Meena M, Sagarka BK, Man MK (2017) Influence of drip irrigation along with nitrogen levels on yield attributes, yield and quality parameters of Rabi drill fennel (*Foeniculum vulgare* Mill). Int J Curr Microbiol Appl Sci 6(5):2115–2121

Michael AM (2008) Irrigation theory and practice, 2nd edn. Vikas Publishing, Delhi

Mubarak MK, Hussain K, Abbas G, Altaf MT, Baloch FS, Ahmad S (2022) Productivity of sorghum (Sorghum bicolar L.) at diverse irrigation regimes and sowing dates in semi-arid and arid environment. Turk J Agri For 46(1):1–18

Mustafa G, Ali MA, Smith D, Masood S, Qayyum MF, Ahmad N, Rehman A, Ahmad S, Hussain S, Arshad M, Muneer S, Khan AHA, Fahad S, Datta R, Iqbal M, Schwinghamer T (2021) Formalin fumigation and steaming of various composts differentially influence the nutrient release, growth and yield of muskmelon (*Cucumis melo* L.). Sci Rep 11:21057

Patil KVO, Das JC (2015) Effect of drip irrigation and fertilizer management on capsicum (*Capsicum annuum* L.). J Agric Vet Sci 8(1):10–13

Smith RJ, Uddin JM, Gillies MH, Moller P, Clurey K (2016) Evaluating the performance of automated bay irrigation. Irrig Sci 34:175–185

Vijayakumar G, Tamilmani D, Selvaraj PK (2010) Maximizing water and fertilizer use efficiencies under drip irrigation in chili crop. J Manag Pub Policy 2(1):85–95

Zakir I, Wahab AA, Abbas T, Haider ST, Irum S, Sabir S, Hussain S, Ahmad S (2023) Cereal-vegetable intercropping: hindrances and strategies to increase intercropping. Turk J Agric For 47(6):1115–1129

Zakir I, Ahmad S, Haider ST, Ahmad T, Hussain S, Saleem MS, Khalid MF (2024) Sweet pepper farming strategies in response to climate change: enhancing yield and shelf life through planting time and cultivar selection. Sustain For 16:6338

Zakir I, Ahmad S, Saleem MS, Hussain S, Nawaz H, Ahmed T, Khalid MF (2025) First insights on sowing time and cultivar selection of sweet corn for optimal yield and shelf life under agro-climatic conditions of Multan. Int J Plant Prod https://link.springer.com/article/10.1007/s42106-025-00329-3

Weed Management in Vegetable Crops 15

Omer Farooq, Rao Muhammad Ikram, Imran Shabbir,
Hulya Dogan, and Shakeel Ahmad

Abstract

With every passing year, population of the world and its associated food demand are increasing. To fulfill this food and nutritional demand, vegetables can play a vital role. Under the situation, vegetable production can be increased by bringing more area under cultivation or by increasing productivity of vegetable species. Among different biotic factors limiting vegetable production, weeds are considered as worse enemies. Weeds interfere with vegetables by competition for resources and by allelopathy, therefore, weeds should be controlled to avoid yield losses and fulfill the production challenges. There are several methods used to control or manage the weeds effectively; each method has its own repercussions. Preventive measures occupy a key position and can play a vital role if fully adopted. Whereas, non-chemical methods are also of prime significance to ensure the production of safe food free from toxic chemicals and their residual impact. For getting long-term and sustainable benefits, agronomic measures are the best options to follow, such as intercropping or rotation with allelopathic crops, managing proper planting geometry and cultural methods like stale seedbed preparation; all these can minimize the occurrence of weeds. Chemical method though provides good and efficient results but growers should adopt this as a last option. Lastly, the modern concept of integrated weed management (IWM) is an ideal way to deal with this ubiquitous enemy.

O. Farooq · I. Shabbir · S. Ahmad (✉)
Institute of Agronomy, Bahauddin Zakariya University, Multan, Pakistan
e-mail: shakeelahmad@bzu.edu.pk

R. M. Ikram
Department of Agronomy, MNS University of Agriculture, Multan, Pakistan

H. Dogan
Department of Agriculture and Food, Hemp Research Institute, Yozgat Bozok University, Yozgat, Türkiye

Z. Iqbal et al. (eds.), *Sustainable and Innovative Vegetable Production in times of Climate Change*, https://doi.org/10.1007/978-981-96-6283-8_15

Keywords

Allelopathy · Cover crops · Crop rotation · Herbicides · Intercropping · Integrated weed management · Mulching · Soil solarization · Stale seedbed

15.1 Introduction

Among horticultural crops, vegetables are much important worldwide, particularly in developing countries. Vegetables being a rich source of vitamins, fibers, and minerals provide healthy food, nutrition security, raw ingredients for industries, and also earn foreign exchange. Burgeoning population of the globe is expected to rise over 09 billion by the end of 2050. Vegetables are considered as the prime source for provision of healthier diets to billions of people on this planet. There are various factors including abiotic, biotic, social, and economic that hinder the successful vegetable production. Climate changes, temperature extremes, drought, and salinity are among abiotic aspects which reduce vegetables' productivity; however, important biotic reasons include insect-pests, disease-pathogens, and weeds that diminish their yield and quality. Stress caused by these factors greatly influences final yield and more importantly quality of vegetables (Bakht and Khan 2014; Channapagoudar and Biradar 2007; Khan et al. 2012; Sanjay et al. 2019; Singh et al. 2005; Syriac and Geetha 2007).

Weeds are damaging biotic elements which ultimately affect vegetable production worldwide. Weeds cause highest productivity loss because of competition for available resources (nutrients, water, light, space, etc.) and allelopathy (release of certain allelochemicals that influence growth and normal functioning of vegetable crops). Weeds are ubiquitous in nature, almost found in every vegetable production system under all weed-controlling practices (Singh et al. 2010). Despite the fact that the farmers have to deal with several production challenges, weed competition is one of the biggest profit-limiting factors for the vegetable growers (Brown et al. 2019), especially for those following organic agriculture and having small-scale vegetable farming due to lack of funds or assistance to invest on labor or other weed control options (McErlich and Boyston 2014).

15.2 Importance of Weed Management in Vegetables

Vegetable production is predicted to decline by 45–95% due to weed growth (Mennan et al. 2020). Weeds not only hinder vegetable production but also reduce their quality, which lowers their market worth. According to Majrashi and Khandaker (2020), increased weed densities can cause a 50% decrease in agricultural productivity.

Weeds delay the maturity of crops and reduce their yield and quality. Types of weeds, the severity and period of weed competition, as well as soil and climate conditions, all affect how much yield is lost. Weeds reduce production potential of different vegetables including size of onion bulbs, lettuce, and cabbage heads (Rodenburg and Johnson 2009). Studies have shown productivity losses of 66% for cabbage, 51% for cauliflower, 70% for pea, 40% for okra, 60% for tomato, 62–82% for potato, 95% for beetroot, 28–78% for carrot, 2–41% for radish, 42% for onion, and 60% for garlic (Sandhu et al. 2002; Kaur et al. 2015).

Weeds are host for numerous pests and illnesses and also create phyto-sanitary issues. Green peach aphid uses *Eleusine indica* as a host and is recognized to spread viral diseases i.e. potato leaf roll and mosaic (Rao 2000). It has been discovered that getting rid of this weed lowers the likelihood of this pest on potatoes. Pests are also carried by weeds from season to season. On some crops, some weeds have allelopathic effects. When picking lettuce, celery, or strawberries, the irritating sting of burning nettle (*Urtica urens*) can significantly reduce workers' productivity. Therefore, there is a critical requirement for effective weed control strategies in these crops if one wants to maximize the returns from inputs applied to these horticulture crops.

15.3 Weed-Vegetable Competition

Weeds compete with vegetable crops for water, nutrition, space, light, and gases, hence, ongoing monitoring and weed-removal efforts are required for a healthy plantation along with higher yield. Most of the vegetables are susceptible to weed competition in their early stages of development, making yield losses up to 70–80% (Rana et al. 2011). However, early weed competition may not have much impact on final production for some crops with non-determinate development habits. The cucurbit family appears to be less vulnerable to weed competition in the early stages. Vegetables-weeds competition periods are mentioned in Table 15.1. Crop harvesting may be impacted or interfered by crop-weed competition. Economy and product quality are also impacted by this competition.

Vegetables have shallow roots and a slow rate of growth, making them poor competitors with weeds. According to Akemo et al. (2000), the weed cover in pea reached 73% in the absence of weed management practices. Competition with *Cyperus rotundus* L. and *Amaranthus palmeri* weeds caused pepper fruit yields to drop up to 94% and 44%, respectively (Norsworthy et al. 2007), while pepper needed weed-free time for 12.2 weeks to minimize productivity losses.

Weeds may improve the habitat for specific infections like downy mildew and *Sclerotinia sclerotiorum* in addition to competing for the fundamental growth requirements. The disease epidemiology is considerably aided by the dense weed growth that restricts airflow.

Table 15.1 Critical period of vegetables-weed competition

Name of vegetable	Competition period	Crop stage	References
Spinach	20–30 days	After transplanting	Shawquat et al. (2009)
Cabbage	2–4 weeks	Post-emergence	Weaver (2006)
Broccoli	15–30 days	After transplanting	Latif et al. (2021)
Brinjal	20–60 days	After sowing	Rana et al. (2011), Bhullar et al. (2015)
Turmeric	30–40 days	Post-planting	Kifelew and Getachew (2017)
Turnip	6–8 weeks	After sowing	Davies (2003)
Radish	3–4 weeks	Post-planting	Harris et al. (2015)
Cucumber	4 weeks	After seeding	Weaver (2006)
Peas	20–70 days	After sowing	Singh et al. (2015a, b)
Lettuce	35–37 days	After transplanting	Riva et al. (2023)
Onion	20–60 days	After transplanting	Chopra and Chopra (2006)
Pepper	08 weeks	After transplanting	Galal et al. (2020)
Potato	19–24 days	Postemergence	Ahmadvand et al. (2009)
Tomato seeded	9 weeks	Post-planting	Weaver (2006)
Okra	2–4 weeks	Post-emergence	Imoloame (2013)
Cauliflower	0–38 days	After transplanting	Qasem (2009)
French bean	11–28 days	Post-emergence	Stagnari and Pisante (2011)

15.4 Weed Control Strategies in Vegetables

15.4.1 Preventive Approaches

In all the weed management systems of vegetables, prevention as a first choice has a keystone position. It is undoubtedly one of the cheapest strategies to adopt for the farmers. Whereas, such strategies are sometimes diverse in nature that include combination of practices and policies that checks weeds entry to a field and further spread of some species to an area that was earlier free from those specific weed species. Precautionary approaches are efficient for all sorts of areas, including a minor vegetable area.

Prevention approach is a vital pillar of integrated weed management (IWM) (Norris et al. 2003) but the approach sometimes requires great attention as in some circumstances, growers are not willing to adopt this especially when land under cultivation is on rent. Following are some key points that should be strictly observed for weeds prevention:

- Field water channels, main water channels, and canal banks should be carefully monitored and weeds should be removed to avoid weed seed entry in the fields.
- Farmhouse equipment must be cleaned before its usage in fields.
- If farm yard manure is used for nutritional purpose, it should be weed seed free.
- Government rules and regulations should be strictly followed concerning entrance and seeds movement or seed materials.

15.4.2 Non-chemical Approaches

Prior to development of selective herbicides, non-chemical approaches were the primary means of controlling weeds. However, wide herbicides usage in the latter part of the twentieth century largely replaced non-chemical approaches. The necessity of non-chemical methods of weed control is escalating due to worries about the harmful effects of herbicides on environment as well as human health. There is a rising demand for organic vegetables globally, and non-chemical weed control methods are essential for their production. The significance of non-chemical weed control is highlighted even more in vegetable production systems due to the emergence of herbicide-resistant weeds (Heap 2014). The non-chemical approaches include preventive measures to curtail weed germination, cultural techniques to enhance crop competition, and direct physical methods to reduce weed survival. The preventive techniques encompass crop rotation, mulching, and cover cropping, which lessen weed germination by hindering light and controlling soil temperature. Cultural methods, like intercropping, crop rotation, and robust soil manipulation like hand-weeding and hoeing are being used over a period.

15.4.2.1 Agronomic/Cultural Measures

Cultural methods for controlling weeds consist of various options, such as crop rotation, enhancing crop competitiveness, seeding earlier or later, flooding, solarization, a stale seedbed, green manure, cover crops, and intercropping. The major goal is to stop prevailing weeds from spreading and to curtail new weed species' introduction. To enhance crop competitiveness against weeds, it is recommended to select suitable crops and cultivars, consider the weed species and climate conditions, ensure prompt and uniform crop emergence via appropriate seedbed preparation, utilize appropriate seed and planting depth, raise planting density, and adjust planting arrangement to suppress weed growth. Adequate and localized resource applications as well as optimum crop management are also key factors in increasing crop competitiveness against weeds (Bhadu and Kaswan 2022). In addition to well-known tactics for choosing competitive cultivars, optimizing seed density, and using fertilizers, several other measures play a significant role in agroecosystem design and are gaining a lot of attention.

15.4.2.1.1 Tillage Measures

Weed emergence and soil seedbank are significantly impacted by tillage. Vertical dissemination of weed's seed in a soil profile is altered by tillage, which also affects soil environment around the seeds, influencing weed seed germination (Chauhan and Johnson 2009). Changes in moisture, fluctuations in temperature, availability of light, and activity of predators can have an impact on weed seeds when they are located at different depths (Singh et al. 2015a, b). In systems with minimal soil disturbance, like zero or minimum cultivation, most weeds seed tend to be concentrated near or on the soil surface. Conversely, traditional tillage methods that rely mainly on plowing tend to distribute weed seeds uniformly throughout tilled soil. Weed species having small seeds, rarely dormant, and with a short lifespan can quickly boost if not adequately managed (Graziani et al. 2012). The plowing tillage strategy is suggested for burying surface weed seed in deeper soil layers, leading to increased mortality, reduced seedling emergence, and a decline in the seed bank. Nonetheless, plowing can conserve weed species whose seeds exhibit long lifespan and secondary dormancy by burying them deeper in the soil (Graziani et al. 2012). Seeds of some weed species germinate 02–03 weeks prior to the emergence or transplant of vegetable crops when seedbed has been prepared earlier and irrigation has been applied (Rasmussen et al. 2011). These weed seedlings should be eradicated through harrowing or other practices before sowing or transplanting the crop. Stale seedbed works well for integrated weed control in onion (Tei et al. 1999), cabbage (Tei et al. 2005), spinach (Peruzzi et al. 2006), and lettuce (Tei et al. 2007).

15.4.2.1.2 Crop Rotation, Cover Crop, Intercropping

Crop rotation is an additional strategy that can help with weed issues. Crop rotation exerts multiple impacts on the weed communities and significantly curtails growth of leading weed's species. Rotation is effective in eradicating a weed community that is weedicide-resistant but may not be effective to control species that are similar to crops. To accomplish this, it is possible to alternate different types of crops, for example, leafy vegetables (such as spinach, lettuce, and cabbage), root vegetables (such as beetroot, carrot, and radish), tuber vegetables (such as potatoes and Jerusalem artichoke), bulb vegetables (such as garlic and onions), and fruit vegetables (such as cucurbits). Whenever possible, cultivate cereals, legumes, and cruciferous plants, which not only aid in weed management, but also enhance soil fertility and minimize soil erosion. Vegetables from the same family shouldn't be grown in the same spot year after year. Once a crop is chosen, select adaptable and robust cultivars that are disease-resistant. Plants with disease cannot successfully compete with weeds. Switching between crops with various crop cycles; avoiding uninterrupted cropping systems; switching between low competition crops (such as lettuce onion, radish, and garlic) and high competition crops (such as cucurbits, potato, etc.); alternate between crops requiring simple, low-cost weed control and crops requiring complex, high-cost weed control are some important strategies for weed control. Brainard et al. (2008) reported reduced weed seedbank in response to crop rotation in snap bean, sweet corn, and cabbage. Larkin (2020) reported that bean yields were 08% greater after squash than after pepper due to the vegetable rotation, whereas squash yields were 15% higher after beans compared to pepper, while pepper yielded 11% more after beans compared to squash. These findings aid in defining crop rotation to enhance production of organic vegetables and offer lucrative returns to the growers.

Various researchers have explored the potential of utilizing cover cropping in conjunction with crop rotations. Crops sown as cover help for green manuring and living or dead mulch. The success of cover crops, including weed control, soil fertility improvement, and erosion reduction, is significantly affected by several factors, like crop selected, planting dates, composition of weed flora, water availability, fertilizer application, as well as their competition with the major crop. Radicetti et al. (2013) reported positive response in pepper yields when oat was used as a cover crop. Among various cover crop species, the most effective for weed mitigation in cabbage were *Sorghum bicolor*, *Sorghum sudanense* (P) Stapf., and *Vicia villosa*, which reduced weed biomass and densities (Mennan et al. 2009). *V. villosa* produced a higher biomass of 7.49 t ha^{-1}, as a cover crop in tomato and the living mulch from this cover crop reduced biomass of some weed species (*Amaranthus retroflexus*, *Digitaria sanguinalis*, *Portulaca oleracea*) by 40%, and weed density by >70% (Campiglia et al. 2010). According to Jabran et al. (2015), usage of crops as cover may result in allelopathy interactions among minor as well as other crops or weeds.

In addition to weed control, intercrops can be superior to solitary crops due to higher productivity and economic return. However, intercropping system must be designed with the best component crops, additional IWM aspects, and environmental variables for better weed control under certain circumstances (Weerarathne et al. 2016). Hairy vetch as a cover crop reduces weed infestation by 70–73%, and can be used for suppressing weeds and enhancing productivity of sweet peppers (Isik et al. 2009) and tomatoes (Campiglia et al. 2010). Fenugreek can act as a living mulch to decrease weed biomass in coriander production (Pouryousef et al. 2015). However, the efficacy of fenugreek as a weed control option is limited and additional weed management strategies such as using a competitive cultivar or a stale seedbed should be employed to attain adequate control throughout the season. Demir and Polat (2011) reported that intercropping broccoli and crisp salad had an LER (Land Equivalent Ratio) value greater than 01 under greenhouse conditions. Yildirim and Guvenc (2005) found that intercropping cauliflower with lettuce, salad, onion, radish, and eggplant with bean, leaf lettuce, and cos lettuce provided similar results.

15.4.2.1.3 Planting Geometry

One effective approach to suppress weeds and increase yields is to increase the density of crop plants in each row (Korres 2018). Proper plant densities and closer row spacing promote quick crop ending. Later-emerging weeds are shaded out by a closed canopy, and weed seeds that need light to germinate are prevented from doing so. Weeds rarely become an issue until the canopy covers. De Cauwer et al. (2021) discovered that using narrow intra-row spacing resulted in lower biomass per plant and greater overall as well as commercial spinach biomass. Other studies have also demonstrated that increasing populations within rows can enhance the competitiveness of crops, even those with slower early growth (Alba et al. 2020). Additionally, localized, and appropriate resource application, as well as optimal crop management, can aid in enhancing crop competitiveness against weeds (Bhadu, and Kaswan, 2022). This technique could be integrated with mechanical weeding treatments among rows for further reducing the weed growth (Gazoulis et al. 2021).

15.4.2.1.4 Other Means

Planting at the appropriate time is a valuable cultural technique to increase crop competitiveness. Depending on the ideal temperature for their growth, vegetables can be classified as warm- or cool-season crops. Emergence time and earlier seedling vigor of crops are influenced by the planting date and these are crucial in determining crop competitiveness.

The most popular technique used by farmers to get rid of weeds in vegetable cowpea is manual weeding. Its effectiveness is limited by high wage rate, unavailability of laborers when needed, and unusual weather conditions. Cowpea pod output decreased significantly and crop plant damage was induced by manual weeding after full ground coverage and during the reproductive stage (Adigun et al. 2014). Hand weeding conducted at 3–6 weeks post-sowing has been demonstrated to be an effective approach for controlling weed growth in cowpea. Kujur et al. (2015) obtained highest pods/plant and yield of 1016.66 kg ha^{-1} of cowpea with hand weeding at 20 and 40 days after sowing (DAS). Grain yield was equivalent to weed-free when hand weeding and intercultural activities were conducted 20 and 40 DAS (Kumar and Singh 2017). Two-hand weeding plus one intercultural operation produced best productivity of cowpea (1581.02 kg ha^{-1}) and increased production by 25.2% over the weedy control. For green gram, the optimal timing for weeding was found to be 02–05 weeks post emergence, resulting in lowest dry weight, highest nodules/plant, and highest yield attributes (Kumar et al. 2017).

(a) **Stale Seedbed Technique**

Pre-sowing irrigation was used to facilitate the growth of weed seeds, which were then eliminated through spraying non-selective weedicides or pre-sowing tillage techniques once they had germinated (Gnanavel and Kathiresan 2014). Effectiveness of the stale seedbed (SSB) method in controlling weeds is dependent on numerous factors such as the preparation method, types of weeds present, technique used to eliminate emerging weeds, environmental conditions, and season (Singh 2014). According to Riemens et al. (2007), the use of mechanical weed control in conjunction with the stale seedbed (SSB) technique proved to be equally effective in reducing weed density as the chemical method in cowpea at 15, 30, and 45 DAS. According to Sinchana (2020), SSB method was found to significantly reduce overall density and weed biomass. Additionally, when combined with mulching using dried banana leaves and quizalofop-p-ethyl herbicide, it demonstrated the lowermost weed index, highest weed management efficiency, along with the highest green pods productivity of vegetable cowpea.

(b) **Selection of Cultivars**

The yield, quality, and resilience to biotic and abiotic stresses are all taken into consideration while choosing crop cultivars. To better compete with weeds, new kinds should have additional traits including root development, quicker beginning growth, large leaf area, and allelopathic action. Despite the observed benefits of using competitive cultivars for weed management in other crops, plant breeders have not paid sufficient attention to the potential of this technique in minor crops, which could deliver safe as well as environment-friendly solution (Mahajan and Chauhan 2013).

In organic farming, where the use of herbicides is limited or prohibited, agronomic techniques like changing plant spatial patterns, adjusting sowing times, and using narrow crop rows can be used to maximize the benefits of competitive cultivars. This can enhance the significance of competitive cultivars as a weed management strategy (Mahajan and Chauhan 2013). In a non-chemical weeding system, the use of competitive cultivars can lead to significant reductions in selection pressure, herbicide application, and labor costs while also enhancing weed control efficacy. To increase competitiveness and improve weed control for minor crops, it is recommended to select crop cultivars with specific characteristics such as droopy leaves, high tiller density, and early-stage biomass accumulation. Other desirable traits include high specific leaf area, high leaf area index during vegetative growth, quick canopy coverage, and early vigor. These characteristics have been observed to promote competitiveness in certain crops; however, using weed-competitive cultivars alone may not be sufficient for effective weed management. Instead, it should be considered as a valuable component of an integrated approach (Chauhan 2012).

15.4.2.2 Mechanical Approaches

New mechanical weed control techniques have been developed as a result of organic farming during the past two decades. Now these techniques are more efficient, quick, and do not leave chemical residues on crops (Pannacci and Tei 2014). Therefore, the primary direct method of weed suppression in organic or low-input systems is mechanical weed management. In conventional cropping systems, mechanical techniques could be employed to decrease dependency on weedicides. However, there are certain disadvantages of mechanical weed control: (1) it takes a lot of time and requires money that may have been used for various other crop works; (2) weather and soil conditions can have a big impact on proper time and efficiency; (3) limited impact on intra-row weeds; (4) high initial cost and high ongoing costs; and (5) need for qualified operators to better tune and operate mechanical weed equipment. Based on how they work on the crop row, mechanical weed instruments can be divided into sub-categories: inter-row and intra-row.

15.4.2.2.1 Inter-row Cultivators

Using instruments that exclusively till between row spaces, mechanical weed management in row crops (like okra, corn) can be done. Strong weed-suppression techniques include cultivation and uprooting, cutting, shredding, tearing, and burying.

However, the type of weed, its stage of development, and the tools being used all affect how effectively the weeds can be removed. The three types of hoes used for mechanical weed control are traditional hoes, rotary hoes, and rotary tillage cultivators. A traditional hoe is towed by a tractor and features a variety of operating implements placed on a fixed or moveable shank (sweep, blade, duck foot, etc.). Using a rotary hoe or more advanced working tools (functioning in inter-row space), including discs as well as spike wheels, enable an uprooting and burying weed management action. A late mechanical weed management is possible with both traditional and rotary hoes only when plants are mature (0.4–0.5 m height and prior to canopy cessation). Also, they cultivate quickly and take little time because of their high operative speed, which ranges from 04 to 08 km h^{-1}. The effectiveness of weeding with a hoe depends on the characteristics of the soil like moisture content and texture, weather after soil cultivation, and characteristics of the working equipment (size, shape, and working depth). Weeding effectiveness can decrease by 30–40% in rainy weather after soil cultivation (Lichtenhahn et al. 2005). Rotary tilling cultivators are employed to control perennial weeds due to their strong activity in soil. Hoeing's primary drawback is its lack of efficiency against intra-row weeds because the instruments are placed far from the row to prevent crop damage (Ascard and Fogelberg 2002).

To maximize the effectiveness of weed management on crop rows, hoeing should always be done in conjunction with ridging. This feature was emphasized for the minor crops including sunflower, maize, and soybean (Pannacci and Tei 2014). The traditional hoe has been successfully supplanted by precision hoe in small-scale crops like radish, carrot, onion, fennel, and spinach grown organically and is also suitable for greenhouse vegetables (Peruzzi et al. 2005). The use of a precision hoe, equipped with a range of tools like flexible teeth, goosefoot, and L-shaped tines can help to reduce weeds density. Pannacci and Tei (2014) reported that up to 65% of chemical input can be saved with this combined approach.

Brush weeders and split-hoe are also used as inter-row weeders. With an inter-row interval ranging from 0.20 to 0.40 m, brush weeders are frequently used for greenhouse and vegetable crops such as lettuce, tomato, chicory, spinach, carrot, etc. (Turner et al. 2007). Low operation capacity (between 0.3 and 0.5 ha h^{-1}), rapid brush deterioration, poor efficiency against established weeds, and effectiveness solely dependent on soil moisture are some of the disadvantages. Soil moisture and weed species directly affect the efficacy of hoes, rotary tilling cultivators, and brush weeders, while split hoes are less reliant on these factors. Further, split hoe has a more potent and non-selective effect on tiny, established, and perennial weeds species (Pannacci and Tei 2014). In comparison to an untreated control, split-hoe has demonstrated efficiency against a weed flora comprising *Solanum nigrum* L., *Chenopodium album* L. as well as *Echinochloa crus-galli* L. (Pannacci and Covarelli 2005). The results of the study emphasized the capability of split-hoe as a mechanical approach for managing weeds in crops that are either transplanted or sown, including onions, carrots, lentils, beans, chickpeas, cabbage, lettuce, and chicory. However, this mechanical technique should be carried out multiple times, and the inter-row spacing should be within the range of 0.20–0.25 m.

15.4.2.2.2 Intra-row Cultivators

Many intra-row mechanical weeding methods with great crop plant selectivity have been created in recent times. Rubber-finger weeders are effective in controlling weeds in vegetable as well as greenhouse crops that have inter-row space ranging 0.25–0.30 m to 0.40–0.50 m. This technique is applied in the inter-row as well as the intra-row spaces for weed management in crops such as onion, lettuce, tomato, chicory, spinach, leek, fennel, radish, and others. Intra-row weeding is carried out using rubber-finger weeders that move along soil surface. This weeding method is recommended when plants have reached a well-established stage, typically after four true leaves have emerged. The working depth range is between 10 and 30 mm. As a result, the instrument can only cultivate a maximum of 06 crop rows at once and operate at a speed of 04–05 km h^{-1}. In clay soils with vegetable crops having 02–04 true leaves, rubber-finger weeders have demonstrated a low effectiveness against weeds and a less crop selectivity (Kurstjens and Bleeker 2000).

When there were 182 plants of *Portulaca oleracea* L. and 27 plants of *Amaranthus retroflexus* per square meter, finger weeding had 50% efficiency against weeds in seeded onions when the crop was at the cotyledonal stage. Same conditions were utilized for the study of spinach where the crop was planted with an inter-row 0.25 m distance, and finger weeding was carried out when the crop had 04–06 true leaves. With an inter-row distance of 01 m in organic transplanted tomatoes, finger weeders can be employed after the plants have 02–03 leaves (Pannacci and Tei 2014).

A very basic instrument known as a "torsion weeder" is made up of two spring tines pairs with a 90 mm diameter that are attached to a strong frame. A precision hoe is mounted on a more advanced machine with a torque weeder. Precision hoe is made up of numerous implements which enhance tilling and finish torsion weeder's activity by focusing the inter-row spacing (Peruzzi et al. 2005). The effectiveness of the torsion weeder has been demonstrated in both horticultural and greenhouse crops, including but not limited to common bean, tomato, cabbage, leek, celery, fennel, and lettuce. Production yield for onions was comparable when employing chemical weed control and torsion weeders (Ascard and Bellinder 1996). Torsion weeders exhibited 80% weed killing in leeks (Kurstjens and Bleeker 2000). The efficacy of the torque weeder was demonstrated in trials conducted on maize as well as sugar beet, particularly for controlling small weeds at cotyledon to two true leaves growth stage. The selectivity of the crop was affected by the growth stage as well as appropriate tool settings (Peruzzi et al. 2006).

Around the crop plants, in a 24 cm radius zone, 62–87% weed reductions were achieved in transplanted cabbage through robocrop (Tillett et al. 2008). For transplanted celery and lettuce, robocrop was used instead of a typical hoe to provide superior intra-row weed control. Melander et al. (2015) found no appreciable differences in the intra-row weed management of transplanted cabbage and onion when using the robovator and other equipment (flex-tine harrow, torsion weeder, and finger weeder).

15.4.2.3 Physical Measures

Physical methods for weed control, such as mulching, solarization, flaming, and steaming, have been widely used in small-scale crop management. However, current-day research is exploring alternative thermal weed managing approaches, including freezing, microwaves, laser, electric as well as gamma ray, as feasible options.

15.4.2.3.1 Mulching

In addition to preventing soil erosion, mulching also reduces weed populations during early crop growth stages by preventing weed emergence and establishment. By obstructing sunlight or establishing conditions that will stop weeds from germinating, emerging, and growing later, mulches control weed growth. Mulches are of two kinds: natural and artificial.

Artificial mulches made of synthetic materials are simple to use, effective at keeping weeds out of the row, moisture-saving, and raising soil temperature. The usage of plastic mulches in the cultivation of vegetables is becoming more and more important. In these kinds of plantings, weed control is different from bare-ground production. Herbicide application beneath the plastic is not necessary when methyl bromide fumigation is utilized.

Organic mulches are more common in organic farming systems. These are crucial in raising agricultural productivity (Mahmood et al. 2015). The addition of organic mulch, such as cereal straw or mush cortex, to the soil can lead to potential improvements in the soil's biological, chemical, and physical characteristics (Wang et al. 2009). Thickness of mulch determines how effective it is (Teasdale and Mohler 2000). However, compared to some translucent mulch, this technique only slightly raises soil temperature. Paper mulches can effectively remove weed plants in watermelons, common beans, and tomatoes (Sanchez et al. 2008) and in processing tomatoes contaminated with *Cyperus rotundus* L. (Cirujeda et al. 2012). According to Mohler and Johnson (2009), cucurbits (i.e., squash, zucchini, watermelon, and melon) tend to become weedy if mulches are not used because it is practically impossible to cultivate or physically weed them once the stems have spread out of the row. Moreover, by the time these crops reach maturity, weeds may have often already produced seeds. To address this issue, a series of short-season plants, such as spinach and lettuce, are grown after the cucurbits to suppress weed growth before they can produce seeds. Yet, for practical, economic, and market reasons, crop rotation is difficult to apply (particularly in a greenhouse cropping system).

Fasna et al. (2022) evaluated six non-chemical methods for weed management in chilli, including stale seedbed method, mulching, and hand weeding. They found that black polythene mulch was the most effective method for weed control, resulting in the highest plant height, fruit yield (22.44 t ha^{-1}), and organic carbon content in soil. Straw mulching was also effective when used alone or followed by one-hand weeding. Straw mulching had the highest values of available phosphorus and potassium, while black polythene mulch had the highest available nitrogen at 60 days after transplanting and stale seedbed afterward one manual weeding had the highest available N after the final harvest. The black polythene mulch had the highest benefit-cost ratio (1.84), followed by straw mulching (1.42).

Gupta et al. (2019) revealed two manual weedings on 20 as well as 40 DAP and straw mulching with one-hand weeding were the most effective non-chemical control methods in reducing weed density (86%) in potato (*Solanum tuberosum* L). Akshatha et al. (2018) reported that mulching with silver-black polythene alone resulted in the highest yield of 16.83 t ha^{-1} by maintaining a weed-free condition in cabbage.

15.4.2.3.2 Soil Solarization

Soil solarization is a non-chemical, hydrothermal approach of soil disinfestation in which soil absorbs solar radiation under a translucent polyethylene sheet (Cohen et al. 2019). When the soil is sufficiently moist inside a polythene sheet, sun radiation heats the soil, pasteurizing it to kill dangerous pests while preserving the population of beneficial soil bacteria that encourage plant growth (Lombardo et al. 2012). Soil solarization is a potential technology to lessen reliance on synthetic pesticides and chemicals for weed management. The efficacy of soil solarization depends on the duration of the process and the temperature. The study found that soil solarization for 08 weeks was most effective in inhibiting weed growth and resulted in the highest productivity for *Abelmoschus esculentus* (L.) Moench. Therefore, using soil solarization in agricultural fields can increase crop productivity (Kapoor 2020).

The method entails placing a clear plastic film or mulch over moist soil and using sunlight to heat it. The mulch must remain on soil surface 04–12 weeks period for maintaining higher temperature. While it is commonly used to treat nematodes and fungal infections, this technique is also effective for managing weeds growth (Gill et al. 2009). This technique makes it possible to keep soil temperatures over 40 °C, which destroys weed seeds and seedlings (Chase et al. 1999). Although solarization has been shown to be beneficial in both outdoor and indoor settings, its usage is mostly limited to vegetable and small-scale crops (Lombardo et al. 2012). Maheswari and Karthik (2019) revealed enhanced carrot yield and weed control when soil was solarized for 03–09 weeks against non-solarized plots.

15.4.2.3.3 Flaming

The most used thermal weed extraction technique is flaming (Ascard et al. 2007). Flaming's effectiveness is influenced by the weed's shape and stage of growth. In comparison to mature plants and grass species, dicotyledonous and young weed plants are more susceptible because their meristems are not shielded (Ascard 1995). Liquid petroleum gas (LPG)-powered flaming equipment can be utilized for various weed management practices, including pre-sowing or pre-transplanting with an old seedbed, pre-emergence, post-emergence using or not using a protecting shield depending on the crop's tolerance to elevated temperatures. It seems that for field crops with wide rows, flaming may be a suitable option than in greenhouse crops. Flaming can be safely used on direct-seeded and bulb onions before and after emergence. The main benefits of flame weeding include the ability to handle herbicide-resistant weeds in conventional systems, manage organic vegetable crop production, and utilize alternative approach against cultivation if the field is wet (Horesh et al. 2019).

15.4.2.3.4 Steaming

Weed seeds and propagules can be rendered inert with the help of steam. Melander et al. (2004) found that steaming at a temperature of 60 °C can minimize emergence by 90% and can eradicate resistant weeds when done for 06–09 s. Using the temperatures of 80 °C for *Echinochloa crus-galli*, and 76 °C for *Setaria viridis*, and 75 °C for *Solanum nigrum* for 02–05 s, devitalization of some weeds can reach 99% (Vidotto et al. 2013). Various continuous and intermittent steaming technologies have been developed to enhance their efficiency, while simultaneously reducing the time and cost involved in their application. Laboratory experiments indicated that steam heat was well-tolerated by leek, onion, and occasionally carrot seeds when sown immediately after steaming. This finding suggests that crop sowing and steaming could be combined, enabling both tasks to be completed simultaneously (Ascard et al. 2007). Carlesi et al. (2021) reported that by decreasing weed density, the use of steam application provided carrots with a competitive edge, resulting in a fresh yield increase from 47% to 92% compared to yields from plots without steam treatment.

15.4.2.4 Exploiting Allelopathic Measures

Adopting natural herbicides is a novel strategy to counteract the harmful effect of synthetic weedicides for vegetable production. One of the main obstacles in achieving effective competition among crops as well as weeds in ecosystems is allelopathy, naturally occurring phenomenon in chemical ecology. Unlike synthetic herbicides, most natural allelochemicals are non-toxic to humans, environment-friendly, and easily biodegradable, making them an excellent weed-managing strategy. Primary use of allelopathy involves identifying phenolic compounds in plant extracts that possess allelochemical activity and using as weedicide. Natural products in plants or weeds with phytotoxic properties can be harnessed for weed management, with certain species like *Chenopodium album, Amaranthus retroflexus,* and *Cynodon dactylon* known for producing allelopathic compounds that lead to crop reductions, with *C. dactylon* being the most harmful. However, some allelochemicals can also stimulate crop growth and/or inhibit weed growth by regulating numerous aspects of growth, like photosynthesis, respiration, transpiration, mineral uptake, and enzyme activity.

Alsharekh et al. (2022) reported that *Conocarpus erectus* extracts were tested at different concentrations against several plant species. The results revealed that *C. erectus* leaves and seeds extracts had a maximum inhibition rate against *Chenopodium murale* and *Amaranthus viridis* but did not exhibit any negative effects on tomato as well as cucumber. Mabele and Ndong'a (2020) used guava extract as an allelopathic chemical and successfully reduced the weed density in tomato crop. Besançon et al. (2021) stated that sorghum-Sudan grass cover crop significantly reduced dry biomass of smooth pigweed and common lambs quarters by 77 and 82% respectively, in cabbage. Likewise, allelopathic potential of sorghum and brassica have been tested against the weeds of kharif season (Farooq et al. 2018).

15.4.3 Chemical Measures

Reduced chemical use is a primary benefit of non-chemical weed control techniques. Chemical herbicides can harm soil health, water quality, and non-target organisms, including beneficial insects and wildlife. Moreover, chemical herbicides can leave residues on crops, endangering humans and animals who encounter them. Non-chemical weed control methods can limit the application of chemical herbicides, resulting in more sustainable soils, purer water, and safer food.

Chemical control of weeds using herbicides is a popular method among vegetable farmers due to its convenience and effectiveness. To avoid competition between crops and weeds, weedicides are typically applied prior to critical period of crop-weed competition, either pre- or post-emergence. The approach provides a favorable environment for the crops to grow and develop during their initial stages. Moreover, herbicides can effectively control a wide range of weeds, including those which are morphologically similar to the crops and those which grow within the rows. For instance, Singh et al. (2008) found that pre-emergence application of butachlor at a rate of 1.5 kg ha^{-1} resulted in better weed management option subsequently with higher root productivity of radish. Another study by Singh et al. (2017) recommended flurochloridone at a rate of 625 g ha^{-1} as pre-emergence herbicide for weeding in vegetables. In addition, halosulfuron methyl at a rate of 52.5 g ha^{-1} as post emergence, effectively controlled sedges and increased bottle gourd productivity (Dash and Mishra 2014). Application of pre- and post-emergence herbicides was found to be an effective strategy for managing weeds in okra as compared to pre-emergence alone (Patel et al. 2017).

15.4.4 Integrated Weed Management (IWM)

IWM is a key strategy to change the crop's competition with weeds. It is superior to any single technique for reducing the growth of weeds in crops (Lamichhane et al. 2017). By using all available weed management methods, a systemic strategy is needed to maintain weed density below economic threshold level (ETL). Daramola et al. (2020) evaluated the efficacy and cost-effectiveness of different weed managing strategies in okra production. Pre-emergence herbicide in combination with manual hoeing improved okra yield and profitability comparatively at a lower expense. De Cauwer et al. (2021) reported that narrow-row spacing combined with post-emergence broadcast harrowing resulted in lowest weeds total dry matter in spinach. Kaur et al. (2020) discovered that utilizing pendimethalin as a pre-emergence herbicide and crop residue mulch, followed by imazethapyr as a post-emergence herbicide, produced pea yields comparable to weed-free control and resulted in higher weed control efficiency, leaf area index, biomass, net returns, and net benefit-cost than most other treatments. They suggested that this herbicide and residue mulch combination could lead to improved weed management and higher pea productivity and profitability without the risk of herbicide residue in the pea and soil, while also reducing the use and intake of residues into the environment. Shafiq and Kaur (2021) used weedicides and paddy straw as IWM in potatoes. They reported that uncontrolled weeds caused a 49.7% yield reduction in potatoes. Metribuzin applied as pre-emergence killed weeds effectively at 30 days after planting but lost effectiveness by harvest. Post-emergence application of clodinafop reduced grasses and a pre-mix of clodinafop and metribuzin killed broadleaf weeds and grasses, resulting in higher tuber yield and economic returns compared to individual applications of metribuzin or clodinafop. The most effective treatment for achieving high tuber yield and 95.3% weed control efficiency at harvest was the integration of paddy straw mulch with post-emergence application of clodinafop plus metribuzin. The second most effective treatment was the application of paddy straw mulch with pre-emergence metribuzin. Faruq et al. (2021) found the highest weed control efficiency and marketable tuber yield with a combination of hand weeding and mulching, followed by herbicide application. They suggested that using a combination of hand weeding and mulching could be a cost-effective solution to weed management and labor shortage.

15.5 Conclusion

Weeds are frequently removed manually from vegetable fields. However, the effectiveness of this method is constrained and the crop's output is impacted by the unavailability of labor at the appropriate time, pay rate increase, and unusual weather patterns. An effective and convenient alternative to handle this scenario is chemical weed control. Due to environmental issues and emergence of weed resistance, it is not recommended to solely rely on herbicides to manage weeds. Hence, other weed management strategies are also necessary for optimal weed management. As a result, a system that incorporates cultivation, proper crop husbandry techniques, and use of herbicides should be adopted. IWM may be simply and successfully implemented by a grower as a component of any comprehensive managing approach. It is a cost-effective, sound, and reliable approach.

References

Adigun J, Osipitan AO, Lagoke ST, Adeyemi RO, Afolami SO (2014) Growth and yield performance of cowpea (*Vigna unguiculata* (L.) Walp) as influenced by row-spacing and period of weed interference in Southwest Nigeria. J Agric Sci 6(4):188–198

Ahmadvand G, Mondani F, Golzardi F (2009) Effect of crop plant density on critical period of weed competition in potato. Sci Hortic 121(3):249–254

Akemo MC, Regnier EE, Bennett MA (2000) Weed suppression in spring-sown rye (Secale cereale)–pea (*Pisum sativum*) cover crop mixes. Weed Technol 14(3):545–549

Akshatha V, Prameela KP, Narayankutty C, Menon MV (2018) Weed management in cabbage (*Brassica oleracea* var. *capitata* L.). J Trop Agric 56(2):187–190

Alba OS, Syrovy LD, Duddu HS, Shirtliffe SJ (2020) Increased seeding rate and multiple methods of mechanical weed control reduce weed biomass in a poorly competitive organic crop. Field Crop Res 245:107648

Alsharekh A, El-Sheikh MA, Alatar AA, Abdel-Salam EM (2022) Natural control of weed invasions in hyper-arid arable farms: allelopathic potential effect of *Conocarpus erectus* against common weeds and vegetables. Agronomy 12(3):703

Ascard J (1995) Effects of flame weeding on weed species at different developmental stages. Weed Res 35:397–411

Ascard J, Bellinder RM (1996) Mechanical in-row cultivation in row crop. In: Proceedings of the second International Weed Control Congress, 25–28 June 1996, Copenhagen, Denmark, pp 1121–1126

Ascard J, Fogelberg F (2002) Mechanical intra-row weed control in organic onion production. In: Proceedings of the 5th EWRS workshop on physical weed control, 11–13 March 2002, Pisa, Italy, p 125

Ascard J, Hatcher PE, Melander B, Upadhyaya MK (2007) Thermal weed control. In: Upadhyaya MK, Blackshaw RE (eds) Non-chemical weed management: principles, concepts and technology. CAB International, Wallingford, pp 155–175

Bakht T, Khan IA (2014) Weed control in tomato (*Lycopersicon esculentum* Mill.) through mulching and herbicides. Pak J Bot 46(1):289–292

Besançon TE, Wasacz MH, Heckman JR (2021) Weed suppression, nitrogen availability, and cabbage production following sunn hemp or sorghum-sudangrass. Hortic Technol 31(4):439–447

Bhadu K, Kaswan B (2022) Organic weed management an important component of organic farming: a review. Pharm Innov J 11(7):1655–1660

Bhullar SM, Kaur T, Kaur S, Yadav R (2015) Weed management in vegetable and flower crop-based systems. Indian J Weed Sci 47(3):277–287

Brainard DC, Bellinder RR, Hahn RR, Shah DA (2008) Crop rotation, cover crop, and weed management effects on weed seedbanks and yields in snap bean, sweet corn, and cabbage. Weed Sci 56(3):434–441

Brown B, Hoshide AK, Gallandt ER (2019) An economic comparison of weed management systems used in small-scale organic vegetable production. Org Agric 9:53–63

Campiglia E, Caporali F, Radicetti E, Mancinelli R (2010) Hairy vetch (*Vicia villosa* Roth.) cover crop residue management for improving weed control and yield in no-tillage tomato (*Lycopersicon esculentum* Mill.) production. Eur J Agron 33:94–102

Carlesi S, Martelloni L, Bigongiali F, Frasconi C, Fontanelli M, Bàrberi P (2021) Effects of band steaming on weed control, weed community diversity and composition and yield in organic carrot at three Mediterranean sites. Weed Res 61:385–395

Channapagoudar BB, Biradar NR (2007) Physiological studies on weed control efficiency in direct sown onion. Karnataka J Agric Sci 20(2):375–376

Chase CA, Sinclair TR, Chellemi DO, Olson SM, Gilreath JP, Locascio SJ (1999) Heat-retentive films for increasing soil temperatures during solarization in a humid, cloudy environment. Hortic Sci 34:1085–1089

Chauhan BS (2012) Weed ecology and weed management strategies for dry-seeded rice in Asia. Weed Technol 26:1–13

Chauhan BS, Johnson DE (2009) Influence of tillage systems on weed seedling emergence pattern in rainfed rice. Soil Tillage Res 106:15–21

Chopra N, Chopra NK (2006) Crop-weed competition and determination of critical period in onion (*Allium cepa*) under north-west plain zone. Indian J Weed Sci 38(1–2):89–91

Cirujeda A, Anzalone A, Aibar J, Moreno MM, Zaragoza C (2012) Purple nutsedge (*Cyperus rotundus* L.) control with paper mulch in processing tomato. Crop Prot 39:66–71

Cohen O, Gamliel A, Katan J, Shubert I, Guy A, Weber G, Riov J (2019) Soil solarization based on natural soil moisture: a practical approach for reducing the seed bank of invasive plants in wetlands. NeoBiota 51:1–18

Daramola OS, Adigun JA, Adeyemi OR (2020) Efficacy and economics of integrated weed management in okra (*Abelmoschus esculentus* (L.) Moench). Agric Trop Subtrop 53(4):199–206

Dash RR, Mishra MM (2014) Bio-efficacy of halosulfuron-methyl against sedges in bottle gourd. Indian J Weed Sci 46(3):267–269

Davies K (2003) Swedes and Turnips-Integrated weed management. Technical note T 547. https://www.sruc.ac.uk/media/bedjnetx/tn547-swedes-turnips-weeds.pdf. Accessed 15 Mar 2025

De Cauwer B, Delanote L, Devos M, De Ryck S, Reheul D (2021) Optimization of weed control in organic processing spinach (*Spinacia oleracea* L.): impacts of cultivar, seeding rate, plant spacing and integrated weed management strategy. Agronomy 11(1):53

Demir H, Polat E (2011) Effects of broccoli-crispy salad intercropping on yield and quality under greenhouse conditions. Afr J Agric Res 6(17):4116–4121

Farooq O, Rehman AU, Sarwar N, Hussain M, Wasaya A, Naeem M, Iqbal MM, Khaliq A (2018) Herbicidal potential of sorghum and brassica against the weeds of cotton. Planta Daninha 36

Faruq MO, Khan MSH, Uddin MR, Alam MR, Hassan ME (2021) Effect of different weed management practices on yield of potato in Chattogram hill districts of Bangladesh. Asian J Plant Soil Sci 6(1):239–244

Fasna P, Menon MV, Sindhu PV, Anitha P, Antony S (2022) Evaluation of non-chemical weed management practices in chilli (*Capsicum annuum* L.). J Trop Agric 59(2):306–311

Galal RM, Hassanein AMA, Fedlallah AM (2020) Determination critical periods of weed competition and weed control influence on yield productivity of sweet pepper (*Capsicum annuum* L.). J Plant Prod 11(2):127–137. https://doi.org/10.21608/jpp.2020.79106

Gazoulis I, Kanatas P, Antonopoulos N (2021) Cultural practices and mechanical weed control for the management of a low-diversity weed community in spinach. Diversity 13(12):616

Gill HK, McSorley R, Treadwell DD (2009) Comparative performance of different plastic films for soil solarization and weed suppression. Hortic Technol 19:769–774

Gnanavel I, Kathiresan RM (2014) Eco-friendly weed control options for sustainable agriculture – a review. Agric Rev 35(3):172–183

Graziani F, Onofri A, Pannacci E, Tei F, Guiducci M (2012) Size and composition of weed seedbank in long-term organic and conventional low-input cropping systems. Eur J Agron 39:52–61

Gupta V, Joshi E, Singh DSSSL, Kasana BS, Singh YK (2019) The effect of chemical and non-chemical control methods on weeds and yield in potato (*Solanum tuberosum* L.) cultivation under potato-based organic cropping system. Int J Curr Microbiol Appl Sci 8(7):2737–2747

Harris KD, Geretharan T, Dilsath MSA, Srikrishnah S, Nishanthi S (2015) Critical period of weed control in radish (*Raphinus sativus* L.). AGRIEAST 10:6–10

Heap I (2014) Herbicide resistant weeds. In: Pimentel D, Peshin R (eds) Integrated pest management: pesticide problems, vol 3. Springer, Dordrecht, pp 281–301

Horesh A, Igbariya K, Peleg Z, Lati RN (2019) LPG flaming – a safe post-emergence weed control tool for direct-seeded and bulb onion. Agronomy 9(12):786

Imoloame EO (2013) Critical period of weed interference in okra (*Abelmoschus esculentus* (L.) Moench) in a humid forest and rainforest-Savanna transition zones of Eastern and Western Nigeria: a review. Int J Agri Sci 3(8):610–614

Isik D, Kaya E, Ngouajio M, Mennan H (2009) Weed suppression in organic pepper (*Capsicum annuum* L.) with winter cover crops. Crop Prot 28:356–363

Jabran K, Mahajan G, Sardana V, Chauhan BS (2015) Allelopathy for weed control in agricultural systems. Crop Prot 72:57–65

Kapoor RT (2020) Effect of soil solarization for weed management in *Abelmoschus esculentus* (L.) Moench. Plant Arch 20(1):641–646

Kaur N, Bhullar MS, Gill G (2015) Weed management options for sugarcane-vegetable intercropping systems in North-Western India. Crop Prot 74:18–23

Kaur R, Das TK, Banerjee T, Raj R, Singh R, Sen S (2020) Impacts of sequential herbicides and residue mulching on weeds and productivity and profitability of vegetable pea in north-western Indo-Gangetic plains. Sci Hortic 270:109456

Khan A, Muhammed S, Hussain Z, Khattak AM (2012) Effect of different weed control methods on weeds and yield of chilies (*Capsicum annuum* L.). Pak J Weed Sci Res 18(1):71–78

Kifelew H, Getachew W (2017) Critical time of weed competition and evaluation of weed management techniques on turmeric (*Curcuma Longa*) at Tepi, South West Ethiopia. Int J Res St Agric Sci 3(7):15–22. https://doi.org/10.20431/2454-6224.0307003

Korres NE (2018) Agronomic weed control: a trustworthy approach for sustainable weed management. In: Jabran K, Chauhan BS (eds) Non-chemical weed control, 1st edn. Academic Press, London, pp 97–114

Kujur A, Bhadauria N, Rajput RL (2015) Effect of weed management practices on seed yield and nutrient (NPK) uptake in cowpea. Legum Res 38(4):555–557

Kumar P, Singh R (2017) Integrated weed management in cowpea (*Vigna unguiculata* (L.) wasp.) under rainfed conditions. Int J Curr Microbiol Appl Sci 6(3):97–101

Kumar N, Hazra KK, Nadarajan N (2017) Efficacy of pre- and post-emergence herbicides in rainy season green gram (Vigna radiata). Indian J Agric Sci 87(9):1219–1224

Kurstjens D, Bleeker P (2000) Optimizing torsion weeders and finger weeders. In: Proceedings of the 4th EWRS workshop on physical weed control, Elspeet, The Netherlands, pp 30–32

Lamichhane JR, Devos Y, Beckie HJ, Owen MDK, Tillie P, Messéan A, Kudsk P (2017) Integrated weed management systems with herbicide-tolerant crops in the European Union: lessons learnt from home and abroad. Crit Rev Biotechnol 37(4):459–475

Larkin RP (2020) Effects of cover crops, rotation, and biological control products on soil properties and productivity in organic vegetable production in the Northeastern US. Org Agric 10(2):171–186

Latif A, Jilani MS, Baloch MS et al (2021) Evaluation of critical period for weed crop competition in growing broccoli crop. Sci Hortic 287(3):110270. https://doi.org/10.1016/j.scienta.2021.110270

Lichtenhahn M, Koller M, Dierauer H, Baumann D (2005) Weed control in organic vegetable cultivation – technical guide. Research Institute of Organic Agriculture (FiBL), CH-Frick. https://orgprints.org/id/eprint/25433/. Accessed 22 Jan 2023

Lombardo S, Longo AMG, Lo Monaco A, Mauromicale G (2012) The effect of soil solarization and fumigation on pests and yields in greenhouse tomatoes. Crop Prot 37:59–64

Mabele A, Ndong'a M (2020) Efficacy of guava (*Psidium guajava*) mulch allelopathy in controlling tomato (*Solanum lycopersicum*) weeds. East Afr J Agric Biotechnol 1(1):7–11

Mahajan G, Chauhan BS (2013) The role of cultivars in managing weeds in dry seeded rice production systems. Crop Prot 49:52–57

Maheswari UM, Karthik TA (2019) Recent techniques of weed management practices in organic farming. Acta Sci Agric 3(2):150–157

Mahmood A, Ihsan MZ, Khaliq A, Hussain S, Cheema ZA, Naeem M, Daur I, Hussain HA, Alghabari F (2015) Crop residues mulch as organic weed management strategy in maize. Clean Soil Air Water 44(3):317–324

Majrashi A, Khandaker MM (2020) Survey of Chenopodiaceae family from Taif, Saudi Arabia Peninsula. Plant Arch 20(2):5958–5964

McErlich AF, Boydston RA (2014) Current state of weed management in organic and conventional cropping systems. In: Young S, Pierce F (eds) Automation: the future of weed control in cropping systems. Springer, Dordrecht, pp 11–32

Melander B, Elsgaard L, Jørgersen MH (2004) Band-steaming reduces laborious hand-weeding in vegetables. Newsletter from Danish Research Centre for Organic Farming No. 3. https://core.ac.uk/download/pdf/10921348.pdf. Accessed 23 Mar 2023

Melander B, Lattanzi B, Pannacci E (2015) Intelligent versus non-intelligent mechanical intra-row weed control in transplanted onion and cabbage. Crop Prot 72:1–8

Mennan H, Ngouajio M, Kaya E, Isik D (2009) Weed management in organically grown kale using alternative cover cropping systems. Weed Technol 23:81–88

Mennan H, Jabran K, Zandstra BH, Pala F (2020) Non-chemical weed management in vegetables by using cover crops: a review. Agronomy 10(2):257

Mohler CL, Johnson SE (2009) Crop rotation on organic farms: a planning manual. Natural Resource, Agriculture, and Engineering Service (NRAES) Cooperative Extension, Ithaca

Norris RF, Caswell-Chen EP, Kogan M (2003) Concepts in integrated pest management. Prentice Hall, Upper Saddle River, 586 pp

Norsworthy JK, Malik MS, Jha P, Riley MB (2007) Suppression of *Digitaria sanguinalis* and *Amaranthus palmeri* using autumn-sown glucosinolate-producing cover crops in organically grown bell pepper. Weed Res 47(5):425–432

Pannacci E, Covarelli G (2005) Mechanical weed control in sunflower. In: Proceedings of 13th EWRS symposium, 19–23 June 2005, Bari, Italy

Pannacci E, Tei F (2014) Effects of mechanical and chemical methods on weed control, weed seed rain and crop yield in maize, sunflower and soybean. Crop Prot 64:51–59

Patel TU, Zinzala MJ, Patel HH, Patel DD, Patel HM, Italiya AP (2017) Summer okra as influenced by weed management. AGRES Int e-J 6(1):129–133

Peruzzi A, Ginanni M, Mazzoncini M, Raffaelli M, Fontanelli M, Di Ciolo S, Verna P, Casaccia D, Recinelli E (2005) La gestione fisica delle infestanti su carota biologica e su altre colture tipiche dell'altopiano del Fucino (a cura di Peruzzi A). Stamperia Editoriale Pisana

Peruzzi A, Ginanni M, Mazzoncini M, Raffaelli M, Fontanelli M, Di Ciolo S, Fantoni E, Costa I, Di Colo M, Boschetti I (2006) Il controllo fisico delle infestanti su spinacio in coltivazione biologica ed integrata nella bassa Valle del Serchio (a cura di Peruzzi A). Stamperia Editoriale Pisana

Pouryousef M, Yousefi AR, Oveisi M, Asadi F (2015) Intercropping of fenugreek as living mulch at different densities for weed suppression in coriander. Crop Prot 69:60–64

Qasem JR (2009) Weed competition in cauliflower (*Brassica oleracea* L. var. *botrytis*) in the Jordan Valley. Sci Hortic 121(3):255–259

Radicetti E, Mancinelli R, Campiglia E (2013) Influence of winter cover crop residue management on weeds and yield in pepper (*Capsicum annuum* L.) in a Mediterranean environment. Crop Prot 52:64–71

Rana MK, Sood S, Sood R (2011) Weed management in vegetable crops. In: Rana MK (ed) Fundamentals of vegetable production. New India Publishing Agency, New Delhi, pp 484–528

Rao VS (2000) Principles of weed science, 2nd edn. Science Publishers, Enfield

Rasmussen J, Henriksen CB, Griepentrog HW, Nielsen J (2011) Punch planting, flame weeding and delayed sowing to reduce intra-row weeds in row crops. Weed Res 51(6):489–498

Riemens MM, van der Weide RY, Bleeker PO, Lotz LAP (2007) Effect of stale seedbed preparations and subsequent weed control in lettuce (cv. Iceboll) on weed densities. Weed Res 47(2):149–156

Riva NB, Biffe DF, Nalin D et al (2023) Weed interference periods in lettuce crop. Hortic Bras 41. https://doi.org/10.1590/s0102-0536-2023-e2566

Rodenburg J, Johnson DE (2009) Weed management in rice-based cropping systems in Africa. Adv Agron 103:149–218

Sanchez E, Lamont WL Jr, Orzolek MD (2008) Newspaper mulches for suppressing weeds for organic high-tunnel cucumber production. Hortic Technol 18(1):154–157

Sandhu KS, Singh D, Sandhu MS, Gill BS, Singh J (2002) Weed management in seed crop of radish. PAU J Res 39(4):504–507

Sanjay MT, Dhanapal GN, Nagarjun Sandeep A (2019) Response of mulching and weed management practices on weed control, yield and economics of garlic. Indian J Weed Sci 51(2):217–219

Shafiq M, Kaur S (2021) Weed control using paddy straw mulch in integration with herbicides in autumn potato in North-West India. Potato Res 64(4):761–773

Shawquat M, Hossain MA, Islam MN et al (2009) Effect of duration of weed competition and weed control on the yield of Indian spinach. Bangladesh J Agric Res 33(4). https://doi.org/10.3329/bjar.v33i4.2306

Sinchana JK (2020) Integrated weed management in bush type vegetable cowpea (*Vigna unguiculata* subsp. *unguiculata* (L.) Verdcourt). M.Sc. Agriculture thesis, Kerala Agricultural University, Thiruvananthapuram, Kerala, 181 pp

Singh R (2014) Weed management in major kharif and rabi crops. National Training on Advances in Weed Management, 14–23 January 2014, Directorate of Weed Science Research, Jabalpur, pp 31–40

Singh VP, Mishra JS, Gogoi AK (2005) Effect of weed interference and fertilizer levels on weeds and productivity of potato. Indian J Weed Sci 37(4):1–18

Singh M, Prabhukumar S, Sairam CV, Kumar A (2008) Effect of different herbicides on weed management in radish. Indian J Weed Sci 40(1–2):96–97

Singh KN, Singh P, Singh P, Alie BA, Bahar FA, Panotra N (2010) Integrated weed management in vegetable crops under temperate conditions. Res J Agric Sci 1(2):160–163

Singh M, Bhullar MS, Chauhan BS (2015a) Seed bank dynamics and emergence pattern of weeds as affected by tillage systems in dry direct-seeded rice. Crop Prot 67:168–177

Singh M, Kumar R, Kumar S, Kumar V (2015b) Critical period for weed control in field pea. Legum Res Int J 39. https://doi.org/10.18805/lr.v0iOF.6787

Singh RS, Bisen N, Rani M (2017) Flurochloridone- a promising herbicide for weed management in carrot. Indian J Weed Sci 49(4):409–410

Stagnari F, Pisante M (2011) The critical period for weed competition in French bean (*Phaseolus vulgaris* L.) in Mediterranean areas. Crop Prot 30(2):179–184

Syriac EK, Geetha K (2007) Evaluation of pre-emergence herbicides and soil solarization for weed management in brinjal (*Solanum melongena* L.). Indian J Weed Sci 39(1&2):109–111

Teasdale JR, Mohler CL (2000) The quantitative relationship between weed emergence and the physical properties of mulches. Weed Sci 48(3):385–392

Tei F, Ascard J, Baumann DT, Caussanel JP, Dobrzanski A, Froud-Williams RJ, Kleifeld Y, Pardo Iglesias A, Rocha F, Ruuttunen P, Rzozi RS, Sanseovic T, Suso L (1999) Weeds and weed management in onion – a review. In: Proceedings of the 11th EWRS symposium, 28 June–01 July 1999, Basel, p 131

Tei F, Pannacci E, Cirujeda A, Dobrzanski A, Froud-Williams RJ, Rocha F, Ruuttunen P, Sanseovic T, Simoncic A, Torma M, van der Weide R (2005) Weeds and weed management in cabbages – a review. In: Proceedings of the 13th EWRS symposium, 19–23 June 2005, Bari, pp 2–5

Tei F, Cirujeda A, Dobrzanski A, Froud-Williams RJ, Mennan H, Neuweiler R, Pannacci E, Qasem J, Rocha F, Simončič A, Verschwele A (2007) Weeds and weed management in lettuce. In: Proceedings of the 14th EWRS symposium, 17–21 June 2007, Hamar, Norway, pp 17–21

Tillett ND, Hague T, Grundy AC, Dedousis AP (2008) Mechanical within-row weed control for transplanted crops using computer vision. Biosyst Eng 99:171–178

Turner RJ, Davies G, Moore H, Grundy AC, Mead A (2007) Organic weed management: a review of the current UK farmer perspective. Crop Prot 26:377–382

Vidotto F, De Palo F, Ferrero A (2013) Effect of short-duration high temperatures on weed seed germination: high temperatures affecting weed seeds. Ann Appl Biol 163:454–465

Wang FX, Feng SY, Hou XY, Kang SZ, Han JJ (2009) Potato growth with and without plastic mulch in two typical regions of Northern China. Field Crop Res 110:123–129

Weaver SE (2006) Critical period of weed competition in 3 vegetable crops in relation to management practices. Weed Res 24(5):317–325

Weerarathne LVY, Marambe B, Chauhan BS (2016) Intercropping as an effective component of integrated weed management in tropical root and tuber crops: a review. Crop Prot 95:89–100

Yildirim E, Guvenc I (2005) Intercropping based on cauliflower: more productive, profitable and highly sustainable. Eur J Agron 22:11–18

Applications of Plant Growth Regulators in Vegetables

16

Ishtiaq Ahmad, Mehmet Ramazan Bozhuyuk, Maryam, Riaz Ahmad, and Muhammad Akbar Anjum

Abstract

Plant growth regulators (PGRs) include both natural as well as synthetic substances that regulate plant growth and development. The use of PGRs in vegetable production has gained tremendous interest in recent years as these can improve crop growth, productivity, quality, and tolerance against biotic and abiotic stresses. This chapter reviews the role of PGRs in the production and postharvest physiology of vegetables. This chapter starts with a discussion of distinct types of PGRs, such as auxins, gibberellins, cytokinins, abscisic acid, and ethylene, as well as their roles in the regulation of plant growth and development. This chapter delves into the specific effects of PGRs on different vegetable crops including tomato, cucumber, pepper, eggplant, and others. For instance, gibberellins have been shown to improve fruit set and size of tomato and cucumber, while cytokinins have been found to delay senescence and improve postharvest quality of leafy vegetables like lettuce and spinach. The chapter also discusses challenges associated with the use of PGRs in vegetable production, such as

I. Ahmad
Department of Horticultural Sciences, The Islamia University of Bahawalpur, Bahawalpur, Pakistan

M. R. Bozhuyuk
Horticulture Department, Igdir University, Igdir, Türkiye

Maryam
Department of Botany, The Govt. Sadiq College for Women University, Bahawalpur, Pakistan

R. Ahmad
Department of Horticulture, The University of Agriculture, D.I. Khan, Pakistan

M. A. Anjum (✉)
Department of Horticulture, Bahauddin Zakariya University, Multan, Pakistan
e-mail: akbaranjum@bzu.edu.pk; anjumbzu@yahoo.com

Z. Iqbal et al. (eds.), *Sustainable and Innovative Vegetable Production in times of Climate Change*, https://doi.org/10.1007/978-981-96-6283-8_16

environmental pollution, toxicity, and resistance development. It then explores the influence of PGRs on postharvest physiology. Overall, the chapter highlights the potential of PGRs in improving vegetable production and postharvest quality. It emphasizes the need for careful selection, application, and monitoring of PGRs to ensure their safe and effective use in vegetable production and postharvest handling.

Keywords

Parthenocarpy · Phytohormones · Postharvest physiology · Seed germination · Sex expression

Abbreviations

ABA	Abscisic acid
PGRs	Plant growth regulators
IAA	Indole acetic acid
NAA	Naphthaleneacetic acid
GA	Gibberellic acid
BA	Benzyl adenine

16.1 Introduction

Plant hormones are biochemical substances which are vital for growth and development of plants as these are involved in many metabolic pathways. Plant hormones are also called as "phytohormones" if these are produced naturally in plants; however, synthetics are usually termed as plant growth regulators (PGRs). Five different types of PGRs exist, i.e., auxins, gibberellins, cytokinins, abscisic acid, and ethylene. Phytohormones are chemical substances that are not nutrients in nature, synthesized in one part of the plant and translocated to other plant parts, where these are responsible for biochemical and physiological activities (Kaur et al. 2018). PGRs are categorized into two groups, i.e., growth promoters (auxins, gibberellins, and cytokinins) and growth retardants (abscisic acid, ethylene, and others). The role of PGRs in different physiological processes of plants is discussed below.

16.2 Role of PGRs in Physiology of Vegetable Crops

16.2.1 Seed Germination

PGRs are involved in many physiological processes and are responsible for the significant increase in vegetable production. PGRs can be used in vegetables to improve seed germination, ensure crop uniformity, ease of harvesting, and extend their storability. Certain vegetable crops are very sensitive to photoperiod for germination and difficult to grow in off-season, i.e., lettuce. This problem can be overcome by seed soaking in indole-3-acetic acid (IAA) (10–40 mg/L) solution overnight, which accelerates seed germination in any photoperiod. The methods of application of PGRs to enhance seed germination of various vegetable crops are presented in Table 16.1.

16.2.2 Flowering

GA_3 is used to manipulate the photoperiod phenomenon as it is quite helpful in inducing flowering in long-day plants under short-day photoperiod. Application of GA_3 @ 50 mg/L to potato plants is helpful for profuse flowering. Further, GA_3 is also helpful to initiate early flowering in lettuce. It has been witnessed that the application of GA_3 in okra induces first flower in minimum of days. Fruit drop or flower drop is a devastating problem in vegetable production, i.e., in capsicum, tomato, and brinjal. However, application of 2,4-D @ 02–05 mg/L before initiation of flowering in solanaceous vegetables results in a tremendous reduction in flower drop (Pal and Bal 2020). Flowering is an important phenomenon in all plants for breeding, seed multiplication, and fruit and seed yields. Application of GA_3 @ 100 mg/L on young leaves, emerged after steckling of non-flowering carrot roots, successfully initiated flowering (Thakur 2022). Moreover, Verma and Sen (2008) reported that carrot root soaking in GA_3 solution @ 400 mg/L produced early flowering and best quality seed. Early crop is always the source of attraction for vegetable markets and guarantees premium income for farmers; foliar sprays of GA_3 and NAA can induce early flowering in tomato (Kumar et al. 2018).

It has been observed that just before the flowering stage, foliar application of GA_3 is quite helpful to manipulate flowering phenomenon in members of the *Brassicaceae* family. A delay in flowering is helpful to increase leaf area by around

Table 16.1 PGRs and their methods of application to enhance seed germination of various vegetable crops

Crops	PGR	Method of Application	References
Eggplant	GA_3 (750 mg/L)	Presoaking of seed	Neto et al. (2017)
Pea	GA_3 (10 mg/L)	Seed soaking	Bahadur and Singh (2014)
Okra	GA_3 (20 mg/L) and IAA (200 mg/L)	Seed priming	Kalariya et al. (2016), Lamichhane et al. (2021), Pal and Bal (2020)
Tomato	GA_3 (0.5 mg/L) Kinetin (10 mg/L)	Seed soaking Seed priming	Bahadur and Singh (2014); Nawaz et al. (2013)
Bitter gourd	NAA (0.9 mg/L)	Seed soaking	Muttaleb et al. (2022)
Capsicum	GA_3 (500 mg/L) + NAA (100 mg/L)	Seed priming	Tombegavani et al. (2020)
Melons	Ethephon (480 mg/L)	Seed soaking	Pal and Bal (2020)

35.5% (Jong et al. 2009). Coriander (*Coriandrum sativum* L.) is a prominent herbaceous plant of Umbelliferae family. Basically, it is grown for its fresh leaves and seeds; however, being an annual plant it bolts quickly under the conditions of hot weather and long photoperiod. Bolting in coriander causes deterioration in flavor and the crop becomes unmarketable which is a significant economic loss to farmers (Laribi et al. 2015). GA_3 is responsible for bolting and with anti-GA activity, bolting can be delayed. It was witnessed that foliar application of prohexadione calcium and paclobutrazol can reduce bolting up to 78% when applied 05–06 weeks after seed germination (Meyering et al. 2020). In another study, application of IAA and kinetin significantly delayed flowering in coriander (Hernandez-Davila et al. 2004). Similarly, in the case of Chinese cabbage, lettuce, and spinach, inhibitors of GA biosynthesis have successfully been utilized to restrict bolting (Abed 2018).

16.2.3 Sex Expression

Monoecious cucurbitaceous vegetables face improper balance in staminate and pistillate flowers which leads to not only loss of early crop potential but also lower fruit yield. PGRs are helpful to balance male and female flower ratio by shifting male flowers to female flowers to produce optimum yield (Mia et al. 2014). It was observed that in bitter gourd (*Momordica charantia* L.), a greater number of male flowers was born as compared to female flowers. Such type of flowering behavior is uneconomical for growers as it results in low fruit set and yield. Expression of flower sex can be manipulated by certain environmental attributes, i.e., temperature, photoperiod, and nutrition. In addition to this, certain PGRs are also quite helpful for this expression (Dalai et al. 2016). These substances are applied to optimize crop growth and yield of crops, as they regulate many morphological and physiological plant processes but desired results depend on PGRs' concentration, growth stage, season, and crop species (Dalai et al. 2015). Among these PGRs, GA_3 has a significant role in sex alteration in many cucurbits (Hilli et al. 2010). Application of GA_3 @ 10 μg/mL resulted in increased female flowers and decreased male flowers in cucumber (Sanusi 2019). Similarly, when ethrel @ 200 mg/L was applied two times, i.e., at the second and fourth leaf stages in cucumber, a significant conversion of male flowers into female flowers was recorded (Jadav et al. 2010). Similarly, when ethrel @ 200 mg/L was applied at 02–04 leaf stage in cucumber, it resulted in the early emergence of the first pistillate flower (Aishwarya et al. 2019). Ansari and Chowdhary (2018) found significant sex alteration in bottle gourd, when plants were sprayed with ethrel @ 100 mg/L at 02–04 leaf stage along with soaking of seed in boron 0.05% solution for 12 h. Further, in another study, various concentrations of maleic hydrazide (MH) at different growth stages of cucumber plants were tested. It was observed that MH @ 150 mg/L when applied at vegetative stage showed significant results related to sex modification and fruit setting attributes (Sarkar et al. 2019). Furthermore, ethrel @ 150 mg/L is reported to induce early pistillate flower in watermelon (Chaurasiya et al. 2016).

16.2.4 Parthenocarpy

Usually after fertilization and fruit set, fruit development proceeds. In tomato (*Solanum lycopersicum* L.), fruit growth occurs in two successive phases, i.e., active cell division and cell expansion. During the process of fruit development, ovary wall grows into a pericarp having three distinctive layers: an exocarp, mesocarp. and endocarp (Gillaspy et al. 1993). In parthenocarpy, ovary wall develops into seedless fruit without pollination and fertilization. In some vegetables, it occurs naturally; however, it can also be induced through exogenous application of PGRs or by increasing their endogenous levels. Parthenocarpy increases productivity, and improves quality and processing aspects of vegetables such as watermelon, cucumber, and eggplant in which seeds are problematic during consumption. This attribute is highly beneficial to develop fruits under the environmental conditions that are hostile for successful pollination and subsequent fertilization, especially in greenhouse grown and particularly in cross-pollinated vegetables. In fact, phytohormones play a vital part in fruit set and development and their genetic manipulation can result in seedless fruits (Dhatt and Kaur 2016). The applications of phytohormones to induce parthenocarpy in various vegetable crops are summarized in Table 16.2.

16.2.5 Yield

Production of vegetables is greatly determined by the genetic makeup of cultivars grown and also affected by various climatic factors and cultural practices. Physiological and hormonal imbalance leads to flower and fruit drop which is responsible for low yield (Tamilselvi et al. 2019). In tomato, increasing dose of PGRs improved flower number per cluster, individual fruit weight, and economic yield. Various concentrations of GA_3 and NAA on cherry tomato were tested to improve yield and quality attributes. Results depicted that the highest fruit yield per plant (6.94 kg) was achieved with application of GA_3 @ 375 mg/L (Baby et al. 2018). Moreover, it was confirmed that seed soaking in naphthoxyacetic acid (NOA) @ 25–50 mg/L had a favorable effect on tomato fruit yields (Pal and Bal 2020).

Yield is directly linked with flowering and fruit-setting phenomena. Flower and fruit drop is a major problem in vegetables causing serious yield losses. Market demand and consumer preference for eggplant depend on fruit color, shape, size, and stage of maturity (Gopalan et al. 2007). Cloxyfonac, 4-CPA, and CPPU are effective PGRs that are helpful in inducing flowers and subsequent fruit set in eggplant under unfavorable conditions i.e. low temperature and reduced light intensity (Lee et al. 2004). It was reported by Choudhury et al. (2013) that application of 4-CPA+ GA_3 @ 20 mg/L each on tomato resulted in the maximum growth and yield contributing parameters, i.e., stem height at 60 DAT (86.01 cm), flower clusters per plant (10.60), flower number per plant (39.69), fruit number per plant (36.54), individual fruit weight (74.01 g), and overall production (28.40 t/ha).

Table 16.2 Applications of hormones to induce parthenocarpy in vegetables

Crop	PGRs	Time of application	References
Chili pepper	2,4-D @ 20 mg/L, NAA @ 0.05%	Prior to anthesis	Kondo et al. (2021), Hu et al. (2019), Heuvelink and Korner (2001)
Cucumber	NAA @ 100 mg/L	1 day before anthesis and at anthesis.	Su et al. (2021), Qian et al. (2018)
Okra	IAA @ 200 mg/L	Flower bud and ovary injection	Hossain et al. (2021)
Eggplant	β-naphthoxyacetic acid @ 60 mg/L	Emasculated flowers	Makrogianni et al. (2018)
Watermelon	Cytokinin class (CPPU) @ 0.5 mL/ flower	During anthesis	Kawamura et al. (2018)
Pumpkin	GA_3 @ 150 ppm	Injection before blossoming	Hossain (2015)
Muskmelon	CPPU @ 10 mg/L and BA @ 10 mg/L	During anthesis to the ovaries	Hayata et al. (2000)
Bottle gourd	CPPU @ 10–100 mg/L	After or 2 days before anthesis	Yu (1999)

16.2.6 Mitigation of Adverse Effects of Environmental Stresses

According to the Food and Agriculture Organization, global production of vegetables was approximately 1.17 billion tons in 2019, showing an increase of 2.2% compared to the previous year (FAO 2019). However, environmental stresses, i.e., drought, flooding, extreme temperatures, soil salinity, heavy metal toxicity, and pests and diseases have negative impacts on vegetable production, resulting in yield losses, inferior produce quality, and high production costs (Bulgari et al. 2019; Anuar et al. 2023). The effects of these stresses are expected to worsen with climate change and increase demand for food (Scheffran and Battaglini 2011).

Application of PGRs can help vegetables to cope with these stress conditions and enhance their growth and yield (Dias 2019). Each of the five main groups of PGRs, i.e., auxins, cytokinins, gibberellins (GA), abscisic acid (ABA), and ethylene, has a different effect on plant growth and development, and they interact with each other to regulate plant growth (Agboola et al. 2014). Auxins are responsible for growth and development of apical meristem. They also regulate root growth and tropisms, and help plants to survive under drought stress conditions (Guo et al. 2015). Cytokinins delay senescence and improve stress tolerance (Gou et al. 2022). GA_3 is involved in cell elongation, seed germination, and flower formation (Binenbaum et al. 2018; Kou et al. 2021). In response to stress, GA_3 can help plants to adjust their growth patterns to better cope with the stressor. For example, when plants are exposed to high salinity, they often experience reduced growth due to osmotic stress. However, GA_3 can promote cell elongation in stem and root, allowing the plant to sustain growth even under stressful environments. ABA is known to play a crucial role in regulating plant responses to abiotic stresses, i.e., drought, salinity, and extreme temperatures (Tuteja 2007; Kohli et al. 2013). When plants are exposed to these abiotic stresses, they produce more ABA, which triggers a cascade of responses aimed at conserving water and nutrients (Czarnocka and Karpiński 2018). ABA can help to close stomata to reduce water loss, promote root growth to increase water uptake, and stimulate the synthesis of stress-related proteins to protect against damage (Gupta et al. 2020). Ethylene regulates several physiological processes in plants, which include fruit ripening, flower senescence, and response to biotic constraints (Kolbert et al. 2019). Under stress conditions, ethylene production may increase, leading to alterations in gene expression and physiological responses. For example, in response to pathogen attack, ethylene can stimulate the biosynthesis of defense-related compounds, such as phytoalexins, and induce closure of stomata to prevent further pathogen entry. Jasmonates (JA), another class of plant hormones, can also mitigate biotic stresses such as pest and disease attacks by activating defense responses like the production of glucosinolates and volatile organic compounds. Table 16.3 shows the role of PGRs in protecting vegetables from biotic and abiotic stresses.

16.2.7 Postharvest Physiology

The postharvest period of vegetables is characterized by the continuation of physiological processes, which ultimately affect quality and shelf life of produce (Kyriacou and Rouphael 2018). Understanding the physiology of vegetables during postharvest is essential to ensure that they maintain their quality and nutritional value until they reach the consumer (Brasil and Siddiqui 2018). The role of PGRs in the postharvest physiology of vegetables is critical (Dias 2019). Auxins can delay senescence in vegetables, leading to extended shelf life and improved quality (Pogson and Morris 2004). These can maintain green color, turgidity, and firmness of leaves by regulating the synthesis of chlorophyll and maintaining the balance between cell division and cell death (Bouzroud et al. 2020). For example, the application of indole-3-butyric acid (IBA) can delay the senescence of lettuce leaves while preserving their color and firmness (Miceli et al. 2021). Naphthalene acetic acid (NAA) treatment can delay the senescence of broccoli and sweet pepper hence leading to improved color, texture, and overall quality of both commodities during storage (Mbandlwa et al. 2020; Li et al. 2022). Auxins also maintain the flavor and aroma of vegetables during storage, regulating the synthesis of volatile compounds and enzymes that contribute to the characteristic taste and odor of vegetables (Xu et al. 2022). For example, indole acetic acid (IAA) treatment of tomato fruits during storage can maintain their lycopene content and improve their overall quality by regulating the activity of enzymes involved in the biosynthesis of lycopene and other flavor and aroma compounds (Ilahy et al. 2019). Cytokinins can be used to delay senescence and prolong the shelf life of vegetables such as lettuce, broccoli, and cabbage (Singh et al. 2018). Cytokinin treatments have been found to reduce yellowing and maintain the quality of vegetables such as broccoli (Wang et al. 2023). Ethylene is commonly used to accelerate the ripening of some vegetables

Table 16.3 Examples of biotic and abiotic stresses in vegetables and the role of endogenous plant growth regulators (PGRs) in stress mitigation

Vegetables	Stress type	Endogenous PGR	Role in mitigation	References
Cucumber	Abiotic (heavy metals)	ABA	Increases metal tolerance by regulating gene expression, ROS scavenging, and ion transport	Li et al. (2016), Wang et al. (2020)
Onion	Abiotic (heat)	ABA	Regulates heat shock protein expression, antioxidant defense, and membrane stability to protect cells from heat damage	Wakchaure et al. (2018), Ghodke et al. (2020)
Eggplant	Abiotic (salinity)	Cytokinins	Regulates ion transport, osmotic adjustment, and antioxidant defense to reduce salt uptake and toxicity	Wu et al. (2012)
Carrot	Abiotic (drought)	ABA	Regulates stomatal closure, water uptake, and antioxidant defense to conserve water and mitigate drought stress	Wang et al. (2021)
Lettuce	Abiotic (UV-B radiation)	ABA	Regulates photoprotection, antioxidant defense, and gene expression to mitigate UV-B damage	Esringu et al. (2016), Aksakal et al. (2017)
Potato	Biotic (pathogens)	Ethylene	Activates defense responses such as production of pathogenesis-related proteins and lignin, and enhances disease resistance	Wiesel et al. (2015), Yogendra et al. (2017)
Broccoli	Biotic (herbivores)	Ethylene	Activates defense responses such as production of glucosinolates and trypsin inhibitors, and regulates herbivore feeding behavior	Grubb and Abel (2006), Rechner et al. (2016)

such as tomatoes (Botton et al. 2019). On the other hand, inhibitors of ethylene synthesis such as 1-Methylcyclopropene (1-MCP) can be used to delay senescence and prolong the shelf life of vegetables (Watkins, 2006). ABA can be used to delay fruit ripening and senescence, which can be beneficial for vegetables such as lettuce and broccoli (Alferez et al. 2021).

16.3 Future Concerns

The use of PGRs has become increasingly important in vegetable production and postharvest management, as these can enhance crop yield and quality, improve stress tolerance, and extend the shelf life of harvested produce. However, as with any agricultural practice, there are future concerns associated with the use of PGRs that need to be addressed. One major concern is their potential for negative effects on the environment. The use of PGRs can result in chemical residues in soil and water, which can have adverse effects on ecosystem. It is important to carefully regulate the use of PGRs and ensure that these are used in a responsible manner to minimize environmental damage.

Another concern is their potential negative effects on human health. PGRs can be harmful if these are ingested or come into contact with human skin, and there is also a risk of residue in food. It is important to carefully monitor the use of PGRs and ensure that these are used in a way that is safe for both producers and consumers.

Additionally, there is a concern that excessive use of PGRs can lead to the development of resistance in pests and diseases, which can ultimately render these PGRs ineffective. This highlights the importance of developing integrated pest management strategies that incorporate multiple approaches, including biological control and cultural practices, to reduce reliance on PGRs.

Finally, there is a concern that the use of PGRs may have negative effects on plant genetic diversity and result in the loss of valuable plant traits. It is imperative to carefully study the long-term effects of PGRs use and ensure that these are used in a way that promotes plant diversity and sustainability.

16.4 Conclusion

In conclusion, PGRs play a significant role in vegetable production and postharvest physiology. They offer a wide range of benefits, including increased yield, improved quality, enhanced stress tolerance, and extended shelf life. However, as discussed in this chapter, their use also raises several concerns which need to be addressed. To take full advantage of PGRs, while reducing their adverse effects, it is essential to use these in a responsible and sustainable manner. This involves carefully regulating their use, monitoring their effects on environment and human health, and developing integrated pest management strategies that reduce reliance on PGRs. It also involves considering the long-term effects of PGR use on plant genetic diversity and sustainability. Research in this field is ongoing, and new developments are emerging all the time. By continuing to study the effects of PGRs on vegetable production and postharvest management, we can improve our understanding of these compounds and optimize their uses to meet the increasing demand for high-quality and sustainable produce.

References

Abed M (2018) Effect of paclobutrazol on growth and delaying bolting of lettuce (*Lactuca sativa* L.) cv Balady. Asian J Agric Hortic Res 1(1):9

Agboola DA, Ogunyale OG, Fawibe OO, Ajiboye AA (2014) A review of plant growth substances: their forms, structures, synthesis and functions. J Adv Lab Res Biol 5:152–168

Aishwarya K, Reddy PSS, Syed S, Ramaiah M, Srinivasa RG (2019) Influence of plant growth regulators and stage of application on sex expression of bitter gourd (*Momordica charantia* L.) cv. VK-1-Priya. Plant Arch 19(2):3655–3659

Aksakal O, Tabay D, Esringu A, Aksakal IF, Esim N (2017) Effect of proline on biochemical and molecular mechanisms in lettuce (*Lactuca sativa* L.) exposed to UV-B radiation. Photochem Photobiol Sci 16:246–254

Alferez F, de Carvalho DU, Boakye D (2021) Interplay between abscisic acid and gibberellins, as related to ethylene and sugars, in regulating maturation of non-climacteric fruit. Int J Mol Sci 22:669

Ansari AM, Chowdhary BM (2018) Effects of boron and plant growth regulators on bottle gourd (*Lagenaria siceraria* (Molina) Standle.). J Pharmacogn Phytochem 7(1S):202–206

Anuar MSK, Hashim AM, Ho CL, Wong MY, Sundram S, Saidi NB, Yusof MT (2023) Synergism: biocontrol agents and biostimulants in reducing abiotic and biotic stresses in crop. World J Microbiol Biotechnol 39(5):123

Baby R, Saravanan S, Prasad VM, Baby S, Geethu BL (2018) Effect of GA3 and NAA on plant growth and yield of cherry tomato (*Lycopersicon esculentum* var. *cerasiforme*) under poly-house condition. Pharma Innov J 7(7):79–82

Bahadur A, Singh NK (2014) Plant growth substances in vegetable production. In: Singh KP, Bahadur A (eds). Olericulture: Fundamentals of vegetable production, Vol I. Kalyani Publishers, Noida, Utter Pradesh, India, pp 178–196

Binenbaum J, Weinstain R. Shani E (2018) Gibberellin localization and transport in plants. Trends Plant Sci 23(5): 410–420

Botton A, Tonutti P, Ruperti B (2019) Biology and biochemistry of ethylene. In: Yahia EM, Carrillo-López A (eds) Postharvest physiology and biochemistry of fruits and vegetables. Woodhead Publishing, Sawston/Cambridge, pp 93–112

Bouzroud S, Gasparini K, Hu G, Barbosa MAM, Rosa BL, Fahr M, Zouine M (2020) Down regulation and loss of auxin response factor 4 function using CRISPR/Cas9 alters plant growth, stomatal function and improves tomato tolerance to salinity and osmotic stress. Genes 11(3):272

Brasil IM, Siddiqui MW (2018) Postharvest quality of fruits and vegetables: An overview. In: Siddiqui MW (ed) Preharvest modulation of postharvest fruit and vegetable quality. Academic Press, Cambridge, MA, pp 1–40

Bulgari R, Franzoni G, Ferrante A (2019) Biostimulants application in horticultural crops under abiotic stress conditions. Agronomy 9(6):306

Chaurasiya J, Verma RB, Ahmad M, Adarsh A, Kumar R, Pratap T (2016) Influence of plant growth regulators on growth, sex expression, yield and quality of muskmelon (*Cucumis melo* L.). Ecol Environ Conserv 22:39–43

Choudhury S, Islam N, Sarkar MD, Ali MA (2013) Growth and yield of summer tomato as influenced by plant growth regulators. Int J Sustain Agric 5(1):25–28

Czarnocka W, Karpiński S (2018) Friend or foe? Reactive oxygen species production, scavenging and signaling in plant response to environmental stresses. Free Radic Biol Med 122:4–20

Dalai S, Singh MK, Kumar M, Singh KV, Kumar V (2016) Growth, flowering and yield of cucumber (*Cucumis sativus* L.) as influenced by different levels of NAA and GA3. J Plant Dev Sci 8(9):445–450

Dalai S, Singh MK, Singh KV, Kumar M, Malik S, Kumar V (2015) Effect of foliar application of GA3 and NAA on growth, flowering yield and yield attributes of cucumber (*Cucumis sativus* L.). Ann Hortic 8(2):181–194

Dhatt AS, Kaur G (2016) Parthenocarpy: a potential trait to exploit in vegetable crops: a review. Agric Rev 37(4):300–308

Dias JPT (2019) Plant growth regulators in horticulture: practices and perspectives. Biotecnol Veg 19:3–14

Esringu A, Aksakal O, Tabay D, Kara AA (2016) Effects of sodium nitroprusside (SNP) pretreatment on UV-B stress tolerance in lettuce (*Lactuca sativa* L.) seedlings. Environ Sci Pollut Res 23:589–597

FAO (2019) FAOSTAT 2019. Food and Agriculture Organization of the United Nations, Rome, Italy. Available at: http.//fao.org/download/Q/Q/E. Accessed on: 28th June 2024.

Ghodke P, Khandagale K, Thangasamy A, Kulkarni A, Narwade N, Shirsat D, Singh M (2020) Comparative transcriptome analyses in contrasting onion (*Allium cepa* L.) genotypes for drought stress. PLoS One 15(8)

Gillaspy G, Ben-David H, Gruissem W (1993) Fruits: a developmental perspective. Plant Cell 5(10):1439

Gopalan C, Rama Sastri BV, Balasubramanian S (2007) Nutritive value of Indian foods. National Institute of Nutrition (NIN), Indian Council for Medical Research (ICMR), Hyderabad, India

Gou T, Su Y, Han R, Jia J, Zhu Y, Huo H, Gong H (2022) Silicon delays salt stress-induced senescence by increasing cytokinin synthesis in tomato. Sci Hortic 293:110750

Guo H, Zhang W, Tian H, Zheng K, Dai X, Liu S, Wang S (2015) An auxin responsive CLE gene regulates shoot apical meristem development in Arabidopsis. Front Plant Sci 6:295

Gupta A, Rico-Medina A, Caño-Delgado AI (2020) The physiology of plant responses to drought. Science 368:266–269

Hayata Y, Niimi Y, Inoue K, Kondo S (2000) CPPU and BA, with and without pollination, affect set, growth, and quality of muskmelon fruit. HortScience 35(5):868–870

Hernández-Dávila J, Zavala-García F, Valdés-Lozano CGS, Salinas-García G, Cárdenas-Cerda E, Montes-Cavazos F, Gámez-González H (2004) Retraso de la floración en cilantro (*Coriandrum sativum* L.) con sustancias reguladoras del crecimiento. Rev Chapingo Ser Hortic 10:51–56

Heuvelink E, Korner O (2001) Parthenocarpic fruit growth reduces yield fluctuation and blossom-end rot in sweet pepper. Ann Bot 88(1): 69–74

Hilli JS, Vyakarnahal BS, Biradar DP, Hunje R (2010) Effect of growth regulators and stages of spray on growth, fruit set and seed yield of ridge gourd (*Luffa acutangula* L. Roxb). Karnataka J Agric Sci 23(2):239–242

Hossain ABMS (2015) Seedless pumpkin vegetable production using gibberellic acid (GA3) as plant hormone and genetically modified technique. Global J Biol Agric Health Sci 4(3):6–8

Hossain AS, Alenazi MM, Taha RM (2021) Seedless okra production by indole 3-acetic acid micro syringe injection on flower bud, ovary and shoot xylem and its vitamin and mineral content development: an innovation. Sci Hortic 283:110010

Hu Z, Lan S, Zhao N, Su N, Xue Q, Liu J, Zhang M (2019) Soft-X-irradiated pollens induce parthenocarpy in watermelon via rapid changes of hormone-signaling and hormonal regulation. Sci Hortic 250:317–328

Ilahy R, Tlili I, Siddiqui MW, Hdider C, Lenucci MS (2019) Inside and beyond color: Comparative overview of functional quality of tomato and watermelon fruits. Front Plant Sci 10:769

Jadav RG, Patel TV, Parmar AB, Saiyad MY (2010) Sex modification of cucumber vegetable through PGRs. J Pure Appl Sci 18:13–14

Jong MD, Mariani C, Vriezen WH (2009) The role of auxin and gibberellin in tomato fruit set. J Exp Bot 60:1523–1532

Kalariya VD, Pamar VK, Patel NM, Sanket M (2016) Impact of seed treatment of plant growth regulators on seed germination and growth of okra (*Abelmoschus esculentus* (L.) Moench) cv. GAO-5 Multilogic Sci 5(15):305–308

Kaur P, Mal D, Sheokand A, Singh L, Datta D (2018) Role of plant growth regulators in vegetable production: a review. Int J Curr Microbiol Appl Sci 7(6):2177–2183

Kawamura S, Ida K, Osawa M, Ikeda T (2018) No effect of seed presence or absence on sugar content and water status of seeded and seedless watermelon fruits. HortScience 53(3):304–312

Kohli A, Sreenivasulu N, Lakshmanan P, Kumar PP (2013) The phytohormone crosstalk paradigm takes center stage in understanding how plants respond to abiotic stresses. Plant Cell Rep 32:945–957

Kolbert Z, Feigl G, Freschi L, Poór P (2019) Gasotransmitters in action: nitric oxide-ethylene crosstalk during plant growth and abiotic stress responses. Antioxidants 8(6):167

Kondo F, Hatakeyama K, Sakai A, Minami M, Nemoto K, Matsushima K (2021) Parthenocarpy induced fluctuations in pungency and expression of capsaicinoid biosynthesis genes in a

Japanese pungency-variable sweet chili pepper 'Shishito' (*Capsicum annuum*). Hortic J 90(1):48–57

Kou E, Huang X, Zhu Y, Su W, Liu H, Sun G, Song S (2021) Crosstalk between auxin and gibberellin during stalk elongation in flowering Chinese cabbage. Sci Rep 11(1):3976

Kumar S, Singh R, Singh V, Singh MK, Singh AK (2018) Effect of plant growth regulators on growth, flowering, yield and quality of tomato (Solanum lycopersicum L.). J Pharmacogn Phytochem 7(1):41–44

Kyriacou MC, Rouphael Y (2018) Towards a new definition of quality for fresh fruits and vegetables. Sci Hortic 234:463–469

Lamichhane A, Mamata KC, Shrestha, M, Baral B (2021) Effect of seed priming on germination of okra (Abelmoschus esculentus). Malays J Sustain Agric 5(2): 111–114

Laribi B, Kouki K, M'Hamdi M, Bettaieb T (2015) Coriander (Coriandrum sativum L.) and its bioactive constituents. Fitoterapia 103: 9–26

Lee JS, Shin YA, Um YC, Lee S (2004) Effect of plant growth regulators on fruit set and yield of eggplant (Solanum melongena L.). Korean J Hortic Sci Technol 4:403–406

Li X, Meng Z, Malik AU, Zhang S, Wang Q (2022) Maintaining the quality of postharvest broccoli by inhibiting ethylene accumulation using diacetyl. Front Nutr 9:1055651

Li Y, Li C, Bai L, He C, Yu X (2016) MicroRNA and target gene responses to salt stress in grafted cucumber seedlings. Acta Physiol Plant 38:1–12

Makrogianni DI, Karapanos IC, Passam HC (2018) Seasonal fluctuations in pollen production and viability in eggplant and the quality of seed-containing and seedless (auxin-set) fruits. J Plant Growth Regul 37:937–946

Mbandlwa NP, Fotouo MH, Maboko MM, Sivakumar D (2020) Preharvest application of naphthalene acetic acid and kelpak® improve postharvest quality and phytonutrient contents of sweet pepper. Int J Vegetable Sci 26(1):3–14

Meyering B, Hoeffner A, Albrecht U (2020) Reducing preharvest bolting in open-field-grown cilantro (*Coriandrum sativum* L. cv. *santo*) through use of growth regulators. Hort. Science 55(1):63–70

Mia MB, Islam MS, Shamsuddin ZH (2014) Altered sex expression by plant growth regulators: An overview in medicinal vegetable bitter gourd (*Momordica charantia* L.). J Medicinal Plants Res 8:361–367

Miceli A, Vetrano F, Moncada A (2021) Influence of Ecklonia maxima extracts on growth, yield, and postharvest quality of hydroponic leaf lettuce. Horticulturae 7:440

Muttaleb QA, Shamukh AF, Mohsen DM (2022) Efficacy of seed priming through plant hormones on the germination of bitter melon (*Momordica charantia* L.). Proc Pakistan Acad Sci 59(3):1–8

Nawaz A, Amjad M, Khan SM, Afzal I, Ahmed T, Iqbal Q, Iqbal J (2013) Tomato seed invigoration with cytokinins. J Animal Plant Sci 23(1):121–128

Neto FJD, Dalanhol SJ, Machry M, Junior AP, Rodrigues JD, Ono EO (2017) Effects of plant growth regulators on eggplant seed germination and seedling growth. Aust J Crop Sci 11(10): 1277–1282

Pal S, Bal S (2020) Plant Growth Regulators: A boon to increase vegetable production. Agriculture & Food e-Newsletter

Pogson BJ, Morris SC (2004) Postharvest senescence of vegetables and its regulation. In: Nooden L (ed) Plant cell death processes. Academic Press, Cambridge, MA, pp 319–329

Qian C, Ren N, Wang J, Xu Q, Chen X, Qi X (2018) Effects of exogenous application of CPPU, NAA and GA on parthenocarpy and fruit quality in cucumber (*Cucumis sativus* L.). Food Chem 243:410–413

Rechner O, Neugart S, Schreiner M, Wu S, Poehling HM (2016) Different narrow-band light ranges alter plant secondary metabolism and plant defense response to aphids. J Chem Ecol 42:989–1003

Sanusi AA (2019) Effect of seed presoaking in gibberellic acid on growth, flowering, and yield of cucumber (*Cucumis sativus* L.) plants. J Sci Agric 3:9–13

Sarkar MD, Moniruzzaman M, Alam MS, Rahman MJ, Quamruzzaman RNR, Subramaniam S (2019) Growth, sex expression and nutrient composition of cucumber (*Cucumis sativus*) as influenced by maleic hydrazide. Pak J Bot 51:117–123

Scheffran J, Battaglini A (2011) Climate and conflicts: the security risks of global warming. Reg Environ Chang 11:27–39

Singh AK, Rai AC, Rai A, Singh M (2018) Applications in postharvest management of vegetable crops. In: Singh B, Singh S, Koley TK (eds) Advances in postharvest technologies of vegetable crops, 1st edn. Apple Academic Press, pp 171–202

Su L, Rahat S, Ren N, Kojima M, Takebayashi Y, Sakakibara H, Qi X (2021) Cytokinin and auxin modulate cucumber parthenocarpy fruit development. Sci Hortic 282:110026

Tamilselvi C, Manimekalai R, Sathish G, Vijayashanthi VA, Yogameenakshi P (2019) Effect of growth regulators on flowering and fruit characters of chillies. Asian J Hortic 14:13–16

Thakur O (2022) Recent advances of plant growth regulators in vegetable production: a review. The Pharma Innovation J 11(3):1726–1731

Tombegavani SS, Zahedi B, Mousavi Fard S, Ahmadpour A (2020) Response of germination and seedling growth of pepper cultivars to seed priming by plant growth regulators. Int J Hortic Sci Technol 7(1):59–68

Tuteja N (2007) Abscisic acid and abiotic stress signaling. Plant Signal Behav 2:135–138

Verma P, Sen NL (2008) The impact of plant growth regulators on growth and biochemical constituents of coriander (*Coriandrum sativum* L.). J Herbs Spices Medicinal Plants 14:144–153

Wakchaure GC, Minhas PS, Meena KK, Singh NP, Hegade PM, Sorty AM (2018) Growth, bulb yield, water productivity and quality of onion (*Allium cepa* L.) as affected by deficit irrigation regimes and exogenous application of plant bioregulators. Agric Water Manag 199:1–10

Wang HT, Ou LY, Chen TA, Kuan YC (2023) Refrigeration, forchlorfenuron, and gibberellic acid treatments differentially regulate chlorophyll catabolic pathway to delay yellowing of broccoli. Postharvest Biol Technol 197:112221

Wang Y, Gong X, Liu W, Kong L, Si X, Guo S, Sun J (2020) Gibberellin mediates spermidine-induced salt tolerance and the expression of GT-3b in cucumber. Plant Physiol Biochem 152:147–156

Wang YH, Que F, Li T, Zhang RR, Khadr A, Xu ZS, Tian YS, Xiong AS (2021) DcABF3, an ABF transcription factor from carrot, alters stomatal density and reduces ABA sensitivity in transgenic Arabidopsis. Plant Sci 302:110699

Watkins CB (2006) The use of 1-methylcyclopropene (1-MCP) on fruits and vegetables. Biotechnol Adv 24:389–409

Wiesel L, Davis JL, Milne L, Redondo Fernandez V, Herold MB, Middlefell Williams J, Morris J, Hedley PE, Harrower B, Newton AC, Birch PRJ, Gilroy EM, Hein I (2015) A transcriptional reference map of defence hormone responses in potato. Sci Rep 5:15229

Wu X, Zhu Z, Li X, Zha D (2012) Effects of cytokinin on photosynthetic gas exchange, chlorophyll fluorescence parameters and antioxidative system in seedlings of eggplant (*Solanum melongena* L.) under salinity stress. Acta Physiol Plant 34:2105–2114

Xu L, Zang E, Sun S, Li M (2022) Main flavor compounds and molecular regulation mechanisms in fruits and vegetables. Crit Rev Food Sci Nutr 1:21

Yogendra KN, Sarkar KK, Kage U, Kushalappa AC (2017) Potato NAC43 and MYB8 mediated transcriptional regulation of secondary cell wall biosynthesis to contain *Phytophthora infestans* infection. Plant Mol Biol Report 35:519–533

Yu JQ (1999) Parthenocarpy induced by N-(2-chloro-4-pyridyl)-N′-phenylurea (CPPU) prevents flower abortion in Chinese white-flowered gourd (*Lagenaria leucantha*). Environ Exp Bot 42(2):121–128

Part IV

Stresses and Constraints

17 Constraints in Vegetables Production Due to Climate Change

Muhammad Saqib, Muhammad Akbar Anjum, Ishtiaq Ahmad, Sami Abou Fayssal, and Riaz Ahmad

Abstract

Vegetable production is facing numerous constraints due to the changing climate. The current chapter provides an overview of the major constraints faced by vegetable farmers and how climate change intensifies these challenges. Increasing temperature can cause heat stress in vegetable crops, leading to decreased yield with poor quality. Heat stress can also affect the timing of bud initiation, flowering, and fruiting, causing management problems in vegetable production. Climate change can result in either water scarcity or excessive rainfall, both of which pose challenges to vegetable production. Water scarcity can lead to inadequate irrigation, affecting crop growth and yield. Excessive rainfall can cause waterlogging and soil erosion, damaging root systems and hindering nutrient uptake. Climate change can also elevate the occurrence of pests and diseases as warmer temperature and irregular rainfall can indulge the explosion of pests and spread of diseases, resulting in decreased marketable yield. Ozone, ultraviolet (UV) radiation, floods, winds, and hail storms can have detrimental effects on

M. Saqib · R. Ahmad (✉)
Department of Horticulture, The University of Agriculture, Dera Ismail Khan, Pakistan
e-mail: riaz.ahmad@uad.edu.pk

M. A. Anjum
Department of Horticulture, Bahauddin Zakariya University, Multan, Pakistan

I. Ahmad
Department of Horticultural Sciences, The Islamia University of Bahawalpur, Bahawalpur, Pakistan

S. A. Fayssal
Faculty of Agronomy, Department of Agronomy, University of Forestry, Sofia, Bulgaria

Department of Plant Production, Faculty of Agriculture, Lebanese University, Beirut, Lebanon

Z. Iqbal et al. (eds.), *Sustainable and Innovative Vegetable Production in times of Climate Change*, https://doi.org/10.1007/978-981-96-6283-8_17

vegetable production. These constraints force the growers to adopt climate-resilient farming practices in vegetable production. This comprises the use of climate-resilient cultivars, and improved irrigation system like drip irrigation to conserve water and sustain optimum soil moisture levels. Integrated pest management approaches and crop diversification can also support in lessening the risks linked with climate change by spreading out the production and reducing susceptibility to explicit weather patterns. Hence, there is a need to implement climate-resilient farming practices to warrant sustainable vegetable production in changing climate scenario.

Keywords

Climate-resilient practices · Irregular weathering patterns · Low yield · Poor flowering and fruiting · Postharvest losses

17.1 Introduction

Climate change refers to the long-term alteration of climatic parameters, characterized by changes in temperature, wind patterns, precipitation, and other climate-related factors. Climate change also refers to the modifications in weather patterns brought out by rises in greenhouse gases. The emission of greenhouse gases (GHGs) into the atmosphere is the main cause of human-induced climate change. Carbon dioxide (CO_2), methane (CH_4), nitrous oxide (N_2O), and fluorinated gases are released as a result of industrial activities, deforestation, and the burning of fossil fuels for energy (Ebi and Hess 2020). These GHGs cause the greenhouse effect, which warms the globe by trapping heat in the atmosphere. Forests act as carbon sinks, absorbing CO_2 from the atmosphere. Urban and industrial expansion contribute to deforestation, which decreases the planet's ability to absorb carbon and raises atmospheric CO_2 levels (Lindgren et al. 2018). Changes in temperature, precipitation, and wind patterns can result from modifying land for industries and urbanization.

Climate change has substantial negative consequences, including global warming, which raises the temperature. The melting of the polar ice caps and glaciers is accelerated by this warming, which raises sea levels and intensifies coastal flooding, endangering low-lying areas and islands (Ahvenainen 1996). Extreme weather events like hurricanes, heat waves, droughts, and heavy rains are becoming more frequent due to climate change, which can seriously harm communities, infrastructure, and ecosystems. Ocean acidification is a result of increasing ocean CO_2 absorption due to higher atmospheric CO_2 levels. Climate changes may disturb ecosystems and alter the distribution and behavior of various plant and animal species. The resilience of ecosystems may be threatened, and biodiversity loss could occur. So, increasing temperatures, altered precipitation patterns, and an increase in extreme weather events are the ways that climate change affects vegetable production (Martí et al. 2016).

Vegetables are essential components of a balanced diet, providing vital nutrients and vitamins necessary for human health. Producing reliable and adaptive vegetables is essential for ensuring food security when climate change disrupts traditional agricultural methods, especially in areas that are more susceptible to its effects. As carbon sinks, plants absorb CO_2 during photosynthesis to help slow down global warming (Schreinemachers et al. 2018). Agroforestry and no-tillage farming are two sustainable vegetable production approaches that can improve carbon sequestration in soil and vegetation. Sustainable vegetable production methods frequently emphasize effective water management, which is crucial in areas where water constraint is a result of varying climatic trends. To reduce the greenhouse gas emissions caused by the long-distance transportation of food, local and seasonal vegetable production should be encouraged (Saqib et al. 2022). Rising temperatures can result in heat stress, reduced yield, and poor crop quality. Changes in precipitation patterns can lead to water scarcity or excessive rainfall, which can hinder irrigation practices causing soil erosion. Moreover, climate change increases the incidence of pests and diseases, affecting crop health and marketable yield. These constraints necessitate the adoption of climate-resilient farming practices, i.e., improved irrigation techniques, integrated pest management, crop diversification, and sustainable vegetable production in the face of climate change (Ahvenainen 1996).

This chapter focuses on assessing the impact of climate change on various aspects of vegetable production, i.e., germination, growth, yield, quality, and post-harvest losses. Further, it also explores the potential adaptation and mitigation strategies that can be employed to overcome these constraints and ensure sustainable vegetable production in the face of climate change.

17.2 Effects of Climate Change on Vegetable Crops

17.2.1 Vegetative Growth (Seed Germination, Growth, and Development)

Vegetable seed germination, growth, and development are severely impacted by climate change, causing serious problems for agriculture and food security. The environmental factors that influence the earliest stages of plant life are altered by warming temperatures, shifting precipitation patterns, and severe weather conditions (Bueren et al. 2011). Rising temperatures and unpredictable precipitation patterns can affect the progress of seed germination. Increased temperatures, decreased water availability, or extended exposure to extreme weather conditions may cause some seeds to lose viability. Temperature and moisture changes in the soil may disturb the timing of germination, which could result in uneven seedlings' emergence (Chapman and Chakraborty 2012). Changes in temperature and moisture cause a significant impact on the vegetative growth of vegetables. Growth of some vegetables can be accelerated by higher temperatures, which result in early flowering and slower vegetative growth (Mahmoud et al. 2020). On the other hand, protracted heat stress can retard growth and lower crop yields. Changes in precipitation patterns can result in waterlogging or drought, both of which have a detrimental effect on plant growth (Dang et al. 2019).

One of the most crucial variables impacting vegetable growth is temperature. The rate of plant development can be accelerated by higher temperatures, reducing the period between planting and harvest. However, heat stress can be brought on by extreme heat and results in poor seed germination, stunted growth, and wilted leaves. Low temperatures can also inhibit seed germination and restrict growth and development in peppers (Quesada-Ocampo et al. 2010). Tomato seed germination was hampered due to low temperatures of about 8–12 °C. As a result of increasing temperature unpredictability, vegetable growers have to adopt suitable measures, i.e., choosing heat-tolerant cultivars and implementing heat stress management techniques such as providing shade or using protective structures (Bazilian et al. 2011).

Fluctuations in precipitation patterns linked with climate change can also influence vegetable growth. Inadequate rainfall and extended droughts can result in water stress, distressing plant vegetative growth (Heureux et al. 2022). Vegetables, i.e., potato, onion, and okra seeds were not germinated under water deficit conditions. On the contrary, unnecessary rainfall or intense storms can cause flooding, soil erosion, and nutrient leaching, which can lead to root injury and reduced plant vigor. Irrigation practices like drip irrigation or efficient water management schemes become decisive in guaranteeing suitable water supply to vegetable crops (Chauhan et al. 2014).

The majority of vegetables, particularly those with shallow roots, are quite vulnerable to flooding. As soil air is displaced by overflowing water, the roots search for oxygen under soggy conditions (Shehzad 2023). Suffocation severely hinders root respiration, which is necessary for them to continue their normal activities of nutrition and water intake. So, poor vegetative growth comes as a result. It has been noted that tomato plants that have been submerged produce endogenous ethylene, which harms the plants' vegetative growth (Akhtar and Masud 2022). During the flooding, it was found that ABA concentrations rose in the roots of pea plants, which led to a build-up of ABA in the leaves, causing the stomata to partially close (Khan et al. 2020). Continuous flooding of asparagus plants for 48 h after 02–03 weeks of transplanting stunted their growth and decreased chances of their survival.

Atmospheric ozone exhaustion and increased revelation to UV radiation have opposing effects on the vegetative growth of vegetables. High levels of UV-B radiation can harm plant cells and tissues, leading to reduced photosynthetic machinery and low growth (Ayyogari et al. 2014). Moreover, UV radiation can modify the nutritional composition of vegetables, reducing their quality and nutritional value. These effects highlight the need for approaches to mitigate ozone depletion and UV radiation acquaintance to confirm sustainable vegetable production. UV-B radiation seems to be more harmful because it disrupts photosynthesis rate, stomatal operation, and carbohydrate metabolism. UV radiation was found to be harmful to the crops, i.e., cabbage, tomato, potatoes, and sugar beets (Ayyogari et al. 2014). The big concerns in vegetables related to elevated levels of ozone are diminished rate of photosynthesis, yellowing of leaves, decreased biomass, reduced leaf size, and stunted plant growth.

Winds and hail storms can have harmful effects on plant growth. Strong winds can uproot plants and break branches. Hail storms can cause physical injury to plant tissues, resulting in lesions. The summertime in the humid tropics and tropical monsoon territories introduces heavy rain, which causes flooding and waterlogging (Abbas et al. 2022).

17.2.2 Reproductive Growth (Flowering, Fruiting, and Seed Production)

Climate change is having a profound impact on the reproductive growth of vegetables. Variations in temperature, precipitation, and CO_2 levels are fluctuating the timing and duration of flowering, disturbing pollination and ultimately manipulating the overall fruiting and seed production. Phenological alterations are one of the major impacts of climate change on flowering (Shehzad 2023). Plants may flower earlier or later than usual due to rising temperatures and shifting weather patterns. Whereas, unpredictably occurring frost incidents in late winter or early spring might harm blooms and impede fruit set. For instance, warmer springs may result in earlier flowering. The availability and abundance of pollinators during crucial flowering times may be affected by these phenological fluctuations, which might set off the timing between flowering plants and their pollinators (Chauhan et al. 2014).

Pollination and fertilization are largely influenced by elevated temperature (Table 17.1). The success of a fruit set and seed production is significantly influenced by the prevailing temperature. Various temperature ranges are necessary for the best pollination, subsequent fruit set, development, and seed production in

Table 17.1 Negative response of vegetable crops to elevated temperatures

Crops	Physiological disorders	References
Tomato	Fruit cracking, sun scald, small fruit and seed size, blossom-end-rot	Rosales et al. (2011)
Lettuce, cauliflower, cabbage, broccoli, and celery	Tip burn, bolting, loose puffy heads	Gora et al. (2019)
Broccoli and cauliflower	Hollow stem, bract buds, leafiness	Gora et al. (2019)
Pea	Sunscald, less fruit, pod number, small seed size	Gonçalves et al. (2022)
Pepper	Sunburn, floral abortion, fruit cracking, blossom-end-rot	Gora et al. (2019)
Potato	Brown fleck disorder	Gora et al. (2019)
Carrot	Less carotene content	Bahadur et al. (2011)
Watermelon	Blossom-end-rot	Gora et al. (2019)

vegetable crops. Heat waves in particular can produce thermal stress, which lowers pollen viability, disrupts pollen tube growth, and hinders fertilization (Hatfield et al. 2018). Poor fruit set and development can result from high temperatures' impact on a plant's female reproductive system. However, low temperatures can lead to poor pollination and floral sterility, which also reduces fruit set (Bhatia and Kumar 2013). The most crucial developmental stage for tomatoes is 08–13 days period before flower opening. Temperature increases around 29 °C for 02 weeks disturbed anthesis process in tomato crop. High temperatures during flowering cause flower decline, malformed flower initiation, and physiological depression in tomato and pepper (Lee et al. 2018). In peppers, elevated temperature at the pre-anthesis stage did not disturb pistil or stamen viability, while elevated temperature distressed fruit set and seed formation after pollination. Hence, peppers are sensitive to elevated temperatures during fertilization.

Flowering, fruit set, and seed production can all be impacted by changes in precipitation patterns brought on by climate change. Drought can be caused due to insufficient precipitation or water scarcity and impacts fruit growth, pollen viability, and flower production in horticultural crops (Table 17.2). Cucumbers, melons, pumpkins, squashes, peas, peppers, and tomatoes are much sensitive to water shortage at flowering, fruiting, and seed development phases. Fruit set on these vegetables can be extremely decreased due to water deficit conditions. Significant precipitation or extended periods of high humidity can raise the danger of fungal diseases and degrade quality of flowers and fruits. Moreover, unreliable rainfall patterns might result in unreliable irrigation methods, which can change the amount of moisture in the soil and influence flower size, fruits drop and quality, and seed formation (Abbas et al. 2022; Akhtar and Masud 2022).

Increased atmospheric CO_2, a major contributor to climate change, can affect blooming and fruit set both favorably and unfavorably. Enhanced flower production can occur from enhanced plant growth and photosynthesis caused by higher CO_2

Table 17.2 Response of vegetable crops to waterlogging and drought conditions

Crops	Water logging	References
Tomato	Accumulation of endogenous ethylene, rapid wilting, plant death	Malhotra and Srivastva (2014)
Onion	Decrease in bulb size, yield loss	Malhotra and Srivastva (2014)
Pepper	Reduced photosynthetic rate and stomatal conductance	Ou et al. (2011)
Lettuce	Decreased chlorophyll and carotenoids	Cabillo (2019)
Drought		
Lettuce	Reduced chlorophyll and biomass	Cabillo (2019)
Amaranth	Decreased leaf area and plant biomass	Liu and Stutzel (2004)
Potato	Small canopy size, lass number of tubers, lower tuber weight	Nasir and Toth (2022)
Tomato	Reduced net photosynthetic rate	Liang et al. (2020)
Pepper	Lower carotenoid and chlorophyll in leaves	Okunlola et al. (2017)

concentrations (Chung et al. 2016). However, it can also change nutritional makeup of pollen, which might impair its viability and quality. Moreover, increased CO_2 levels may interfere with pollinators' feeding patterns, which could reduce pollination efficiency, fruit set, and seed production (Chung et al. 2017).

High ozone and UV radiation are serious threats to pollination, flowering, fruiting, and seed production (Table 17.3). Inhibiting photosynthesis and nutrient uptake, ozone can harm plants by causing noticeable symptoms including poor pollination, flowers and fruit drop, low fruit quality, and poor seed production. Plants that are exposed to UV radiation may have aberrant fruit growth, decreased pollen viability and germination, and other reproductive process disruptions (Dong et al. 2020).

Flowering, fruiting, and economic yields are negatively impacted by hailstorms, strong winds, and floods. Storms with hail can obliterate early flowers and fruits, reducing the overall production of vegetable crops. Strong winds can uproot plants, break branches, and interfere with pollination processes, which reduce fruit set and seed production.

Table 17.3 Response of vegetable crops to higher CO_2 concentration, UV-B radiation, and higher O_3 level

CO_2 concentration		
Crops	Responses	References
Potato	Increase in photosynthesis, tuber initiation, flowering, tuber number, weight, and productivity	Katny and Hoffmann-Thoma (2005)
Onion	Higher mass proportion in leaves, stems, and bulbs, increase in photosynthesis at bulb development	Gora et al. (2019)
Asparagus and broccoli	Decrease in respiration rate	Peppelenbos and van't Leven (1996)
Tomato	Reduction in respiration, increase in fruit number and yield	Gora et al. (2019)
UV-B radiation		
Cucumber	Reduction in the early stage of germination of cotyledons	Ayyogari et al. (2014)
French bean	Burnishing and browning of leaves, increase in vulnerability to virus attack	Ayyogari et al. (2014)
Elevated O_3 level		
Tomato	Delay in the ripening time, a temporary increase in carotenoids	Calvo et al. (2007)
French bean	Decrease in photosynthetic activity	Flowers et al. (2007)
Broccoli and carrot	Increase in respiration rate	Song et al. (2001)
Cucumber	Delay in the ripening time	Calvo et al. (2007)
Lettuce, turnip, and spinach	Necrosis, yellowing of leaves, reduction in yield	Saqib et al. (2022)
Radish and cabbage	Severe reduction in yield	Malhotra and Srivastva (2014)

17.2.3 Yield and Produce Quality

Vegetable crops around the world have been significantly affected by climate change. As a result of agricultural ecosystems being disrupted by rising temperatures, changing precipitation patterns, and extreme weather, crop productivity and quality are reduced. Vegetable growth and development can be significantly impacted by heat stress, which can result in lower yields and a higher vulnerability to diseases and pests. Crop production changes are made worse by shifting rainfall patterns, including droughts and floods, which alter crop water availability and nutrient uptake. Moreover, changes in temperature and humidity foster an environment that is conducive to pests and illnesses, which further lowers vegetable yield. Frost damage can significantly diminish yield or quality of vegetable crops (Kumar et al. 2012). Bolting or the premature generation of seed heads is brought on by sudden rises in temperature during the developing phase of lettuce and celery. This causes poor-quality heads and decreased yields. Diamondback moth survival rates rise in warm climates. Poor germination, decline in growth, pollination distress, flowers and fruit impairment of cellular integrity, diminished crop yield as well as death of the plants, are all signs of a plant's response to salinity (Koyama et al. 2012).

Reduced production and quality, as a result of drought stress, occur when storage organs are growing. Due to their great sensitivity to drought, potatoes may suffer tuber production losses even under modest water stress (Flowers et al. 2007). The generation and transport of carbohydrates from leaf to root or bulb determine crop yields of root and bulb crops including potatoes, carrots, and onions. Since leafy greens are naturally succulent, drought state causes the water content of produce to decrease, resulting in poor yield and quality. Vegetative growth significantly declines under severe water stress, and plants grown under such conditions produce little fruits with lower quality (Gautam et al. 2013). Floods can result in waterlogging, which suffocates root system, causes nutritional deficiencies, and increases the risk of diseases. Flooding throughout bulb development causes yield reductions that can reach 30–40% in onions (Kumar et al. 2012). Annual rains lower garlic and carrot production in simulated potential trends (Potopová et al. 2016). Floods can aid in the dispersion of water-borne pathogens, while droughts and heat waves can make plants more infected, and during and after flooding, winds can boost plant pathogens' spore dispersal (Songy et al. 2019).

The pollination mechanism is disrupted by changes in temperature, precipitation patterns, CO_2 levels, and phenological alterations, which also harm fruit yield and quality (Lee et al. 2018). Lower yields are directly related to decreased flower production, poor pollination, and decreased fruit set (Gora et al. 2019). Further lowering the marketable yield is the possibility of fruit with irregular forms, reduced size, or poor fruit development as a result of poor pollination (Hatfield et al. 2011). However, profitable vegetable production is possible under the problems posed by climate change, and sustainable and fruitful agriculture can be assured in the future through the adoption of climate-resilient agricultural practices, improvements in plant breeding, and conservation of pollinators (Settele et al. 2016).

17.2.4 Shelf Life and Postharvest Losses

The postharvest life of vegetables is determined by their quality at harvest. The production of plant pigments, i.e., chlorophylls, carotenoids, xanthophyll, bioactive photo-chemicals, and carbohydrates, is responsible for this postharvest quality (Kiaya 2014). This pigment and carbohydrate biosynthesis occurs via photosynthetic activity, which is inextricably linked to environmental conditions or pre-harvest environments during production and harvesting stages. Multiple pre-harvest factors influence vegetable quality, including environmental as well as genetic variables (Atanda and Pessu 2011). The climate during production phase is one of the most important factors influencing quantity as well as quality of the vegetables produced.

Changing climate patterns like elevated temperature, erratic rainfall trends, raised CO_2 intensity, irregular tropical storms, salinity, scarcity of water, biotic stress, and so on influence vegetable quality directly and indirectly through modifying maturation period, oxygen levels, stomatal conductance, development of biochemicals, flavor components, and antioxidants, among other factors (Parajuli et al. 2019). Changes in postharvest quality caused by climate change could be either positive or negative. Cultivation of improved crop cultivars can increase vegetable production while maintaining the utmost postharvest life (Magan and Aldred 2007). Enhanced ozone concentrations disrupted tomato ripening as well as decreased crop sugar content (Moretti et al. 2010). Depending on the species or growth stage, ozone prompted leaf yellowing, cell death, and yield decrease in turnip, tomato, lettuce, and spinach (Thwe et al. 2015). C_4 plants were found to be more tolerant of ozone than C_3 plants. The interaction of ozone and elevated CO_2 levels affects the foliage effectiveness and starch composition of potato tubers (Hauglustaine and Brasseur 2001).

17.3 Tolerance Mechanism in Response to Climate Change

Phytohormones, i.e., abscisic acid (ABA), salicylic acid (SA), ethylene, etc., are critical for regulating numerous signaling pathways as well as responses under different kinds of abiotic stresses. According to crosstalk, ABA plays a significant part in the control of signaling pathways through interplay with several other hormones (Pangga et al. 2013). ABA is a particularly critical and essential hormone for trying to regulate climatic stresses in plants. Plant growth is seriously slowed in extreme drought, and ABA content in cells rises. During drought, ABA buildup regulates transpiration through stomatal disclosure. ABA also stimulates numerous physiological effects in plants during this period, i.e., regulates stomata closure, and generates numerous stress-responsive genes (Zhang et al. 2020). In recent times, the process of operation of ABA signaling machinery has been illustrated (Mou et al. 2016). SA has recently been studied for its effects on fruit productivity, legume nutrient uptake, temperature resistance, stomatal closure, respiration, senescence-related genes, and cell growth (Ullah et al. 2018). SA regulation may be secondary in these occurrences because it regulates the output of additional plant stress-responsive hormonal changes (Zhang et al. 2020).

Phytohormones have a significant impact on stress reactions by modulating various signal transduction mechanisms in the face of climatic variability. Ethylene is one of the most crucial phytohormones. Ethylene was discovered over a century ago and several research investigations have taken place since then to uncover its biosynthesis. Under various abiotic and biotic stresses, ethylene regulates seed germination, ripening, leaf expansion, and senescence. It is thought that ethylene serves as a signaling pathway between plant growth as well as condition (Ahmad et al. 2023). Ethylene response factors (ERFs) in plants are members of a large transcription factor family and are stimulated in reactions to different physiological as well as environmental hardships. There have been detailed researches on ERFs that propose its responsibility in abiotic stresses, but there is no indication of any specific signaling cascade under abiotic stresses (Ahmad and Anjum 2023). In a recent study, ERFs in tomatoes were exposed to drought, salinity, heat, cold, and extreme water circumstances to profile their expression (Klay et al. 2018).

Phytohormones play a crucial role in regulating plant growth and development, and their response to climate change in vegetable crops is of great significance (Malhotra and Srivastva 2014). Climate change, including rising temperatures and altered precipitation patterns can affect synthesis, transport, and signaling of phytohormones, leading to significant impacts on plant growth and productivity (Lansink and Bezlepkin 2003; Ahmad and Anjum 2023; Ahmad et al. 2023). Higher temperatures can disrupt hormone balance, resulting in reduced seed germination, poor root development, and altered flowering patterns. Changes in rainfall patterns can influence hormone distribution, affecting plant water relations and nutrient uptake. Understanding the intricate relationship between phytohormones and climate change is crucial for developing strategies to enhance the resilience of vegetables (Liu and Stützel 2004). This information can support in developing crop varieties with improved tolerance to climate stressors and in optimizing cultivation practices to mitigate the negative effects of climate change on vegetable production.

17.4 Adaptive Measures Against Climate Change

Adopting appropriate measures is essential to defend against the adverse outcomes of climate change on vegetable production. The adaptability of vegetable farming can be improved by using sustainable agricultural practices such as crop diversification, water-efficient methods, and climate-resilient crop types. Reducing greenhouse gas emissions from transportation is another benefit of promoting seasonal and local production. Promoting agroforestry and soil carbon sequestration also helps to mitigate climate change. Together, these actions protect vegetable production, guaranteeing food security despite the difficulties caused by climate change (Bisbis et al. 2018). Different mitigation and adaptation measures to reduce the adverse effects of climate change on vegetables are listed in Fig. 17.1.

Fig. 17.1 Different mitigation and adaptation measures to reduce the adverse effects of climate change on vegetables

17.4.1 Conventional Approaches

To lessen the risk and uncertainty associated with climate change, farmers should implement crop diversification strategies by cultivating a wide range of vegetable crops. By using this strategy, the chance of complete crop loss brought on by unexpected climate-related issues can be reduced. Furthermore, various kinds of crops help to strengthen ecosystem resilience and lessen its susceptibility to pests and diseases (Ayyogari et al. 2014). Vegetable intercropping is a more effective approach and a less risky technology that can be used to deal with these issues. Due to their short lifespan and N-fixing capacity, the use of legumes is the most prevalent in intercropping systems. In the cultivation of vegetables, intercropping with legumes has become more predictable and stable than sole cropping patterns. Intercropping of baby corn with chili, brinjal, and okra during summer and with brinjal, tomato, pea, and chili during autumn–winter is a well-known profitable as well as productive pattern than sole crop (Olesen et al. 2011).

Crop rotation techniques and establishment of cover crops can improve soil quality, lessen soil erosion, and aid in the control of pests and diseases. During times of drought or heavy rain, this method helps vegetable crops by maximizing fertilizer use and improving water retention. Legumes can help in fixing atmospheric nitrogen through crop rotation, which reduces the need to consume fossil fuels to make chemical fertilizers. An efficient way to increase soil organic carbon storage and carbon sequestration is to cultivate cover crops. Moreover, by restricting moisture loss via evaporation, cover crops help to conserve soil moisture, making them a key component of drought adaptation strategies (Min et al. 2016; Chepkoech et al. 2023).

Water management is a crucial adaptation method under an array of climatic situations. Water is one of the most essential ingredients for crop production and a crucial part of all biological systems (White et al. 2011). Climate change immediately affects water sources and decreases the amount of water available. The monsoons are influenced by climate change, which also frequently causes crop failure. Low crop yields and a lack of water are the results of a delayed or failed monsoon. Critical aspects of irrigation management under climate change include appropriate irrigation and soil moisture retention. The benefits of cover crops, mulching. and precision farming in preserving soil moisture are also necessary. Sprinkler and drip irrigation are micro-irrigation methods that have already been successfully used to conserve water (Abbaspour et al. 2009).

Vegetables are shielded from heavy rains, scorching heat, and flooding by mulching, which also helps to retain soil moisture and minimizes soil degradation (Spaldon et al. 2015). Under stress, vegetable crops like brinjal, okra, round melons, ridge gourds, bottle gourds, and sponge gourds are grown using both organic as well as inorganic mulches. Mulching controls soil temperature, inhibits weed growth, and lessens soil runoff and erosion. Mulching also minimizes soil moisture evaporation (Chen et al. 2019). As they are allowed to decompose after usage, organic mulches, i.e., fenugreek, rice straw, cluster beans, and grasses serve to improve soil fertility and add organic carbon to the soil. The warm-weather tomato crop is benefited from rice straw mulching. Due to higher fruit production and reduced initial input requirements throughout the summer, rice straw mulching in a okra crop displays a maximum benefit-cost ratio (Tiwari et al. 1998). Plastic or inorganic mulches fail to enrich the soil with organic materials, but they do keep the soil moist and inhibit weed development (Zhang et al. 2017).

Vegetable crops can be protected during production from harsh weather and copious rain using greenhouses, polytunnels, and shade nets. These protective structures offer a controlled environment that enables growers to modify temperature, humidity, and other elements to maximize crop development and safeguard against climatic changes (Gruda et al. 2019). Several technologies, including heating and cooling systems, CO_2 emission and absorption systems, automated need-based irrigation, and fertilizer-providing systems, can be used to control climate inside greenhouses. Weeds, salinity, alkalinity, acidity, and diseases transmitted through contaminated soil are also avoided through soilless farming. Vegetables that have been harvested can be sold at a good price (Bisbis et al. 2018).

17.4.2 Molecular Approaches

Genetic improvement primarily functions as an adaptation method since it helps crops to adjust against expected climatic conditions in the future. For their profitable cultivation and food security, crop plants need to be genetically improved so that they can survive the negative effects of climate change. Due to intricacy of abiotic stress resilience, absence of criteria for selection, and lack of accurate knowledge regarding genetics of stress resistance, breeding for stress resilience is challenging. A simple but effective technique for breeding vegetables is to choose resistant lines from existing populations and produce new types from their offsprings (Kwak 2019).

Clustering of existing germplasm into discrete groups is aided by genetic diversity (Abdelrahman et al. 2022). It aids in locating several parents for hybridization. As variability and acquisition of superior variants in segregating stages will be larger, more diverse parents appear. The transfer of abiotic stress tolerance genes from resistant genotypes to agronomically better cultivars is facilitated by hybridization or heterosis breeding (Raza et al. 2019). Cassava and other tropical tuber crops may quickly recover from drought and can survive moisture stress. In vegetable crops, efforts might be made to locate and characterize genes that are responsible for drought tolerance. For drought tolerance in cassava, genetic markers including 800 simple sequence repeat, 3000 restriction fragment length polymorphisms, 09 isozymes, and 120 random amplified polymorphic DNA have been identified (Naqvi et al. 2022).

The parthenocarpic fruit growth in tomatoes at high temperatures is controlled by *Pat-2* gene. Once temperatures are high and normal fruit set is compromised, this feature will help tomatoes produce more fruits. A few wild tomato relatives can withstand environmental conditions. Both *Solanum cheesmaniae* and *S. pimpinellifolium* can withstand heat and salt, respectively. The extended primary root and large secondary root system of *S. chilence* and dense cuticle, waxy leaves of *S. pennellii* enable conservation of leaf water in dry soils, respectively (Koundinya et al. 2018). Hence, genome selection, molecular markers, DNA fingerprinting, genome editing, genetic transformation, markers-assisted selection, genome mapping, and genome sequencing are some approaches that can be utilized for the development of climate-resilient germplasm.

17.4.3 Forecasting

Innovations have been developed to enhance accuracy and use of data on vegetable production against climate change. Forecasting is the process of predicting future value using data from the past. The benefit of daily weather forecasting via remote sensing, validating various land-use products, and information transmission are offered by weather prediction models (Kearney and Wintle 2010). The crop growth simulation models use a variety of parameters, such as future weather possibilities predicted by meteorologists or global circulation models, to forecast crop growth as well as yield during future climatic conditions. The potential effects of climate change on crop output can be foreseen using these, and they can also assist in formulating the necessary adaptation measures (Min et al. 2016). To enable the implementation of preventive measures, many pest and disease models for projections are also available that can forecast development of pests and illnesses in advance. To predict potato yield in India, three global climate models (HadCM3Q, EH5OM, and CCAM-Mark 3.5) are being employed (Koundinya et al. 2018).

17.5 Conclusion

Climate change poses significant constraints to crop production. Rising temperatures, erratic rainfall patterns, and increased frequency of extreme weather events have adverse effects on crop growth, yield, and quality. Heat stress can stunt plant growth and reduce marketable yield, while droughts can lead to water scarcity and inadequate irrigation. Prolonged periods of heavy rainfall can cause soil erosion and waterlogging, impacting root health and nutrient absorption. Moreover, changing pest and disease patterns challenge crop resilience. To mitigate these constraints, innovative agricultural practices, i.e., precision irrigation, crop diversification, and development of heat- and drought-tolerant cultivars are crucial. The most emphasis needed is to investigate the genetic premise governing these processes to comprehend plant responses under various abiotic conditions. Collaborative efforts between researchers, farmers, and policymakers are needed to ensure sustainable vegetable production in the face of climate change.

References

Abbas Q, Han J, Bakhsh K, Ullah R et al (2022) Adaptation to climate change risks among dairy farmers in Punjab, Pakistan. Land Use Policy 119:106184

Abbaspour KC, Faramarzi M et al (2009) Assessing the impact of climate change on water resources in Iran. Water Resour Res 45:10434

Abdelrahman M, Tran LSP et al (2022) Physiological and molecular perspectives of stress tolerance in vegetables. Front Plant Sci 13:1004093

Ahmad R, Anjum MA (2023) Mineral nutrition management in fruit trees under salt stress: a review. Erwerbs-obstbau 65(2):397–405

Ahmad R, Manzoor M, Muhammad (2023) Exogenous melatonin spray enhances salinity tolerance in Zizyphus germplasm: a brief theory. Life 13:493

Ahvenainen R (1996) New approaches in improving the shelf life of minimally processed fruit and vegetables. Trends Food Sci Technol 7:179–187

Akhtar R, Masud MM (2022) Dynamic linkages between climatic variables and agriculture production in Malaysia: a generalized method of moments approach. Environ Sci Pollut Res 29:41557–41566

Atanda SA, Pessu PO (2011) The concepts and problems of postharvest food losses in perishable crops. Afr J Food Sci 5:603–613

Ayyogari K, Sidhya P et al (2014) Impact of climate change on vegetable cultivation – a review. Int J Agric Environ Biotechnol 7:145–155

Bahadur A, Chatterjee A et al (2011) Physiological and biochemical basis of drought tolerance in vegetables. Veg Sci 38:1–16

Bazilian M, Rogner H, Howells et al (2011) Considering the energy, water and food nexus: Towards an integrated modelling approach. Energy Policy 39:7896–7906

Bhatia A, Kumar V (2013) Effect of elevated ozone and carbon dioxide interaction on growth and yield of maize. Maydica 58:291–298

Bisbis MB, Gruda N, Blanke M et al (2018) Potential impacts of climate change on vegetable production and product quality – a review. J Clean Prod 170:1602–1620

Bueren VEL, Jones SS et al (2011) The need to breed crop varieties suitable for organic farming, using wheat, tomato and broccoli as examples: a review. NJAS Wageningen J Life Sci 58:193–205

Cabillo CM (2019) Biomass production of lettuce (*Lactuca sativa* L.) under water stress. Int J Sci Eng Res 10:928–935

Calvo E, Martin C, Sanz M (2007) Ozone sensitivity differences in five tomato cultivars: visible injury and effects on biomass and fruits. Water Air Soil Pollut 186:167–181

Chapman SC, Chakraborty S (2012) Plant adaptation to climate change – opportunities and priorities in breeding. Crop Pasture Sci 63:251–268

Chauhan BS, Mahajan G et al (2014) Global warming and its possible impact on agriculture in India. Adv Agron 123:65–121

Chen H, Li L, Luo X et al (2019) Modeling impacts of mulching and climate change on crop production and N2O emission in the Loess Plateau of China. Agric For Meteorol 268:86–97

Chepkoech W, Stöber S et al (2023) What drives diversity in climate change adaptation strategies for African indigenous vegetable production in Kenya? Econ Anal Policy 77:716–728

Chung BN, Canto T, Tenllado F et al (2016) The effects of high temperature on infection by Potato virus Y, Potato virus A, and Potato leafroll virus. Plant Pathol J 32:321

Chung BN, Koh SW et al (2017) Temperature and CO2 level influence Potato leafroll virus infection in *Solanum tuberosum*. Plant Pathol J 33:522

Dang HL, Li E et al (2019) Factors influencing the adaptation of farmers in response to climate change: a review. Clim Dev 11:765–774

Dong J, Gruda N, Li X et al (2020) Sustainable vegetable production under changing climate: The impact of elevated CO2 on yield of vegetables and the interactions with environments – a review. J Clean Prod 253:119920

Ebi KL, Hess JJ (2020) Health risks due to climate change: inequity in causes and consequences. Health Aff 39:2056–2062

Flowers MD, Fiscus EL et al (2007) Photosynthesis, chlorophyll fluorescence, and yield of snap bean (*Phaseolus vulgaris* L.) genotypes differing in sensitivity to ozone. Environ Exp Bot 61:190–198

Gautam HR, Bhardwaj et al (2013) Climate change and its impact on plant diseases. Curr Sci 105:1685–1691

Gonçalves L, Rubiales D et al (2022) Grass pea (*Lathyrus sativus* L.) – a sustainable and resilient answer to climate challenges. Agronomy 12:1324

Gora JS, Verma AK et al (2019) Climate change and production of horticultural crops. Agric Impacts Clim Change 1:45–61

Gruda N, Bisbis M, Tanny J (2019) Influence of climate change on protected cultivation: Impacts and sustainable adaptation strategies – a review. J Clean Prod 225:481–495

Hatfield JL, Boote KJ, Kimball BA et al (2011) Climate impacts on agriculture: implications for crop production. Agron J 103:351–370

Hatfield JL, Wright-Morton L, Hall B (2018) Vulnerability of grain crops and croplands in the Midwest to climatic variability and adaptation strategies. Clim Chang 146:263–275

Hauglustaine DA, Brasseur GP (2001) Evolution of tropospheric ozone under anthropogenic activities and associated radiative forcing of climate. J Geophys Res Atmos 106:32337–32360

Heureux AMC, Alvar-Beltrán J et al (2022) Climate trends and extremes in the Indus River Basin, Pakistan: implications for agricultural production. Atmos 13:378

Katny MAC, Hoffmann-Thoma G (2005) Increase of photosynthesis and starch in potato under elevated CO2 is dependent on leaf age. J Plant Physiol 162:429–438
Kearney MR, Wintle BA (2010) Correlative and mechanistic models of species distribution provide congruent forecasts under climate change. Conserv Lett 3:203–213
Khan I, Lei H, Shah IA, Ali I et al (2020) Farm households' risk perception, attitude and adaptation strategies in dealing with climate change: promise and perils from rural Pakistan. Land Use Policy 91:104395
Kiaya V (2014) Postharvest losses and strategies to reduce them. Action Contre la Faim 25:1–25
Klay I, Gouia S, Liu M et al (2018) Ethylene response factors (ERF) are differentially regulated by different abiotic stress types in tomato plants. Plant Sci 274:137–145
Koundinya AVV, Kumar PP et al (2018) Adaptation and mitigation of climate change in vegetable cultivation: a review. J Water Clim Change 9:17–36
Koyama R, Itoh H et al (2012) Augmentation of antioxidant constituents by drought stress to roots in leafy vegetables. HortTechnology 22:121–125
Kumar R, Solankey SS et al (2012) Breeding for drought tolerance in vegetables. Veg Sci 39:1–15
Kwak SS (2019) Biotechnology of the sweetpotato: ensuring global food and nutrition security in the face of climate change. Plant Cell Rep 38:1361–1363
Lansink AO, Bezlepkin I (2003) The effect of heating technologies on CO2 and energy efficiency of Dutch greenhouse firms. J Environ Manag 68:73–82
Lee SG, Kim SK et al (2018) Impact of moderate and extreme climate change scenarios on growth, morphological features, photosynthesis, and fruit production of hot pepper. Ecol Evol 8:197–206
Liang G, Liu J et al (2020) Effects of drought stress on photosynthetic and physiological parameters of tomato. J Am Soc Hortic Sci 145:12–17
Lindgren E, Harris F et al (2018) Sustainable food systems – a health perspective. Sustain Sci 13:1505–1517
Liu F, Stützel H (2004) Biomass partitioning, specific leaf area, and water use efficiency of vegetable amaranth (*Amaranthus* spp.) in response to drought stress. Sci Hortic 102:15–27
Magan N, Aldred D (2007) Postharvest control strategies: minimizing mycotoxins in the food chain. Int J Food Microbiol 119:131–139
Mahmoud MT, Mohammed SA et al (2020) Impact of topography and rainfall intensity on the accuracy of immerge precipitation estimates in an arid region. Remote Sens 13:13
Malhotra SK, Srivastva AK (2014) Climate smart horticulture for addressing food, nutritional security and climate challenges. ASM Foundation, New Delhi, pp 83–97
Martí R, Roselló S et al (2016) Tomato as a source of carotenoids and polyphenols targeted to cancer prevention. Cancers 8:58
Min J, Lu K, Sun H et al (2016) Global warming potential in an intensive vegetable cropping system as affected by crop rotation and nitrogen rate. Clean Soil Air Water 44:766–774
Moretti CL, Mattos LM et al (2010) Climate changes and potential impacts on postharvest quality of fruit and vegetable crops: a review. Food Res Int 43:1824–1832
Mou W, Li D, Bu et al (2016) Comprehensive analysis of ABA effects on ethylene biosynthesis and signaling during tomato fruit ripening. PLoS One 11
Naqvi RZ, Siddiqui HA et al (2022) Smart breeding approaches in post-genomics era for developing climate-resilient food crops. Front Plant Sci 13:972164
Nasir MW, Toth Z (2022) Effect of drought stress on potato production: a review. Agronomy 12:635
Okunlola GO, Olatunji OA et al (2017) Physiological response of the three most cultivated pepper species (*Capsicum* spp.) in Africa to drought stress imposed at three stages of growth and development. Sci Hortic 224:198–205
Olesen JE, Trnka M et al (2011) Impacts and adaptation of European crop production systems to climate change. Eur J Agron 34:96–112
Ou LJ, Dai XZ et al (2011) Responses of pepper to waterlogging stress. Photosynthetica 49:339–345
Pangga IB, Hanan J et al (2013) Climate change impacts on plant canopy architecture: implications for pest and pathogen management. Eur J Plant Pathol 135:595–610

Parajuli R, Thoma G et al (2019) Environmental sustainability of fruit and vegetable production supply chains in the face of climate change: a review. Sci Total Environ 650:2863–2879

Peppelenbos HW, van't Leven J (1996) Evaluation of four types of inhibition for modelling the influence of carbon dioxide on oxygen consumption of fruits and vegetables. Postharvest Biol Technol 7:27–40

Potopová V, Štěpánek P et al (2016) Drought stress impact on vegetable crop yields in the Elbe River Lowland between 1961 and 2014. Cuadernos Investig Geogr 42:127–143

Quesada-Ocampo LM et al (2010) Resistance in tomato and wild relatives to crown and root rot caused by *Phytophthora capsici*. Phytopathology 100:619–627

Raza A, Razzaq A, Mehmood SS, Zou X, Zhang X, Lv Y, Xu J (2019) Impact of climate change on crops adaptation and strategies to tackle its outcome: a review. Plan Theory 8:34

Rosales MA, Cervilla LM et al (2011) The effect of environmental conditions on nutritional quality of cherry tomato fruits: evaluation of two experimental Mediterranean greenhouses. J Sci Food Agric 91:152–162

Saqib M, Anjum MA et al (2022) Horticultural crops as affected by climate change. In: Jatooi WN, Mubeen M, Ahmad A, Cheema MA, Lin Z, Hashmi MZ (eds) Building climate resilience in agriculture. Springer, Cham, pp 95–109

Schreinemachers P, Simmons EB et al (2018) Tapping the economic and nutritional power of vegetables. Glob Food Secur 16:36–45

Settele J, Bishop J et al (2016) Climate change impacts on pollination. Nat Plants 2:1–3

Shehzad K (2023) Extreme flood in Pakistan: is Pakistan paying the cost of climate change? A short communication. Sci Total Environ 880:162973

Song J, Fan L et al (2001) Effect of ozone treatment and controlled atmosphere storage on quality and phytochemicals in high blueberries. Acta Hortic 600:417–424

Songy A, Fernandez O et al (2019) Grapevine trunk diseases under thermal and water stresses. Planta 249:1655–1679

Spaldon S, Samnotra et al (2015) Climate resilient technologies to meet the challenges in vegetable production. Int Res Curr Acad Rev 3:28–47

Thwe AA, Vercambre G et al (2015) Effects of acute ozone stress on reproductive traits of tomato, fruit yield and fruit composition. J Sci Food Agric 95:614–620

Tiwari KN, Mal PK et al (1998) Response of okra (*Abelmoschus esculentus* (L.) Moench.) to drip irrigation under mulch and non-mulch conditions. Agric Water Manag 38:91–102

Ullah A, Manghwar H et al (2018) Phytohormones enhanced drought tolerance in plants: a coping strategy. Environ Sci Pollut Res 25:33103–33118

White JW, Hoogenboom G et al (2011) Methodologies for simulating impacts of climate change on crop production. Field Crop Res 124:357–368

Zhang F, Zhang W et al (2017) Is crop biomass and soil carbon storage sustainable with long-term application of full plastic film mulching under future climate change? Agric Syst 150:67–77

Zhang S, Dai J, Ge Q (2020) Responses of autumn phenology to climate change and the correlations of plant hormone regulation. Sci Rep 10:1–10

Understanding Drought Stress, Its Effects, and Tolerance Mechanism for Sustainable Vegetable Production

18

Ali Asad Bahar, Hafiz Nazar Faried, Sami Ullah, and Ahmad Ali

Abstract

Climate change, water shortage, food insecurity, and population pressure are the challenges being faced in today's world. Drought stress poses a great constraint and is becoming a constant challenge in the way of sustainable agriculture. It changes the normal behavior of plants via altering its morphology, physiology, biochemical attributes, and nutrients uptake. Therefore, in this chapter, we highlight the impacts of water scarcity on different phases of plants' life cycle, i.e., seed germination, plant development, photosynthesis, nutrients uptake, antioxidants, ROS balance, and yield. In addition to this, we also summarize drought tolerance induction mechanism especially in vegetable crops by application of plant growth regulators, potassium, selenium, chitosan, microbes, and by grafting and innovative irrigation scheduling, which would facilitate plant scientists and vegetable growers to lessen the damaging effects of drought. The alleviation of these effects of drought is of great importance in sustaining vegetable production.

Keywords

Antioxidant enzymes · Drought tolerance induction · Innovative irrigation scheduling · Water stress

A. A. Bahar · H. N. Faried (✉) · S. Ullah
Department of Horticulture, MNS University of Agriculture, Multan, Pakistan
e-mail: nazar.farid@mnsuam.edu.pk

A. Ali
Department of Soil, Water and Ecosystem Sciences, Institute of Food and Agricultural Sciences, University of Florida, Gainesville, FL, USA

Z. Iqbal et al. (eds.), *Sustainable and Innovative Vegetable Production in times of Climate Change*, https://doi.org/10.1007/978-981-96-6283-8_18

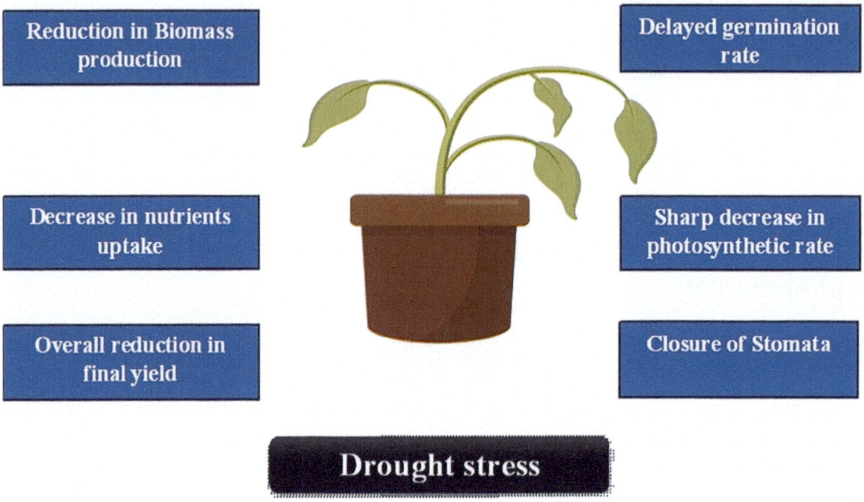

Fig. 18.1 Negative onsets of water stress on plants' growth and productivity

18.1 Introduction

Productivity of vegetable crops is constrained by various abiotic stresses, including cold, heat, drought, and salinity thereby disturbing their growth and life cycle (Xu et al. 2014; Negrão et al. 2017; Nadeem et al. 2018). Among these abiotic constraints, drought is the utmost challenging aspect in crop production (Sehgal et al. 2018). Moreover, due to global climate change, it is expected that in future water scarcity will increase in many parts of the globe as an outcome of reduced rainfall and increased evaporation (Dey et al. 2019).

Commonly, drought is referred to "as a stage of below-average rainfall or as an event without consequential rainfall" (Verslues et al. 2006). Globally, water shortage is an enormous problem, thus posing a huge threat to agriculture in terms of food security and stability. Overall, water stress causes several adverse effects on crop plants (Fig. 18.1). By viewing the climate change scenario and future onsets of water scarcity on plants, it is the need of time to adopt strategies to cope with extreme weather events particularly drought. Therefore, this chapter is aimed to present a complete picture of water scarcity effects on plants along with various management approaches to induce drought tolerance in vegetable crops.

18.2 Changes Induced by Water Scarcity in Plants

18.2.1 Seed Germination

Germination of seed, the first phase toward plant growth, is quite sensitive to low moisture. Seed germination and its establishment are affected severely due to low

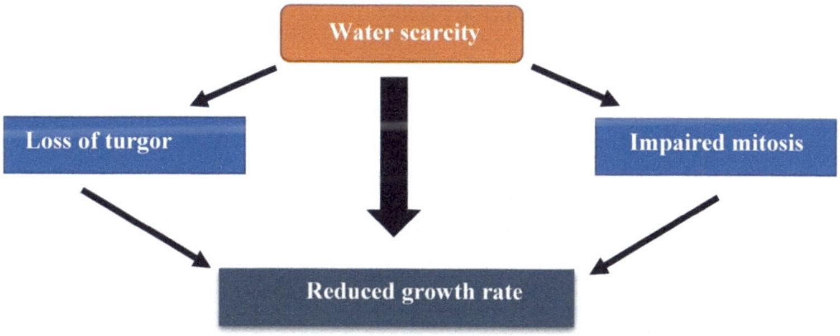

Fig. 18.2 Mechanism of water stress contributing toward reduced plant growth

soil moisture (Ali et al. 2020b; Saux et al. 2020). Initially, it decreases length and dry weight of plumule due to constrained translocation of nutrients from cotyledons or endosperms to growing embryo. Seed germination % declines with the increase of PEG-6000 concentration. At −0.3 MPa (5% PEG), it has a stimulating influence on seed germination, while at −0.9 MPa (15% PEG), it vividly reduces the germination (Li et al. 2013). In purslane (*Portulaca oleracea*), it was observed that seed germination was delayed when treated with polyethylene glycol (PEG) at the stress level of −0.4 MPa (Rahdari et al. 2012). As a consequence of water stress, reduced germination and impaired plant growth have been recorded in several vegetables including tomato (Eke et al. 2019). Several scientists discovered that impaired seed germination and increase in mean germination time (days) were highly linked with osmotic stress (Liu et al. 2013; Guo et al. 2017).

18.2.2 Morphology

Plant growth is accomplished by numerous events mainly via cell division, expansion, and differentiation. Drought causes alterations/changes in plant cell events (Fig. 18.2); it hampers mitosis and cell expansion process and thereby reduces plant growth (Hussain et al. 2008). Drought also triggers the process of turgor loss which is another reason for the slow growth rate (Taiz and Zeiger 2006; Seleiman et al. 2019). For instance, water scarcity causes negative impacts on plant growth pattern (Table 18.1).

18.2.3 Physiology

Generally, the first and foremost drought influence on crop plants is the limitation of photosynthetic rate (*Pn*) (Germ et al. 2007; Zhao et al. 2020). This occurs due to decline in leaf extension, early leaf senescence, reduction in photosynthetic mechanism as well as in the production of food (Farkas et al. 2020). Decline in photosynthesis is because of stomatal closure and reduction in leaf area that leads toward

Table 18.1 Influence of water stress on crop plants

Crop	Parameters	Impact	References
Solanum scabrum	Shoot height, shoot dry weight, and leaf area	Decrease (↓)	Muthomi and Musyimi (2009)
Melons	Plant height, diameter of stem, fresh and dry weights of shoot, leaf area, and number of leaves	↓	Kuşvuran et al. (2011)
Tomato	Plant height, root length, shoot fresh and dry weights, leaf area, and number	↓	Rezaei et al. (2012)
Peppermint	Plant height, plant fresh and dry weights, leaf number, and root dry weight	↓	Zade et al. (2019)
Mint	Plant height, root dry weight, leaf area, and number	↓	Elansary et al. (2020)

Table 18.2 Effect of water stress on plants physiology

Crop	Parameters	Impact	References
Potato	Photosynthetic rate (*Pn*) and stomatal conductance (*gs*)	Decrease (↓)	Liu et al. (2006)
Pepper	Chlorophyll fluorescence	↓	Guang-cheng et al. (2011)
Egg plant	*gs*	↓	Zhang et al. (2014b)
Tomato	*Pn*	↓	Liu et al. (2015)
Faba bean	Chlorophyll content and *Pn*	↓	Siddiqui et al. (2015)
Watermelon	Chlorophyll fluorescence	↓	Mo et al. (2016)
Tomato	Chlorophyll content	↓	Hassnain et al. (2020)

higher yield losses (Chen and Hao 2015). Closing of stomata occurs likely by signaling of abscisic acid (ABA) that prevents inflow of carbon dioxide (Farooq et al. 2009) and thereby becomes the prior reason for declination in photosynthesis rate (Anjum et al. 2003). In this respect, stomatal closure decreases transpiration rate and hence increases leaf temperature (Arbona et al. 2013). Indeed, higher leaf temperature disturbs various processes, i.e., denatures proteins, enzymes, chlorophyll synthesis, as well as reduces membrane permeability (Rahbarian et al. 2011). Additionally, it was also found that water scarcity affected photosynthesis working due to a degeneration of Rubisco activity (Bota et al. 2004; Farooq et al. 2009). This disruption in Rubisco activity directly inhibits protein synthesis, causes anatomical alterations in leaf structure, and thus limits conductance of mesophyll cells (Tränkner et al. 2018). These changes induce disruption in respiration and photosynthesis, amino acids and proteins synthesis, and minerals nutrition uptake (Tiwari and Yadav 2020). Conclusively, water stress disrupts normal functioning of plants and has been reported in many crops (Table 18.2).

18.2.4 Nutrients Availability

Plant growth and productivity heavily rely on the availability of nutrients. Overall, there are 17 essential nutrients that plants need for their existence. However, their absorption is highly dependent on soil moisture because water acts as a transport medium and its acquisition is restricted under water-deficient soil (Usmani et al. 2020). Diminution in transpiration rate, imbalance in membrane permeability, and decreased active transport are the main pillars for this disruption (Hu et al. 2007). Several investigations have been carried out in several crops to investigate the availability of different nutrients under water stress conditions.

18.2.4.1 Nitrogen (N)

N is the most widely used mineral in agricultural production. Plant development is highly correlated with the application of nitrogenous fertilizers. N requisite for optimal plant growth usually ranges 2–5% of dry weight depending upon the nature of crop and its developmental phase (Marschner 2012). It is involved in leaf expansion and also delays the aging of leaves. It takes part in improving the photosynthetic rate and quality of produce (Leghari et al. 2016). However, as mentioned earlier, its uptake is restricted under low level of soil moisture. It has been reported that plants of both potato and pepper were unable to absorb available N under low moisture (Dalla Costa et al. 1997; Dalla Costa and Gianquinto 2002). That's why a crop with optimum soil moisture gets more benefit from the applied nutrients. Results of an experiment conducted on five cultivars of cherry tomato indicated a clear reduction in uptake of macro-nutrients specifically N under moderately water-stressed conditions (Sánchez-Rodríguez et al. 2010). Moreover, due to dehydrated soil, N availability becomes the minimum, and thus N deficiency occurs. Alteration in the uptake of nitrogenous compounds has also been testified in sweet potato plants when they were exposed to water scarcity (Ekanayake and Collins 2004).

18.2.4.2 Phosphorus (P)

P is a vital and major part for all vegetable crops. It induces early canopy development and improves root cell division, crop quality, and number of tubers in potato (Rosen et al. 2014). It is also involved in enzyme activation as well as carbohydrate metabolism (Razaq et al. 2017). Plants absorb this mineral nutrient in the form of anion orthophosphate ($H_2PO_4^-$). According to Wang and Zhang (2012), greater bioavailability of P was ensured by maintaining relatively high soil water content through frequent irrigation. Hence, ample amount of water is necessary for the plants to absorb phosphate ions. The phosphate ions flow through soils via diffusion process and if there is low soil moisture content, the diameter of water-filled pores shrinks, tortuousness improves, and thus movement of the phosphate ions decreases (Faye et al. 2006). This situation leads toward reduction in both absorption and transportation of phosphate ions. In an experiment, in tomato crop, P absorption was greatly reduced under deficit irrigation as compared to partial drying of root zone (Wang et al. 2012). Similarly, in another experiment conducted on potato crop to analyze the extent of P uptake under water scarcity, P uptake gradually decreased with increase in the intensity of water scarcity (Liu et al. 2017).

18.2.4.3 Potassium (K)

In order to complete the life span by plants, K is a vital mineral. It improves plant growth and productivity by stimulating more than 60 enzymes (Abdel-Ati et al. 2007). Additionally, it induces tolerance against abiotic and biotic constraints (Wang et al. 2013). It creates tolerance in crop plants against drought stress by maintaining stomatal regulation, energy condition, osmoregulation, and energy balance (Hasanuzzaman et al. 2018b). Moreover, it lessens soil moisture loss by decreasing transpiration rate and increasing water retention in the plant body (Umar and Moinuddin 2002). However, its availability depends upon plant species, application timing, duration, and severity of drought. The uptake of K under water scarcity environments has already been illustrated in various vegetable crops (Table 18.3).

18.2.5 Antioxidant Enzymes

Antioxidant is a stable molecule that has the ability to scavenge free radicals (Yashin et al. 2017). A free radical is an unstable and highly reactive molecule. It produces deleterious impacts on carbohydrates, proteins, lipids, and DNA (Evers et al. 2010; Gill and Tuteja 2010). Water scarcity disturbs the equilibrium between antioxidants and reactive oxygen species (ROS), thereby leads toward higher generation of ROS. The most noticeable oxygen-containing free radicals are superoxide anion (O^{2-}), hydroxyl radical, oxygen singlet, hydrogen peroxide (H_2O_2), nitric oxide, and peroxynitrite radical (Cruz de Carvalho 2008). Additionally, it also results in destructing the defensive system of enzymatic and non-enzymatic antioxidants, thus causing a failure in eliminating O^{2-} (Zhang et al. 2014a). Earlier investigations have revealed that water scarcity succeeds toward higher ROS generation and has been witnessed in several plant species such as in potato (Dalla Costa et al. 1997; Martínez-Gutiérrez et al. 2012), sweet potato (Lin et al. 2006), and tomato (Jangid and Dwivedi 2017). Furthermore, ROS damage is more when its production increases in comparison to antioxidative activities. Therefore, plants detoxify free radicals via activating antioxidant activities (Nazar et al. 2011). Enzymatic

Table 18.3 Impact of water scarcity on potassium uptake in various crops

Crop	K^+ uptake	References
Tomato (cherry)	Reduced	Sánchez-Rodríguez et al. (2010)
Maize	Reduced	Ge et al. (2012)
Egg plant	Reduced	El-Miniawy (2015)
Tomato	Reduced	Azad et al. (2016)
Potato	Reduced	Bahar et al. (2021)
Potato	Reduced	Zaki and Radwan (2022)

	Sprouting & Emergance (15-20 Days)	Vegetative Growth (15-20 Days)	Tuber initiation (15-20 Days)	Tuber Bulking (45-55 Days)	Maturity (20-25 Days)
Impact of Water Deficit	• Reduction in root and plant development	• Restricted stolon initiation • Reduced nutrient accumulation	• Delayed tuber initiation • Reduced tuber number	• Poor tuber size and quality • Distorted tuber shape • Promotes early senescence	• Limited tuber size with poor shelf life

Fig. 18.3 Impact of water scarcity on different growth stages of potato

antioxidants are peroxidase (POD), catalase (CAT), superoxide dismutase (SOD), glutathione peroxidase (GPx), and glutathione reductase (GR) (Willcox et al. 2004). These enzymes take part in counteracting the ROS. SOD is regarded as the primary defense line because it transforms O^{2-} radical into H_2O_2. CAT or GPx converts the oxidant formed H_2O_2 into water and oxygen. GR also exerts a powerful defense against oxidants as it oxidizes glutathione and limits peroxide radicals (Dröge 2002; Halliwell 2007).

18.2.6 Drought and Crop Yield

Worldwide, drought causes up to 50% losses in average crop yield (Wang et al. 2003). In fact, water scarcity is a multifaceted phenomenon, which is determined by several factors, i.e., genetic potential of the plant, its stage of growth (Fig. 18.3), and period and intensity of the stress (Zhu 2002). Under water scarcity, negative onsets in crop yield (Table 18.4) occur due to reduction in photosynthetic active emission, radiation use efficacy, and harvest index (Earl and Davis 2003).

18.3 Drought Tolerance Induction

Plants employ several strategies and adaptations to alleviate the drastic effects of drought stress. Additionally, scientists are utilizing different methodologies for drought tolerance induction in crop plants. Among these breeding, grafting, and applications of plant growth regulators and other chemicals are of greater value to induce drought tolerance at specific plant stage.

Table 18.4 Yield losses (↓) in some vegetable crops due to water scarcity

Crop	Yield reduction (%)	References
Potato	11	Lahlou et al. (2003)
	13	Kawakami et al. (2006)
Egg plant	60	Karam et al. (2011)
Tomato	47	Wahb-Allah et al. (2011)
Pepper	69.5	Mardani et al. (2017)
Watermelon	38	Mandizvo et al. (2022)

18.3.1 Plant Growth Regulators (PGRs)

Application of PGRs (synthetic or natural) mitigates water stress and induces tolerance in crop plants (Moumita et al. 2019). Gibberellic acid (GA_3) works as a protectant thereby activating secondary metabolites lowers ROS generation (Khan et al. 2018b). GA_3 application regulates photosynthetic machinery, improves chlorophyll content, water use efficiency (WUE), and nutrients uptake as well as ameliorates drought adverse impacts; hence improves plant productivity (Hasanuzzaman et al. 2018a). PGRs, i.e., jasmonic acid and abscisic acid increase crop performance under water deficit condition (Ahmad et al. 2018). Similarly, other hormones such as benzyl adenine, (a cytokinin), uniconazole, and brassinolide induce drought tolerance mechanism in various crops by improving protein content and activating enzymatic antioxidants activities (POD, SOD, and CAT) (Jan et al. 2019).

Salicylic acid is an effective plant hormone that controls several biochemical and physiological activities especially under drought conditions (Zamaninejad et al. 2013). Exogenous application of salicylic acid has demonstrated reduction in drought susceptibility in tomato crop (Fan et al. 2022). Moreover, foliar application of salicylic acid enhances accumulation of proline in sweet basil (Zulfiqar et al. 2021). It also induces drought tolerance by deferring leaf rolling (Gomes et al. 2004). In this way, salicylic acid enhances plants' defense system to deal with biotic and abiotic constraints including drought.

18.3.2 Grafting

Grafting is a tissue fusion of two plants together for better plant development. The upper part (scion) contributes stems, leaves, flowers, and fruits, while the bottom part (rootstock) of the plant contributes roots and upholds plant (Lee et al. 2010). Grafted plants have better capability to survive in areas where water insufficiency is a problem (Tu et al. 2018). Grafted plants have deeper root system, which enables them to acquire water from deep soil layer (Jin et al. 2016). A robust and deeper root system of pumpkin rootstock was the main contributor in drought tolerance of grafted "Charleston Gray" watermelon (Poor 2015). Likewise, Sánchez-Rodríguez et al. (2014) and López-Marín et al. (2017) observed improved root growth, dry biomass, and metabolic activity under deficit irrigation in grafted plants in contrast to non-grafted ones. In addition to this, the most beneficial impacts of grafting observed in crops were improvement in photosynthetic machinery, water use efficiency (WUE), and nutrients accumulation. Tolerant rootstock grafted plants have enhanced absorption, accumulation, and upward transfer of nutrients under water deficit condition (Rouphael et al. 2008). Improved WUE of cucumber grafted on luffa rootstock was also noticed in comparison to non-grafted plants. Such improvement in grafted plants was primarily a result of higher CO_2 intake, accompanied by improved ability of attaining more water and nutrients from the deep layer of soil (Liu et al. 2016). Besides this, grafting also plays an integral part in activating plants defense system by regulating antioxidant enzymatic activities. Increased antioxidant enzymes activities (SOD, CAT) and decreased oxidative stress indicators, i.e., H_2O_2 and malondialdehyde (MDA) were recorded in grafted tomato plants as compared with un-grafted control plants (Sánchez-Rodríguez et al. 2016). Such processes directly act as ROS scavenger, via binding to antioxidant enzymes (Groppa and Benavides 2008). Overall, grafting can enhance plant development, increase antioxidant enzymes activities, reduce oxidative stress, and regulate water homeostasis in plants to induce drought tolerance.

18.3.3 Selenium and Chitosan

The plants subjected to water deficit condition generate ROS which results in oxidative damage to the biomolecules, i.e., lipids, proteins, carbohydrates, and nucleic acid, thereby inhibit the normal functioning of plant by reducing photosynthesis and respiration rates, and growth of plants (Dar et al. 2017). Application of selenium and chitosan is helpful in creating tolerance against drought (Table 18.5). Under water deficit, selenium supplementation can cause buildup of compatible solutes and hence, helpful in ameliorating the oxidative damage in plants (Ebeed et al. 2017). Delay in senescence is another potent characteristic of selenium which means that it maintains water status in plant body (Bocchini et al. 2018). Irrespective to this, chitosan application on plants can create a permeable film that may alter leaf internal environment resulting in improved stomatal conductance and reduced transpiration rate (El-Saadony et al. 2021). Moreover, it was observed that chitosan possesses effective antioxidant and antimicrobial activities that play a key role in plants' survival under stressful environment (Rabêlo et al. 2019). In conclusion, both selenium and chitosan activate protective enzymatic activities such as SOD, thereby reducing the ROS formation. Such anti-oxidative protectant role of selenium and chitosan has a considerable impact on plants in protecting their water status under water deficit conditions.

18.3.4 Microbes

Plant microbiomes play an integral part in the adaptability of plants against drought conditions via phytohormone production (Azeem et al. 2022). Various *plant*

Table 18.5 Drought tolerance induction in plants via selenium and chitosan application

Crops	Mechanism	Reference
Selenium-treated crops		
Tomato	Improved growth and yield, suppressed oxidative stress markers with improved antioxidant defense system, and maintained photosynthetic efficiency, membranes integrity, and cell water status	Rady et al. (2020)
Potato	Higher efficacy of energy conversion in PS-II and higher tuber mass	Germ et al. (2007)
Lettuce	Increased fresh and dry biomass, upregulated antioxidant system and leaf soluble sugars with decrease in electrolyte leakage and improved free proline content and activities of SOD, POD, and CAT in leaf tissues	Tufail et al. (2023)
Chitosan-treated crops		
Pea	Enhanced antioxidative enzymes (SOD, CAT, and POD) activities	Rasheed et al. (2020)
Faba bean	Stimulated plant height and seed yield and improved quality attributes, i.e., protein, carbohydrates, and amino acid contents	Fouda et al. (2021)
Tomato	Improved chlorophyll content, relative water content, and yield attributes	Hassnain et al. (2020)
Cucumber	Improved shoot length, root length, and chlorophyll content	Ali et al. (2020a)

growth-promoting rhizobacteria (PGPR), isolated from different sources, are helpful in inducing drought tolerance, i.e., *Bacillus* and *Paenibacillus* from maize and sunflower (Vardharajula et al. 2011) and *Paenibacillus, Planococcus*, and *Staphylococcus* from wheat rhizosphere (Verma et al. 2016). In fact, PGPRs mitigate drought stress, as they colonize in the rhizosphere, thereby enhancing uptake of nutrients that induces drought tolerance in plants (Khan et al. 2018a). Another key role of PGPRs is production of phytohormones, 1-aminocyclopropane-1-carboxylate (ACC) deaminase, hydrogen cyanide, ammonia, siderophores, and secondary metabolites (Fig. 18.4). These not only induce drought tolerance in crops but also aid in improving soil fertility status via nutrient cycling (Thilagar et al. 2018).

18.3.5 Potassium (K): A Shield Against Drought

K is a vital plant nutrient, which takes part in several physiological and metabolic activities (Ul-Allah et al. 2020). It has a spectacular role in turgor pressure maintenance as well as cell expansion in plants. This impact favors water uptake and phloem performance with respect to nutrients translocation and movement of assimilates to the whole plant body (Guo et al. 2019). Due to its high solubility, it is extremely mobile in plants' body. It improves and stimulates photosynthesis rate and thereby responsible for generating high energy status in ATP form. This energy is consumed by plant for stimulating its morpho-physiological activities resulting in improved plant productivity (Ul-Allah et al. 2020). In this way, K application

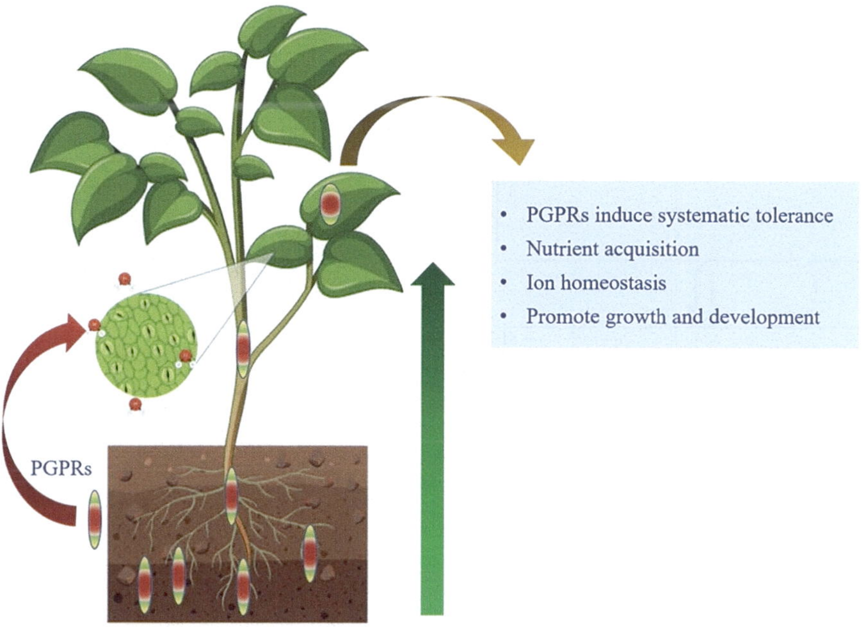

Fig. 18.4 Role of PGPRs in promoting plant growth and development

alleviates drought drastic impacts (Fig. 18.5). Under water deficit condition, K regulates stomatal conductance, enhances antioxidant activities in plants, and hence makes them adaptive to drought condition (Hasanuzzaman et al. 2018b).

Adequate K availability under drought conditions lessens photosynthesis prohibition via extenuating ROS formation (Bahar et al. 2021). That's why, adequate K is required for plants for regulating their physiological and biochemical activities (Shehzad et al. 2020). Further, low yield of crops particularly cultivated under drought stress could be enhanced via exogenous K application (Zelelew et al. 2016; Mattar et al. 2020). Conclusively, K application keeps the plants normal even under drought stress and thereby helpful in attaining optimal yield and quality (Kumar et al. 2019).

18.3.6 Innovative Irrigation Scheduling

Agriculture is the largest consumer for water resources. In future, as discussed earlier, the demand for fresh water will increase both for domestic as well as for agriculture, but water status will be at the brick of risk. That's why irrigation management must be shifted from conventional so-called "Full Irrigation (FI)" to innovative irrigation, i.e., "Partial Root Irrigation (PRI)" (Fig. 18.6). PRI is an effective approach for efficient usage of water. It includes two types: fixed and alternate. In fixed PRI, only one-half of the rhizosphere is irrigated while the other half remains dry during

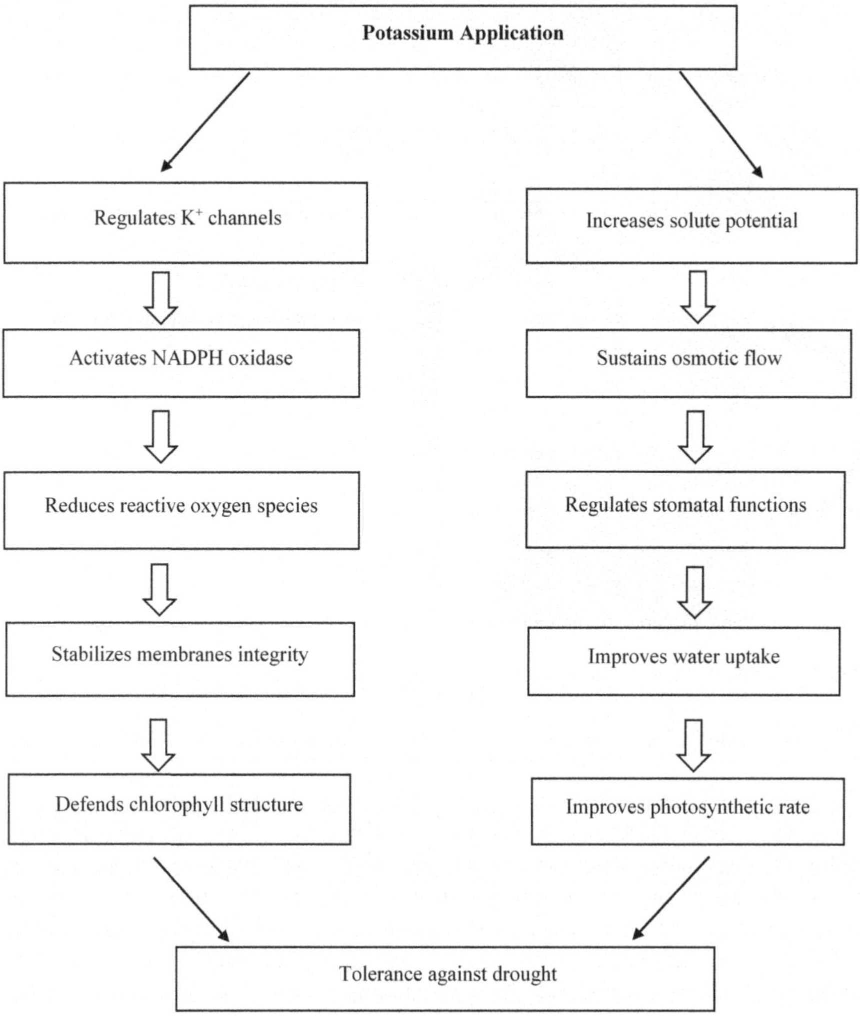

Fig. 18.5 Mechanism of potassium in inducing drought tolerance

the growing season. In alternate PRI, watered and dry sides of the rhizosphere are rotated, which allows the irrigated side of the rhizosphere to dry out and dry side to become wet (Jovanovic and Stikic 2018). Further, improvement in yield by PRI has been recorded in several crops, i.e., pepper (Sezen et al. 2014), eggplant (Howladar 2018), tomato (Mattar et al. 2020), and potato (Bahar et al. 2021). Moreover, antioxidant enzyme activities are also stimulated by PRI and have been witnessed in several crops, i.e., potato (Jovanovic et al. 2010; Bahar et al. 2021) and tomato (Alves et al. 2018).

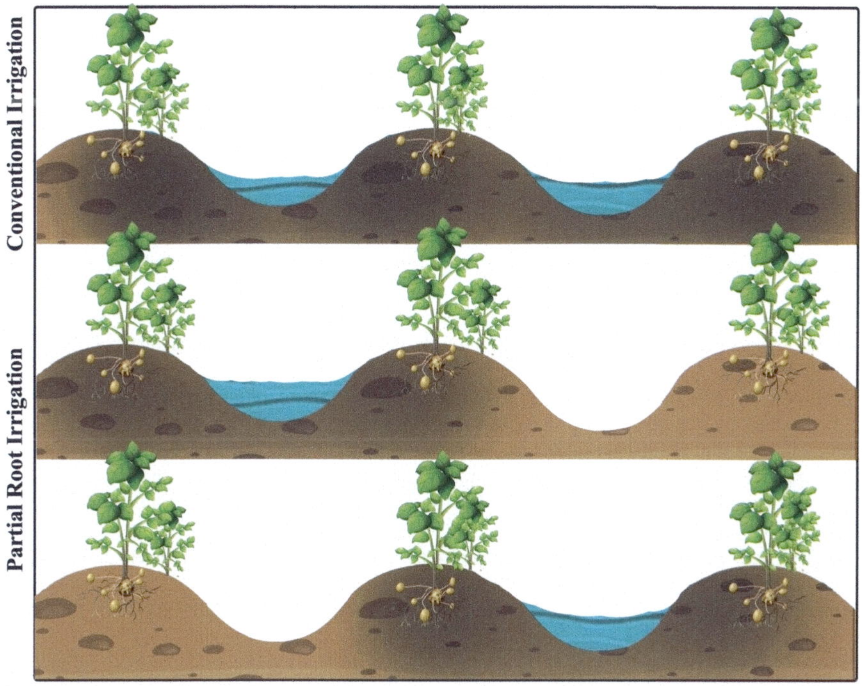

Fig. 18.6 Conventional and partial irrigation scheduling

18.4 Conclusions

Current climate change, inconsistent productivity, and food insecurity are challenging issues globally being faced in the modern era due to the increased prevalence of abiotic stresses. Among these stresses, drought is gaining intense attention, as it limits plants' growth and causes significant yield losses in vegetable crops. In conclusion, drought is causing abrupt changes in normal life cycles of plants, i.e., from germination to harvesting. To combat with water deficit conditions, plants evolve certain mechanisms to induce drought tolerance. However, it is necessary to exploit the unexplored adaptation traits found in numerous plants, which can be incorporated into the sensitive genotypes which might be able to tolerate negative impacts of drought stress and improve their productivity. The breeding aspects have a diverse potential for inducing drought tolerance, however, it is a time-taking process. Therefore, short-term strategies such as applications of PGPRs, grafting techniques, minerals, and innovated irrigation scheduling have gained more attention for drought tolerance induction particularly under arid and semi-arid conditions. Accumulation of proline, compatible solutes, leaf rolling, root increment, delayed senescence, boost in transpiration and water use efficiency and hormonal and osmotic adjustment are the approaches adopted by plants under drought conditions. Apart from drought tolerance induction mechanism, there is a great need to identify drought-tolerant genes and their insertion into vegetable crops, particularly those having a great potential in ensuring food security.

References

Abdel-Ati Y, El-Maziny MY et al (2007) Effect of water stress and potassium fertilization on yield quantity and quality of potato. Afr Crop Sci Soc 8:445–455

Ahmad I, Kamran M, Ali S et al (2018) Seed filling in maize and hormones crosstalk regulated by exogenous application of uniconazole in semiarid regions. Environ Sci Pollut Res 25(33):33225–33239

Ali M, Ayyub CM et al (2020a) Optimization of chitosan level to alleviate the drastic effects of heat stress in cucumber (*Cucumis sativus* L.). J Pure Appl Agric 5(1):30–38

Ali Q, Perveen R et al (2020b) Low doses of *Cuscuta reflexa* extract act as natural bio stimulants to improve the germination vigor, growth, and grain yield of wheat grown under water stress: photosynthetic pigments, antioxidative defense mechanisms, and nutrient acquisition. Biomol Ther 10(9):1212

Alves RC, de Medeiros AS et al (2018) The partial root-zone saline irrigation system and antioxidant responses in tomato plants. Plant Physiol Biochem 127:366–379

Anjum F, Yaseen M et al (2003) Water stress in barley (*Hordeum vulgare* L.). I Effect on morphological characters. Pak J Agric Sci 40(1–2):43–44

Arbona V, Manzi M et al (2013) Metabolomics as a tool to investigate abiotic stress tolerance in plants. Int J Mol Sci 14(3):4885–4911

Azad HN, Poor AH et al (2016) Physiological responses of two tomato (*Lycopersicum esculentum* M.) cultivars to Azomite fertilizer under drought stress. Iran J Plant Physiol 6(4):1833–1843

Azeem M, Haider MZ et al (2022) Drought stress amelioration in maize (*Zea mays* L.) by inoculation of *Bacillus* spp. strains under sterile soil conditions. Agriculture 12(1):50

Bahar AA, Faried HN et al (2021) Potassium-induced drought tolerance of potato by improving morpho-physiological and biochemical attributes. Agronomy 11(12):2573

Bocchini M, D'Amato R, Ciancaleoni S et al (2018) Soil selenium (Se) biofortification changes the physiological, biochemical and epigenetic responses to water stress in *Zea mays* L. by inducing a higher drought tolerance. Front Plant Sci 9:389

Bota J, Medrano H, Flexas J et al (2004) Is photosynthesis limited by decreased Rubisco activity and RuBP content under progressive water stress? New Phytol 162(3):671–681

Chen X, Hao M-D (2015) Low contribution of photosynthesis and water-use efficiency to improvement of grain yield in Chinese wheat. Photosynthetica 53(4):519–526

Cruz de Carvalho MH (2008) Drought stress and reactive oxygen species: production, scavenging and signaling. Plant Signal Behav 3(3):156–165

Dalla Costa L, Gianquinto G (2002) Water stress and water table depth influence yield, water use efficiency, and nitrogen recovery in bell pepper: lysimeter studies. Aust J Agric Res 53(2):201–210

Dalla Costa L, Delle Vedove G et al (1997) Yield, water use efficiency and nitrogen uptake in potato: influence of drought stress. Potato Res 40(1):19–34

Dar MI, Naikoo MI et al (2017) An introduction to reactive oxygen species metabolism under changing climate in plants. In: Khan MIR, Khan NA (eds) Reactive oxygen species and antioxidant systems in plants: role and regulation under abiotic stress. Springer, Singapore, pp 25–52

Dey R, Lewis SC, Arblaster JM et al (2019) A review of past and projected changes in Australia's rainfall. Wiley Interdiscip Rev Clim Chang 10(3):Article e577

Dröge W (2002) Free radicals in the physiological control of cell function. Physiol Rev 82(1):47–95

Earl HJ, Davis RF (2003) Effect of drought stress on leaf and whole canopy radiation use efficiency and yield of maize. Agron J 95(3):688–696
Ebeed HT, Hassan NM et al (2017) Exogenous applications of polyamines modulate drought responses in wheat through osmolytes accumulation, increasing free polyamine levels and regulation of polyamine biosynthetic genes. Plant Physiol Biochem 118:438–448
Ekanayake IJ, Collins W (2004) Effect of irrigation on sweet potato root carbohydrates and nitrogenous compounds. J Food Agric Environ 2:243–248
Eke P, Kumar A, Sahu KP et al (2019) Endophytic bacteria of desert cactus (*Euphorbia trigonas* Mill) confer drought tolerance and induce growth promotion in tomato (*Solanum lycopersicum* L.). Microbiol Res 228:126302
Elansary HO, Mahmoud EA et al (2020) Effects of water stress and modern bio stimulants on growth and quality characteristics of mint. Agronomy 10(1):6
El-Miniawy SM (2015) Growth and yield of eggplant grown under drought stress conditions and different potassium fertilizer rates. Middle East J Agric Res 4(4):1113–1124
El-Saadony MT, Khalil OS et al (2021) Bioactive peptides supplemented raw buffalo milk: Biological activity, shelf life and quality properties during cold preservation. Saudi J Biol Sci 28(8):4581–4591
Evers D, Lefevre I, Legay S et al (2010) Identification of drought-responsive compounds in potato through a combined transcriptomic and targeted metabolite approach. J Exp Bot 61(9):2327–2343
Fan S, Wu H, Gong H, Guo J (2022) The salicylic acid mediates selenium-induced tolerance to drought stress in tomato plants. Sci Hortic 300:111092
Farkas Z, Varga-Laszlo E et al (2020) Effects of waterlogging, drought and their combination on yield and water-use efficiency of five Hungarian winter wheat varieties. Water 12(5):1318
Farooq M, Wahid A et al (2009) Plant drought stress: effects, mechanisms and management. Agron Sustain Dev 29:185–212
Faye I, Diouf O et al (2006) Characterizing root responses to low phosphorus in pearl millet [*Pennisetum glaucum* (L.) R. Br.]. Agron J 98(5):1187–1194
Fouda SE, El-Saadony FM et al (2021) Improving growth and productivity of faba bean (Vicia faba L.) using chitosan, tryptophan, and potassium silicate anti-transpirants under different irrigation regimes. Saudi. J Biol Sci 29(2):955–962
Ge TD, Sun NB et al (2012) Effects of drought stress on phosphorus and potassium uptake dynamics in summer maize (*Zea mays*) through the growth cycle. Acta Physiol Plant 34:2179–2186
Germ M, Kreft I, Stibilj V, Urbanc-Berčič O (2007) Combined effects of selenium and drought on photosynthesis and mitochondrial respiration in potato. Plant Physiol Biochem 45(2):162–167
Gill SS, Tuteja N (2010) Reactive oxygen species and antioxidant machinery in abiotic stress tolerance in crop plants. Plant Physiol Biochem 48(12):909–930
Gomes AMM, Lagoa AMM et al (2004) Interactions between leaf water potential, stomatal conductance and abscisic acid content of orange trees submitted to drought stress. Braz J Plant Physiol 16:155–116
Groppa M, Benavides M (2008) Polyamines and abiotic stress: recent advances. Amino Acids 34(1):35–45
Guang-cheng S, Rui-qi G et al (2011) Photosynthetic, chlorophyll fluorescence and growth changes in hot pepper under deficit irrigation and partial root zone drying. Afr J Agric Res 6(19):4671–4679
Guo Q, Wang Y, Zhang H et al (2017) Alleviation of adverse effects of drought stress on wheat seed germination using atmospheric dielectric barrier discharge plasma treatment. Sci Rep 7(1):16680
Guo J, Jia Y, Chen H et al (2019) Growth, photosynthesis, and nutrient uptake in wheat are affected by differences in nitrogen levels and forms and potassium supply. Sci Rep 9(1):1248
Halliwell B (2007) Biochemistry of oxidative stress. Biochem Soc Trans 35(5):1147–1150
Hasanuzzaman M, Al Mahmud J et al (2018a) Drought stress tolerance in wheat: omics approaches in understanding and enhancing antioxidant defense. In: Zargar S, Zargar M (eds) Abiotic

stress-mediated sensing and signaling in plants: an omics perspective. Springer, Singapore, pp 267–307

Hasanuzzaman M, Bhuyan M et al (2018b) Potassium: a vital regulator of plant responses and tolerance to abiotic stresses. Agron 8(3):31

Hassnain M, Alam I et al (2020) Efficacy of chitosan on performance of tomato (*Lycopersicon esculentum* L.) plant under water stress condition. Pak J Agric Res 33:27–41

Howladar SM (2018) Potassium humate improves physio-biochemical attributes, defense systems activities and water-use efficiencies of eggplant under partial root-zone drying. Sci Hortic 240:179–185

Hu Y, Burucs Z et al (2007) Short-term effects of drought and salinity on mineral nutrient distribution along growing leaves of maize seedlings. Environ Exp Bot 60(2):268–275

Hussain M, Malik M et al (2008) Improving drought tolerance by exogenous application of glycinebetaine and salicylic acid in sunflower. J Agron Crop Sci 194(3):193–199

Jan S, Abbas N, Ashraf M, Ahmad P (2019) Roles of potential plant hormones and transcription factors in controlling leaf senescence and drought tolerance. Protoplasma 256(2):313–329

Jangid KK, Dwivedi P (2017) Physiological and biochemical changes by nitric oxide and brassinosteroid in tomato (*Lycopersicon esculentum* Mill.) under drought stress. Acta Physiol Plant 39(3):73

Jin Y, Zhang C, Liu W et al (2016) The alcohol dehydrogenase gene family in melon (*Cucumis melo* L.): bioinformatic analysis and expression patterns. Front. Plant Sci 7:670

Jovanovic Z, Stikic R (2018) Partial root-zone drying technique: from water saving to the improvement of a fruit quality. Front Sustain Food Syst 1:3

Jovanovic Z, Stikic R et al (2010) Partial root-zone drying increases WUE, N and antioxidant content in field potatoes. Eur J Agron 33(2):124–131

Karam F, Saliba R et al (2011) Yield and water use of eggplants (*Solanum melongena* L.) under full and deficit irrigation regimes. Agric Water Manage 98(8):1307–1316

Kawakami J, Iwama K, Jitsuyama Y (2006) Soil water stress and the growth and yield of potato plants grown from micro tubers and conventional seed tubers. Field Crop Res 95(1):89–96

Khan N, Bano A et al (2018a) Interaction between PGPR and PGR for water conservation and plant growth attributes under drought condition. Biologia 73(11):1083–1098

Khan N, Bano A, Zandi P (2018b) Effects of exogenously applied plant growth regulators in combination with PGPR on the physiology and root growth of chickpea (*Cicer arietinum*) and their role in drought tolerance. J Plant Interact 13(1):239–247

Kumar A, Nayak A et al (2019) Effects of water deficit stress on agronomic and physiological responses of rice and greenhouse gas emission from rice soil under elevated atmospheric CO_2. Sci Total Environ 650:2032–2050

Kuşvuran Ş, Daşgan H, Abak K (2011) Responses of different melon genotypes to drought stress. Yüzüncü Yil Univ J Agric Sci 21(3):209–219

Lahlou O, Ouattar S et al (2003) The effect of drought and cultivar on growth parameters, yield and yield components of potato. Agronomie 23(3):257–268

Lee J-M, Kubota C et al (2010) Current status of vegetable grafting: diffusion, grafting techniques, automation. Sci Hortic 127(2):93–105

Leghari SJ, Wahocho NA et al (2016) Role of nitrogen for plant growth and development: a review. Adv Environ Biol 10(9):209–219

Li H, Li X, Zhang D et al (2013) Effects of drought stress on the seed germination and early seedling growth of the endemic desert plant *Eremosparton songoricum* (Fabaceae). EXCLI J 12:89–101

Lin KH, Chao PY et al (2006) The effects of flooding and drought stresses on the antioxidant constituents in sweet potato leaves. Bot Stud 47(4):417–426

Liu F, Shahnazari A et al (2006) Effects of deficit irrigation (DI) and partial root drying (PRD) on gas exchange, biomass partitioning, and water use efficiency in potato. Sci Hortic 109(2):113–117

Liu H, Zhang YH et al (2013) Alginate oligosaccharides enhanced *Triticum aestivum* L. tolerance to drought stress. Plant Physiol Biochem 62:33–40

Liu J, Wang W, Wang L, Sun Y (2015) Exogenous melatonin improves seedling health index and drought tolerance in tomato. Plant Growth Regul 77(3):317–326

Liu S, Li H, Lv X et al (2016) Grafting cucumber onto luffa improves drought tolerance by increasing ABA biosynthesis and sensitivity. Sci Rep 6(1):20212

Liu C, Liu F et al (2017) Impact of wood biochar and its interactions with mycorrhizal fungi, phosphorus fertilization and irrigation strategies on potato growth. J Agron Crop Sci 203(2):131–145

López-Marín J, Gálvez A et al (2017) Selecting vegetative/generative/dwarfing rootstocks for improving fruit yield and quality in water stressed sweet peppers. Sci Hortic 214:9–17

Mandizvo T, Odindo AO et al (2022) Drought tolerance assessment of citron watermelon (*Citrullus lanatus* var. *citroides* (LH Bailey) Mansf. ex Greb.) accessions based on morphological and physiological traits. Plant Physiol Biochem 180:106–123

Mardani S, Tabatabaei SH et al (2017) Physiological responses of pepper plant (*Capsicum annuum* L.) to drought stress. J Plant Nutr 40(10):1453–1464

Marschner H (2012) Marschner's mineral nutrition of higher plants, vol 89. Academic, London

Martínez-Gutiérrez R et al (2012) Exogenous H2O2 in phytoplasma-infected potato plants promotes antioxidant activity and tuber production under drought conditions. Am J Potato Res 89(1):53–62

Mattar MA, El-Abedin TKZ et al (2020) Soil water status and growth of tomato with partial root-zone drying and deficit drip irrigation techniques. Irrig Sci 38(2):163–176

Mo Y, Yang R, Liu L et al (2016) Growth, photosynthesis and adaptive responses of wild and domesticated watermelon genotypes to drought stress and subsequent re-watering. Plant Growth Regul 79(2):229–241

Moumita, Al Mahmud J et al (2019) Exogenous application of gibberellic acid mitigates drought-induced damage in spring wheat. Acta Agrobot 72(2):1776

Muthomi J, Musyimi D (2009) Growth responses of African nightshades (Solanum scabrum Mill) seedlings to water deficit. ARPN J Agric Biol Sci 4(5):24–31

Nadeem M, Li J et al (2018) Unraveling field crops sensitivity to heat stress: mechanisms, approaches, and future prospects. Agronomy 8(7):128

Nazar R, Iqbal N et al (2011) Salicylic acid alleviates decreases in photosynthesis under salt stress by enhancing nitrogen and sulfur assimilation and antioxidant metabolism differentially in two mungbean cultivars. J Plant Physiol 168(8):807–815

Negrão S, Schmöckel S, Tester M (2017) Evaluating physiological responses of plants to salinity stress. Ann Bot 119(1):1–11

Poor RE (2015) Investigating the effect of grafted watermelon on tolerance to drought and salinity. J Novel Appl Sci 4:670–673

Rabêlo VM, Magalhães PC et al (2019) The foliar application of a mixture of semisynthetic chitosan derivatives induces tolerance to water deficit in maize, improving the antioxidant system and increasing photosynthesis and grain yield. Sci Rep 9(1):8164

Rady MM, Belal HE et al (2020) Selenium application in two methods promotes drought tolerance in *Solanum lycopersicum* plant by inducing the antioxidant defense system. Sci Hortic 266:109290

Rahbarian R, Khavari-Nejad R et al (2011) Drought stress effects on photosynthesis, chlorophyll fluorescence and water relations in tolerant and susceptible chickpea (*Cicer arietinum* L.) genotypes. Acta Biol Cracov Bot 53(1):47–56

Rahdari P, Hosseini SM, Tavakoli S (2012) The studying effect of drought stress on germination, proline, sugar, lipid, protein and chlorophyll content in purslane (*Portulaca oleracea* L.) leaves. J Med Plants Res 6(9):1539–1547

Rasheed R, Ashraf MA et al (2020) Interactive effects of chitosan and cadmium on growth, secondary metabolism, oxidative defense, and element uptake in pea (*Pisum sativum* L.). Arab J Geosci 13:847

Razaq M, Zhang P, Shen HL (2017) Influence of nitrogen and phosphorus on the growth and root morphology of Acer mono. PLoS One 12(2):e0171321

Rezaei MA, Jokar I et al (2012) Morpho-physiological improving effects of exogenous glycine betaine on tomato (*Lycopersicum esculentum* Mill.) cv. PS under drought stress conditions. Plant Omics 5(2):79–86

Rosen CJ, Kelling KA et al (2014) Optimizing phosphorus fertilizer management in potato production. Am J Potato Res 91(2):145–160

Rouphael Y, Cardarelli M et al (2008) Yield, mineral composition, water relations, and water use efficiency of grafted mini-watermelon plants under deficit irrigation. Hortic Sci 43(3):730–736

Sánchez-Rodríguez E et al (2010) Study of the ionome and uptake fluxes in cherry tomato plants under moderate water stress conditions. Plant Soil 335(1):339–347

Sánchez-Rodríguez E et al (2014) How does grafting affect the ionome of cherry tomato plants under water stress? Soil Sci Plant Nutr 60(2):145–155

Sánchez-Rodríguez E et al (2016) Accumulation of free polyamines enhances the antioxidant response in fruits of grafted tomato plants under water stress. J Plant Physiol 190:72–78

Saux M, Ponnaiah M et al (2020) A multiscale approach reveals regulatory players of water stress responses in seeds during germination. Plant Cell Environ 43(5):1300–1313

Sehgal A, Sita K et al (2018) Drought or/and heat-stress effects on seed filling in food crops: impacts on functional biochemistry, seed yields, and nutritional quality. Front Plant Sci 9:1705

Seleiman MF, Refay Y et al (2019) Integrative effects of rice-straw biochar and silicon on oil and seed quality, yield and physiological traits of *Helianthus annuus* L. grown under water deficit stress. Agronomy 9(10):637

Sezen SM, Yazar A et al (2014) Evaluation of crop water stress index (CWSI) for red pepper with drip and furrow irrigation under varying irrigation regimes. Agric Water Manag 143:59–70

Shehzad MA et al (2020) Protective effect of potassium and chitosan supply on growth, physiological processes and antioxidative machinery in sunflower (*Helianthus annuus* L.) under drought stress. Ecotoxicol Environ Saf 187:109841

Siddiqui MH et al (2015) Response of different genotypes of faba bean plant to drought stress. Int J Mol Sci 16(5):10214–10227

Taiz L, Zeiger E (2006) Plant Physiology, 4th edn. Sinauer Associates Inc. Publishers, Sunderland

Thilagar G et al (2018) Bacillus sonorensis, a novel plant growth promoting rhizobacterium in improving growth, nutrition and yield of chilly (Capsicum annuum L.). Proc Natl Acad Sci India Sect B Biol Sci 88(2):813–818

Tiwari YK, Yadav SK (2020) Effect of high-temperature stress on ascorbate–glutathione cycle in maize. Agric Res 9(2):179–187

Tränkner M, Tavakol E, Jákli B (2018) Functioning of potassium and magnesium in photosynthesis, photosynthate translocation and photoprotection. Physiol Plant 163(3):414–431

Tu Y, Wang R, Zhang Y, Wang J (2018) Progress and expectation of atmospheric water harvesting. Joule 2(8):1452–1475

Tufail B, Ashraf K et al (2023) Effect of selenium on growth, physio-biochemical and yield traits of lettuce under limited water regimes. Sustain For 15(8):6804

Ul-Allah S, Ijaz M et al (2020) Potassium application improves grain yield and alleviates drought susceptibility in diverse maize hybrids. Plan Theory 9(1):75

Umar S, Moinuddin (2002) Genotypic differences in yield and quality of groundnut as affected by potassium nutrition under erratic rainfall conditions. J Plant Nutr 25(7):1549–1562

Usmani MM et al (2020) Sulfate-mediated drought tolerance in maize involves regulation at physiological and biochemical levels. Sci Rep 10(1):1147

Vardharajula S, Ali et al (2011) Drought-tolerant plant growth promoting *Bacillus* spp.: effect on growth, osmolytes, and antioxidant status of maize under drought stress. J Plant Interact 6(1):1–14

Verma P, Yadav AN et al (2016) Molecular diversity and multifarious plant growth promoting attributes of Bacilli associated with wheat (*Triticum aestivum* L.) rhizosphere from six diverse agro-ecological zones of India. J Basic Microbiol 56(1):44–58

Verslues PE et al (2006) Methods and concepts in quantifying resistance to drought, salt and freezing, abiotic stresses that affect plant water status. Plant J 45(4):523–539

Wahb-Allah MA et al (2011) Drought tolerance of several tomato genotypes under greenhouse conditions. World Appl Sci J 15(7):933–940

Wang Y, Zhang Y (2012) Soil inorganic phosphorus fractionation and availability under greenhouse subsurface irrigation. Commun Soil Sci Plant Anal 43(3):519–532

Wang W, Vinocur B, Altman A (2003) Plant responses to drought, salinity and extreme temperatures: towards genetic engineering for stress tolerance. Planta 218(1):1–14

Wang Y, Liu F, Jensen CR (2012) Comparative effects of partial root-zone irrigation and deficit irrigation on phosphorus uptake in tomato plants. J Hortic Sci Biotechnol 87(6):600–604

Wang M, Zheng Q et al (2013) The critical role of potassium in plant stress response. Int J Mol Sci 14(4):7370–7390

Willcox JK, Ash SL, Catignani GL (2004) Antioxidants and prevention of chronic disease. Crit Rev Food Sci Nutr 44(4):275–295

Xu J, Yuan Y, Xu Y et al (2014) Identification of candidate genes for drought tolerance by whole-genome resequencing in maize. BMC Plant Biol 14(1):83

Yashin A, Yashin Y, Xia X, Nemzer B (2017) Antioxidant activity of spices and their impact on human health: a review. Antioxidants 6(3):70

Zade NSE, Sadeghi A, Moradi P (2019) Streptomyces strains alleviate water stress and increase peppermint (*Mentha piperita*) yield and essential oils. Plant Soil 434(1):441–452

Zaki HE, Radwan KS (2022) Response of potato (*Solanum tuberosum* L.) cultivars to drought stress under in vitro and field conditions. Chem Biol Technol Agric 9:1–19

Zamaninejad M et al (2013) Effect of salicylic acid on morphological characteristics, yield and yield components of corn (*Zea mays* L.) under drought condition. Eur J Exp Biol 3(2):153–161

Zelelew DZ, Lal S et al (2016) Effect of potassium levels on growth and productivity of potato varieties. Am J Plant Sci 7(12):1629–1638

Zhang N, Liu B, Ma C et al (2014a) Transcriptome characterization and sequencing-based identification of drought-responsive genes in potato. Mol Biol Rep 41(1):505–517

Zhang Q, Wu S, Chen C et al (2014b) Regulation of nitrogen forms on growth of eggplant under partial root-zone irrigation. Agric Water Manag 142:56–65

Zhao W, Liu L, Shen Q, Yang J et al (2020) Effects of water stress on photosynthesis, yield, and water use efficiency in winter wheat. Water 12(8):2127

Zhu JK (2002) Salt and drought stress signal transduction in plants. Annu Rev Plant Biol 53(1):247–273

Zulfiqar F, Chen J et al (2021) Application of trehalose and salicylic acid mitigates drought stress in sweet basil and improves plant growth. Plan Theory 10(6):1078

Toxicity of Heavy Metals in Vegetables and Their Impacts on Human Health

19

Muhammad Kabir, Natasha Aziz, Muhammad Hammad, Muhammad Zafar Iqbal, and Muhammad Athar Tariq

Abstract

Vegetables are an essential share of human nutrition globally as they provide essential nutrients, metabolites, and antioxidants. In recent decades, the deleterious impacts of different heavy metals (HMs) contamination on harvests and human fitness are a matter of thoughtful concern. Most prevalent causes of HMs exposure in humans are vegetables, responsible for 90% of metal intake. HMs such as chromium (Cr), cadmium (Cd), lead (Pb), mercury (Hg) copper (Cu), and iron (Fe) are dangerous to both floras and faunas and pose main undesirable impacts on human health. They create severe ecological harms when they are existing in soil. HMs concentration in environment is due to the activities like mining, urbanization, industrialization, transportation, and agronomic operations. They produce a negative effect on the soil's chemical composition and vegetable nutrients as soil acts as elementary part for plants growing and improvement. Vegetables absorb these dangerous chemicals from the soil along with essential nutrients. There is also a considerable deviation in concentration of the HMs in different regions of the plants and even in several varieties of same species. The feeding of these polluted vegetables inserts severe threats to human health as their persistent and non-degradable nature results in disturbed human metabolism and serious fitness concerns such as cancer, nervous disorders, and

M. Kabir (✉) · N. Aziz · M. Hammad
Department of Biological Sciences, Thal University, Bhakkar, Pakistan
e-mail: m.kabir@tu.edu.pk

M. Z. Iqbal
Department of Botany, University of Karachi, Karachi, Pakistan

M. A. Tariq
California Department of Food and Agriculture, Sacramento, CA, USA

Z. Iqbal et al. (eds.), *Sustainable and Innovative Vegetable Production in times of Climate Change*, https://doi.org/10.1007/978-981-96-6283-8_19

renal damage. This chapter effectively explores the lethality of HMs-polluted soils on vegetables and social healthiness as contamination of HMs is a pressing ecological and health issue. Assessment of the contamination of HMs in humans requires a focused investigation. Detection of the heavy metals' contaminated areas is essential to reduce their negative impact otherwise they may cause dreadful problems in the future. Hyper accumulators like root and stem tubers also need to be regulated by a specific policy to limit the accumulation of HMs in different vegetables. Suitable strategies like pH modifications and phytoremediation technologies should be implemented in agricultural practices to minimize the absorption of heavy metals by vegetables growing in heavy metal-contaminated soils.

Keywords

Agronomic operations · Non-degradable nature · Heavy metals contamination · Disturbed human metabolism · Phytoremediation

19.1 Introduction

It's projected that cumulative yearly economic impact of heavy metals is more than 10 billion dollars each year (He et al. 2015). The continental mantle is naturally home to heavy metals that are dangerous when present in concentrations above an acceptable level to plant and human health as well. As a major contributor to pollution, contaminated heavy metals pose a risk to the environment and community health worldwide and can cause diseases in both humans and animals when they consume polluted vegetables. Vegetables are a crucial component of a typical diet because they include nutrients that are essential for human health. Vegetables cultivated in heavy metal polluted areas can build up greater intensities of the HMs in them as compared to those developed in healthy soil (Figs. 19.1 and 19.2). Roots of plants absorb heavy metals from polluted soils, which then enter the edible parts of plant tissues. Heavy metals typically lead to harmful biological consequences since they are bio-accumulative. Their high concentrations can make plants toxic by triggering oxidative stress and by upsetting the balance of antioxidants inside plant cells, which sequentially leads to problems like lipid oxidation and denaturation of cell and membrane structure ultimately causing a programmed cell death.

The triangle of mercury, lead, and chromium is known as "the big three" as of their ecological hazards (Sandeep et al. 2019). The most emerging ecological issue on the planet is the presence and contamination of HMs in soil that reduce the efficiency of plants (Jadid et al. 2017; Manzoor et al. 2018). The release of different heavy metals from different sources in all the components of environment is shown in Fig. 19.3.

Fig. 19.1 Healthy vegetables growing in unpolluted soil. (**a**) Garlic; (**b**) Broccoli; (**c**) Broccoli and Cabbage

19.2 Toxicity of Heavy Metals

19.2.1 Lead Toxicity Stress on Plants and Human Health

Excessive lead usage causes serious harm to environment and also human beings. Lead forms intricate mix of compounds depending on situation and it begins to tarnish when exposed to air (Sharma and Dubey 2005). Lead is frequently occurring

Fig. 19.2 Unhealthy vegetables growing in polluted soil

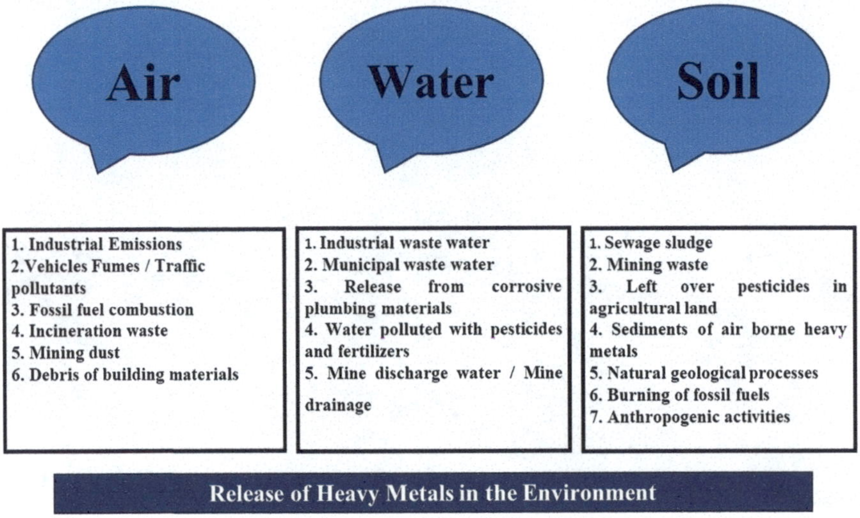

Fig. 19.3 Different sources for releasing heavy metals in all the components of environment

in aquatic ecosystem and in polluted soil because of the activities such as smelting, scorching of fossil, and excessive fertilizers utility (Jadid et al. 2022). The manufacturing of reactive oxygen species is accelerated by high concentration of lead which may disturb the chlorophyll structure and photosynthetic process and also destroy the structure of lipid membrane (Najeeb et al. 2017). Foods kept in metallic cans run the risk of wrapping, covering, and corroding, particularly if the pH of the meal is acidic. When heavy metals including cadmium and lead migrate to stored products through corroded metal packaging, their concentration in food increases (Li et al. 2021). Lead poisoning in plants inhibits growth, disrupts mitosis, reduces seed viability, produces chlorosis, interferes with mineral absorption, alters the permeability and structure of cell membranes, and reduces photosynthetic efficiency. These processes may cause metabolic imbalances, which eventually result in cell death. The impact, however, varies from plant to plant (Hadi and Aziz 2015; Zhou et al. 2016).

19.2.2 Chromium Addition in Plants and Human Health Impacts

Chromium is the most prevalent element that is present on earth in two common forms as Cr^{+3} and Cr^{+6}; both are extremely poisonous to faunas and floras as well. Humans release chromium in environment through fertilizers and sewage (Monalisa and Kumar 2013). The wastewater irrigation is a serious threat to agricultural lands which disturbs the fertility of soil, yield of crops, and quality of vegetables to be consumed by human beings. The phytotoxicity of chromium causes yellowing of leaves and reduces seed germination and growth of roots (Duan et al. 2010; Hosseini and Mousavi 2021; Liu et al. 2021) with some resistance (Ullah and Haq 2022). The extreme absorption of chromium reduces the uptake of vital minerals, leaving bad effects on natural resources and commercially important crops (Kabata-Pendias and Szteke 2015; Chigonum et al. 2019; Wakeel and Xu 2020; Kapoor et al. 2022). Chromium (VI) is highly toxic due to its high oxidation state and increased solubility, which enhances its bioavailability and reactivity (Singh et al. 2022).

19.2.3 Impact of Mercury on Plants and Human Health

Mercury is an extremely perilous heavy metal, commonly present in environment which is deadly in the form of a vapor, the biological mercury. By municipal wastewater treatment, burning fossil fuels, producing heat in industrial furnaces, or ejecting material from active volcanoes, mercury, and its associated mixtures are spread, persist in soil, and released in the atmosphere (Chen et al. 2012; Rafati-Rahimzadeh et al. 2014). Humans are specifically exposed to mercury by eating mercury-contaminated food, particularly marine seafood and water, and undergoing dental care treatments (Sabarathinam et al. 2016; Tan et al. 2018). Soil pollution occurs due to the occurrence of mercury in the manures and fertilizers. The widespread occurrence of contamination due to mercury through the food chain is caused by the significant intake of mercury into the soils. Plant morphology and photosynthetic procedures are harmfully impacted by the existence of lead (Asati et al. 2016).

19.2.4 Cadmium Stress on Plants and Human Health

Cadmium lingers in soils for a very long period. As indicated in different studies, cadmium has poisoned more than 11,000 ha of the land and the industrial waste that is released in environment yearly contains 680 tons of cadmium (Han et al. 2009; Mutlu et al. 2012). Due to the rapid frequency of soil-to-plant transmission, Cd is primarily identified in vegetables and fruits. It has a negative impact on cells' enzymatic systems causing nutritional deficiencies in plants and humans and oxidative stress (Satarug et al. 2010; Irfan et al. 2013; Chen et al. 2016; Chellaiah 2018; Kubier and Pichler 2019). Cd toxicity also causes chlorosis and reduces the growth of plants (Jali et al. 2016; Abbas et al. 2017; Xu et al. 2017; Abedi and Mojiri 2020). The toxicity of Cd damages different body parts in humans but mostly causes kidney, liver, and bones' disorders (Mahajan and Kaushal 2018; Lata et al. 2019; Hussain et al. 2021). Chemical fertilizers' usage enhances the cadmium's absorption in vegetation (Mamun et al. 2021). Transportation and consumption of numerous components and water balance have all been demonstrated to be laden by cadmium. By reducing the rate of activity of nitrate reductase within buds, cadmium also lowers the nitrate transfer between roots and shoots.

19.2.5 Impacts of Iron on Plants and Human Health

On the surface of planet, iron is the second most prevalent metal (USEPA 1993). Exposure to total iron concentrations as low as 1 mg/L has been shown to inhibit the growth and development of certain marine reed species, highlighting the potential toxicity of iron to aquatic plants (Phippen et al. 2008). When the consumed iron doesn't fix to the protein, a diversity of lethal free radicals is produced, containing harmful influence on the application of iron in mammalian cells and fleshly solutions. Heart, liver, and brain cells are all accessed by these free irons because of the interruption and occurrence of lipid peroxidation, which seriously harms cellular organelles such as mitochondria and microsomes (Albretsen 2006; Grazuleviciene et al. 2009).

19.2.6 Copper Toxicity and Its Impacts

Copper is an important metal for humans and plants' growth on the Earth (Karlin and Tyeklar 2012). In nature, it is present in air, water, and rocks while anthropogenically, it is produced by fertilizers, municipal waste, and fungicides (Leygraf et al. 2019). Copper, as a micronutrient, is needed for plant development, however, high levels via sensitive oxygen species formation create oxidative stress (Ameh and Sayes 2019; Nazir et al. 2019; Shabbir et al. 2020). Increase in Cu level becomes harmful to environment (Shahid et al. 2018; Antoniadis et al. 2019; Khalid et al. 2019; Amen et al. 2020; Shahid et al. 2020). Additionally, excessive copper toxicity to cellular macromolecules can occur when the manufacture of reactive oxygen species (ROS) is increased by copper toxicity (Jaime-Perez et al. 2019). A considerable amount of Cu is released in environ via different agronomic actions such as application of algaecides, bactericides, and fungicides (Iwinski et al. 2017). The most common type of copper is Cu^{2+} that is taken up by plants through soil because it has high affinity to bind with OM (Ogunkunle et al. 2019).

19.3 Heavy Metals Toxicity Stress on Leafy Vegetables

HMs accumulation in the surroundings is accumulative due to different natural events, usage of pesticides, coal combustion, and industrial pollutants in city and suburban areas (Ferri et al. 2015; Li et al. 2015; Fu Dietrich et al. 2017; Karimyan et al. 2020). Humans and animal species may suffer from the poisonous impacts of heavy metals exposure over time (Xu et al. 2013; Itoh et al. 2014; Rose et al. 2015; Shahid et al. 2015). The ingestion of heavy metals from polluted vegetables is the main source of exposure for humans. The toxicity of HMs or non-biodegradable elements has been the subject of latest published research because of their propensity to be absorbed from the environment (Singh et al. 2014; Pan et al. 2016; Hu et al. 2017; Zwolak et al. 2019). HMs can be taken up by the human body even through inhalation, ingestion, and skin contact. HMs contamination in soil can endanger human health through tainted fruits, vegetables, and crops (Khan et al. 2015). Mainly Pb, Hg, and Cd can destroy biological activity when consumed in excess (Sevindik et al. 2017; Korkmaz et al. 2018). Leafy vegetables can absorb ambient Hg and Pb through their leaves, due to their exposure to the environment. Leafy vegetables generally have more Hg and Pb when related to non-leafy vegetables (Ao et al. 2017; Obrist et al. 2017). Different ripening phases of tomato and vegetative and reproductive phases of lady's finger are shown in Figs. 19.4 and 19.5, respectively.

Fig. 19.4 Different ripening phases of tomato in unpolluted soil

Fig. 19.5 Growth and reproductive phases of okra in unpolluted soil

19.4 Occurrence, Distribution, and Health Effects of HMs in Locally Marketed Vegetables

Because of diverse anthropogenic activities, 500 Mha lands worldwide are contaminated with a range of HMs and metalloids (Liu et al. 2018; Ali et al. 2019; Li et al. 2019). Compared to industrialized nations, the issue is worst in some developing and closely inhabited countries where industrialization, urbanization, and substantial agriculture have all been occurring quickly (Rajindiran et al. 2015; Islam et al. 2018). The ingesting of polluted vegetables can cause severe health issues, including neurological, reproductive, cancerous, and developmental (Mandal et al. 2017). An investigation carried out in Nigeria, China, and Iran revealed that vegetables from local markets enclosed high levels of lead and chromium (Nwachukwu et al. 2013; Wang et al. 2014; Nabizadeh et al. 2014; Ogunbileje et al. 2020) In Egypt, the use of contaminated vegetables was linked with an increased threat of kidney and liver damage (Gouda et al. 2019).

The growers often prefer wastewater utilization for irrigation purpose because of its nutrients' availability and some beneficial effects on crops but the reuse of industrial wastewater releases harmful substances into the environment (Chaoua et al. 2019; Mallick et al. 2019; Haroon et al. 2019; Salam et al. 2020). Furthermore, the wastewater can penetrate soil and seep to nearby places both of which can have harmful impacts on society (Sharma et al. 2016).

Vegetables are called protective food for humans because of their high nutritional value. These are essential for human beings for their survival and protection from various diseases (Heiner et al. 2012). Root crops have a lower concentration of heavy metals and leaves show larger concentration because organic ligands increase the bioavailability (Schiptsova et al. 2020). According to the current studies, Cd had

Table 19.1 Presence of HMs in unpolluted topsoil and plant tissues, as related to standards selected as inception of hyper-accumulator plant

S. No.	Metals	Average range in soil[a] (mg kg^{-1} dry weight)	Average range in plant tissues[b] (mg kg^{-1} dry weight)	Threshold level for hyperaccumulators[c] (mg kg^{-1} dry weight)	Average range in soil[d] (mg kg^{-1} dry weight)
1	Chromium	Not determined	0.3—1	1000–0.1%	05—1000
2	Cadmium	01—02	0.03—0.5	100—0.01%	
3	Copper	02—60	02—20	1000—0.1%	
4	Iron	Not determined	Not determined	Not determined	
5	Lead	10—150	0.1—05	10,000—0.1%	

[a] Lepp (1981), [b]Markert (1996), [c]Mclntyre (2003), [d]Adriano (2001)

highest soil root transfer value due to its great mobility than other metals. Spinach and lettuce absorb more HMs through roots than shoot. The transfer of copper and zinc through root and shoot is more than other heavy metals. Such kinds of vegetables which are irrigated by sewage water accumulate greater amount of HMs in their leaves that are not suitable for human and animals' use (Eissa and Negim 2018). Plants are considered as an important source of pollutant sink as they help in decreasing the accumulation of pollutants. Considering the importance of better and clear environment, plants growing in the polluted environment help us in understanding the current status of pollutants in the environment. In fact, the toxicity of all the crucial trace elements follows a general trend: under supply leads to deficiency, most favorable (optimum) supply helps in strong growth, and over supply leads to toxicity and death of organisms. Translocation factor (TF) is a measure of the extent to which heavy metals are transferred from soil to the above-ground parts of a plant, providing insight into the plant's ability to accumulate and transport these metals (Table 19.1).

19.5 Evaluation of Health Hazards Linked to Heavy Metal Contamination

Numerous studies have confirmed that such metal contaminants can be engrossed and accumulated by plant roots before slowly making their way to portions that are edible, potentially posing a risk to the health of people by means of the food chain. Residents' health may be seriously threatened by the direct ingestion of leafy vegetables. Prolonged exposure to Pb and Cd can cause a number of diseases, including fractures in the bones, kidney damage, skin cancer, and nervous system problems. The most significant source of human exposure to toxic substances that may endanger people's health, is contaminated food. Due to physiological behaviors like rubbing eyes, young children are typically more sensitive to harmful substances than

Table 19.2 Heavy metals' impacts on human health

Metals toxicity	References
Exposure to high chromium levels can inhibit erythrocyte glutathione reductase, hindering the conversion of methemoglobin to hemoglobin in humans.	Koutras et al. (1965)
Methyl mercury is a neurotoxic toxin that harms mitochondria, microtubules, and lipids.	Patrick (2002)
Cadmium binding to metallothionein or cysteine-rich proteins leads to hepatotoxicity in the liver, followed by nephrotoxicity in the kidneys due to the accumulation of the cysteine-metallothionein complex in renal tissue.	Castagnetto et al. (2002)
Chromate chemicals can harm DNA in a number of ways including chromosomal abnormalities and alterations in DNA transcription.	Matsumoto et al. (2006)
Organic mercury exposure can lead to neurological symptoms including depression, memory impairment, tremors, fatigue, headaches, and alopecia (hair loss).	Martin and Griswold (2009)
C*admium stores* in larger amount in the epithelial *cells* of the *proximal tubule* causing reabsorptive *dysfunction.* In chronic cadmium poisoning, *hypercalciuria may occur leading to renal stone formation.* Cadmium can also cause many other health issues including cancer, bone disease, flu, lung damage, and pneumonitis.	Nordberg et al. (2007), Bernard (2008) and Prozialeck and Edwards (2012)
A prolonged Pb exposure causes neurodegenerative disorders and additionally may contribute to the development of deep vein clotting in the lower limbs.	Ferrante et al. (2017)
Consuming contaminated vegetables can lead to heavy metal exposure, causing adverse health effects including neurological damage, reproductive issues, and increased cancer risk.	Mandal et al. (2017)
Lead in blood has been linked to changes in the brain in the general population including a decline in IQ.	Karri et al. (2018).
Cr (VI) toxicity is related to its ability to cross cell membranes and enter cells. Subsequently, intracellular Cr (VI) reduces to Cr (III) generating ROS as a reaction intermediate.	Rubiya et al. (2018) and Singh et al. (2022)
The soil contaminated with lead causes carcinogen and multiple organs failure in humans like heart and kidney abnormalities.	Meena et al. (2020)

adults when they consume these contaminated foods. Local children who reside close to mining areas might suffer from negative health effects. Even though certain HMs like zinc (Zn), selenium (Se), and copper (Cu) are necessary for human beings along with other living things in small quantities, high doses can have adverse impacts disrupting the normal functioning of various body parts (Table 19.2).

The potential of vegetables to absorb heavy metals varies, and in some vegetables, there is an increased chance of accumulation of a large amount of heavy metals than others. Frequently grown vegetables such as lettuce, carrots, cabbage, tomatoes, and spinach, which are irrigated by municipal effluents, absorb heavy metals in high quantities. The utilization of plants to remove such hazardous metals from the environment could result in a cleaner planet for all of us (Shiyab 2018).

19.6 Heavy Metal Impacts on Microorganisms

Fungi and bacteria are common microorganisms and both are considered decomposers. These microorganisms release some secondary metabolites such as polysaccharides, organic acids, and glycoprotein that make cell resistant to the effect of heavy metals. These metabolites also help to withstand stress by joining with heavy metal ions. Frequently heavy metal ions harm the cells by creating oxidative stress and damage and lead to generating oxygen free radicals. These radicals disrupt many processes happening in the organism (Zhang et al. 2018). Microbiological humiliation of organic pollutants is reduced due to the trace metals in the soil (Mondal et al. 2020). Enhanced concentrations and HM availability in soil change the behavior of soil-dwelling microorganisms, which is crucial for plant growth (Bakht et al. 2016). The long-term soil contamination also creates an unfavorable condition for the growth of plants and the activities of microbes by interfering their functions, damaging cell membrane, and disturbing protein synthesis (Lwin et al. 2018). Depletion of soil organic carbon and change in the activity of microbes in soil due to HMs contamination lead to poor physical properties of soil such as lower water holding capacity and soil porosity (Ramamoorthy 2015; USDA-NRCS 2015).

19.7 Heavy Metal-Induced Lower Water Quality

The survival of human, wildlife, and animals depends on freshwater. However, despite an annual rise in global water demand, pollution has harmed viable water supplies (UN-Water 2018). One of the major risks to humans is the contamination of water with heavy metals (Burgess et al. 2015; Saha and Paul 2019). This is a serious issue in semi-arid and arid regions where there is a shortage of water (Mokarram et al. 2020). Some organic molecules may interact with pollutants under certain circumstances and produce more dangerous complex organic metal pollutants (Chowdhury et al. 2016). Canal water has higher heavy metals concentration, and soil it irrigates shows poor condition (Sahare 2014).

19.8 Conclusion

From this literature review, it is inferred that various heavy metals are released into different components of environment because of natural, anthropogenic, and industrial activities. This affects food security and food production by entering in the food chain especially in vegetables. Plants are producers on the earth's crust but polluted constituents of environment are also affecting different growth, development, and other metabolic processes of different plant species. Although vegetables are an important component of our food and necessary for maintaining healthy life of human and other organisms, polluted vegetables cannot perform these functions properly. There is a need to develop strategies for clean and green spaces and heavy metals free environment where vegetables could be grown on surface of the earth for healthy life of all living organisms.

References

Abbas T, Rizwan M, Ali S, Adrees M, Zia-ur-Rehman M, Qayyum MF, Ok YS, Murtaza G (2017) Effect of biochar on alleviation of cadmium toxicity in wheat (*Triticum aestivum* L.) grown on Cd-contaminated saline soil. Environ Sci Pollut Res 25:25668–25680

Abedi T, Mojiri A (2020) Cadmium uptake by wheat (*Triticum aestivum* L.): an overview. Plan Theory 9:500

Adriano DC (2001) Trace elements in terrestrial environments: biochemistry, bioavailability and risks of metals, 2nd edn. Springer, New York, p 867. https://doi.org/10.1007/978-0-387-21510-5

Albretsen J (2006) The toxicity of iron, an essential element. Veterinary Medicine-Bonner Springs Then Edwardsville 101(2):82

Ali H, Khan E, Ilahi I (2019) Environmental chemistry and ecotoxicology of hazardous heavy metals: environmental persistence, toxicity, and bioaccumulation. J Chem 2019:1–14

Ameh T, Sayes CM (2019) The potential exposure and hazards of copper nanoparticles: a review. Environ Toxicol Pharmacol 71:103220

Amen R, Bashir H, Bibi I, Shaheen SM, Niazi NK, Shahid M, Hussain MM, Antoniadis V, Shakoor MB, Al-Solaimani SG, Wang H, Bundschuh J, Rinklebe J (2020) A critical review on arsenic removal from water using biochar-based sorbents: the significance of modification and redox reactions. Chem Eng J 396:125195

Antoniadis V, Shaheen S, Levizou E, Shahid M, Niazi N, Vithanage M, Ok Y, Bolan N, Rinklebe J (2019) A critical prospective analysis of the potential toxicity of trace element regulation limits in soils worldwide: are they protective concerning health risk assessment? – a 1194 review. Environ Int 127:819–847

Ao M, Meng B, Sapkota A, Wu YG, Qian XL, Qiu GL, Zhong SQ, Shang LH (2017) The influence of atmospheric Hg on Hg contaminations in rice and paddy soil in the Xunyang Hg mining district, China. Acta Geochim 36:181–189

Asati A, Pichhode M, Nikhil K (2016) Effect of heavy metals on plants: an overview. Int J Appl Innov Eng Manag 5(3):56–66

Bakht J, Khan L, Shafi M (2016) Phytoaccumulation of heavy metals and protein expression by different vegetables collected from various parts of Khyber Pukhtunkhawa Province, Pakistan. Sains Malaysiana 45(2):167–176

Bernard A (2008) Cadmium & its adverse effects on human health. Indian J Med Res 128(4):557–564

Burges A, Epelde L, Garbisu C (2015) Impact of repeated single-metal and multi-metal pollution events on soil quality. Chemosphere 120:8–15

Castagnetto JM, Hennessy SW, Roberts VA, Getzoff ED, Tainer JA, Pique ME (2002) MDB: the metalloprotein database and browser at the Scripps Research Institute. Nucleic Acids Res 30(1):379–382

Chaoua S, Boussaa S, El Gharmali A, Boumezzough A (2019) Impact of irrigation with wastewater on accumulation of heavy metals in soil and crops in the region of Marrakech in Morocco. J Saudi Soc Agric Sci 18(4):429–436

Chellaiah ER (2018) Cadmium (heavy metals) bioremediation by *Pseudomonas aeruginosa*: a mini-review. Appl Water Sci 8:154–164

Chen CW, Chen CF, Dong CD (2012) Distribution and accumulation of mercury in sediments of Kaohsiung River Mouth, Taiwan. APCBEE Proc 1:153–158

Chen B, Stein AF, Castell N, Gonzalez-Castanedo Y, De-La Campa AS, De La Rosa JD (2016) Modeling and evaluation of urban pollution events of atmospheric heavy metals from a large Cu-smelter. Sci Total Environ 539:17–25

Chigonum WJ, Ikenna OC, John CU (2019) Dynamic impact of chromium on nutrient uptake from soil by fluted pumpkin (*Telfairia occidentalis*). Am J Biosci Bioeng 7:1–9

Chowdhury S, Mazumder MJ, Al-Attas O, Husain T (2016) Heavy metals in drinking water: occurrences, implications, and future needs in developing countries. Sci Total Environ 569:476-488

Duan N, Wang XL, Liu XD, Lin C, Hou J (2010) Effect of anaerobic fermentation residues on a chromium-contaminated soil-vegetable system. Procedia Environ Sci 2:1585–1597

Eissa MA, Negim OE (2018) Heavy metals uptake and translocation by lettuce and spinach grown on a metal-contaminated soil. J Soil Sci Plant Nutr 18(4):1097–1107

Ferrante M, Fiore M, Conti GO, Fiore V, Grasso A, Copat C, Signorelli SS (2017) Transition and heavy metals compared to oxidative parameter balance in patients with deep vein thrombosis: a case-control study. Mol Med Rep 15(5):3438–3444

Ferri R, Hashim D, Smith R, Guazzetti S, Donna F, Ferretti E, Curatolo M, Moneta C, Beone GM, Lucchini RG (2015) Metal contamination of home garden soils and cultivated vegetables in the province of Brescia Italy: implications for human exposure. Sci Total Environ 518-519:507–517

Fu Dietrich M, Huling J, Krekeler MPS (2017) Metal pollution investigation of Goldman Park, Middletown Ohio: evidence for steel and coal pollution in a high child use setting. Sci Total Environ. https://doi.org/10.1016/j.scitotenv.2017.09

Gouda AA, El-Baz FK, Abdel-Mageed WS, Taha TA (2019) Health risk assessment of heavy metals in vegetables from local markets in Alexandria Egypt. Environ Monit Assess 191(3):164

Grazuleviciene R, Nadisauskiene R, Buinauskiene J, Grazulevicius T (2009) Effects of elevated levels of manganese and iron in drinking water on birth outcomes. Pol J Environ Stud 18(5):819–825

Hadi F, Aziz T (2015) A mini review on lead (Pb) toxicity in plants. J Biol Life Sci 6:91–101

Han JX, Shang Q, Du Y (2009) Effect of environmental cadmium pollution on human health. Health 1(03):159

Haroon B, Ping A, Pervez A, Irshad M (2019) Characterization of heavy metal in soils as affected by long-term irrigation with industrial wastewater. J Water Reuse Desalin 9(1):47–56

He Z, Shentu J, Yang X, Baligar VC, Zhang T, Stoffella PJ (2015) Heavy metal contamination of soils: sources indicators and assessment. J Environ Indic 9:17–18

Heiner B, Angela B, Bub A, Ellinger S, Haller D, Leschik-Bonnet KA, Muller E, Oberritter MJ, Schulze H, Stehle M (2012) Critical review: vegetables and fruit in the prevention of chronic diseases. Eur J Nutr 51(6):637–663

Hosseini H, Mousavi SM (2021) Bacterial cellulose/polyaniline nanocomposite aerogels as novel bio adsorbents for removal of hexavalent chromium: experimental and simulation study. J Clean Prod 278:123817

Hu W, Huang B, Tian K, Holm PE, Zhang Y (2017) Heavy metals in intensive greenhouse vegetable production systems along Yellow Sea of China: levels transfer and health risk. Chemosphere 167:82–90

Hussain B, Ashraf MN, Abbas A, Li J, Farooq M (2021) Cadmium stress in paddy fields: effects of soil conditions and remediation strategies. Sci Total Environ 754:142188

Irfan M, Hayat S, Ahmad A, Alyemeni MN (2013) Soil cadmium enrichment: allocation and plant physiological manifestations. Saudi J Biol Sci 20(1):1–10

Islam MM, Karim M, Zheng X, Li X (2018) Heavy metal and metalloid pollution of soil water and foods in Bangladesh: a critical review. Int J Environ Res Pub Health 15(12):2825. https://doi.org/10.3390/ijerph15122825

Itoh H, Iwasaki M, Sawada N, Takachi R, Kasuga Y, Yokoyama S, Onuma H, Nishimura H, Kusama R, Yokoyama K, Tsugane S (2014) Dietary cadmium intake and breast cancer risk in Japanese women: a case–control study. Int J Hyg Environ Health 217:70–77

Iwinski KJ, Rodgers JH, Kinley CM, Hendrikse M, Calomeni AJ, McQueen AD, Geer TD, Liang J, Friesen V, Haakensen M (2017) Influence of CuSO4 and chelated copper algaecide exposures on biodegradation of microcystin-LR. Chemosphere 174:538–544

Jadid N, Maziyah R, Nurcahyani DD, Mubarokah ND (2017) Growth and physiological responses of some *Capsicum frutescens* varieties to copper stress. AIP Conf Proc 2017 1854:020018

Jadid N, Safitri CE, Jannah AL, Muslihatin W, Purwani KI, Masud F (2022) Genetic diversity and growth responses of Indonesian tomato (*Solanum lycopersicum* L) genotypes under lead stress. Sci Prog 105(3):00368504221122364

Jaime-Perez N, Kaftan D, Bína D, Bokhari SNH, Shreedhar S, Kupper H (2019) Mechanisms of sublethal copper toxicity damage to the photosynthetic apparatus of *Rhodospirillum rubrum*. Biochimica et Biophy Acta (BBA)-Bioenerg 1860(8):640–650

Jali P, Pradhan C, Das AB (2016) Effects of cadmium toxicity in plants: a review article. Schol Acad J Biosci 4:1074–1081. https://doi.org/10.21276/sajb20164123

Kabata-Pendias A, Szteke B (2015) Trace elements in abiotic and biotic environments. CRC Press, Boca Raton

Kapoor RT, Mfarrej MFB, Alam P, Rinklebe J, Ahmad P (2022) Accumulation of chromium in plants and its repercussion in animals and humans. Environ Pollut 301:119044

Karimyan K, Alimohammadi M, Maleki A, Yunesian M, Nodehi RN, Foroushani AR (2020) Human health and ecological risk assessment of heavy metal (loid) s in agricultural soils of rural areas: a case study in Kurdistan Province Iran. J Environ Health Sci Eng 18:469–481

Karlin KD, Tyeklár Z (2012) Bioinorganic chemistry of copper springer science. Business Media

Karri V, Ramos D, Martinez JB, Odena A, Oliveira E, Coort SL, Kumar V (2018) Differential protein expression of hippocampal cells associated with heavy metals (Pb, As and MeHg) neurotoxicity: deepening into the molecular mechanism of neurodegenerative diseases. J Proteome 187:106–125

Khalid S, Shahid MBM, Bibi I, Natasha NM, Niazi N (2019) A critical review of different factors governing the fate of pesticides in soil under biochar application. Sci Total Environ 711:134645

Khan A, Khan S, Khan MA, Qamar Z, Waqas M (2015) The uptake and bioaccumulation of heavy metals by food plants their effects on plants nutrients and associated health risk: a review. Environ Sci Pollut Res Int 22:13772–13779

Korkmaz Aİ, Akgul H, Sevindik M, Selamoglu Z (2018) Study on determination of bioactive potentials of certain lichens. Acta Aliment 47(1):80–87

Koutras GA, Schneider AS, Hattori M, Valentine WN (1965) Studies on chromated erythrocytes mechanisms of chromate inhibition of glutathione reductase. Br J Hematol 11(3):360–369

Kubier A, Pichler T (2019) Cadmium in groundwater—a synopsis based on a large hydrogeochemical data set. Sci Total Environ 689:831–842

Lata S, Kaur HP, Mishra T (2019) Cadmium bioremediation: a review. Int J Pharmacol Sci Res 10:4120–4128

Lepp NW (1981) Effect of heavy metal pollution on plants effects of trace metal pollution on plant function, vol 1. Applied Science Publishers, London, pp 352–367

Leygraf C, Chang T, Herting G, Odnevall WI (2019) The origin and evolution of 1602 copper patina colour. Corros Sci 157:337–346

Li N, Kang Y, Pan W, Zeng L, Zhang Q, Luo J (2015) Concentration and transportation of heavy metals in vegetables and risk assessment of human exposure to bio accessible heavy metals in soil near a waste-incinerator site South China. Sci Total Environ 521–522:144–151

Li C, Zhou K, Qin W, Tian C, Qi M, Yan X, Han W (2019) A review on heavy metals contamination in soil: effects, sources and remediation techniques. Soil Sediment Contam 28(4):380–394. https://doi.org/10.1080/15320383.2019.1592108

Li C, Li C, Yu H, Cheng Y, Xie Y, Yao W, Guao W, Qian H (2021) Chemical food contaminants during food processing: sources and control. Crit Rev Food Sci Nutr 61(9):1545–1555

Liu LW, Li W, Song WP, Guo MX (2018) Remediation techniques for heavy metal-contaminated soils: principles and applicability. Sci Total Environ 633:206–219. https://doi.org/10.1016/j.scitotenv.2018.03.161

Liu X, Pang H, Liu X, Li Q, Zhang N, Mao L, Qiu M, Hu B, Yang H, Wang X (2021) Orderly porous covalent organic frameworks-based materials: superior adsorbents for pollutants removal from aqueous solutions. The Innov 2(1):100076

Lwin CS, Seo BH, Kim HU, Owens G, Kim KR (2018) Application of soil amendments to contaminated soils for heavy metal immobilization and improved soil quality – a critical review. Soil Sci Plant Nutr 64(2):156–167

Mahajan P, Kaushal J (2018) Role of phytoremediation in reducing cadmium toxicity in soil and water. J Toxicol 101155(2018/4864365):4864365–4864381

Mallick SR, Proshad R, Islam MS, Sayeed A, Uddin M, Gao J, Zhang D (2019) Heavy metals toxicity of surface soils near industrial vicinity: a study on soil contamination in Bangladesh. Arch Agric Environ Sci 4(4):356–368

Mamun SA, Sultana N, Hasan M, Latif MB, Islam A, Tusher TR (2021) Phytoaccumulation of cadmium in leafy vegetables grown in contaminated soil under varying rates of compost and phosphate fertilizer application. Commun Soil Sci Plant Anal 52(18):2161–2176

Mandal S, Datta S, Das AK, Pal S (2017) Heavy metal contamination of vegetables and crops from Kolkata metropolitan city of India and associated health risk assessment. Bull Environ Contam Toxicol 98(4):554–561

Manzoor J, Sharma M, Wani KA (2018) Heavy metals in vegetables and their impact on the nutrient quality of vegetables: a review. J Plant Nutr 41(13):1744–1763

Markert B (1996) Phyto technologies solutions for sustainable land management – inorganic contaminants: occurrence and significance of chemical element with special reference to plants encyclopedia of life support system (EOLSS) Fliederwerg 17 D – 49733 Haren Germany p 7

Martin S, Griswold W (2009) Human health effects of heavy metals. Environ Sci Technol Br Citiz 15:1–6

Matsumoto ST, Mantovani MS, Malaguttii MIA, Dias AL, Fonseca IC, Marin-Morales MA (2006) Genotoxicity and mutagenicity of water contaminated with tannery effluents as evaluated by the micronucleus test and comet assay using the fish Oreochromis niloticus and chromosome aberrations in onion root-tips. Genet Mol Biol 29:148–158

McIntyre T (2003) Phytoremediation of heavy metals from soils. Adv Biochem Eng/BioTechnol 78:97–123

Meena V, Dotaniya ML, Saha JK, Das H, Patra AK (2020) Impact of lead contamination on agroecosystem and human health. In: Lead in plants and the environment, pp 67–82

Mokarram M, Saber A, Sheykhi V (2020) Effects of heavy metal contamination on river water quality due to release of industrial effluents. J Clean Prod 277:123380

Monalisa M, Kumar PH (2013) Effect of ionic and chelate assisted hexavalent chromium on mung bean seedlings (*Vigna radiata* L Wilczek var k-851) during seedling growth. J Stress Physiol Biochem 9(2):232–241

Mondal SC, Sarma B, Farooq M, Nath DJ, Gogoi N (2020) Cadmium bioavailability in acidic soils under bean cultivation: role of soil additives. Int J Environ Sci Technol 17:153–160

Mutlu A, Lee BK, Park GH, Yu BG, Lee CH (2012) Long-term concentrations of airborne cadmium in metropolitan cities in Korea and potential health risks. Atmos Environ 47:164–173

Nabizadeh R, Esmaili-Sari A, Mehrizi EA (2014) Determination of cadmium and lead in vegetables cultivated in suburb areas of shiraz Iran. Bull Environ Contam Toxicol 92(6):677–681

Najeeb U, Ahmad W, Zia MH, Zaffar M, Zhou W (2017) Enhancing the lead phytostabilization in wetland plant *Juncus effusus* L through somaclonal manipulation and EDTA enrichment. Arab J Chem 10:S3310–S3317

Nazir F, Hussain A, Fariduddin Q (2019) Hydrogen peroxide modulate photosynthesis and antioxidant systems in tomato (*Solanum lycopersicum* L) plants under copper stress. Chemosphere 230:544–558

Nordberg GF, Nogawa K, Nordberg M, Friberg LT (2007) Cadmium. In: Nordberg GF, Fowler BA, Nordberg M, Friberg LT (eds) Handbook on the toxicology of metals, 3rd edn. Academic, pp 445–486. https://doi.org/10.1016/B978-012369413-3/50078-1

Nwachukwu E, Egwuonwu C, Okorie O, Ezejiofor T (2013) Heavy metal content of selected green leafy vegetables sold in a typical Nigerian urban market. Afr J Food Agric Nutr Dev 13(1):7261–7275

Obrist D, Agnan Y, Jiskra M, Olson CL, Colegrove DP, Hueber J, Moore CW, Sonke JE, Helmig D (2017) Tundra uptake of atmospheric elemental mercury drives arctic mercury pollution. Nature 547:201–204

Ogunbileje TO, Eletta OAA, Oluyemi EA, Adeyemi OI, Adeleye IO, Adegoke OA (2020) Assessment of heavy metal concentrations in selected vegetables in urban farms and market gardens of Lagos Nigeria. Food Sci Qual Manag 97:55–62

Ogunkunle CO, Bornmann B, Wagner R, Fatoba PO, Frahm R, Lutzenkirchen-Hecht D (2019) Copper uptake tissue partitioning and biotransformation evidence by XANES in cowpea (*Vigna unguiculata* L) grown in soil amended with nano-sized copper particles. Environ Nano Technol Monit Manag 12:100231

Pan XD, Wu PG, Jiang XG (2016) Levels and potential health risk of heavy metals in marketed vegetables in Zhejiang China. Sci Rep 6:20317. https://doi.org/10.1038/srep20317

Patrick L (2002) Mercury toxicity and antioxidants: part I: role of glutathione and alpha-lipoic acid in the treatment of mercury toxicity. Altern Med Rev 7(6):456–472

Phippen B, Horvath C, Nordin R, Nagpal N (2008) Ambient water quality guidelines for iron: overview. Ministry of Environment Province of British Columbia

Prozialeck WC, Edwards JR (2012) Mechanisms of cadmium-induced proximal tubule injury: new insights with implications for biomonitoring and therapeutic interventions. J Pharmacol Exp Ther 343(1):2–12. https://doi.org/10.1124/jpet.110.166769

Rafati-Rahimzadeh M, Rafati-Rahimzadeh M, Kazemi S, Moghadamnia AA (2014) Current approaches of the management of mercury poisoning: need of the hour. DARU J Pharmacol Sci 22:46

Rajindiran S, Dotaniya ML, Coumar MV, Panwar NR, Saha JK (2015) Heavy metal polluted soils in India: Status and countermeasures. JNKVV Res J 49(3):320–337

Ramamoorthy D (2015) Assessment of heavy metal pollution and its impacts on soil physical chemical properties and β-glucosidase activities in agricultural lands Puducherry region. Project report submitted to Department of Science, Technology and Environment Government of Puducherry

Rose M, Fernandes A, Mortimer D, Baskaran C (2015) Contamination of fish in UK freshwater systems: risk assessment for human consumption. Chemosphere 122(2):183–189

Rubiya S, Dhriti K, Bhat AA (2018) Heavy metal toxicity in plants: a review. Plant Archiv 18(2):1229–1238

Sabarathinam J, Vishnu Riya V, Gayathri R (2016) Mercury poisoning and management: a systematic review. Asian J Pharmacol Clin Res 9:8–12

Saha P, Paul B (2019) Assessment of heavy metal toxicity related with human health risk in the surface water of an industrialized area by a novel technique. Human and Ecol Risk Assess An Int J 25(4):966–987

Sahare D (2014) Long term impact of industrial effluents on agricultural soil. Recent Res Sci Technol 6(1):253–257

Salam SA, Javed MS, Toor MD, Adnan M, Awais M, MMU D, Tampubolon K (2020) Influence of industrial wastewater on soil and plants: a review. Curr Res Agric Farm 1(4):19–23

Sandeep G, Vijayalatha KR, Anitha T (2019) Heavy metals and its impact in vegetable crops. Int J Chem Study 7(1):1612–1621

Satarug S, Garrett SH, Sens MA, Sens DA (2010) Cadmium environmental exposure and health outcomes. Environ Health Persp 118(2):182–190

Schiptsova N, Larionov G, Vasilyev O, Fadeeva N, Terentyeva M (2020) Effect of sewage sludge application on heavy metals contamination in soil and carrot. In: IOP conference series: earth and environmental science, vol 604 No 1. IOP Publishing, p 012034

Sevindik M, Akgul H, Akata I, Selamoğlu Z (2017) *Geastrum Pectinatum* as an alternative antioxidant source with some biochemical analysis. Med Mycol Open Access 3(2:25):2471–8521

Shabbir Z, Sardar A, Shabbir A, Abbas G, Shamshad S, Khalid S, Shahid M (2020) Copper uptake essentiality toxicity detoxification and risk assessment in soil-plant environment. Chemosphere 259:127436

Shahid M, Khalid S, Abbas G, Shahid NM (2015) Heavy metal stress and crop productivity. In: Hakeem KR (ed) Crop production and global environmental issues SE–1. Springer, pp 1–25

Shahid M, Niazi NK, Dumat C, Naidu R, Khalid S, Rahman MM, Bibi I (2018) A meta-analysis of the distribution sources and health risks of arsenic-contaminated groundwater in Pakistan. Environ Pollut 242:307–319

Shahid M, Niazi NK, Rinklebe J, Bundschuh J, Dumat C, Pinelli E (2020) Trace elements-induced phytohormesis: a critical review and mechanistic interpretation. Crit Rev Environ Sci Technol 50(19):1984–2015

Sharma P, Dubey RS (2005) Lead toxicity in plants Brazilian. J Plant Physiol 17:35–52

Sharma A, Katnoria JK, Nagpal AK (2016) Heavy metals in vegetables: screening health risks involved in cultivation along wastewater drain and irrigating with wastewater. Springer Plus 5(1):1–16

Shiyab S (2018) Phytoaccumulation of copper from irrigation water and its effect on the internal structure of lettuce. Agriculture 8(2):29

Singh KP, Bhattacharya S, Sharma P (2014) Assessment of heavy metal contents of some Indian medicinal plants. Am-Eurasian J Agric Environ Sci 14:1125–1129

Singh V, Singh N, Verma M, Kamal R, Tiwari R, Sanjay Chivate M, Rai SN, Kumar A, Singh A, Singh MP, Vamanu E, Mishra V (2022) Hexavalent-chromium-induced oxidative stress and the protective role of antioxidants against cellular toxicity. Antioxidants (Basel) 11(12):2375. https://doi.org/10.3390/antiox11122375

Tan BL, Norhaizan ME, Liew WP, Rahman HS (2018) Antioxidant and oxidative stress: a mutual interplay in age-related diseases. Front Pharmacol 9:1–28

Ullah H, Haq ZU (2022) Effect of industrial wastewater on local soil quality and vegetables in industrial estate Hattar Pakistan. Int J Waste Res 12(11):1000497

United State Environmental Protection Agency (USEPA) (1993) Standard methods for the examination of water and wastewater; US: American Pub Health Assoc 1993

UN-Water (2018) The United Nations world water development report 2018: nature-based solutions for water 139 pp

USDA-NRCS 2015: Soil quality indicators Sheets Nat Resour Conservation Serv Available at: http://www.nrcs.usda.gov/wps/portal/nrcs/detail/soils/health/assessment/?cid=stelprdb1237387

Wakeel A, Xu M (2020) Chromium morpho-phytotoxicity. Plants 9(5):564

Wang X, Sato T, Xing B, Tao S (2014) Health risks of heavy metals to the general public in Tianjin China via 350:28–37

Xu D, Zhou P, Zhan J, Gao Y, Dou C (2013) Assessment of trace metal bioavailability in garden soils and health risks via consumption of vegetables in the vicinity of Tongling mining area. China Ecotoxicol Environ Safety 90:103–111

Xu ZM, Li QS, Yang P, Ye HJ, Chen ZS, Guo SH, Wang LL, He BY, Zeng EY (2017) Impact of osmoregulation on the differences in Cd accumulation between two contrasting edible amaranth cultivars grown on Cd-polluted saline soils. Environ Pollut 224:89–97. https://doi.org/10.1016/jenvpol201612067

Zhang BW, Zhao JT, Wu EW (2018) Water pollution characteristics of heavy metals in Fuyang River system and their toxicity to *Vibrio qinghaiensis* sp Q67 and *Scenedesmus obliquus*. Asian J Ecotoxicol 13(1):179–189

Zhou H, Yang WT, Zhou X (2016) Accumulation of heavy metals in vegetable species planted in contaminated soils and the health risk assessment. Int J Environ Res Public Health 13:289

Zwolak A, Sarzynska M, Szpyrka E, Stawarczyk K (2019) Sources of soil pollution by heavy metals and their accumulation in vegetables: a review. Water Air Soil Pollut 230:1–9

Part V

Genetics and Biotechnology

Applications of Biotechnology in Vegetable Crops

20

Riaz Ahmad, Sezai Ercisli, Arslan Akbar, and Muhammad Akbar Anjum

Abstract

Sustainable vegetable production is mainly based on strong linkages among horticulturists, plant breeders, biotechnologists, and other stakeholders. Molecular markers, genome sequencing, gene editing, genetic transformation, and tissue culture are efficient biotechnological tools that are assumed to be useful for plant breeders for crop improvement purposes. Biotechnological approaches have the potential to improve tolerance in the vegetables growing under biotic and abiotic stresses. Since morphological and biochemical markers are highly affected by environmental influences, their applications have been very limited. Many molecular markers and sequencing techniques are regularly utilized in phylogenetic studies, evaluation of genetic variability and evolutionary relationships, precise taxonomy-based classification, identification of diverse germplasm, and high-yielding hybrids. Further, various biotechnological tools are being used to identify desired and undesirable attributes that are regulated by numerous genes. Over the past 20 years, many plant species have been successfully characterized by using these well-established molecular techniques. Recent molecular marker approaches provide an improved genotypic platform for the development of high-yielding cultivars. The present study explores deep insights into several molecular markers, functional genomics, genome editing, and genetic transformation approaches that have been used for numerous genomic analyses of veg-

R. Ahmad (✉)
Department of Horticulture, The University of Agriculture, Dera Ismail Khan, Pakistan
e-mail: riaz.ahmad@uad.edu.pk

S. Ercisli
Department of Horticulture, Faculty of Agriculture, Ataturk University, Erzurum, Türkiye

A. Akbar · M. A. Anjum
Department of Horticulture, Bahauddin Zakariya University, Multan, Pakistan

Z. Iqbal et al. (eds.), *Sustainable and Innovative Vegetable Production in times of Climate Change*, https://doi.org/10.1007/978-981-96-6283-8_20

etable crops. CRISPR/Cas-based genome editing tools are now widely used in vegetable crops. The current review validates the adoption of molecular technologies for genome identification, sequencing, editing, transformation, re-sequencing, barcoding, and QTLs mapping. Diverse genotypes can be identified and further utilized for the development of high-yielding germplasm with resilience to biotic and abiotic challenges.

Keywords

Genetic transformation · Genome editing · Molecular markers · Tissue culture

20.1 Introduction

Plant improvement has been carried out for a very long time to get higher yields. Plant genome is being altered through various methods, i.e., domestication, genomic transformation, and traditional breeding to improve a single trait (Fita et al. 2015). The most popular technique for crop enhancement is traditional breeding. In some circumstances, these techniques must be combined with biotechnological techniques, either to increase their efficacy or to achieve a goal that is impossible to achieve with conventional methods (Bartlett et al. 2022).

Biotechnological methods involving plant tissue culture provide a reliable method for identifying a wide range of variations and exploiting the natural diversity for crop improvement programs. These somaclonal variations are as a result of culture conditions, medium constituents, and added plant growth regulators (PGRs) (Gaba 2005) or can be induced in cultures by in *vitro* mutagenesis through the use of mutagens. Through in vitro culture, unexpectedly rich and innovative modifications can be generated, giving rise to a new and significant diversity in revived plants. One opportunity to use this uniqueness for the expansion of new groups is mending of hereditary transformation in plants that results through tissue culture. Through the use of tissue culture, new disease-resistant cultivars have been developed, providing access to a wide range of plants that are resistant to diseases caused by viruses, bacteria, and fungi (Aftab and Roychoudhury 2021). Massive innovations have been developed in the tissue culture methods to multiply plants over the past decade and produce plants with distinctive characteristics, i.e., high yield, and resistance to biotic and abiotic stresses (Khan et al. 2019).

Traditional breeding takes a long time to start and requires several generations of crossbreeding and selection to introduce a valuable genotype. Approaches using test tubes are unquestionably more advanced than conventional methods (Ahmar et al. 2020). As long as any molecular action would be considered biotechnology, the term is seen as being broad. It stands to reason that it would be utilized not only in crop production and improvement but in food processing for brewing and bread making as well (Rommens et al. 2007).

It has been evaluated that biotechnological tools are effective for sustainable vegetable production. There are two key biotechnology strategies for increasing vegetables yield (i) using it to support conventional methods of propagation, and (ii) using it to transfer genes across different organisms. Chromosome-carried genes control morpho-physiological characteristics of an organism. Plant breeders greatly benefit from being able to examine if such genes are present or absent in plants (Hou et al. 2014). This is accomplished through the use of distinctive gene sequences, molecular markers, the concerned gene, or gene-related DNA fragments. The monitoring of such markers in numerous independent individuals, such as new lines that develop from cross-pollination between two diverse plants, has been made possible by biotechnology (Al-Khayri et al. 2015). Hence, biotechnological techniques are fantastic for the development of potential cultivars. Genetic engineering and gene transformation are biotechnological tools which can be used to save time and are best to transfer genetic materials in a particular plant in a single step by avoiding the use of traditional breeding methods (Rommens 2007). Traditional breeding, of course, is restricted to the introduction of genes against plants of the same species or very close relatives. By using genetic engineering techniques, it is possible to incorporate unique genes from different plants, animals, bacteria, or fungi within a crop plant. Genes are made up of two segments: the coding site, which determines what the gene product will be, and the non-coding region. Genes comprise two regions, the promoter, which determines when, where, and how much a gene is expressed, and the coding site, which defines what gene is explored (Caruso et al. 2020). Coding regions and promoters from different genes can be combined at once in the laboratory to develop genes containing effective features (recombinant DNA) (Wissuwa et al. 2009). The recombinant genes can subsequently be modified inside crop plants using genomic transformations. The acquired genes incorporate into the plant genome normally and are transferred into the offsprings. It is possible to improve biological properties in this way and add new features to a crop (Schaart et al. 2016).

The exploration of biotechnology is essential for sustainable vegetable production with the development of excellent germplasm with desired characteristics. The main mechanisms and methods of heredity regulation, from gene selection and separation to safety situations like the stability of transgenes, have been briefly explored for plants (Dietz-Pfeilstetter 2010). The hereditary information can be utilized for improvements of yield contributing traits. Hereditary changes in agronomic characteristics, i.e., resilience to environmental constraints and pests, herbicide tolerance, and increased yield with improved quality are revealed in genetic changes in vegetable crop parts (Rejeb et al. 2014). It is also becoming important to study how genes can be used via biotechnology to increase the post-harvest survival of fruits and vegetables (König 2003). The use of molecular inheritance was used to control food behaviors including nutritional quality as well as sensory features like color and flavor (Dangi et al. 2018). The nutritional content of plant foods can increase the nutritional quality of plant-based products (Eede et al. 2004). Utilization of breeding approaches for sustainable vegetable production is presented in Fig. 20.1.

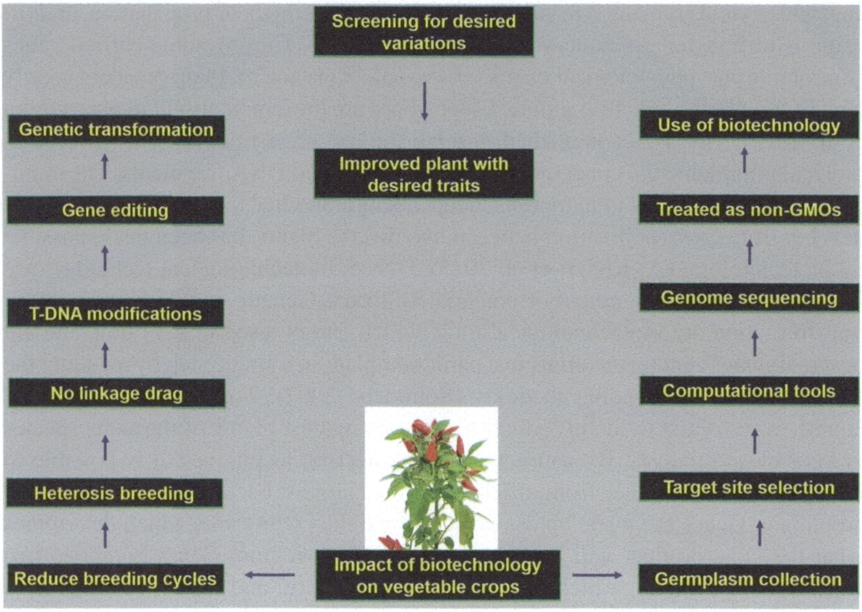

Fig. 20.1 Utilization of breeding approaches for sustainable vegetable production

20.2 Significance of Biotechnology

Mostly, cultivated crops are a result of early farmers' conscious or unintentional selection of plants with better traits that were passed down through their DNA to succeeding generations (Golldack et al. 2011). By the time of historic agricultural civilizations like the Egyptian, Chinese, or Mayan, the majority of the variations in crops were due to domestication such as seed retention just on the plant, easier cultivation, a larger size of harvested parts, variation in plant form, and the decline of bitter and toxic compounds. Breeders had a far better understanding of how to recombine genes by sexual crosses to increase crop yields or change plant features. Biotechnology as a science attained good attention at the beginning of this century (Saito and Matsuda 2010). Genetic modification of cultivars in association with good agronomic practices resulted in higher productivity, yield, and quality in the last few decades (Yadav 2020).

Although our domesticated crops no longer exhibit the majority of their wild attributes, ongoing genetic enhancement attempts are crucial. In addition to continued advancements in the quality of produce, yield, and geographical performance, new pests and disease threats are always emerging, prompting the development of new and improved cultivars (Miller and Gross 2011). Therefore, crop genetic enhancement cannot be achieved all at once. However, maintaining the quality and quantity of vegetable crops requires a systematic discovery of genomic combinations that can give rise to new demand or production features. For breeders to make these breakthroughs, modern biotechnology has produced innovations to complement selection strategies and new resources of genomic combinations (Koinange et al. 1996).

To add novel traits to several crops, breeders have largely depended on introducing genes from similar interfertile wild species (Smýkal et al. 2018). Commercial cultivars of tomatoes and potatoes have been bred from related wild ancestors found in South America that are resistant to diseases like *Fusarium* and *Verticillium* as well as pests like root-knot nematode (Thrall et al. 2011; Boyd et al. 2013). The majority of our domesticated crops particularly vegetables rely on genes obtained from a diverse range of genotypes, or genetically unique variants, of each species. The parent alleles segregate or randomly spread in succeeding generations upon a sexual cross made between wild and domesticated species under genetics laws (Singh et al. 2022). Breeders are interested in tracking these genes and learning which offsprings have inherited particular genes. Generally, this is accomplished by visual evaluation, disease or pest screens, chemical analyses, or some other criteria. However, a visual feature (phenotype) is frequently absent or only detectable in fully grown plants (Sarker et al. 2018).

The interdisciplinary science of biotechnology focuses on several aspects of life, including production of energy and food and medicine that has been genetically modified (Sarker et al. 2017). Although planting materials like potato tubers and banana suckers cannot be used for a long time, even these are inexpensive and disease-free. Cryopreservation allows for the preservation of genomic resources for a long time and their further use in breeding programs (Keller 2002). In addition to broadening the range of traits that can be addressed, plant biotechnology has sped up the event for growing vegetables (Ruta et al. 2020). Conventional breeding barriers, such as relationship drag challenge, vast cross-sexual obstacle, nutritional avoidance problem, etc. can be commendably and quickly handled using a biotechnology method. Recent developments in molecular biology, including genetic engineering (cisgenesis, transgenic, and intragenesis), DNA sequencing, cloning, molecular markers, and in vitro mutagenesis, provide new interpretations, contemporary perspectives, and innovative potential of antiquated biotechnology (Grout and Crisp 1995). Moreover, it also offers an up-to-date outlook on microbial involvement in agricultural activities, including bio-control agents, bio-fertilizers, and several other microbiological commodities employed in the present-day agricultural sector (Rao et al. 2014). Thus, the development of vegetable crops may be significantly impacted by contemporary biotechnology advancements.

Vegetables diversity confronted the problems of genomic losses because of commercial farming of high-yielding identical landraces and eradication of natural habitats because of industrialization and urbanization expansion. The phenotypic characterization is very difficult because of hybrid origins and evolutionary characters in crop plants, for example in potato as studied by Raker and Spooner (2002). For molecular characterization, 77 ISSRs (inter-simple sequence repeats) were applied to 28 potato genotypes collected from various fields (Bornet et al. 2002). Moreover, SSRs offer the most genomic information, and are highly repeatable and easy to use for the evaluation of potato genetic resources (Ghislain et al. 2004). The characterization of new commercial cultivars is necessary for sustainable crop production. So, application of molecular tools is imperative for genomic studies of developed germplasm (Frary et al. 2005). Similarly, Chen et al. (2009) also investigated the genetic variants in 216 tomato cultivars, landraces, and hybrids by using two molecular markers systems, i.e., SNPs (single nucleotide polymorphisms) and SSRs (simple sequence repeats). To examine the genetic diversity and genomic relationships among 48 genotypes of peppers collected from nine different countries, Rai et al. (2013) employed 106 SSRs and 17 RAMPs (random amplified microsatellite polymorphism). They validated the evidence of natural interspecific hybridization and proposed that natural crossing was the reason of their origin. So, it is imperative to examine the genetic diversity of vegetables germplasm to preserve genetic resources for breeding programs and to determine how quickly breeding material can be verified and documented through different biotechnological tools.

20.3 Trending Implications in Vegetable Crops

Numerous biotechnology techniques and approaches have been used to improve production and quality of vegetables. Several plant researchers were prompted by DNA-based expertise to successfully extract the specific gene of interest and transfer and assimilate it into the target species. These efforts were made possible due to DNA-based expertise. The conventional crossing was restricted to closely related species alone, but advancements in plant biotechnology have made it possible to improve in good ways beyond the limits (Yadav et al. 2015). Below, some highlighted techniques and strategies have been briefly explained that have been or being utilized for sustainable vegetable production.

20.3.1 Genomic Editing and Genetic Transformation for Biotic Stresses Tolerance

The crop yield losses of approximately 40% are caused by biotic stresses, i.e., insects, parasites, viruses, bacteria, fungi, and weeds (Ahmad and Anjum 2023). To reduce these losses, different strategies have been adopted, including applications of pesticides, phytohormones, etc. However, utilization of molecular tools is much effective in development and speedy characterization of tolerant germplasm against these biotic stresses (Aghdam et al. 2015). Several techniques can be used to alter or transfer genes in plants. *Agrobacterium tumefaciens*, a bacterial pathogen, is used in one exciting method to introduce the required DNA into the plant (Kumar and Srivastava 2016). The bacterium spontaneously inserts a portion of its DNA into the chromosomes of the plant, where it triggers the synthesis of substances that the bacterium consumes. Genome editing and genetic transformation approaches are extensively used for evolving vegetable crops with unique characteristics, i.e., colors, fragrances, flavors, sizes, harvest durability, and tolerance against insects and pests (Kumar and Srivastava 2020). The advent of transgenic technologies brings a greater revolution for the improvement of vegetable germplasm with higher yields. So, it has been evaluated that genome editing and genetic transformations are imperative approaches for sustainable vegetable production.

RNA silencing network improves gene expression and tolerance against vital diseases. Tomato mosaic virus (TMV) coat protein (CP) gene-modified (GM) vegetables were resistant to TMV infestation, and this CP-moderate immunity is widely used to protect many crops from a wide range of viruses (Mushegian and Shepherd 1995). China was the main producer of commercially viable virus-immune GM crops, and finally, virus-immune tomato, potato, watermelon, and squash plants were developed. Through a flaw in the gene arrangement for viral CP, Monsanto strengthened the GM potato's resistance to the potato virus Y, which prevents the virus from replicating. Transplanting viral CP that was resistant to the cucumber mosaic virus greatly improved transgenic tomato and cucumber types. The RNA silencing technique enhanced the resistance of transgenic common beans to bean golden mosaic virus (Lindbo and Falk 2017).

After maturing, GM tomato and brinjal were tested in various nations' agricultural conditions (Shelton 2012). For the development of insect-immune vegetable crops, more insecticidal proteins, such as antibodies, lectins, protease inhibitors, microbial insecticides, wasp, spider, and insect peptide hormones are now cost-effective alternatives (Maheswari et al. 2008). *Bt* brinjal was developed and field assessed by the joint analysis within the US-based company Monsanto and Maharashtra Hybrid Seeds Company (Mahyco), India, by insertion of a crystal gene (*Cry1Ac*) extracted from *Bacillus thuringiensis* (Kumar et al. 2011). By inactivating the insect's serine proteinase enzymes, proteinase inhibitors have proven to be more effective at controlling the growth and development of several insects (Sun et al. 2016). Several classes of proteinase inhibitors, including potato protease inhibitors II and cowpea trypsin inhibitors (*CpTi*), have been demonstrated in agricultural plants. By cloning genes expressing proteinase inhibitors and amylase inhibitors, many insect-immune plant species have thrived (Zhu-Salzman et al. 2003). Relocation of the *cry1Ab* gene to the bottom of a granule-bound starch synthase promoter potentially developed a *Bt* transgenic potato plant that was resistant to the potato tuber moth (Pouvreau et al. 2001). Cucumber is regarded as an ideal plant for genetic study due to its small genome (about 367 Mb) with a higher level of gene expression and short life span among all the species of the Cucurbitaceae family (Lv et al. 2012). Breeders around the world can develop cultivars that can be tolerant against insects and diseases, boosting production and enhancing quality (Yuan et al. 2008). Farmers can be secure against biotic constraints by using the tolerant germplasm for crop improvement (Govindaraj et al. 2015).

20.3.2 Biotechnological Approaches to Improve Abiotic Stress Tolerance

Abiotic stresses negatively impact the production of vegetables all over the globe. Different management practices can be utilized for the amelioration of harmful effects of abiotic stresses (Shahzad et al. 2021; Ahmad et al. 2023). The improved antioxidant potential of plants also improves their yields under stressful conditions. It is possible to identify salt-resistant genes and assess the tolerance mechanisms of crops using several molecular approaches (Machado and Serralheiro 2017). Thus, genetic modification is a successful method for developing genotypes that are salt tolerant. A main challenge to the development of cultivars with high salt tolerance is their polygenic nature. However, by using various physiological, biochemical, and molecular markers, one or more traits can be improved (Ahmad et al. 2020). Moreover, it is yet unknown how certain vegetable cultivars tolerate abiotic stresses. Genome linkage maps can be helpful to overcome this problem in vegetable crops. There is a limited study on vegetable crops to develop tolerant genotypes and evaluate antioxidant mechanisms governing stress tolerance. Field testing for genotypic competence under stressful conditions is difficult, time-consuming, and unreliable. However, short-term evaluation using molecular markers is possible for genotypes that are stress-tolerant or sensitive (Aleem et al. 2020). The emergence of stress-tolerant genotypes through QTL and molecular markers occurs quite slowly in vegetable crops (Zhuang et al. 2014). Conventional breeding practices have several restrictions and do not produce reliable results. The ability to identify genes using molecular techniques and clone them to improve tolerance against various abiotic stresses is growing exponentially. It is urgently required to conduct molecular research for various genetic analyses to better understand the crop plants (Kok et al. 2010).

Proteomics is a useful tool for assessing a specific species or cultivars and their unique genes. Proteins have a role in the defensive mechanisms of plants as well as in many biotic and abiotic stresses. Proteomic strategies can be utilized to choose cultivars which are tolerant to biotic and abiotic constraints. These strategies play a direct role in the selected plant's response to a stress (Zhuang et al. 2014). Through proteomic techniques, a large number of proteins involved in the control of a plant stress may be encoded. Evaluation of new genes involved in biotic and abiotic stresses' tolerance can be made using transcriptional studies in the vegetables (Agregán et al. 2021).

20.3.3 Quantitative Trait Locus (QTL) Mapping and Marker-Assisted Breeding (MAB)

A statistical study known as QTL mapping can be used to analyze complex traits or determine which molecular markers result in a quantitative change of a specific trait. QTL mapping investigations are used to recognize the genes and DNA markers related to the quantitative characters. Thus, QTL mapping serves as the building block for the construction of marker-assisted selection (MAS) (Lian et al. 2021). In backcross breeding, markers can be utilized at the following three levels of selection. i) Target features that require time-consuming phenotypic assessment. ii) Recombinant selection, which involves selecting backcross progeny with the target gene and closely spaced flanking markers to decrease linkage drag. iii) Choosing backcross progeny (already chosen for the target trait) with "background" markers. Multiple genes and QTLs can be combined into a single genotype using the gene pyramiding strategy (Pereira et al. 2018).

Marker-assisted breeding (MAB) is a method for choosing traits in plants that are beneficial to farmers. The capacity to incorporate single or multiple combinations of attributes into an existing cultivar or elite breeding line distinguishes MAB from traditional breeding. The successful commercial lines that farmers desire and demand may be quickly reinvented to deliver improved agronomic or consumer market features with a minimum reinvestment in field studies. Approximately, 08–10 years are required for conventional breeding. So, value-added attributes are being introduced into current cultivars in development spans of just 02–03 years due to MAB, while still maintaining the qualities that made them economical to grow (Simko et al. 2021).

Insect pests and fungal tolerance have been studied using these QTL and MAB approaches. Moreover, these strategies are also much effective for evaluation of qualitative attributes. Genetic diversity, population relatedness, MAS, and discovery of genes important to plant breeders can all be examined using molecular techniques. MAS is the most promising method for cultivar development purposes. MAS relates to the use of closely linked DNA markers to address loci in replacement of or in conjunction with phenotypic evaluation (Ahmad and Anjum 2018). Molecular characterization of cucumber was also conducted by using SSRs, QTLs and whole genome mapping, MAS, evolutionary studies, and taxonomic investigations (Dar et al. 2017).

20.3.4 Molecular Characterization of Vegetable Crops Germplasm

Molecular markers act as tags that help to locate and identify particular genes on chromosomes. The expression of a gene can be verified and its location on the chromosome can even be determined using some different molecular techniques. The markers allow plant researchers to identify specific plants that contain beneficial gene combinations and to construct genomic maps of the placement of genes on chromosomes. Proteins and enzymes have long been utilized to reveal heredity for several significant features. Despite needing to rely on transcriptional activation, molecular markers for DNA enable biologists to directly see the genes and their sequence on the chromosomes (Ahmad and Anjum 2018).

The insertion of root-knot nematode resistance gene (*Mi*) into commercial tomato cultivars is an illustration of the use of biochemical markers in vegetable breeding. This gene was first identified in a wild tomato ancestor before being introduced into cultivated tomatoes through traditional sexual crosses. However, the wild tomato parent gave rise to some distinct undesirable features, including undersized fruit, subpar color, and bitter or toxic substances (Fassuliotis, and Bhatt 1982). Apart from a little region encoding the *Mi* gene, most of the wild-type chromosomes have been deleted after several generations (greater than 10 years) of selection. One of the newest biotechnological techniques, DNA-based markers, can reduce or perhaps completely remove the restrictions that occur in breeding programs. Regarding the *Mi* gene, DNA sequences that overlap with portions of the target gene have been identified and are, therefore, more likely to be inherited across it. To make it easier to transfer this desirable characteristic to other types and even other species, biologists are also utilizing these patterns to extract the *Mi* gene directly (Devran and Söğüt 2010). In other situations, copies of the desired gene have been extracted or cloned and are ready to be used as the marker. These markers have the benefit of identifying DNA of the gene itself as opposed to the production of enzymes or proteins. The protein could hardly be detected because it is only expressed in very minute levels, or it could be influenced by other genes or the habitat in which the plant is grown (Djian-Caporalino et al. 2011). Moreover, while a potentially limitless amount of DNA-based biomarkers can be developed to distinguish between even closely related cultivars, there is relatively little diversity among enzymes or proteins to serve as tags to identify various lines in many plants, i.e., lettuce and tomato (Devran and Söğüt 2014).

Multiple markers or barcodes are especially helpful if the breeder wants to track the inheritance of many genes at once (Park et al. 2021). For example, about 15 genes have the potential to provide tolerance to different strains of lettuce downy mildew (Govindarajulu et al. 2015). The seedlings can be examined utilizing molecular techniques which are specific for every gene rather than having to test each one against every strain of the pathogen, which is challenging to do, and only those that pass the test are grown for the following generation (Paran and Michelmore 1993). Similar to how specific targets can be chosen, many markers closely connected with desired genes can be assessed for features that depend upon the number of genes. This procedure has the advantage because it can analyze huge data quickly (Huibers et al. 2013). Biotechnological tools implemented for genomic examination of vegetable crops are given in Table 20.1.

Vegetables' germplasm resources must be accurately documented and classified aimed to improve genotypes by using biotechnological tools (Hale et al. 2006). Little diversity and comparatively higher similarity revealed a narrow genetic base in broccoli germplasm (Louarn et al. 2007). So, a narrow genetic base is the main problem in the improvement of vegetables germplasm. Therefore, it is necessary to characterize the tolerant/sensitive germplasm. However, the introduction of new cultivars is much imperative for the bordering of the gene pool. Most germplasm within a country is developed through natural and artificial selection processes. So, biotechnological tools can be utilized for characterization of high-yielding germplasm or for other desired traits (Gichuki et al. 2003). Moulin et al. (2012) examined the genetic diversity among 81 cultivars of sweet potato using both RAPD (Random Amplified Polymorphic DNA) and ISSR (Inter Simple Sequence Repeat) markers. According to genetic relationships rather than agronomic features, concise relationships between sweet potato cultivars were explored by Qiang et al. (2009). This allowed identification and selection of prospective parental pairs for breeding of high-quality edible cultivars. However, ISSRs in ginger are more accurate than RAPDs (Goulao and Oliveira 2001). Similarly, Isshiki et al. (2008) examined 18

Table 20.1 Biotechnological tools implemented for genomic examination of vegetable crops

Software name	Operating system	Description	Hyperlink
POPGENE version 1.32	Windows	Effective number of alleles was calculated through dominant and co-dominant nature markers from the studied germplasm	https://mybiosoftware.com/popgene-1-32-population-genetic-analysis-2.html
MEGA version 11-5.2	Windows and MacOS	Phylogenetic and evolutionary relationships of diverse genotypes can be evaluated	https://www.megasoftware.net/
XLSTAT, 2023	Windows	Genetic variation, principal component analysis, clustering of traits, correlation matrix, and genotypes clustering were calculated	https://www.xlstat.com/en/
NTSYSpc Version 2.0, 2.1 & 2.2	Windows	Cluster analysis for germplasm evaluation of similarity/dissimilarity.	http://www.appliedbiostat.com/ntsyspc/update_NTSYSpc.html
Statistix version 8.1 & 8.1.1	Windows	Evaluation of genetic variation, mean data, correlation, regression, and standard error of the collected genotypes and their traits	https://statistix.informer.com/8.1/
Structure version 2.3.4	Windows and UNIX	Assessment of allelic admixture within the species or their individuals	https://web.stanford.edu/group/pritchardlab/structure_software/release_versions/v2.3.4/html/structure.html
RStudio desktop	Windows, NT, Ubuntu, And UNIX	Trait association, principal component analysis, and mean data analyses can be estimated in the genotypes.	https://posit.co/download/rstudio-desktop/
JBrowse 1.10.11 & 1.10.12	Windows	Visualizing the detected structural variants	https://jbrowse.org/blog/2013/12/10/jbrowse-1-10-12-index.html
SnpEff and SnpSift	Windows	Whole genome sequencing and mapping of species and their cultivars	http://pcingola.github.io/SnpEff/
Clustal W / Clustal X	Windows, MacOS, And UNIX	Offers multiple alignments of DNA sequences in numerous species.	http://www.clustal.org/clustal2/
Minitab, 2023	Windows	Phylogenetic relationship can be determined.	https://www.minitab.com/en-us/

(continued)

Table 20.1 (continued)

Software name	Operating system	Description	Hyperlink
Microarray probe	Windows, MacOS, and UNIX	The statistical tool is used for finding the differences arising between numerous genetic resources.	https://pubmed.ncbi.nlm.nih.gov/30252889/
AliGROOVE	Windows, MacOS, and UNIX	Meditation of heterogeneous classification deviation within multiple sequence arrangements and discovery of inflated division support.	https://en.wikipedia.org/wiki/List_of_phylogenetics_software
APE	Windows, MacOS, and UNIX	Investigations of phylogenetic and evolutionary relationships in the plants.	https://en.wikipedia.org/wiki/List_of_phylogenetics_software
LisBeth	Windows, MacOS, and UNIX	Evaluation of phylogenetic and evolutionary relationships in the plants.	http://www.infosyslab.fr/?q=en/resources/software/lisbeth

eggplant cultivars using 100 ISSRs, to assess the evolutionary relationships and identify cultivars. Thus, to evaluate the genetic diversity of vegetables for the goal of improving crops, a huge set of biotechnological tools can be utilized.

20.3.5 Tissue Culture, Micro-propagation, and Somatic Hybridization

To produce plants that are genetically true to type, meristem explant cultures use somatic meristematic tissues as explants. Planting thc buds, or eyes, that are found on the tubers is the main method of potato propagation. The new plants are similar to their parents due to vegetative propagation. To produce sufficient amounts of potato "seed" for commercialization, numerous rounds of seed increase are required because one tuber can only produce a few young plants, one from each eye. So, traditional propagation methods are quite slow and expensive requiring long duration (Butt et al. 2015).

A bio-technique known as micro-propagation, which involves producing a large number of identical plants from tiny buds of the parent plant, can be used to produce disease-free plants or to eliminate pathogens from the progeny plants while still maintaining the benefits of vegetative propagation. Similar to this, a technique called cell culture can be used to separate, multiply, and regenerate plant cells into complete plants. Through this technique, it is possible to produce thousands of disease-free replicas of a single plant. These technologies are useful for plant reproduction and are crucial to more sophisticated approaches of genetic manipulation (Basarygina et al. 2014).

Somatic hybridization is a technique to make hybrids between distantly related plants. Somatic hybrids and cytoplasmic hybrids (cybrids) are produced by using protoplast fusion. Another culture makes use of pollen as explants. It is cheap and less time-consuming to generate many haploid plants using anther culture (Scholze et al. 2010; Irikova et al. 2011). Inbred homozygous lines have been created through anther culture in many vegetable crops, i.e., onion, carrot, cabbage, and cauliflower (Cao et al. 1994). By using PEG-induced symmetric and asymmetric protoplast fusions, somatic hybrids have been created in order to transfer resistance to *Alternaria brassicicola*, *A. brassicae*, *Phoma lingam*, *Plasmodiophora brassicae,* and *Turnip mosaic virus* (TuMV) into *Brassica oleracea* var. *capitata* (cv. "Toskama") and var. *botrytis* (cv. "Korso") (Scholze et al. 2010).

20.4 Conclusions and Future Concerns

The world population is going to increase and food demand is increasing day by day. Biotechnological implications are essential for producing vegetable crops on a sustainable basis. The development of high-yielding and nutritionally rich germplasm is the need of time globally. The use of nutraceuticals as antioxidants for the management of malnutrition as well as the prevention and treatment of diseases have attained high attention in recent times. In conclusion, biotechnology also brings a revolution in the production of vegetables by developing tolerant genotypes against harsh climatic conditions. In the field of vegetable science, the more interdisciplinary effort is needed to evolve new, highly productive, and fast-growing genotypes with high nutritional value.

References

Aftab T, Roychoudhury A (2021) Crosstalk among plant growth regulators and signaling molecules during biotic and abiotic stresses: molecular responses and signaling pathways. Plant Cell Rep 40:2017–2019

Aghdam MS, Sevillano L et al (2015) The contribution of biotechnology to improving postharvest chilling tolerance in fruits and vegetables using heat-shock proteins. J Agric Sci 153:7–24

Agregán R, Echegaray N et al (2021) Proteomic advances in cereal and vegetable crops. Molecules 26:4924

Ahmad R, Anjum MA (2018) Applications of molecular markers to assess genetic diversity in vegetable and ornamental crops – a review. J Hortic Sci Technol 1:1–7

Ahmad R, Anjum MA (2023) Mineral nutrition management in fruit trees under salt stress: a review. Erwerbs-obstbau 65:397–405

Ahmad R, Anjum MA, Balal RM (2020) From markers to genome-based breeding in horticultural crops: an overview. Phyton 89:183

Ahmad R, Manzoor M et al (2023) Exogenous melatonin spray enhances salinity tolerance in Zizyphus germplasm: a brief theory. Life 13:493

Ahmar S, Gill RA et al (2020) Conventional and molecular techniques from simple breeding to speed breeding in crop plants: recent advances and future outlook. Int J Mol Sci 21:2590

Aleem S, Sharif I et al (2020) Heat tolerance in vegetables in the current genomic era: an overview. Plant Growth Regul 92:497–516

Al-Khayri JM, Jain SM, Johnson DV (2015) Breeding strategies. In: Jain SM, Johnson DV (eds) Al-Khayri JM. Springer, Advances in plant breeding strategies: fruits, pp 113–114

Bartlett ME, Moyers BT et al (2022) The power and perils of de novo domestication using genome editing. Annu Rev Plant Biol 74:113–121

Basarygina EM, Panova RI et al (2014) Environmentally sound plant protection in hydroponic vegetable production. Central Sci Agric Libr 67:5–8

Bornet B, Goraguer F, Joly G, Branchard M (2002) Genetic diversity in European and Argentinian cultivated potatoes (*Solanum tuberosum* subsp. *tuberosum*) detected by inter-simple sequence repeats (ISSRs). Genome 45:481–484

Boyd LA, Ridout C et al (2013) Plant-pathogen interactions: disease resistance in modern agriculture. Trends Genet 29:233–240

Butt SJ, Varis S et al (2015) Micro propagation in advanced vegetable production: a review. Adv Life Sci 2:48–57

Cao MQ, Li Y et al (1994) Application of anther-culture and isolated microspore culture to vegetable crop improvement. Hortic Sci Biotechnol 392:27–38

Caruso M, Smith MW et al (2020) Traditional breeding. In: Talon M, Caruso M, Gmitter FG Jr (eds) The genus citrus. Woodhead Publishing, pp 129–148

Chen J, Wang H et al (2009) Genetic variation in tomato populations from four breeding programs revealed by single nucleotide polymorphism and simple sequence repeat markers. Sci Hortic 122:6–16

Dangi AK et al (2018) Combinatorial interactions of biotic and abiotic stresses in plants and their molecular mechanisms: systems biology approach. Mol Biotechnol 60:636–650

Dar AA, Mahajan R, Lay P, Sharma S (2017) Genetic diversity and population structure of *Cucumis sativus* L by using SSR markers. 3 Biotech 7:307

Devran Z, Söğüt MA (2010) Occurrence of virulent root-knot nematode populations on tomatoes bearing the Mi gene in protected vegetable-growing areas of Turkey. Phytoparasitica 38:245–251

Devran Z, Söğüt MA (2014) Response of heat-stable tomato genotypes to Mi-1 virulent root-knot nematode populations. Türkiye Entomoloji Dergisi 38:229–238

Dietz-Pfeilstetter A (2010) Stability of transgene expression as a challenge for genetic engineering. Plant Sci 179:164–167

Djian-Caporalino C et al (2011) The reproductive potential of the root-knot nematode *Meloidogyne incognita* is affected by selection for virulence against major resistance genes from tomato and pepper. Eur J Plant Pathol 131:431–440

Eede VDG, Aarts H et al (2004) The relevance of gene transfer to the safety of food and feed derived from genetically modified (GM) plants. Food Chem Toxicol 42:1127–1156

Fassuliotis G, Bhatt DP (1982) Potential of tissue culture for breeding root-knot nematode resistance into vegetables. J Nematol 14:10–14

Fita A, Rodríguez-Burruezo A et al (2015) Breeding and domesticating crops adapted to drought and salinity: a new paradigm for increasing food production. Front Plant Sci 6:978

Frary A, Xu Y, Liu J et al (2005) Development of a set of PCR-based anchor markers encompassing the tomato genome and evaluation of their usefulness for genetics and breeding experiments. Theor Appl Genet 111:291–312

Gaba VP (2005) Plant growth regulators in plant tissue culture and development. In: Trigiano RN, Gray DJ (eds) Plant development and biotechnology. CRC Press, Boca Raton, pp 87–99

Ghislain M, Spooner DM, Rodriguez F et al (2004) Selection of highly informative and user-friendly microsatellites (SSRs) for genotyping of cultivated potato. Theor Appl Genet 5:881–890

Gichuki ST, Berenyi M et al (2003) Genetic diversity in sweet potato [*Ipomoea batatas* (L.) Lam.] in relationship to geographic sources as assessed with RAPD markers. Genet Res Crop Evol 4:429–437

Golldack D, Lüking I, Yang O (2011) Plant tolerance to drought and salinity: stress regulating transcription factors and their functional significance in the cellular transcriptional network. Plant Cell Rep 30:1383–1391

Goulao L, Oliveira CM (2001) Molecular characterization of cultivars of apple (Malus × domestica Borkh.) using microsatellite (SSR and ISSR) markers. Euphytica 122:81–89

Govindaraj M, Vetriventhan M, Srinivasan M (2015) Importance of genetic diversity assessment in crop plants and its recent advances: an overview of its analytical perspectives. Genet Res Int 2015:431487

Govindarajulu M et al (2015) Host-induced gene silencing inhibits the biotrophic pathogen causing downy mildew of lettuce. Plant Biotechnol J 13:875–883

Grout BWW, Crisp PC (1995) Cryopreservation of germplasm of tomato. In: Bajaj YPS (ed) Cryopreservation of plant germplasm I – biotechnology in agriculture and forestry. Springer, Berlin/Heidelberg, pp 371–380

Hale AL, Farnham MW, Menz MA (2006) Effectiveness of PCR-based markers for differentiating elite broccoli inbreds. J Am Soc Hortic Sci 131:418–423

Hou H, Atlihan N, Lu ZX (2014) New biotechnology enhances the application of cisgenesis in plant breeding. Front Plant Sci 5:389

Huibers RP, Loonen AEHM et al (2013) Powdery mildew resistance in tomato by impairment of SlPMR4 and SlDMR1. PLoS One 8:e67467

Irikova T, Grozeva S, Rodeva V (2011) Anther-culture in pepper (*Capsicum annuum* L.) in vitro. Acta Physiol Plant 33:1559–1570

Isshiki S, Iwata N, Khan MR (2008) ISSR variations in eggplant (*Solanum melongena* L.) and related *Solanum* species. Sci Hortic 117:186–190

Keller ERJ (2002) Cryopreservation of *Allium sativum* L. (garlic). In: Towill LE, Bajaj YPS (eds) Cryopreservation of plant germplasm II. Biotechnology in agriculture and forestry. Springer, Berlin/Heidelberg, pp 37–47

Khan N, Bano A, Babar MA (2019) Metabolic and physiological changes induced by plant growth regulators and plant growth promoting rhizobacteria and their impact on drought tolerance in *Cicer arietinum* L. PLoS One 14:e0213040

Koinange EM, Singh SP, Gepts P (1996) Genetic control of the domestication syndrome in common bean. Crop Sci 36:1037–1045

Kok DTM, de Waard P et al (2010) Antioxidative and antigenotoxic properties of vegetables and dietary phytochemicals: the value of genomics biomarkers in molecular epidemiology. Mol Nutr Food Res 54:208–217

König A (2003) A framework for designing transgenic crops—science, safety and citizen's concerns. Nat Biotechnol 21:274–279

Kumar P, Srivastava DK (2016) Biotechnological advancement in genetic improvement of broccoli (*Brassica oleracea* L. var. *italica*), an important vegetable crop. Biotechnol Lett 38:1049–1063

Kumar P, Srivastava DK (2020) Molecular and biotechnological interventions for improving brassicaceae crops for abiotic stress tolerance. In: Hasanuzzaman M (ed) The plant family brassicaceae. Springer, Singapore, pp 437–450

Kumar S, Misra A et al (2011) Bt Brinjal in India: a long way to go. GM Crops 2:92–98

Lian Q, Fu Q, Xu Y, Hu Z, Zheng J, Zhang A, He Y, Wang C, Xu C, Chen B, Garcia-Mas J (2021) QTLs and candidate genes analyses for fruit size under domestication and differentiation in melon (*Cucumis melo* L.) based on high resolution maps. BMC Plant Biol 21:1–13

Lindbo JA, Falk BW (2017) The impact of "coat protein-mediated virus resistance" in applied plant pathology and basic research. Phytopathology 107:624–634

Louarn S, Torp AM, Holme IB, Andersen SB, Jensen BD (2007) Database derived microsatellite markers (SSRs) for cultivar differentiation in *Brassica oleracea*. Genet Res Crop Evol 54:1717–1725

Lv J, Qi J et al (2012) Genetic diversity and population structure of cucumber (*Cucumis sativus* L.). PLoS One 7:e46919

Machado RMA, Serralheiro RP (2017) Soil salinity: effect on vegetable crop growth. Management practices to prevent and mitigate soil salinization. Horticulturae 3(2):30

Maheswari R, Ashok KR et al (2008) Precision farming technology, adoption decisions and productivity of vegetables in resource-poor environments. Agr Econ Res Rev 21:415–424

Miller AJ, Gross BL (2011) From forest to field: perennial fruit crop domestication. Am J Bot 98:1389–1414

Moulin MM, Rodrigues R et al (2012) A comparison of RAPD and ISSR markers reveals genetic diversity among sweet potato landraces (*Ipomoea* batatas (L.) Lam.). Acta Sci Agron 34:139–147

Mushegian AR, Shepherd RJ (1995) Genetic elements of plant viruses as tools for genetic engineering. Microbiol Rev 59:548–578

Paran I, Michelmore RW (1993) Development of reliable PCR-based markers linked to downy mildew resistance genes in lettuce. Theor Appl Genet 85:985–993

Park G, Sim SC et al (2021) Development of genome-wide single nucleotide polymorphism markers for variety identification of F1 hybrids in cucumber (*Cucumis sativus* L.). Sci Hortic 285:110173

Pereira L, Ruggieri V et al (2018) QTL mapping of melon fruit quality traits using a high-density GBS-based genetic map. BMC Plant Biol 18:1–17

Pouvreau L et al (2001) Relative abundance and inhibitory distribution of protease inhibitors in potato juice from cv. Elkana. J Agric Food Chem 49:2864–2874

Qiang LI, Li XY et al (2009) Selection of parents for breeding edible varieties of sweet potato with high carotene content. Agric Sci China 8:1166–1173

Rai VP, Kumar R et al (2013) Genetic diversity in capsicum germplasm based on microsatellite and random amplified microsatellite polymorphism markers. Physiol Mol Biol Plants 19:575–586

Raker CM, Spooner DM (2002) Chilean tetraploid cultivated potato, *Solanum tuberosum*, is distinct from the Andean populations. Crop Sci 42:1451–1458

Rao KM, Singh PK et al (2014) Use of bio-fertilizers in vegetable production. Indian Hortic J 4:73–76

Rejeb BI, Pastor V, Mauch-Mani B (2014) Plant responses to simultaneous biotic and abiotic stress: molecular mechanisms. Plan Theory 3:458–475

Rommens CM (2007) Intragenic crop improvement: combining the benefits of traditional breeding and genetic engineering. J Agric Food Chem 55:4281–4288

Rommens CM, Haring MA et al (2007) The intragenic approach as a new extension to traditional plant breeding. Trends Plant Sci 12:397–403

Ruta C, Lambardi M, Ozudogru EA (2020) Biobanking of vegetable genetic resources by in vitro conservation and cryopreservation. Biodivers Conserv 29:3495–3532

Saito K, Matsuda F (2010) Metabolomics for functional genomics, systems biology, and biotechnology. Annu Rev Plant Biol 61:463–489

Sarker U, Islam MT, Rabbani MG, Oba S (2017) Genotypic diversity in vegetable amaranth for antioxidant, nutrient and agronomic traits. Indian J Genet Plant Breed 77:173–176

Sarker U, Islam MT, Rabbani MG, Oba S (2018) Phenotypic divergence in vegetable amaranth for total antioxidant capacity, antioxidant profile, dietary fiber, nutritional and agronomic traits. Acta Agric Scand 1:67–76

Schaart JG, van de Wiel CCM et al (2016) Opportunities for products of new plant breeding techniques. Trends Plant Sci 21:438–449

Scholze P, Krämer R et al (2010) Somatic hybrids of vegetable brassicas as source for new resistances to fungal and virus diseases. Euphytica 176:1–14

Shahzad S, Ali S et al (2021) Foliar application of silicon enhances growth, flower yield, quality and postharvest life of tuberose (*Polianthes tuberosa* L.) under saline conditions by improving antioxidant defense mechanism. SILICON 14:1511–1518

Shelton AM (2012) Genetically engineered vegetables expressing proteins from *Bacillus thuringiensis* for insect resistance: successes, disappointments, challenges and ways to move forward. GM Crops Food 3:175–183

Simko I, Jia M et al (2021) Genomics and marker-assisted improvement of vegetable crops. Crit Rev Plant Sci 40:303–365

Singh D, Dhillon T et al (2022) Exploring the genetic diversity of carrot genotypes through phenotypically and genetically detailed germplasm collection. Agronomy 12:1921

Smýkal P et al (2018) The impact of genetic changes during crop domestication. Agronomy 8:119

Sun Q, Zhang B et al (2016) Comparative analysis on the distribution of protease activities among fruits and vegetable resources. Food Chem 213:708–713

Thrall PH, Oakeshott JG et al (2011) Evolution in agriculture: the application of evolutionary approaches to the management of biotic interactions in agro-ecosystems. Evol Appl 4:200–215

Wissuwa M et al (2009) Novel approaches in plant breeding for rhizosphere-related traits. Plant Soil 321:409–416

Yadav AN (2020) Plant microbiomes for sustainable agriculture: current research and future challenges. In: Yadav AN, Singh J, Rastegari AA, Yadav N (eds) Plant microbiomes for sustainable agriculture. Sustainable development and biodiversity. Springer, Cham, pp 475–482

Yadav UP, Ayre BG, Bush DR (2015) Transgenic approaches to altering carbon and nitrogen partitioning in whole plants: assessing the potential to improve crop yields and nutritional quality. Front Plant Sci 6:275

Yuan XJ, Pan JS et al (2008) Genetic mapping and QTL analysis of fruit and flower related traits in cucumber (*Cucumis sativus* L.) using recombinant inbred lines. Euphytica 164:473–491

Zhuang J, Zhang J et al (2014) Transcriptomic, proteomic, metabolomic and functional genomic approaches for the study of abiotic stress in vegetable crops. Crit Rev Plant Sci 33:225–237

Zhu-Salzman K et al (2003) Cowpea bruchid *Callosobruchus maculatus* uses a three-component strategy to overcome a plant defensive cysteine protease inhibitor. Insect Mol Biol 12:135–145

Hybrid Breeding in Vegetable Crops

21

Saeed Rauf, Taiyyibah Basharat, and Zafar Iqbal

Abstract

Vegetables are known for their nutritional and medicinal value, and are widely cultivated to meet the food demands globally. Availability of high-quality seeds, with potential to give higher yields to the growers, is the basic necessity for profitable olericulture in all parts of the world. This chapter reviews the progress of hybrid seed development in various vegetable crops. Several factors contributing to higher herbage and fruit or seed yield are discussed including the manipulation of heterosis, hybrid vigor, and utilization of male sterility. Recurrent selection method for population improvement program which may eventually generate breeding lines with superior combining ability is also described in the chapter. Overall, this chapter emphasizes the need of manipulation of hybrid vigor to improve the economical yield in vegetable crops.

Keywords

Combining ability · Genetic male sterility · Heterosis · Inbred lines · Molecular markers · Recurrent selection

S. Rauf (✉) · T. Basharat
Department of Plant Breeding & Genetics, College of Agriculture, University of Sargodha, Sargodha, Pakistan
e-mail: saeedbreeder@hotmail.com

Z. Iqbal
Faculty of Agriculture, University of Sargodha, Sargodha, Pakistan

Z. Iqbal et al. (eds.), *Sustainable and Innovative Vegetable Production in times of Climate Change*, https://doi.org/10.1007/978-981-96-6283-8_21

21.1 Introduction

Vegetables belong to a diverse group of herbaceous plants, which are consumed raw or after cooking on a daily basis and constitute an integral part of the human diet due to a good source of daily calories, vitamins, and antioxidants. There are plants of about 35–40 genera that are cultivated on a global scale for human foods (Kapur 2013). Demand for vegetables as a cheaper source of the human diet is ever increasing due to the continuous increase in global population, which ultimately exerts pressure on major vegetable production zones of the world. Moreover, climate changes are further challenging global annual production, threatening sustainable supply to meet the ever-increasing demand for vegetables. Recent advances in vegetable raising such as vegetable grafting, vertical gardening, and horizontal greenhouse production have improved the supply of various seasonal and non-seasonal vegetable crops. However, a large proportion of vegetable production is still obtained in developing countries from open farms. Therefore, there is a constant need to breed varieties with improved and uniform production potential and high resilience to environment throughout the growth season. Vegetable seed industry has shown about 10 times growth in the last 40 years. The size of the global vegetable seed market reached USD 8.14 billion in 2023 (Fortune 2024).

Vegetable production and resilience to climate changes can be improved through the exploitation of heterosis or hybrid vigor in F_1 generation (hybrid). Development of hybrid cultivars with improved and uniform marketable yield is one of the prime objectives of vegetable breeding. Hybrid cultivars have been considered superior to the traditional ones due to a greater degree of hybrid vigor. Hybrid is a cross between two or more inbred lines or populations and designated as F_1 generation. Hybrids are characterized as inbred hybrids, population hybrids, and single or double cross hybrids depending upon the degree of homozygosity of parents and the number of parental lines used in each hybrid. Parental lines or populations are carefully selected through mating schemes, and less commonly through molecular markers (Kawamura et al. 2017). Generally, parental lines which are genetically distinct and good combiner produce superior hybrids. Over dominance type of gene action is required for commercial exploitation of heterosis. Superiority of F_1 hybrids over its parental genotypes is unknown but F_1 hybrids are heterozygous having a broader genetic base.

Despite its wide application in various crop species including vegetables, the term heterosis was originally donated by Shull in 1914 (Shull 1948). The effects of self and cross-pollination in the vegetable kingdom were described by Darwin. Heterosis was discovered in maize and later on exploited and applied in numerous species irrespective of the mode of pollination or genetic background. Hybrid cultivars in vegetable crops such as peas have higher pod yield and number of pods plant^{-1}, and lesser days to first harvest (Sharma and Bora 2013). Benefits of exploitation of heterosis to achieve diversified breeding objectives are mentioned in Table 21.1.

Table 21.1 Manipulation of heterosis in various vegetable crops to achieve diversified breeding objectives

Vegetable crop	Heterosis manipulation	References
Peas (*Pisum sativum*)	Seed yield plant^{-1} (11 to 145%), number of seeds pod^{-1} (−10 to 34), pod mass plant^{-1} (−24 to 54%)	Askander and Osman (2018)
Okra (*Abelmoschus esculentus*)	Fruit weight (23 to 54%), number of pods plant^{-1} (19 to 64%), fruit yield plant^{-1} (62 to 95%)	Amutha et al. (2007)
Tomato (*Lycopersicon esculentum*)	Number of fruits plant^{-1} (−53 to +59%), fruit diameter (−43 to +42%), fruit yield plant^{-1} (−56 to +30%)	Yadav et al. (2013)
Chili pepper (*Capsicum annuum*)	Fruits per plant (17 to 103%), fruit weight (10 to 58%), green fruit yield (29 to 143%).	Aiswarya et al. (2020)
Eggplant (*Solanum melongena*)	Fruit size (22 to 46%), weight (20%), fruit yield (52–62%), days to first picking (0.25%)	Kumar et al. (2020) Mistry et al. (2018)
Cauliflower (*Brassica oleracea* var. *botrytis*)	Positive heterosis was noted for various pigments and molecules such as ascorbic acid (24%), anthocyanins (0.25%), and carotenoids (32%)	Dey et al. (2014)
Cucumber (*Cucumis sativus*)	Fruit length (−42 to 30%), diameter (−38 to 27%), days to first male or female flower (−18%), fruit weight (−60 to 136%), number of fruit vine^{-1} (−63 to 63%)	Singh et al. (2015)
Cucumber (*Cucumis sativus*)	Number of fruits per plant (48 to 77%), marketable yield (37 to 148%)	Kumar et al. (2017)
Pumpkin (*Cucurbita moschata*)	Flesh thickness, total carotenoids, and ascorbic acid content	Pandey et al. (2010)

21.1.1 Heterosis and Hybrid Vigor

Heterosis is a biometrical term used to calculate the superiority of F_1 hybrids over the mid parental value or superior parents. It is called as mid parental heterosis which is calculated as:

$$\text{Heterosis} = \left(F_1 - MP\right) / MP,$$

where MP is the mid parental value $(P_1 + P_2)/2$. Economical heterosis is calculated by the formula = $(F_1\text{-}BC)/BC \times 100$, where BC = best commercial cultivar. Hybrids are generally recommended for cultivation when they show significant economical heterosis across a range of environment. Contrastingly, inbreeding depression is calculated as:

$$\text{Inbreeding depression}(\%) = \left(F_1 - F_2\right) \times 100 / F_1$$

where F_1 is the mean performance of the F_1 hybrid and F_2 is the mean performance of F_2 population.

Hybrid vigor is a botanical term used to donate the vigorous performance of F_1 hybrids for growth, development, and yield attributes. Heterosis or hybrid vigor may be required in negative terms when traits such as early maturity and dwarf growth habits are sought or required.

Several studies indicated the use of genetic distance between parental lines as a predictor of heterosis (Kawamura et al. 2016). For instance, in Chinese cabbage (*Brassica rapa*), CAPS (cleaved amplified polymorphic sequences) and SSR (simple sequence repeat) markers were used to estimate the genetic distance between 32 selected inbred lines. However, molecular markers-based genetic distance was not significantly correlated to mid or best parental heterosis. In contrast, leaf size at 21 days after sowing was moderately correlated with mid parent heterosis but not with best parent heterosis. In cabbage, heterosis was associated with the specific combining ability of inbred lines (Dey et al. 2014). Significant heterosis was estimated for total and marketable yield in carrots. Molecular genetic distance calculated by AFLP (amplified fragment length polymorphism) and RAPD (random amplified polymorphic DNA) was significantly related to the total yield of carrot hybrids (Jagosz 2011). A genetic distance of 0.056–0.179 was noted among inbred lines of wax gourd {*Benincasa hispida* (Thunb.) Cogn.}, which correlated with hybrid performance or heterosis but was not useful enough to predict hybrid performance. Hybrid showed significant heterosis for yield and related traits including zinc and iron contents in wax gourd (Sureja et al. 2006). Similarly, in brinjal (*Solanum melongena*), genetic distance was correlated with heterosis of fruit yield plant^{-1} but regression analysis failed to predict performance of hybrid through genetic distance between parental lines (Hazra et al. 2010). Heterosis was significantly related to non-additive gene action estimated in 120 hybrids obtained by crossing 20 *Ogura*-based cytoplasmic male sterile (CMS) lines and 06 doubled haploid (DH) testers. Genetic distance based on SSR and expressed sequence tags (EST) also showed some relationship with heterotic effects (Singh et al. 2019). The magnitude of heterosis in various vegetable crops is mentioned in Table 21.1, while the characterization of vegetable crops on the basis of inbreeding depression is given in Table 21.2.

21.1.2 Genetic Factors Contributing to Heterosis

Exact genetic mechanism of heterosis induction in F_1 is not known. However, several theories such as dominance, over dominance, and epistasis hypothesis are well explained. Heterosis may be due to the presence of dominant alleles which may have masking effects on deleterious recessive alleles. It may also be due to the increased degree of heterozygosity which may lead toward diversified alleles' products across environment and thus hybrids show superior performance than parents. Superiority of hybrids may result from inter-allelic or intergenic interaction known as epistatic interaction. Quantitative trait locus (QTL) mapping has identified regions which are associated with the heterosis or hybrids. Moreover, transcriptomic analysis has led to the identification of genes related to the heterosis. Epigenomic changes in the genome of inter or intraspecific hybrids have also been observed which may reduce the genetic load of the hybrids (Fujimoto et al. 2018).

Table 21.2 Characterization of vegetable crops with respect to inbreeding depression

Vegetable crop	Pollination	Degree of inbreeding depression	References
Pea (*Pisum sativum*)	Self-pollinated	40 to 52% (medium)	Sood et al. (2006)
Okra (*Abelmoschus esculentus*)	Self-pollinated	-1 to 40% (medium)	Amutha et al. (2007)
Potato (*Solanum tuberosum*)	Self-pollinated	Severe	Zhang et al. (2019)
Tomato (*Lycopersicon esculentum*)	Self-pollinated	−6 to 12% (low)	Kumar and Singh (2016)
Eggplant (*Solanum melongena*)	Self-pollinated	−36 to 1.59% (low)	Mistry et al. (2018)
Onion (*Allium cepa*)	Cross-pollinated	Bulb weight (18%), S1 (single generation) (high)	Porta et al. (2014)
Leek (*Allium ampeloprasum* ssp. *porrum*)	Cross-pollinated	Severe	Smith and Crowther (1995)
Cucumber (*Cucumis sativus*)	Cross-pollinated	32 to 86% (high)	Singh et al. (2015)

Inbreeding depression is a term which arises due to loss of hybrid vigor in F_2 and subsequent generations. It is defined as loss of fitness in the progeny when compared with their parents (Zhang et al. 2019). Inbreeding depression varies with crop species and it results due to uncovering of deleterious alleles in segregating generation (Table 21.2). However, the intensity of inbreeding depression decreased in subsequent generation like it decreased from 45% in S_1 to −6% in S_4 generation in summer squashes. Genomic causes of the inbreeding depression have been the subject of many publications (Paige 2010; Zhang et al. 2019).

In potato, 15 genomic regions were identified with severe segregation distortion. The deleterious recessive alleles which induce inbreeding depression were related to the fertility of plants and located in genomic region which undergoes high recombination frequency. One of the recessive alleles uncovered in inbred lines was related to the disruption of embryo development (Zhang et al. 2019).

21.2 Basic Requirements for Hybrid Breeding: Botanical and Morphological Features

The success and performance of hybrids are dependent on genetic, botanical, and agronomic factors. Genetic factors include the magnitude of commercial heterosis for particular cross combination. All crosses may not show high degree of heterosis to surpass its parents and successful elite cultivars. Crosses showing high degree of heterosis also had high degree of over dominance. Over dominance type of gene action is the only case where hybrids are recommended over the open pollinated

Table 21.3 Fertility status and hybrid seed yield in various vegetables species

Species	Fruit yield (g)	No. of seed fruit^{-1}	Factors	References
Cucurbita maxima × *Cucurbita moschata*	26 to 145	116 to 365	Interspecific hybridization	Karaağaç and Balkaya (2013)
Lycopersicon esculentum	0.11 to 0.14	41 to 53	Line × tester	Yesmin et al. (2014)
Momordica charantia	9	47	Insect proof net house vs. open field	Jat et al. (2016)
Abelmoschus esculentus	3.5	61	Timing of pollination	Mallikarjun (2006)

cultivars. Heterosis is also dependent over specific combining ability of parental lines and to a lesser extent to genetic divergence between breeding lines. Hybrids utilize both additive and non-additive gene actions during performance. Inbred lines are based on additive gene action where traits such as days to flowering, plant height, fruit length, and quality are fixed through pure line selection and non-additive gene action is used to maximize the heterosis or hybrid vigor (Saleem et al. 2009). Phenological synchronization between male and female lines is key for the success of hybrids and dictates the amount of seed obtained from a particular cross. In onion, it was noted that male lines were earlier than CMS lines by 12–25 days, moreover, duration of pollen availability was also less. Therefore, pollen availability was a limited factor in hybrid seed production (Padmini et al. 2007). Delayed sowing of male lines for about a week resulted in synchronized flowering of both male and female lines. If the two parental lines do not have synchronized flowering, they may not be able to yield large amount of seed, thus increasing the cost of hybrid seed. Storage of onion bulbs at 9–14 °C could also yield synchronized flowering of male and female lines (Gupta and Singh 2016). Spraying of gibberellic acid on plants may also cause synchronous flowering. Moreover, fertility of parental lines and number of seeds per fruit are also important factors, which finally affect the yield of hybrid seed. Loss of fertility over several generation of inbreeding is a limiting factor for successful maintenance of female and male lines. Moreover, the amount of seed obtained per cross also helps to determine the success of hybrid seed production on a commercial scale (Table 21.3).

21.3 Development of Inbred Lines

Inbred lines are generally developed by pure line selection methods such as the pedigree method. Plant progenies are grown and among them poor progenies are discarded. Best plants within the best progenies are self-pollinated to induce homozygosity. The pedigree record of selected plants is maintained. This process continues for up to six generations.

Sometimes early combining ability test is conducted by crossing some florets within a single plant to a weak tester. Plants with good general combining ability are further selected as candidate for the inbred lines. High degree of homozygosity is achieved in plants with lesser inbreeding depression upon self-pollination, while it cannot be achieved in populations with higher degree of inbreeding depression (Table 21.2). Reciprocal recurrent selection has been recommended to improve the performance of specific combining ability and expand the magnitude of heterosis (Gwanama et al. 2001).

Doubled haploid plants can be regenerated through in vitro techniques in plant species where regeneration protocols have been optimized (Bhatia et al. 2018). Haploid tissues such as immature anthers or ovules carrying unfertilized eggs are dissected from mother tissues and inoculated over the suitable growth medium to regenerate haploid plants which are further treated with anti-tubulin chemicals such as colchicine to develop doubled haploid plants. In cabbage (*Brassica oleracea* var. *capitata*), several factors including immature bud size (4.5 to 5.0 μm), temperature shock ranging between 25 and 32.5 °C for 48 h, and uni- to late bi-nucleate stage, had the highest success for induction of haploid microspore culture. There was positive relationship between the viability of microspores and percentage of haploid plants. The highest viability of microspores was induced in immature bud of 4.0 μm, whereas microspore density of 4×10^4 mL had the highest haploid embryo induction rate. Flow cytometry showed that 42% were diploid spontaneously, while 29% were haploid followed by 11% tetraploid (Bhatia et al. 2018). Selection of buds pre-treated at 4 °C for 1 day, with 50% late uni-nucleate microspores, was considered optimum to produce 4–7 haploid plants per bud in hot pepper (*Capsicum annuum*) (Supena et al. 2006). Doubled haploid plants can be candidate for inbred lines. In some species, haploid inducer lines have been developed which can successfully induce haploid seeds and may be treated to develop doubled haploid lines. In vitro technologies such as somatic cell hybridization have also been recommended to widen the genetic variation and develop crosses between diverse species. Some examples are presented in Table 21.4 while examples on interspecific hybrids are presented in Table 21.5.

21.3.1 Recurrent Selection: A Tool to Remove Specific Defects in Inbred Lines

Recurrent selection has been recommended to improve combining ability and specific defects in inbred lines. Recurrent selection for specific combining ability is made by self-pollinating some selected plants and crossing few florets to an inbred line which is used as tester. Half sib progenies are grown and evaluated. Plants are selected on the basis of performance of half sib progenies. Selected plants are crossed in all possible combinations to develop another base population for another cycle of selection. Reciprocal recurrent selection is performed to improve both general and specific combining ability. In this way, two base populations are developed which also act as testers for each other. Reciprocal full sib selection was performed in base populations obtained in *Cucumis sativus* for fruit size and fruit number plant^{-1}. However, single selection cycle did not show substantial improvement in

Table 21.4 Somatic cell hybridization for development of wide hybrids

Somatic hybrids	Traits	Identification markers	References
Panax ginseng × *Daucus carrota*	Roots and medicinal value	Morphology, chromosome count, isozyme, SDS-page	Yue-bing et al. (2012)
Brassica oleracea var. *capitata* (cabbage) × *Brassica campestris* ssp. *pekinensis*	Plant morphology intermediate between species	Genomic in situ hybridization	Lian et al. (2001)
Leaf mustard (*Brassica juncea*) × broccoli (*Brassica oleracea*)	Viable hybrid obtained	RAPD markers, CMS-based markers	Lian et al. (2011)
Brassica oleracea var. *botrytis* and *B. nigra*	DNA bands of both parents Recombination between mitochondria of two species Chloroplast DNA of *B. Nigra*	Sequence-related amplified polymorphism (SRAP), CAPS, and RFLPs	Wang et al. (2011)

Table 21.5 Interspecific hybrid among various vegetable species

Interspecific hybrids	Techniques	Traits	References
Chinese cabbage × broccoli	Embryo rescue and colchicine doubling	High nutritional value than both parents	Wei et al. (2018)
Cucurbita maxima × *C. moschata*	Interspecific hybridization	Early and compact, high frequency of pistillate flowers	Uretsky and Loy (2017)
Cucurbita pepo L. × sweet gourd (*C. moschata* Duch.)	Interspecific hybridization	High heterosis for fruit yield and quality	Davoodi et al. (2016)
Allium fistulosum × *Allium schoenoprasum*	Embryo rescue	07 times higher yield of edible parts than both parents and high carotenes.	Umehara et al. (2006)
Brassica napus × *B. campestris*	Hybridization	High heterosis for vegetative parts and may be utilized for vegetables and forage	Yun et al. (2003)
Brassica campestris × *Brassica oleracea* var. *acephala*	Hybridization	High heterosis for vegetative parts	Zhou et al. (2006)

both traits as compared to self-pollinated plants (S_1). It was concluded that reciprocal recurrent selection was not suitable for traits showing additive type of gene action but suitable for traits predominantly controlled by non-additive type of gene action such as epistasis or dominance (Nienhuis and Lower 1988). Particular type of gene action was the outcome of particular populations and environments. However, sampling errors may lead to erroneous conclusions.

Recurrent selection was practiced to reduce the pungency of onions to make suitable for salad purpose (Yoo et al. 2020). Base populations, i.e., "T81079" (n = 754) and "T81082" (n = 347) which had pungency determined by the pyruvic acid concentration of 4.1 and 3.5 mM, respectively were subjected to three recurrent selections. Bulbs of selected plants were divided into five groups based on their pungency level. Mean values of each selected group were shifted toward selection differentials and there was high relationship between selection differentials and response to selection for lower pungencies. Mean pungency level decreased to 1.5 mM after three selection cycles. It was concluded that selection for low pungency in onion was possible through 10 cycles of selection (Yoo et al. 2020).

Genotypic recurrent selection for traits such as fruit yield, earliness, and fruit shape was practiced in various populations of *Cucumis sativus* for 9 selection cycles. About 200 to 355 half sib progenies were evaluated in two replication blocks. Overall, selected populations showed an improvement of 37% over unselected populations, while the highest genetic gain was noted in days to first harvest which showed an improvement of 63% over unselected populations (Wehner and Cramer 1996). Phenotypic recurrent selection was practiced to improve disease resistance against *Phytophthora capsici* in pepper which had the polygenic basis of inheritance (Thabuis et al. 2004).

21.4 Factors Affecting Seed Quality

Seeds should be harvested upon the physiological maturity of fruits/pods. Upon maturity, pods may change their color specific to species. However, delayed harvesting may cause shattering of seed. GA_3 sprays have been recommended to improve bolting or alter the sex ratio in various vegetable crops. Produce is dried and cured under shade for sometimes and then seeds are threshed, cleaned, and again dried to remove the extra moisture recommended for specific crop species. In case of tomato, seeds are surrounded by mucilaginous material, which makes seed separation to be difficult. Therefore, seeds from mucilage are extracted by using fermentation, acid extraction, alkali extraction, or mechanical extraction methods. Moisture contents in tomato seed should not exceed 8%. Under optimum storage conditions of temperature and humidity, seed may retain viability for 2–3 years (Cheema and Dhaliwal 2005). In cabbage, pods turn yellowish brown and seeds become reddish brown at maturity. Parameters for commercial hybrid seed production have been presented in detail in Table 21.6.

21.5 Molecular Markers for Validation of Hybrid Seed

Morphological traits are used to differentiate between true to types and off types in grow out test as traditional method for genetic purity analysis of the hybrids. However, it is difficult, laborious, and time-consuming in most cases. On the other hand, molecular markers may also be exploited to estimate the genetic purity of the

Table 21.6 Commercial hybrid seed production of various vegetable crop species

Vegetable crops	Female: male ratio	Male sterility	Isolation distance	Pollinator	Seed yield	References
Onion	4:1, 8:2	CMS	100 m	8–12 hives ha^{-1}	400–600 kg ha^{-1}	Gupta and Singh (2016)
Cabbage	1:1, 2:2	GMS, SI	2000 m	15 bee hives ha^{-1}	–	Zhiyuan et al. (2000)
Radish	2:1, 3:1, 4:1	Ogura-CMS	1200 m	15–18 bee hives ha^{-1}	600–900 kg ha^{-1}	Gupta and Singh (2016)
Cole	1:1, 2;1	SI	1200 m	15–18 bee hives ha^{-1}	400–600 kg ha^{-1}	Gupta and Singh (2016)
Egg plant	4:1	Emasculation	200 m	Manual	100–120 kg ha^{-1}	Sidhu et al. (2005)
Chilies	2:1	GMS	400 m	5–8 bee hives ha^{-1}	200–250 kg ha^{-1}	Hundal and Dhall (2005)
Tomato	3:1	Hand emasculation, GMS	100 m	Manual	210–310 kg ha^{-1}	Cheema and Dhaliwal (2005)
Cauliflower	2:1	SI, CMS	Isolation cages	Bumble bees	1.8–2.3 g $plant^{-1}$	Kucera et al. (2006)
Cauliflower	3:1	SI, GMS	1000 m	Manual		Tripathi and Singh (2000)
Okra	8:1	Emasculation	400 m	Manual	1200–1500 kg ha^{-1}	Mallikarjun (2006)

hybrids. Co-dominant markers may be able to discriminate heterozygous or homozygous genotypes, thus both parents and F_1 hybrids may be confirmed through the use of specific markers (Table 21.7). Hybrids seed may also be contaminated through seed mixtures during seed handling. Therefore, molecular markers may be good tools to judge seed purity before marketing and probe seed adulteration after marketing, especially in developing countries where the quality of hybrid seeds may be compromised at various handling steps. Moreover, molecular markers are also recommended in handling court cases regarding the genetic purity of hybrid seed. Development of male and female specific markers requires screening of the large number of markers which may show specific bands for both parents and helpful to authenticate the genetic purity of hybrids. In *Brassica oleracea*, 325 SSR, ISSR, RAPD, and SRAP markers were screened, and only 7 primers showed male or female-specific markers. These markers showed 97% genetic purity in commercial hybrid "Sugan 21" (Ye et al. 2013).

Table 21.7 Some useful markers for the probing of the genetic purity of hybrids

Vegetable crops	Hybrids	Molecular markers	Genetic purity	References
Tomato (*Lycopersicon esculentum*)	Hezuo 903, and Sufen No. 8	NAURP1123 (a) NAURP403 (b) NAUSSRTO1 (a) NAUSSRTO2 NAUSSRTO22 (a) NAUSSRTOM47	96.2%, and 93.8%	Liu et al. (2007)
Tomato (*Lycopersicon esculentum*)	TMS1 × Money Maker	SSR-22	–	Hameed et al. (2017)
Eggplant (*Solanum melongena*)	PH-5 and DBHL-20	eme08D09 emg11I04	Both hybrids and their parents identified	Mangal et al. (2016)
Cabbage (*Brassica oleracea* var. *capitata*)	Sugan 21	NAUSSR11, NAUSSR45 and NAUSSR46	97.69%	Ye et al. (2013)
Cabbage (*Brassica oleracea* var. *capitata*)	35 commercial hybrids	PS5/15 and PS3/21 (RFLP primer set), HVR2-F/R (class-I SLG) and 60-F/40-R (class-II SP11)	Identified all 35 hybrids	Kawamura et al. (2017)
Cauliflower (*Brassica oleracea* var. *botrytis*	NBH Tania-815	BrgMS565	High accuracy after mixing 5% adulteration	Pattanaik et al. (2018)
Carrot (*Daucus carota*)	Experimental F1 hybrids	14 DcMTD markers	Potential for hybrid testing	Macko and Grzebelus (2008)

21.6 Methods of Male Sterility Induction

Male sterility is induced in female lines to avoid self-pollination or pollen contamination. Male sterility is characterized by the inability to produce functional pollens in lines to be exploited as female during hybrid seed production (Table 21.8). Male sterility may be induced through genetic or non-genetic procedures. Non-genetic procedures include manual removal of immature anthers or using physical factors such as hot or boiling water which may only damage the androecium before it gets mature. Male sterility may also be induced through various chemicals or growth regulators which may be sprayed over the flower and are generally referred as chemical hybridizing agents or gametocides. These biochemicals may enhance female parts or desiccate the anthers before getting matured. There is a long list of potential gametocides such as maleic hydrazide, gibberellins, ethephon, tri-iodobenzoic acid, naphthalene acetic acid, etc. Generally, these biochemical compounds are applied as aqueous spray at bud initiation stage. Some uncommon methods include injecting through micro syringe or dipping floral parts in the solution. An ideal gametocide may be able to induce 100% male sterility without affecting the female organs of a flower. Gametocides may be effective at different reproductive phases. Moreover, it should not induce undesirable effects on plant phenology and mutations. Further, the chemical should neither be a potential pollutant nor hazardous to human health.

Generally, chemical hybridization agent (CHA) affects the pollen production by disrupting the normal meiotic process in microsporocytes. It may also affect the gene encoding pole and pole determinants, resulting in the failure of spindle fiber formation leading to the arrest of meiotic division. CHAs are also known to hinder the exine differentiation due to disrupting of exine-related gene or affecting

Table 21.8 Utilization of various chemical hybridizing agents and their concentrations in various vegetable crops

Vegetable crops	Chemical hybridizing agents	References
Coriander (*Coriandrum sativum*)	Maleic hydrazide (125 mg L^{-1}) at 25 days after sowing	Kalidasu et al. (2009)
17 Cruciferous species	10 mL of 0.2 or 0.4 mg L^{-1} of tribenuron-methyl (repeated 10–12 times), first spray before vacuolated microspores	Yu et al. (2017)
Okra (*Abelmoschus esculentus*), tomato (*Lycopersicon esculentum*), and pepper (*Capsicum annuum*)	Detergent "surf excel" (1.0 and 1.5%)	Chauhan and Agnihotri (2005)
Rapeseed (*Brassica napus*)	0.1 μg mL^{-1} monosulfuron ester sodium (MES) solution, applied 15 mL $plant^{-1}$ at the bolting stage with flower bud <2 mm	Cheng et al. (2013)
Okra (*Abelmoschus esculentus*)	Maleic hydrazide (200 mg L^{-1}) at 20–40 days after sowing	Deepak (2007)

carbohydrate metabolism which starves the tapetum cell causing the arrest of meiotic divisions. Some CHAs also greatly affect DNA synthesis during S-phase and delay condensation of chromosome which hinders prophase-1 and reduces pollen production.

21.6.1 Genetic Male Sterility

The genetic male sterility may be characterized into two major groups, i.e., nuclear or cytoplasmic male sterility based on the organelles in which residing genes are targeted. Nuclear male sterility is recessive trait (ms ms) governed by two recessive alleles residing in the nucleus of a plant. Male sterility alleles have displayed a 1:3 ratio with male fertile plants in F_2 generation and considered to be inherited sporophytically. Male-sterile plants can be differentiated from segregating population at seedling stage by using linked genetic markers. In tomato, 3 types of functional male sterility have been reported including exserted stigma (*ex*) sterility which is induced by the long protruding stigma due to excessive length of style making self-pollination difficult because of unequal length of stigma and anthers. Interspecific hybridization has been done to *ex* type of functional male sterility (Atanassova 2000). Other type of functional sterility includes positional sterility (*ps*) which is induced due to the lateral growth of petals forming connate or psedu-connate, which causes considerable construction of anthers. One of the limitations is the requirement for the removal of anthers which may be done in single operation. On the other hand, in positional sterility gene 2 (*ps2*), the pollens are functional but anthers are indehiscent (Table 21.9). This type of functional male sterility was identified as spontaneous mutation in a Czech cultivar of tomato (Atanassova 2000). It was induced due to structural alteration in the anther dehiscence zone. This type of genetic sterility was not stable across a range of environments and may shed pollens to cause undesirable self-pollination.

Table 21.9 Genetic male sterility in various vegetable crops

Crops	Genes	Sources	References
Tomato	55 recessive genes, *ps*, *ps-2* (positional sterile) *ex* (exserted stigma sterility), ms 10, *7B-1* (photoperiod sensitive male sterility), $ms10^{35}$, SlSTR1 (stamen specific gene)	Yeast artificial chromosome, *Lycopersicon esculentum* × *Lycopersicon pimpinellifolium*, Spontaneous mutation, CRISPR/Cas9	Atanassova (2000)
Tomato	$BC_{10}P^2$,. *Lycopersicom peruvianum* × *L. pennellii*), its complex hybrids × *L. esculentum*	Cytoplasmic male sterility	Petrova et al. (1999)
Onion	S and T cytoplasms	Cytoplasmic male sterility	Havey (2000)

Tomato genetic male sterile plant carrying ms10^{35} alleles had dysfunctional meiosis and abnormal tapetum during anthers' development. Mutant plants encoded a helix-loop-helix transcript, while transcriptomic analyses revealed 246 genes that affect various stages of anther development. However, gene Ms10^{35} was found to have key role in male sterility of tomato (Jeong et al. 2014). Male sterility characterized by cytological studies indicated several defects in *7B-1* tomato mutant for anther development such as nonsynchronous anther maturity, abnormal tapetum development, and poor callose degradation. Several genes linked with genetic male sterility of *7B-1* were identified (Omidvar et al. 2017). Clustered regularly interspaced short palindromic repeat (CRISPR)/CRISPR-associated protein 9 (Cas9) was exploited to induce genetic modification in stamen-specific gene (SlSTR1) which produced genetic male sterility in tomatoes. Male sterility was restored through transformation used in fertility gene (Du et al. 2020). Transcription factor bHLH encoded by SlLS10 gene was knocked out through CRISPR/Cas9 to produce male sterile tomato plants. bHLH was known to affect meiosis and tapetum cell death (Jung et al. 2020). The same technology was also used to co-knock out two genes Ms10^{35} and GSTAA to develop genetic male sterility line of tomato. Biallelic mutation was transferred into woolly tomato to link male sterility with non-woolly tomato (Liu et al. 2021).

21.6.2 Cytoplasmic Male Sterility

Cytoplasmic male sterility (CMS) systems have been extensively exploited in various vegetable crop species such as *Brassica*. CMS is dictated by factors residing in the mitochondrial organelles and thus has maternal inheritance. CMS system has been introduced into cultivated species by crossing them to wild or unrelated species. Hybrid seed production through CMS systems is based on ABR lines, where A line carries cytoplasmic male sterility source which is maintained by B line. B line has the capacity to produce normal pollen but unable to restore fertility of A line. In a single cross-hybrid system, A line is crossed with R line which carries dominant fertility restorer gene (*Rf*) residing in the nucleus.

Spontaneous sources of cytoplasm grouped as S or T were identified in various parts of the world, i.e., USA, Japan, India, and the Netherlands, however, molecular markers revealed high similarities among CMS sources despite being identified independently in geographically diverse regions (Havey 2000).

21.7 Conclusion and Future Prospects

This chapter reviews the progress of hybrid seed production which is a mean for obtaining crops with uniform growth and maturity and also higher yields in all kinds of vegetable crops. The intensity of hybrid vigor depends upon the combination of particular parents due to their genetic distance and specific combining ability. Hybrid vigor is biometrically determined to estimate the increase of vigor over mid parental value. It may vary between crosses and species. Superior inbred lines may be selected through pure line methods and their breeding value may be determined by crossing them to a common male tester parent. In order to speed up the breeding program, in vitro techniques such as haploid development and embryo rescue or somatic cell hybridization methods have been exploited to develop pure lines with superior uniformity and genetic base. Molecular markers are used to predict the genetic distance between accessions. The accessions with high genetic distance are assumed to show higher heterosis which has not been proven in many studies. Association mapping may predict the breeding value with high success in crop plants as it is based on high-density linkage mapping.

References

Aiswarya CS, Vijeth S, Sreelathakumary I, Kaushik P (2020) Diallel analysis of chili pepper (*Capsicum annuum* L.) genotypes for morphological and fruit biochemical traits. Plan Theory 9(1):1. https://doi.org/10.3390/plants9010001

Amutha R, Venkatesan M, Kumar NS, Thangavel P (2007) Hybrid vigor and inbreeding depression in Bhendi (*Abelmoschus esculentus* (L.) Moench). Agric Sci Dig 27(2):131–133

Askander HS, Osman KF (2018) Heterosis and combining ability effects for some traits of pea (*Pisum sativum* L.). Mesop J Agric 46(4):435–450

Atanassova B (2000) Functional male sterility in tomato (*Lycopersicon esculentum* Mill) and its application in hybrid seed production. Acta Physiol Plant 22(3):221–225

Bhatia R, Dey SS, Parkash C, Sharma K, Sood S, Kumar R (2018) Modification of important factors for efficient microspore embryogenesis and doubled haploid production in field grown white cabbage (*Brassica oleracea* var *capitata* L.) genotypes in India. Sci Hortic 233:178–187

Chauhan SVS, Agnihotri DK (2005) Detergent induced pollen sterility in some vegetable crops. Int J Hortic Sci 11(1):85–88

Cheema DS, Dhaliwal MS (2005) Hybrid tomato breeding. J New Seeds 6:1–14

Cheng Y, Wang Q, Li Z, Cui J, Hu S, Zhao H, Chen M (2013) Cytological and comparative proteomic analyses on male sterility in *Brassica napus* L. induced by the chemical hybridization agent monosulphuron ester sodium. PLoS One 8(11):e80191

Davoodi S, Olfati JA, Hamidoghli Y, Sabouri A (2016) Standard heterosis in *Cucurbita moschata* and *Cucurbita pepo* interspecific hybrids. Int J Veg Sci 22(4):383–388

Deepak, KD (2007) Chemical induction of male sterility and histological studies in okra (*Abelmoschus esculentus* (L.) Monech). Doctoral dissertation, University of Agricultural Sciences, Dharwad, India

Dey SS, Singh N, Bhatia R, Parkash C, Chandel C (2014) Genetic combining ability and heterosis for important vitamins and antioxidant pigments in cauliflower (*Brassica oleracea* var. *botrytis* L.). Euphytica 195(2):169–181

Du M, Zhou K, Liu Y, Deng L, Zhang X, Lin L, Li C (2020) A biotechnology-based male-sterility system for hybrid seed production in tomato. Plant J 102(5):1090–1100

Fortune (2024) Vegetable Seeds Market Size, Share & Industry Analysis, By Type (Open Pollinated Varieties and Hybrid), Crop Type (Solanaceae, Root & Bulb, Cucurbit, Brassica, Leafy, and Others), Cultivation Method (Protected and Open Field), and Regional Forecast, 2024–2032. https://www.fortunebusinessinsights.com/vegetable-seed-market-103066. Accessed 14 Dec 2024

Fujimoto R, Uezono K, Ishikura S, Osabe K, Peacock WJ, Dennis ES (2018) Recent research on the mechanism of heterosis is important for crop and vegetable breeding systems. Breed Sci 68(2):145–158

Gupta AJ, Singh BK (2016) Development of hybrids and hybrid seed production in onion. Training Manual on "Principles and Production Techniques of Hybrid Seeds in Vegetables", organized by ICAR-IIVR, Varanasi, pp 101–111

Gwanama C, Botha AM, Labuschagne MT (2001) Genetic effects and heterosis of flowering and fruit characteristics of tropical pumpkin. Plant Breed 120(3):271–272

Hameed A, Saleem MY, Akhtar KP, Shoaib M, Iqbal Q, Asghar M (2017) Molecular confirmation of intraspecific tomato (*Solanum lycopersicum*) hybrids and their evaluation against late blight and cucumber mosaic virus. Mol Biotechnol 59(6):234–240

Havey MJ (2000) Diversity among male-sterility-inducing and male-fertile cytoplasm of onion. Theor Appl Genet 101(5–6):778–782

Hazra P, Sahu PK, Roy U, Dutta R, Roy T, Chattopadhyay A (2010) Heterosis in relation to multivariate genetic divergence in brinjal (*Solanum melongena*). Indian J Agric Sci 80(2):234–243

Hundal JS, Dhall DK (2005) Breeding for hybrid hot pepper. J New Seeds 6:31–50. https://doi.org/10.1300/J153v06n02_03

Jagosz B (2011) The relationship between heterosis and genetic distances based on RAPD and AFLP markers in carrot. Plant Breed 130(5):574–579

Jat GS, Singh B, Tomar BS, Singh J, Ram H, Kumar M (2016) Seed yield and quality as influenced by growing conditions in hybrid seed production of bitter gourd (*Momordica charantia* L.) cv. Pusa Hybrid-1. J Appl Nat Sci 8(4):2111–2115

Jeong HJ, Kang JH, Zhao M, Kwon JK, Choi HS, Bae JH, Kang BC (2014) Tomato male sterile 10 35 is essential for pollen development and meiosis in anthers. J Exp Bot 65(22):6693–6709

Jung YJ, Kim DH, Lee HJ, Nam KH, Bae S, Nou IS, Kang KK (2020) Knockout of SlMS10 gene (Solyc02g079810) encoding bHLH transcription factor using CRISPR/Cas9 system confers male sterility phenotype in tomato. Plan Theory 9(9):1189

Kalidasu G, Sarada C, Reddy PV, Yellam T (2009) Use of male gametocide: an alternative to cumbersome emasculation in coriander (*Coriandrum sativum* L). J Hortic For 1(7):126–132

Kapur A (2013) Development of vegetable hybrids for climate change scenarios. In: Climate-resilient horticulture: adaptation and mitigation strategies. Springer, India, pp 103–111

Karaağaç O, Balkaya A (2013) Interspecific hybridization and hybrid seed yield of winter squash (*Cucurbita maxima* Duch) and pumpkin (*Cucurbita moschata* Duch) lines for rootstock breeding. Sci Hortic 149:9–12

Kawamura K, Kawanabe T, Shimizu M, Nagano AJ, Saeki N, Okazaki K, Fujimoto R (2016) Genetic distance of inbred lines of Chinese cabbage and its relationship to heterosis. Plant Gene 5:1–7

Kawamura K, Shimizu M, Kawanabe T, Pu Z, Kodama T, Kaji M, Okazaki K (2017) Assessment of DNA markers for seed contamination testing and selection of disease resistance in cabbage. Euphytica 213(1):28

Kucera V, Chytilová V, Vyvadilová M, Klíma M (2006) Hybrid breeding of cauliflower using self-incompatibility and cytoplasmic male sterility. Hortic Sci 33(4):148–152

Kumar C, Singh SP (2016) Heterosis and inbreeding depression to identify superior F_1 hybrids in tomato (*Solanum lycopersicum* L) for the yield and its contributing traits. J Appl Nat Sci 8(1):290–296

Kumar S, Kumar R, Kumar D, Gautam N, Singh N, Parkash C, Shukla YR (2017) Heterotic potential, potence ratio, combining ability and genetic control of yield and its contributing traits in cucumber (*Cucumis sativus* L.). NZ J Crop Hortic Sci 45(3):175–190

Kumar A, Sharma V, Jain BT, Kaushik P (2020) Heterosis breeding in eggplant (*Solanum melongena* L): gains and provocations. Plan Theory 9(3):403

Lian YJ, Lin GZ, Zhao XM, Lim HT (2001) Production and characterizations of somatic hybrids between *Brassica campestris* L. ssp. *pekinensis* and *Brassica oleracea* var. *capitata*. J Plant Biotech 31(2001):33–38

Lian YJ, Lin GZ, Zhao XM, Lim HT (2011) Production and genetic characterization of somatic hybrids between leaf mustard (*Brassica juncea*) and broccoli (*Brassica oleracea*). In Vitro Cell Dev Biol Plant 47(2):289–296

Liu LW, Wang Y, Gong YQ, Zhao TM, Liu G, Li XY, Yu FM (2007) Assessment of genetic purity of tomato (*Lycopersicon esculentum* L) hybrid using molecular markers. Sci Hortic 115(1):7–12

Liu J, Wang S, Wang H, Luo B, Cai Y, Li X, Wang X (2021) Rapid generation of tomato male-sterile lines with a marker use for hybrid seed production by CRISPR/Cas9 system. Mol Breed 41(3):1–12

Macko A, Grzebelus D (2008) DcMaster transposon display markers as a tool for diversity evaluation of carrot breeding materials and for hybrid seed purity testing. J Appl Genet 49(1):33–39

Mallikarjun B (2006) Studies on hybrid seed production of bhendi (*Abelmoschus esculentus*). Thesis submitted to University of Agriculture Sciences, Dhrawad, India

Mangal M, Upadhyay P, Kalia P (2016) Characterization of cultivated and wild genotypes of brinjal (*Solanum melongena* L.) and confirmation of hybridity using microsatellite markers. Vegetos 29:2

Mistry CR, Kathiria KB, Sabolu S, Kumar S (2018) Heterosis and inbreeding depression for fruit yield attributing traits in eggplant. Curr Plant Biol 16:27–31

Nienhuis J, Lower RL (1988) Comparison of two recurrent selection procedures for yield in two pickling cucumber populations. J Am Soc Hortic Sci 113(2):272–276

Omidvar V, Mohorianu I, Dalmay T, Zheng Y, Fei Z, Pucci A, Fellner M (2017) Transcriptional regulation of male-sterility in 7B-1 male-sterile tomato mutant. PLoS One 12(2):e0170715

Padmini K, Gowda RV, Naik LB (2007) Studies on parental synchronization in flowering for hybrid seed production in onion (*Allium cepa* L). J Hortic Sci 2(1):47–49

Paige KN (2010) The functional genomics of inbreeding depression: a new approach to an old problem. Biosci 60(4):267–277

Pandey S, Jha A, Kumar S, Rai M (2010) Genetics and heterosis of quality and yield of pumpkin. Ind J Hortic 67(3):333–338

Pattanaik A, Reddy DL, Ramesh S, Chennareddy A (2018) Comparison of traditional grow-out test and DNA-based PCR assay to estimate F_1 hybrid purity in cauliflower. Curr Sci 115(11):2095

Petrova M, Vulkova Z, Gorinova N, Izhar S, Firon N, Jacquemin JM, Stoeva P (1999) Characterization of a cytoplasmic male-sterile hybrid line between *Lycopersicon peruvianum* Mill × *Lycopersicon pennellii* Corr and its crosses with cultivated tomato. Theor Appl Genet 98(5):825–830

Porta B, Rivas M, Gutiérrez L, Galván GA (2014) Variability, heritability, and correlations of agronomic traits in an onion landrace and derived S1 lines. Crop Breed Appl Biotechnol 14(1):29–35

Saleem MY, Asghar M, Haq MA, Rafique T, Kamran A, Khan AA (2009) Genetic analysis to identify suitable parents for hybrid seed production in tomato (*Lycopersicon esculentum* Mill). Pak J Bot 41(3):1107–1116

Sharma VK, Bora L (2013) Studies on genetic variability and heterosis in vegetable pea (*Pisum sativum* L) under high hills condition of Uttarakhand. India Afr J Agric Res 8(18):1891–1895

Shull GH (1948) What is "Heterosis"? Genetics 33:439–446

Sidhu AS, Bal SS, Behera TK, Rani M (2005) An outlook in hybrid eggplant breeding. J New Seeds 6(2–3):15–29

Singh SK, Singh SV, Srivastava JP (2015) Studies on heterosis and inbreeding depression in cucumber (*Cucumis sativus* L.). Agriways 3(2):107–111

Singh S, Dey SS, Bhatia R, Kumar R, Sharma K, Behera TK (2019) Heterosis and combining ability in cytoplasmic male sterile and doubled haploid based *Brassica oleracea* progenies and prediction of heterosis using microsatellites. PLoS One 14(8):e0210772

Smith BM, Crowther TC (1995) Inbreeding depression and single cross hybrids in leeks (*Allium ampeloprasum* ssp. *porrum*). Euphytica 86(2):87–94

Sood M, Kalia P, Gautam G (2006) Heterosis, inbreeding depression and residual heterosis for pod yield, component traits and protein content in garden pea. Indian J Hortic 63(4):460–463

Supena EDJ, Suharsono S, Jacobsen E, Custers JBM (2006) Successful development of a shed-microspore culture protocol for doubled haploid production in Indonesian hot pepper (*Capsicum annuum* L.). Plant Cell Rep 25(1):1–10

Sureja AK, Sirohi PS, Behera TK, Mohapatra T (2006) Molecular diversity and its relationship with hybrid performance and heterosis in ash gourd [*Benincasa hispida* (Thunb) Cogn]. J Hortic Sci Biotech 81(1):33–38

Thabuis A, Lefebvre V, Bernard G, Daubèze AM, Phaly T, Pochard E, Palloix A (2004) Phenotypic and molecular evaluation of a recurrent selection program for a polygenic resistance to *Phytophthora capsici* in pepper. Theor Appl Genet 109(2):342–351

Tripathi SK, Singh PK (2000) Hybrid seed production of cauliflower. J New Seeds 2(4):43–49

Umehara M, Sueyoshi T, Shimomura K, Iwai M, Shigyo M, Hirashima K, Nakahara T (2006) Interspecific hybrids between *Allium fistulosum* and *Allium schoenoprasum* reveal carotene-rich phenotype. Euphytica 148(3):295–312

Uretsky J, Loy JB (2017) Evaluation of morphological traits associated with productivity in F1 interspecific (*Cucurbita maxima* Duch x C. *moschata* Duch) hybrid processing squash. HortSci 529:1156–1163

Wang GX, Tang Y, Yan H, Sheng XG, Hao WW, Zhang L, Liu F (2011) Production and characterization of interspecific somatic hybrids between *Brassica oleracea* var. *botrytis* and *B. nigra* and their progenies for the selection of advanced pre-breeding materials. Plant Cell Rep 30(10):1811–1821

Wehner TC, Cramer CS (1996) Ten cycles of recurrent selection for fruit yield, earliness, and quality in three slicing cucumber populations. J Am Soc Hortic Sci 121(3):362–366

Wei Y, Zhu M, Qiao H, Li F, Zhang S, Zhang S, Sun R (2018) Characterization of interspecific hybrids between flowering Chinese cabbage and broccoli. Sci Hortic 240:552–557

Yadav SK, Singh BK, Baranwal DK, Solankey SS (2013) Genetic study of heterosis for yield and quality components in tomato (*Solanum lycopersicum*). Afr J Agric Res 8(44):5585–5591

Ye S, Wang Y, Huang D, Li J, Gong Y, Xu L, Liu L (2013) Genetic purity testing of F1 hybrid seed with molecular markers in cabbage (*Brassica oleracea* var. *capitata*). Sci Hortic 155:92–96

Yesmin L, Islam MS, Rahman MM, Uddin MN, Ahmad S (2014) Inbred and hybrid seed production potentiality of tomato (*Lycopersicon esculentum*) genotypes and their yield performance during summer. Bang J Agric Res 39(1):13–21

Yoo KS, Pike LM, Patil BS, Lee EJ (2020) Developing sweet onions by recurrent selection in a short-day onion breeding program. Sci Hortic 266:109269

Yu CY, Dong JG, Hu SW, Xu AX (2017) Exposure to trace amounts of sulfonylurea herbicide tribenuron-methyl causes male sterility in 17 species or subspecies of cruciferous plants. BMC Plant Biol 17(1):1–11

Yue-bing L, Ying Z, Yi W, Yu W, Mei-ping Z (2012) Identification of somatic hybrids between *Panax ginseng* and *Daucus carota*. Ginseng Res 1:24–229

Yun W, Kun C, Song LZ (2003) Studies on the utilization of vegetative heterosis of intergeneric hybrids in *Brassica*. Acta Agron Sin 1:119–123

Zhang C, Wang P, Tang D, Yang Z, Lu F, Qi J, Huang S (2019) The genetic basis of inbreeding depression in potato. Nat Genet 51(3):374–378

Zhiyuan F, Wang X, Dongyu GL (2000) Hybrid seed production in cabbage. J New Seeds 1:109–129

Zhou QY, Yin JM, Cui C, Lin N, Chen L, Tang ZL, Li JN (2006) Characters of interspecific hybrids between *Brassica campestris* and *B. oleracea* var. *acephala*. J Southwest Agric Univ 1:116–119

CRISPR-Based Gene Tweaking for Improvement in Vegetable Crops

22

Muhammad Faheem, Anam Niamat, Sameer H. Qari, Nayla Munawar, and Aftab Ahmad

Abstract

In an era of increasing global population and environmental instability, improving vegetable crops is a critical agricultural priority. Vegetable crops are necessary for human nutrition and health because they include a wide range of essential nutrients. They are high in crucial vitamins like vitamins C, K, and B. Vegetables also contain high amounts of minerals including potassium, magnesium, and folate. CRISPR-based gene tweaking offers a possible avenue for improving vegetable crops by integrating accuracy and efficiency into the world of genome editing. Increased yield, disease resistance, climate resilience, and nutritional content are all important areas for improvement in genome editing through the CRISPR/Cas system. CRISPR technology is a revolutionary technique for precise gene editing. This groundbreaking tool is powered by the customized Cas9 protein, which is also known as molecular scissors due to its ability to cut DNA at particular spots. The importance of these breakthroughs stems from their ability to improve global food security, promote sustainable farming methods, and decrease malnutrition in populations. The objective of this chapter is to highlight the understanding, working, and role of CRISPR/Cas9 to improve the crucial

M. Faheem · A. Niamat
Department of Biochemistry, Faculty of Sciences, University of Agriculture, Faisalabad, Pakistan

S. H. Qari
Department of Biology, AlJumum University College, Umm Al-Qura University, Makkah, Saudi Arabia

N. Munawar
Department of Chemistry, United Arab Emirates University, Al-Ain, United Arab Emirates

A. Ahmad (✉)
Natural and Medical Science Research Center (NMSRC), University of Nizwa, Nizwa, Oman
e-mail: aftab.ahmad@uaf.edu.pk

Z. Iqbal et al. (eds.), *Sustainable and Innovative Vegetable Production in times of Climate Change*, https://doi.org/10.1007/978-981-96-6283-8_22

traits in vegetables including yield, nutritional content, disease resistance, and environmental resilience. The role of genetic advancements, and their applications, and significance of CRISPR in agriculture have been discussed. In the end, the prospects of CRISPR in improving the traits of vegetables have also been detailed.

Keywords

CRISPR-Cas9 · Gene Editing · Vegetable Improvement · CRISPR-based gene editing

22.1 Introduction

Vegetable crops are necessary for human nutrition and health because they include a wide range of essential nutrients. They are high in crucial vitamins like vitamins C, K, and B. Vegetables are also high in minerals including potassium, magnesium, and folate. Additionally, vegetables contain fiber and antioxidants such as carotenoids and flavonoids, which have been linked to a variety of health advantages. Thus, consuming a variety of vegetables in the diet is an easy method to guarantee that the body receives a diversity of nutrients (Buturi et al. 2021).

Vegetable crops provide culinary value in addition to nutritional benefits. Because of their varied flavors, textures, and colors, they make excellent complements to meals. The variety of vegetables available, ranging from leafy greens to root vegetables to bright bell peppers and tomatoes, adds to the sensory appeal of recipes (Yahia et al. 2019). This variety not only makes meals more enjoyable but also encourages people to consume a wider variety of nutrients. A varied vegetable diet can help prevent vitamin shortages and maintain a balanced nutritional profile.

Vegetable crops are critical to global food security. They help meet the nutritional needs of billions of people as a staple in diets around the world. Their cultivation is critical not just for individual well-being, but also for national economic progress and stability (Sweta and Singh 2022). Producing vegetable crops in areas with growing populations and increasing urbanization can help to ensure that people have access to fresh and healthy food, ultimately contributing to global food security. Many vegetable crops, such as corn as well as nitrogen-fixing legumes, lend themselves nicely to local and sustainable agricultural methods. They are suitable for growing on tiny areas of land and incorporative for crop rotations. These methods eliminate the demand for synthetic chemicals while also improving soil health (Goddard et al. 2022).

Furthermore, vegetable production allows small-scale and family farmers to generate funds and improve their living conditions. Sustainable vegetable farming strategies aid in the conservation of natural resources and the promotion of environmental responsibility (Tripathi et al. 2019). When compared to other crops, some vegetable crops, such as drought-resilient corn and salt-tolerant wheat, are more resilient to climate change because of their improved growth cycles and tolerance to different climates such as high salinity and water scarcity. This adaptation contributes to a steady food supply in areas prone to unpredictable weather events, ultimately leading to improved agricultural resilience in the face of climate change.

Vegetable gardening is not only a source of nutrition, but it also contributes significantly to the economy. Vegetable gardening could become commercially viable due to the high demand for vegetables in both domestic and international markets (Bisbis et al. 2018). Vegetable gardening supports a wide spectrum of livelihoods, from small-scale farmers to agribusinesses involved in vegetable production, distribution, and marketing. Vegetable crops add to economic value because the raw material of some vegetables such as tomato paste can be processed by industries, benefiting economies as well as communities. Cultivation of various vegetable crops brings agricultural biodiversity as these crops come from a variety of genetic origins and species, which are critical for agricultural adaptability and resilience. Genetic diversity protects against risks of pests, diseases, and environmental changes. Thus, producing and protecting many varieties of vegetables is essential for a robust and sustainable food supply (Ebert 2020). When kept and processed appropriately, several vegetable crops including garlic, onion, dry beans, and potatoes are ideal for long-term preservation. As a consequence, they are critical for food security, particularly in areas where crop production varies seasonally. Vegetables can be preserved using techniques such as freezing, canning, or pickling, maintaining a stable food supply even during times of scarcity. This helps to increase food security and sustainability.

Vegetables have a distinct position in culinary traditions throughout the world, in addition to their nutritional and economic importance. They are an important part of many civilizations, displaying culinary tradition and diversity. The distinct aromas, textures, and cuisines made with vegetables reflect the ingenuity and traditions of various regions. Vegetables are frequently used as the focal point of cultural events and feasts, helping to strengthen the cultural fabric and promote gastronomic diversity (Ghoshal 2018).

The research on vegetable crops has resulted in agricultural and biotechnological advancements. Researchers have generated new cultivars with improved qualities, such as disease resistance or increased nutritional content, using procedures such as genome editing. Agriculture has seen a dramatic transformation in recent years as a result of innovative biotechnology breakthroughs. Furthermore, current advances in genetic engineering i.e., the use of technologies such as Clustered Regularly Interspaced Short Palindromic Repeats (CRISPR), have the potential to revolutionize vegetable crop cultivation. These inventions provide solutions to current agricultural and food production difficulties, ensuring the continued importance of vegetable crops in our changing world (Zulfiqar et al. 2019). CRISPR, a groundbreaking technique that has significantly altered the landscape of gene editing, is at the leading edge of this revolution.

The Clustered Regularly Interspaced Short Palindromic Repeats (CRISPR), along with the CRISPR-associated protein 9 (Cas9) system, is at the heart of gene editing. This revolutionary technique has provided genome editing with an extraordinary precision, efficiency, and versatility. CRISPR-Cas9, which was inspired by the microbial immune system, enables scientists to alter particular DNA sequences inside the genome of an organism with remarkable precision. CRISPR-Cas9 has far-reaching implications beyond science, affecting industries as diverse as medicine, biotechnology, and agriculture. CRISPR-Cas9 shows enormous promise for increasing crop yields, improving disease resistance, and reinforcing nutritional value in vegetable crops (Kumar et al. 2023). We may go on a path to build more robust, nutritious, and sustainable vegetable crops by understanding and leveraging the power of this breakthrough gene-editing technology, which will not only contribute to global food security but also establish a new standard for the future of agriculture.

In this chapter we have discussed the exciting world of CRISPR-based gene editing for vegetable crop development, examining the potential and opportunities that this technology provides (Gupta et al. 2019). This chapter emphasizes the use of CRISPR-based gene editing in the improvement of vegetable crops. The major goal is to glean insights into the possibilities of CRISPR technology in agriculture and provide a comprehensive analysis of its implications for enhancing vegetable crop traits. This chapter introduces CRISPR-Cas9 technology, addresses the need for crop improvement in vegetable crops, highlights specific target features, provides case studies, and discusses the methodologies and protocols involved. It also explores potential difficulties, limitations due to regulation, and ethical concerns, providing a thorough overview of the subject. The chapter serves as an introduction to the fascinating possibilities that CRISPR technology offers in the realm of vegetable crop development.

22.2 Understanding CRISPR-Cas9

The CRISPR-Cas9 system is a revolutionary breakthrough that has transformed the area of genome editing. CRISPR is a family of DNA sequences, discovered in bacteria and archaea genomes that work as a defense mechanism against viral infections. When a virus infects, bacteria grab a fragment of viral DNA and incorporate it into their genome, forming a genetic memory of the invader and is utilized to kill the virus using this system, if it gets entry into bacteria in future. This system consists of Cas9 protein and gRNA (Hossain 2021).

In this system, the Cas9 protein, an enzyme, functions similarly to molecular scissors. It is responsible for accurately binding and cutting DNA at memorized regions in the genome of the virus. Cas9 is programmed by the bacteria to target certain genes by feeding it a guide RNA molecule, which is a specific nucleotide sequence that binds with the complementary sequence in the viral genome. When the Cas9 protein, led by the RNA molecule, comes into contact with the target DNA, it cuts the DNA and prevents bacteria from viral attack (Mushtaq et al. 2021). This method has a lot of potential in agriculture and crop enhancement. CRISPR-Cas9 is being used by scientists to precisely modify the genes of vegetable crops, adding desired features such as disease resistance, increased yield, and greater nutritional value. This capability provides an unprecedented level of precision and control in plant breeding.

22.3 Working of CRISPR Technology

CRISPR technology is a revolutionary technique for precise gene editing. This groundbreaking tool is powered by the customized Cas9 protein, which is also known as molecular scissors due to its ability to cut DNA at particular spots. The procedure starts with the discovery of a target gene within an organism's DNA, with applications ranging from strengthening disease resistance in vegetables to increasing nutritional value. After identifying the target gene, a guide RNA molecule is precisely designed to match the sequence of the gene in issue. This guide RNA serves as a GPS for the Cas9 protein, directing it to the precise place on the DNA strand where the target gene is located (Moon et al. 2019).

The Cas9 protein, now guided by the RNA molecule, transforms into a precision instrument, scanning the DNA for the precise sequence corresponding to the target gene. Once this region has been determined, Cas9 acts as a molecular scissor, producing a precise cut in the DNA. This activates the inherent repair mechanisms of the cell, which include Non-Homologous End Joining (NHEJ) and Homology-Directed Repair (HDR) (Mushtaq et al. 2021). NHEJ can be error-prone, introducing tiny insertions or deletions that may affect the target gene's function (Mushtaq et al. 2021). In contrast, HDR allows for the insertion of particular sequences, ensuring precise gene editing. These genetic changes can result in a variety of results, ranging from gene knockout, which renders the function of the target gene inactive to the addition of new genetic material, which recovers the normal functions of genes. CRISPR technology's adaptability and precision provide several advantages and make it a potent tool for upgrading vegetable crops, improving features such as disease resistance, nutritional content, and shelf life, and playing a critical part in agriculture's future (Janik et al. 2020).

22.4 Advantages and Potential Applications of CRISPR in Agriculture

CRISPR-Cas9 technology has been praised for its precision. It enables researchers to precisely identify and change particular genes inside an organism's DNA. This precision is critical for introducing desirable traits while avoiding unwanted alterations, which is especially important in agriculture, where predictability is critical for crop enhancement. Traditional breeding procedures frequently rely on crossbreeding and the selection of desired features across numerous generations, which can be time-consuming. CRISPR-Cas9 speeds up this process by allowing direct modifications to genes involved in desirable characteristics. This drastically minimizes the amount of time needed for breeding and developing new crop varieties (Nidhi et al. 2021).

CRISPR technology has the potential to improve disease resistance in vegetable crops. Scientists have strengthened a plant's natural defenses by precisely changing specific genes, making it more resistant to common infections (Wada et al. 2020). Several genes have been targeted with the CRISPR Cas9 system to increase disease resistance. An example includes the production of late blight-resistant potatoes. In this example, susceptibility genes such as St DND1 and St CHL were knocked out using the CRISPR cas9 system. These genes were major contributors to the proliferation of pathogen *Phytophthora infestans*. Thus, knocking out of these genes resulted in improved disease resistance in potatoes (Kieu et al. 2021). This not only increases the disease resistance but can also help to lessen reliance on chemical pesticides while also promoting healthier, more sustainable farming practices.

CRISPR can greatly increase the nutritional value of vegetable crops. Crops can be modified to deliver increased quantities of vital vitamins and minerals by altering genes responsible for nutrient synthesis and content. This has far-reaching consequences for combating malnutrition and encouraging healthier dietary choices. CRISPR technology can be used to extend the shelf life of harvested crops. Vegetables can be kept fresh for longer periods by modifying genes related to senescence and decay. An example of increased shelf life is the production of genetically modified tomatoes that have greater shelf life as compared to traditional varieties. The Pectate lyase (PL) gene has been double knocked out using CRISPR. Pectate lyase is an enzyme that is responsible for degrading pectin. Pectin is an essential component of the tomato cell wall and provides structural strength to tomatoes. The knout out of the PL gene resulted in the increased shelf life of tomato as it removed the degrading impact of pectate lyase (Wang et al. 2019). This not only lowers food waste but also improves food security by preserving the quality of harvested products over longer periods.

Climate change causes substantial problems to agriculture, influencing elements such as temperature, precipitation, and insect dynamics. CRISPR has the potential to play a critical role in the development of crops that are more resilient and adaptable to changing climatic conditions. The practical example of CRISPR-based climate resilience is the production of genetically modified rice that can tolerate water deficiencies and efficiently use water in severe conditions. In recent research, the OsEPFL9 gene is knocked out using CRISPR. This gene is responsible for regulating the stomatal density on the inner surface of rice leaves. Successful knockout of OsEPFL9 resulted in improved water using efficiency of rice under controlled conditions, ensuring steady food production even in the face of climate-related uncertainty (Wang et al. 2021).

CRISPR technology leads to a lower environmental impact in agriculture by reducing the usage of chemical pesticides and promoting sustainable agricultural methods. This has the potential to minimize the harmful consequences of agricultural pollution on ecosystems and biodiversity. CRISPR technology enables crop customization to meet regional and commercial demands (Thakur et al. 2023). Flavor, texture, and appearance can all be customized to fit consumer preferences. This amount of personalization improves crop marketability and can lead to higher consumer satisfaction.

Fig. 22.1 CRISPR-based improvements of vegetable crops

Since it produces predictable and precise genetic modifications, CRISPR is widely regarded as a more ethically acceptable approach to genetic modification. This feature can speed up regulatory approval and increase public acceptance of genetically modified crops. CRISPR-Cas9 technology brings about a new era in agriculture. Its precision, adaptability, and sustainability make it an invaluable tool for developing more robust, nutritious, and ecologically friendly vegetable crops. CRISPR technology has numerous applications in agriculture, adding to the promise of a more environmentally friendly and effective food production system in the future (Piergentili et al. 2021) (Fig. 22.1).

22.5 The Significance of Crop Improvement

Vegetable crops, like all agricultural initiatives, face a variety of problems that affect their growth, productivity, and long-term viability. Pests, diseases, and the effects of climate change are among the most serious threats to vegetable crops. Vegetable crops are vulnerable to a variety of pests, including insects, nematodes, and other herbivores. Pests that feed on plants can cause severe harm, resulting in lower agricultural yields and quality. Aphids, caterpillars, and beetles are common vegetable crop pests (Elias et al. 2019). Pesticides have long been used to control pests, but misuse can result in resistance and environmental damage. Integrated Pest Management (IPM) techniques, which mix biological, chemical, and cultural practices, are becoming more prevalent to reduce pest impact while limiting environmental harm.

Vegetable crops are also susceptible to a variety of diseases caused by fungi, bacteria, and viruses. These infections can spread quickly in the field, producing wilting, discoloration, and, eventually, crop loss. Disease spread can be regulated by factors such as moisture, temperature, and the presence of sensitive host plants. Crop rotation, disease-resistant cultivars, and sanitation methods are routinely employed to manage disease outbreaks in vegetable crops (Baker et al. 2020).

Climate change poses a significant threat to vegetable crop cultivation. Rising temperatures, changing precipitation patterns, and an increase in extreme weather events can all interrupt growing seasons and reduce crop yields. Droughts that last for an extended period can induce water stress, while excessive rains can bring flooding and soil saturation. Inconsistent weather patterns can also foster the growth of pests and diseases. To address these issues, researchers are working to create vegetable types that are more resistant to climate change. Precision agriculture, smart irrigation systems, and the adoption of temperature-resistant rootstocks are among the practices being used to adapt to the effects of climate change (Skendžić et al. 2021).

22.6 The Role of Genetic Advancements in Addressing Problems

In vegetable crop agriculture, genetic improvement is critical in tackling the challenges posed by pests, diseases, and climate change. Genetic advancements help to improve the resilience and productivity of vegetable crops through a variety of breeding strategies, including classical breeding and modern biotechnology. Here's how genetic advancements can help address these issues. Plant breeders can create cultivars that prevent or repel common pests by discovering and introducing certain resistance genes. This eliminates the demand for chemical pesticides while also reducing environmental effects. Incorporating insect-resistance genes into vegetable crops such as tomatoes and peppers, has proven helpful in minimizing pest damage (Singh et al. 2020).

Genetic improvement, like pest resistance, results in vegetable crop variants with higher disease resistance. Plants can better fight themselves against bacterial, fungal, and viral pathogens by incorporating disease-resistant genes into their genome. This eliminates the need for chemical treatments while also protecting agricultural production (Swarup et al. 2021). Disease-resistant tomato cultivars are an excellent example of how genetic modification might help to alleviate disease-related problems.

Crop agriculture faces additional problems as a result of climate change, such as altered growth seasons and extreme weather occurrences. Genetic improvement can aid in the development of vegetable crop types that are more adaptable to changing environmental conditions. Traits such as heat tolerance, drought resilience, and adaptability to changing temperature and precipitation patterns can be introduced by using biotechnological techniques and selective breeding. This guarantees that crops can thrive in a variety of climates (Swarup et al. 2021).

Genetic enhancement can be used to improve the nutritional value of vegetable crops in addition to pest and disease resistance. Breeders can design crops with higher levels of vital vitamins and minerals by identifying genes responsible for nutrient production and content. This addresses the issue of providing nutritious food to the world's rising population (Simko et al. 2021).

Vegetable crops' shelf life can be prolonged by genetic modification. Crops can be kept fresher for longer periods by altering genes related to senescence and decay. This solves the difficulty of reducing food waste while still ensuring that harvested crops remain edible. Combining genetic innovation with sustainable agriculture techniques provides a holistic strategy for tackling the issues that vegetable crops encounter. It not only improves crop resilience and productivity, but it also helps to construct ecologically friendly and sustainable agricultural systems. As research and innovation progress, genetic improvement's role in addressing these difficulties becomes increasingly important in guaranteeing food security and sustainable agriculture in the face of changing agricultural landscapes and global climate trends (Asrey et al. 2020).

22.7 Target Traits in Vegetable Crops

Disease Resistance is a crucial trait in the improvement of vegetable crops. Many vegetables are susceptible to pathogen-induced diseases that are mostly caused by fungi, bacteria, and viruses. Improving disease resistance is critical for lowering yield losses and dependency on chemical pesticides. Researchers and breeders can design vegetable crops that are better suited to defend against common infections, by finding and introducing disease-resistant genes, boosting overall crop health and production stability (Li et al. 2017).

Pest Resistance, a critical trait, tackles the problem of safeguarding vegetable crops against herbivorous insects and pest invaders. Such invaders can cause significant damage, resulting in decreased crop quality and yield. A technique for mitigating the impact of these pests is to develop pest-resistant crop cultivars through genetic advancement. It not only maintains crop quality but also decreases the need for chemical pest control procedures, contributing to environmentally friendly and sustainable farming practices (Divekar et al. 2022).

Yield and productivity are critical aspects of agricultural farming. Increasing vegetable crop yield and overall productivity is a long-term aim in agriculture since it has a direct impact on food security and economic sustainability. Yield-related characteristics, such as the number and size of fruits or vegetables produced per plant, are critical for increasing the efficiency and profitability of vegetable agriculture. These qualities can be optimized by genetic enhancement, leading to increased agricultural yields and economic benefits for farmers (Kashyap et al. 2022).

Nutritional content is a characteristic with far-reaching public health implications. Breeders can address nutritional deficits and encourage healthy diets by enhancing the nutritional value of vegetable crops. This entails breeding for higher levels of important vitamins, minerals, and antioxidants, adding to customers' nutritional well-being. Vegetables high in these nutrients can help to combat malnutrition and diet-related health problems around the world. Challenges posed by climate change including drought and heat tolerance are becoming increasingly important traits. A variety of vegetable crops, that can grow under heat and drought stress, is critical for crop resilience in the face of shifting climatic patterns. Crops that are bred for these features are better able to survive water scarcity and high temperatures, increasing their resilience to an ever-changing environment. Early maturity is a feature that extends the growing season and reduces the risk of bad weather conditions. Early-maturing crops can be made via genetic modifications, planted, harvested, and marketed faster, making them less vulnerable to weather-related disturbances.

Shelf life and postharvest quality are critical factors in reducing food wastage. Longer shelf life and postharvest quality are qualities that have a direct impact on the marketability and consumer attractiveness of vegetable crops. Efforts in genome editing to improve traits can focus on features that postpone senescence and decay, ensuring that vegetables with high quality stay fresh for longer periods, eventually minimizing food waste (Hoppu et al. 2020).

Flavor and taste characteristics include the sensory qualities that impact consumers' acceptance and preference. Improving flavor and taste characteristics contributes to vegetable crops' marketability and popularity, making them more appealing to customers. Uniformity and appearance are two characteristics that have a substantial impact on consumer preferences and marketability. Vegetable crops with uniform and consistent size, shape, and appearance are more likely to match customer expectations and requests. To meet the taste of customers as well as business demands, breeders and farmers must ensure that vegetables are consistent in these aspects (Hoppu et al. 2020).

The identification of these crucial traits provides the base for future genetic enhancement efforts in vegetable crop development. The selection of specific traits for genetic improvements is influenced by a variety of factors, including crop type, regional climate conditions, customer preferences, and crop improvement program aims. By addressing these traits through modern biotechnology such as CRISPR-Cas9, the potential for developing vegetable crop varieties that are more resilient, productive, and nutritionally rich, emerges, ultimately contributing to global food security and agricultural sustainability (Hemalatha et al. 2023).

22.8 Practical Examples of Improved Traits

Increased yield is an important characteristic of tomato breeding. Breeders have successfully increased production by selecting tomato varieties with stronger fruit-setting potential or breeding for larger fruit sizes. Indeterminate tomato cultivars with prolific fruit-bearing features have greatly increased tomato yield per plant, benefiting both farmers and customers (Iqbal et al. 2019).

Late blight is a disease that mostly affects potatoes. Potato varieties with increased resistance to late blight have been developed as a result of genetic improvement. Breeders have generated potatoes that need fewer fungicides and suffer fewer disease losses by transferring specific resistance genes from wild potato species. Cucumbers are frequently sensitive to downy mildew, which can reduce output. Cucumber varieties with genetic resistance to downy mildew have been developed effectively by breeders, lowering the requirement for fungicides in cucumber production (Ayenan et al. 2019).

Breeders have worked on increasing the nutritional value of carrots by increasing beta-carotene content, which is a precursor to vitamin A. Carrot variants with higher levels of beta-carotene have been generated through genetic modifications. These "orange" carrots have more nutritional value since beta-carotene is important for eye health and immunological function. Spinach is a good source of iron, but certain spinach cultivars have been engineered to have even more iron. These iron-enriched spinach types have increased nutritional value, contributing to individuals' dietary needs, particularly those at risk of iron deficiency (Elgin Karabacak & Karabacak 2019).

These examples show how various properties of vegetable crops, such as yield, disease resistance, and nutritional value are improved by genetic modifications. These advancements are crucial for satisfying the needs of a growing population and addressing issues such as food security and health while minimizing agriculture's environmental effects (Higgins et al. 2022).

22.9 CRISPR Applications in Vegetable Crop Improvement

CRISPR-Cas9 is utilized to target a specific gene, R2R3, MYB, and SlAN2, in tomatoes. This gene controls the formation of anthocyanin, a natural pigment that gives foods their color and flavor. Scientists successfully generated tomatoes with a deeper purple color and a richer, more attractive flavor by increasing the expression of this gene using CRISPR (Menconi et al. 2023). CRISPR technology is also implemented to improve the taste of tomatoes while also potentially increasing their nutritional value.

In the realm of fungus, CRISPR-Cas9 was used to create disease-resistant white button mushrooms. Researchers altered the mushroom's genome by targeting a Poly Phenol Oxidase (PPO) gene to make it more resistant to browning, a typical problem in mushroom farming. This advancement not only increased the shelf life of white button mushrooms but also reduced food waste, resulting in more sustainable practices in agriculture. Late blight is a well-known problem with potato agriculture. CRISPR-Cas9 was used by scientists to target a susceptibility gene such as St DMR1 and St DND1 in potatoes, making them immune to the pathogen (*Phytophthora infestans*) that causes late blight (Fig. 22.2). This discovery has the potential to reduce the demand for chemical fungicides while also improving potato production sustainability, addressing both economic and environmental concerns (Adaskaveg and Blanco-Ulate 2023). Furthermore, the StERF3 gene is also knocked out by CRISPR-mediated dual mRNA targeting, resulting in late blight resistance in potatoes (Razzaq et al. 2022).

The CRISPR technique has also been used to improve disease resistance in cabbage crops. Researchers altered key genes related to disease susceptibility, yielding

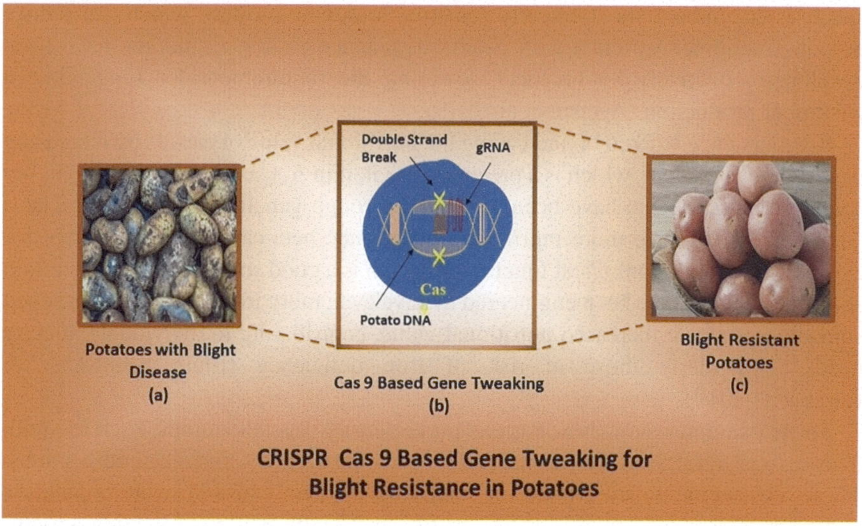

Fig. 22.2 CRISPR-Cas9-based gene tweaking to increase late blight resistance in potatoes

cabbage cultivars that are less susceptible to pathogen infections such as black rot and downy mildew. This genetic improvement not only increased crop output and quality but also reduced the need for chemical treatments, enabling more sustainable and environmentally friendly cabbage cultivation.

CRISPR-Cas9 is also implemented to boost the iron content in lettuce, treating nutritional deficits and promoting healthier eating habits. Researchers altered genes involved in iron intake and storage in lettuce, resulting in iron-rich cultivars. This application of CRISPR technology has the potential to help those who are iron-deficient and contribute to the production of more nutritious vegetable crops. These case studies demonstrate CRISPR technology's adaptability and potential for vegetable crop enhancement. CRISPR offers unique methods to improve the quality, sustainability, and nutritional value of vegetable crops by targeting specific genes linked with flavor, disease resistance, shelf life, and nutritional content, with promising implications for agriculture and food production (Ma et al. 2019; Damerum et al. 2020).

22.10 Regulation of CRISPR Edited Crops

Regulatory bodies are critical in evaluating and certifying genetically modified crops, particularly those generated with CRISPR technology. These authorities, like the USDA and EFSA, conduct extensive safety evaluations to guarantee that CRISPR-modified crops are safe to eat and do not pose any environmental problems (Abd-Elsalam and Ahmad 2024). The regulation process varies by location, but it always includes severe examinations to protect the public and the environment. Following are some important considerations of regulations.

Labeling genetically modified (GM) items clearly and accurately is an ethical factor that aligns with customers' right to make informed decisions. It is critical to require adequate labeling of CRISPR-modified crops to ensure customer understanding and choice in the marketplace. Transparent information enables customers to make decisions that are consistent with their preferences and values. The possible environmental impact of CRISPR-modified crops is an ethical problem. Responsible crop production must assess how these crops may interact with local ecosystems and affect non-target species and biodiversity. It is an ethical imperative to ensure that these crops do not unintentionally harm the environment (Kuzma 2018).

Intellectual property rights, including patents, govern ownership and control of CRISPR-mediated genetically modified crop innovations and have caused an ethical issue, particularly in areas where food security is a concern. Striking a balance between intellectual property rights and the global demand for enhanced crops is an ethical challenge. Using CRISPR technology for agricultural enhancement can have socioeconomic consequences. Access to sophisticated biotechnologies may be unequal, potentially leading to discrepancies in crop development and enhanced variety access. Ethical considerations concentrate on resolving these discrepancies and ensuring that technology's benefits are widely distributed. The potential for off-target consequences in gene editing is a significant regulatory consideration. If unintentional alterations in the genome occur, resulting in unanticipated repercussions in modified crops, ethical problems arise. To alleviate these issues, regulators must ensure the accuracy and predictability of CRISPR alterations (Lianos 2018).

Ethical considerations also extend to how plants or organisms employed in research are treated. It is critical to ensure that the application of CRISPR technology is done ethically and humanely. Ethical norms and oversight are required to support responsible research techniques and ensure that study subjects are treated ethically. Both ethical and regulatory judgments should include public participation and inclusion. Giving the public, stakeholders, and affected communities the opportunity to engage in decision-making processes is critical for guaranteeing a more democratic and ethical approach to CRISPR-based agricultural development (Lianos 2018). Involving a varied range of opinions and perspectives helps to produce well-rounded as well as informed decisions.

In an increasingly globalized environment, ethical and regulatory concerns should extend to international collaboration and conventions (Islam et al. 2020). Collaborative efforts are required to solve cross-border difficulties and unify CRISPR-based agricultural enhancement regulations and ethical standards. International collaboration guarantees that ethical and regulatory norms are uniformly followed and are adaptable to changing biotechnological settings.

22.11 Techniques and Protocols for Efficient CRISPR-Based Genome Editing

The process begins with the careful selection of certain qualities in the target crop that require improvement. These characteristics might range from disease resistance and greater productivity to improved nutritional content. Once these features have been found, the genes responsible for them become the prime targets for CRISPR editing. The construction of CRISPR constructs, which include guide RNA (gRNA) sequences and the Cas9 protein, is the next essential step. The gRNA is precisely constructed to match the target gene present at a specific chromosomal position. This design precision is critical to ensuring that the CRISPR system edits the correct gene while reducing off-target consequences. By combining the intended gRNAs with the Cas9 protein, the CRISPR-Cas9 system is created. Typically, this is performed by cloning the gRNA into an appropriate vector alongside the Cas9 gene. The vector should contain the regulatory elements required for the efficient expression of CRISPR components within the plant. The precise gene editing process is dependent on the proper synthesis of CRISPR-Cas9 constructs (Gupta et al. 2021).

Following construct development, the desired plant material is transformed with CRISPR constructs, which might comprise plant cells, tissues, or complete plants (Matson et al. 2019). Agrobacterium-mediated transformation, biolistics (gene gun), and protoplast transformation are among the transformation technologies that are easily accessible (Atiq et al. 2019). This process infuses the plant material with the CRISPR components required for gene editing. After successful transformation, the next step is to regenerate transgenic plants from the genetically modified cells or tissues. This usually entails cultivating the material on selective media to promote the formation of plantlets with the required genetic modifications (Atiq et al. 2019).

Various molecular approaches are used to confirm the presence of CRISPR-induced genetic changes. To confirm the presence of desired changes in the plant genome, polymerase chain reaction (PCR) and DNA sequencing are routinely used. This phase is critical for ensuring that the desired changes are accurate (Gupta et al. 2021).

Transgenic plants with CRISPR-induced alterations are grown under controlled conditions and their phenotypic features are assessed. This step comprises a thorough evaluation of whether the expected gains, such as disease resistance or increased nutritional content, were realized. The examination of phenotypes provides vital information into the practical impact of gene editing. Before proceeding to field testing, full environmental and biosafety studies are performed. These assessments try to detect and mitigate any environmental and non-target species implications of the genetically modified plants. These evaluations ensure appropriate management of genetically modified plants by analyzing ecological considerations and safety implications (Liao et al. 2019).

Field trials are an important step in evaluating the real-world performance of CRISPR-edited plants in agricultural settings. These trials entail growing modified crops in a variety of field conditions and climates. Researchers regularly monitor crop growth, yield, and interactions with local ecosystems. Field experiments give critical information about the practical feasibility of genetically modified crops in agricultural settings (Long et al. 2020).

Compliance with applicable regulatory bodies is a non-negotiable feature of crop gene editing. Regional and international norms and guidelines must be followed by researchers and organizations involved in the process. As the criteria vary by location, this compliance may include acquiring permissions or authorization for performing field testing or the possible commercial release of the modified crops (Vlčko & Ohnoutková 2020).

Following successful evaluations, altered crops may be scaled up. In this stage, the altered crops' cultivation and seed production are expanded to permit their wider use in agriculture and food production. Collaboration with seed firms and the agricultural business is frequently essential to this scaling-up process, which has the potential to bring the benefits of gene-edited crops to a broader agricultural setting. Key features of CRISPR/Cas9 in vegetables are detailed in Table 22.1.

Table 22.1 Key features of CRISPR/Cas9 in vegetables

Aspect	Description	References
Genome editing precision	CRISPR/Cas9 allows for precision vegetable DNA alteration, targeting specific genes for trait enhancement while reducing unwanted modifications.	Gupta et al. (2019)
Improved crop traits	CRISPR/Cas9 is used by scientists to improve desired features including yield, nutritional value, illness and pest resistance, and stress resistance.	Mushtaq et al. (2021)
Reduced chemical usage	CRISPR-edited crops with greater disease resistance require fewer chemical pesticides and fungicides, which benefits the environment.	Wang et al. (2021)
Accelerated vegetable breeding	CRISPR speeds up the production of superior vegetable varieties compared to traditional breeding approaches.	Hemalatha et al. (2023)
Sustainability	CRISPR aids in the development of resource-efficient vegetables that use less water and fertilizer, helping to sustainability.	Singh et al. (2020)
Addressing nutritional deficiencies	Genome editing has the potential to strengthen vegetables with critical vitamins and minerals, thereby treating malnutrition in specific areas.	Wang et al. (2021)
Reduced food waste	Postharvest trait editing increases shelf life and decreases spoiling, resulting in less food waste.	Simko et al. (2021)
Public and ethical concerns	CRISPR in agriculture raises issues of ownership of intellectual property, balanced technology access, and ethical concerns.	Eş et al. (2019).
Regulatory challenges	The implementation and commercialization of CRISPR-edited vegetables are hampered by complex laws and public views.	Lindberg et al. (2023)

22.12 Challenges and Limitations in Applying CRISPR to Vegetables

While CRISPR technology has enormous potential for crop improvement, especially vegetable development, it also has several hurdles and constraints that researchers and agriculturalists must consider. Here are some of the major obstacles and limits of using CRISPR on veggies.

The risk of off-target effects is one of the key challenges connected with CRISPR technology. Unintended genetic alterations can occur in sections of the genome that are similar to the intended target gene, despite ongoing attempts to construct highly precise guide RNAs (gRNAs). Precision and off-target consequences are ongoing challenges in CRISPR-based genetically modified vegetable crop development. It is not always easy to transport CRISPR components into plant cells, and the mechanism of transformation varies amongst vegetable crops. Some vegetables may be more resistant to transformation than others, complicating the deployment of CRISPR technology to a wide range of crops (Eş et al. 2019).

It is still difficult to optimize the delivery process for various vegetable species. The genetic background or genotype of the vegetable plant can influence the success of CRISPR-based gene editing. Some kinds may be more amenable to gene editing, whereas others may be resistant or have low transformation efficiency. Because of this genotype dependence, CRISPR's applicability to specific vegetable cultivars and strains is limited, making it critical to select appropriate kinds for successful gene editing (Eş et al. 2019).

The regulatory landscape for genetically modified organisms (GMOs) differs greatly by area, making it difficult to navigate complex regulatory systems. Obtaining regulatory permissions for CRISPR-edited vegetable crops can be time-consuming and expensive, discouraging some researchers as well as companies from pursuing these breakthroughs. Ethical and societal considerations can influence the public acceptability of gene-edited crops, especially vegetables. Transparency, labeling, and customer perception can all pose difficulties in bringing CRISPR-edited veggies to market. Engaging the public and addressing these ethical concerns is critical to the technology's acceptability (Das et al. 2023).

Evaluating the potential environmental interactions and ecological effects of CRISPR-edited crops is a difficult task. Researchers are trying to figure out how these crops affect non-target animals and ecosystems, as well as the long-term consequences on local biodiversity. Environmental issues and responsible management are important topics of continuing research. While CRISPR is a highly precise gene-editing technique, attaining the required results across a wide range of vegetable crops remains a goal. Researchers are still working to improve the editing process to ensure consistent and efficient results, while technological hurdles related to precision and efficiency remain. Researchers and the agricultural business are working to find solutions to these problems as they aim to fully utilize CRISPR in crop enhancement (Kumari et al. 2022).

22.13 Prospects and Challenges for the Future

CRISPR and gene editing technologies have the potential to guide in a new era of agriculture by providing novel answers to some of the most critical food production concerns. This technology enables precise plant genome alteration, allowing for the production of crops with improved features. Gene editing can help to address global food security issues by allowing the production of crops that can live in harsh environments and produce better yields. Furthermore, crop nutrition can be improved, potentially reducing malnutrition in vulnerable people. This technique promises not only enhanced agricultural output but also a road to more sustainable and environmentally friendly farming practices by eliminating the need for chemical inputs (Lindberg et al. 2023).

While CRISPR and gene editing have significant potential benefits in agriculture, various problems and obstacles must be overcome before they can be widely adopted. Regulatory complexity is a substantial barrier, with different regions having varied requirements for gene-edited crop licensing and commercialization. Global regulation and uniformity are critical for preventing trade obstacles and encouraging innovation. Another major issue is public acceptance, as attitudes toward genetically modified crops and genome editing vary greatly. To overcome this barrier, it is critical to educate the public and develop faith in the safety and benefits of these technologies. Smaller agricultural enterprises may face hurdles due to intellectual property rights and access to technology. Access to gene editing breakthroughs must be equitable to ensure a fair and inclusive agricultural economy. The ethical implications and public image of agricultural gene editing must not be disregarded. Transparency is essential in the development and deployment of gene-edited crops. Open and unambiguous information regarding the procedures, potential hazards, and benefits of these crops should be prioritized in ethical frameworks (Zhu et al. 2020). Environmental and social fairness should be prioritized, with gene editing promoting sustainability and benefiting all, especially small-scale farmers and people with limited resources. Long-term safety must be properly evaluated, and responsible innovation is a basic ethical precept. To guarantee that gene editing conforms with societal values and serves the greater good, decision-making procedures should be guided by inclusivity and public engagement.

22.14 Conclusion

The chapter emphasizes the critical function of CRISPR-based gene editing in improving vegetable crop yield. CRISPR technology enables accurate and efficient genetic modification of vegetable crops to obtain desired features. There are numerous potential benefits, including increased agricultural output, disease resistance, climatic resilience, and nutritional enhancement. These developments are crucial for resolving global food security challenges and developing more environmentally friendly agricultural techniques. CRISPR technology offers the potential to minimize reliance on chemical inputs, reduce farming's environmental imprint, and increase the nutritional content of crops. While the potential benefits are substantial, there are still problems and complications to overcome, such as regulatory hurdles, public perception, and ethical concerns. To fully achieve CRISPR's promise in agriculture, it is critical to overcome these issues, align worldwide standards, and participate in open communication with the public. Researchers, legislators, and stakeholders must collaborate to advance the use of CRISPR technology for the benefit of vegetable crops and the agricultural landscape as a whole.

References

Abd-Elsalam KA, Ahmad A (2024) Regulations of genome-edited plants: notes from editors. Elsevier eBooks, pp 1–20. https://doi.org/10.1016/b978-0-443-18444-4.00007-7

Adaskaveg JA, Blanco-Ulate B (2023) Targeting ripening regulators to develop fruit with high quality and extended shelf life. Curr Opin Biotechnol 79:102872. https://doi.org/10.1016/j.copbio.2022.102872

Asrey R, Barman K, Prajapati U, Sharma S, Yadav A (2020) Genetically modified fruit and vegetable – an overview on senescence regulation, postharvest nutraceutical quality preservation and shelf life extension. J Hortic Sci Biotechnol 96(3):271–287. https://doi.org/10.1080/14620316.2020.1845986

Atiq G, Khan NN, Kanwal S, Raheem MAR, Iqbal RK (2019) Plant transformation in biotechnology. Middle East J Appl Sci Tech (MEJAST), 2(3): 103–123. Available at: https://papers.ssrn.com/sol3/papers.cfm?abstract_id=3446375. Accessed 15 Jan 2025

Ayenan MAT, Danquah A, Hanson P, Ampomah-Dwamena C, Sodedji FAK, Asante IK, Danquah EY (2019) Accelerating breeding for heat tolerance in tomato (*Solanum lycopersicum* L.): an integrated approach. Agronomy 9(11):720. https://doi.org/10.3390/agronomy9110720

Baker BP, Green TA, Loker AJ (2020) Biological control and integrated pest management in organic and conventional systems. Biol Control 140:104095. https://doi.org/10.1016/j.biocontrol.2019.104095

Bisbis MB, Gruda N, Blanke M (2018) Potential impacts of climate change on vegetable production and product quality–a review. J Clean Prod 170:1602–1620. https://doi.org/10.1016/j.jclepro.2017.09.224

Buturi CV, Mauro RP, Fogliano V, Leonardi C, Giuffrida F (2021) Mineral biofortification of vegetables as a tool to improve human diet. Food Secur 10(2):223. https://doi.org/10.3390/foods10020223

Damerum A, Chapman MA, Taylor G (2020) Innovative breeding technologies in lettuce for improved post-harvest quality. Postharvest Biol Technol 168:111266. https://doi.org/10.1016/j.postharvbio.2020.111266

Das T, Anand U, Pal T, Mandal S, Kumar M, Radha, Gopalakrishnan AV, de la Lastra JMP, Dey A (2023) Exploring the potential of CRISPR/Cas genome editing for vegetable crop improvement: an overview of challenges and approaches. Biotechnol Bioeng 120:1215. https://doi.org/10.1002/bit.28344

Divekar PA, Narayana S, Divekar BA, Kumar R, Gadratagi BG, Ray A, Singh AK, Rani V, Singh V, Singh AK, Kumar A, Singh RP, Meena RS, Behera TK (2022) Plant secondary metabolites as defense tools against herbivores for sustainable crop protection. Int J Mol Sci 23(5):2690. https://doi.org/10.3390/ijms23052690

Ebert AW (2020) The role of vegetable genetic resources in nutrition security and vegetable breeding. Plan Theory 9(6):736. https://doi.org/10.3390/plants9060736

Elgin Karabacak Ç, Karabacak H (2019) Factors affecting carotenoid amount in carrots (*Daucus carota*). Ecol life Sci 14(2):29–39

Elias EH, Flynn R, Idowu OJ, Reyes J, Sanogo S, Schutte BJ, Smith R, Steele C, Sutherland C (2019) Crop vulnerability to weather and climate risk: analysis of interacting systems and adaptation efficacy for sustainable crop production. Sustain For 11(23):6619. https://doi.org/10.3390/su11236619

Eş I, Gavahian M, Marti-Quijal FJ, Lorenzo JM, Mousavi Khaneghah A, Tsatsanis C, Kampranis SC, Barba FJ (2019) The application of the CRISPR-Cas9 genome editing machinery in food and agricultural science: current status, future perspectives, and associated challenges. Biotechnol Adv 37(3):410–421. https://doi.org/10.1016/j.biotechadv.2019.02.006

Ghoshal G (2018) Emerging food processing technologies. In: Food processing for increased quality and consumption. Hand book of food bioengineering. Academic, pp 29–65. https://doi.org/10.1016/B978-0-12-811447-6.00002-3

Goddard T, Kassam A, Mkomwa S (2022) Moving paradigms – conservation agriculture with alternative agronomics to minimize inputs. CABI eBooks, pp 189–205. https://doi.org/10.1079/9781789245745.0010

Gupta D, Bhattacharjee O, Mandal D, Sen MK, Dey D, Dasgupta A, Kazi TA, Gupta R, Sinharoy S, Acharya K, Chattopadhyay D, Ravichandran V, Roy S, Ghosh D (2019) CRISPR-Cas9 system: a new-fangled dawn in gene editing. Life Sci 232:116636. https://doi.org/10.1016/j.lfs.2019.116636

Gupta S, Kumar A, Patel R, Kumar V (2021) Genetically modified crop regulations: scope and opportunity using the CRISPR-Cas9 genome editing approach. Mol Biol Rep 48:4851. https://doi.org/10.1007/s11033-021-06477-9

Hemalatha P, Abda EM, Shah S, Prabhu SV, Jayakumar M, Karmegam N, Kim W, Govarthanan M (2023) Multi-faceted CRISPR-Cas9 strategy to reduce plant-based food loss and waste for sustainable bio-economy—a review. J Environ Manag 332:117382. https://doi.org/10.1016/j.jenvman.2023.117382

Higgins DS, Goldenhar KE, Kenny GG, Perla DM, Hausbeck MK (2022) An evaluation of year-to-year fungicide efficacy and cultivar resistance combined with fungicide programs to manage cucumber downy mildew. Crop Prot 168:106176. https://doi.org/10.1016/j.cropro.2022.106176

Hoppu U, Puputti S, Sandell M (2020) Factors related to sensory properties and consumer acceptance of vegetables. Crit Rev Food Sci Nutr 61:1–11. https://doi.org/10.1080/10408398.2020.1767034

Hossain MA (2021) CRISPR-Cas9: a fascinating journey from bacterial immune system to human gene editing. Prog Mol Biol Transl Sci 178:63–83. https://doi.org/10.1016/bs.pmbts.2021.01.001

Iqbal R, Saeed K, Khan A, Noreen I, Bashir R (2019) Tomato (*Lycopersicum esculentum*) fruit improvement through breeding. Inno Scholar J Appl Sci Res 2(7):7–21. http://innovationinfo.org/articles/SJASR/SJASR-1-243.pdf

Islam MA, Rony SA, Rahman MB, Cinar MU, Villena J, Uddin MJ, Kitazawa H (2020) Improvement of disease resistance in livestock: application of immunogenomics and CRISPR/Cas9 technology. Animals 10(12):2236. https://doi.org/10.3390/ani10122236

Janik E, Niemcewicz M, Ceremuga M, Krzowski L, Saluk-Bijak J, Bijak M (2020) Various aspects of a gene editing system—CRISPR–Cas9. Int J Mol Sci 21(24). https://doi.org/10.3390/ijms21249604

Kashyap P, Prusty AK, Panwar AS, Paramesh V, Natesan R, Shamim M, Verma N, Jat PC, Singh MP (2022) Achieving food and livelihood security and enhancing profitability through an integrated farming system approach: a case study from western plains of Uttar Pradesh, India. Sustainability 14(11):6653. https://doi.org/10.3390/su14116653

Kieu NP, Lenman M, Wang ES, Petersen BL, Andreasson E (2021) Mutations introduced in susceptibility genes through CRISPR/Cas9 genome editing confer increased late blight resistance in potatoes. Sci Rep 11(1):4487. https://doi.org/10.1038/s41598-021-83972-w

Kumar R, Singh S, Kumar S, Tewarson V, Hakim M, Kaushik K, Kumar S, Kumar B (2023) A new era of clustered regularly interspaced short palindromic repeats/CRISPR-associated protein 9 gene editing technology in cardiovascular diseases: opportunities, challenges, and perspectives. Heart Views 24(4):201–201. https://doi.org/10.4103/heartviews.heartviews_49_23

Kumari C, Sharma M, Kumar V, Sharma R, Kumar V, Sharma P, Kumar P, Irfan M (2022) Genome editing technology for genetic amelioration of fruits and vegetables for alleviating post-harvest loss. Bioengineering 9(4):176. https://doi.org/10.3390/bioengineering9040176

Kuzma J (2018) Regulating gene-edited crops. Issues Sci Technol 35(1):80–85. https://www.jstor.org/stable/26594291

Li MY, Hou XL, Wang F, Tan GF, Xu ZS, Xiong AS (2017) Advances in the research of celery, an important Apiaceae vegetable crop. Crit Rev Biotechnol 38(2):172–183. https://doi.org/10.1080/07388551.2017.1312275

Lianos I (2018) Agro-chemical mega-mergers and innovation—between competition law, regulation and IP rights. In: Muscolo G, Tavassi M (eds) The interplay between competition law and IP: an international perspective, p 49. https://papers.ssrn.com/sol3/papers.cfm?abstract_id=3265685

Liao S, Qin X, Luo L, Han Y, Wang X, Usman B, Nawaz G, Zhao N, Liu Y, Li R (2019) CRISPR/Cas9-Induced Mutagenesis of Semi-Rolled Leaf1,2 Confers Curled Leaf Phenotype and Drought Tolerance by Influencing Protein Expression Patterns and ROS Scavenging in Rice (*Oryza sativa* L.). Agronomy 9(11):728. https://doi.org/10.3390/agronomy9110728

Lindberg S, Bain C, Selfa T (2023) Regulating gene editing in agriculture and food in the European Union: disentangling expectations and path dependencies. Sociologia Ruralis 63(2):348–369. https://doi.org/10.1111/soru.12429

Long KC, Alphey L, Annas GJ, Bloss CS, Campbell KJ, Champer J, Chen C-H, Choudhary A, Church GM, Collins JP, Cooper KL, Delborne JA, Edwards OR, Emerson CI, Esvelt K, Evans SW, Friedman RM, Gantz VM, Gould F, Hartley S (2020) Core commitments for field trials of gene drive organisms. Science (New York, N.Y.) 370(6523):1417–1419. https://doi.org/10.1126/science.abd1908

Ma C, Liu M, Li Q, Si J, Ren X, Song H (2019) Efficient BoPDS gene editing in cabbage by the CRISPR/Cas9 system. Hortic Plant J 5(4):164–169. https://doi.org/10.1016/j.hpj.2019.04.001

Matson AW, Hosny N, Swanson ZA, Hering BJ, Burlak C (2019) Optimizing sgRNA length to improve target specificity and efficiency for the GGTA1 gene using the CRISPR/Cas9 gene editing system. PLoS One 14(12):e0226107. https://doi.org/10.1371/journal.pone.0226107

Menconi J, Perata P, Gonzali S (2023) Novel R2R3 MYB transcription factors regulate anthocyanin synthesis in aubergine tomato plants. BMC Plant Biol 23(1):148. https://doi.org/10.1186/s12870-023-04153-7

Moon SB, Kim DY, Ko JH, Kim YS (2019) Recent advances in the CRISPR genome editing tool set. Exp Mol Med 51(11):1–11. https://doi.org/10.1038/s12276-019-0339-7

Mushtaq M, Dar AA, Skalicky M, Tyagi A, Bhagat N, Basu U, Bhat BA, Zaid A, Ali S, Dar TUH, Rai GK, Wani SH, Habib-Ur-Rahman M, Hejnak V, Vachova P, Brestic M, Çığ A, Çığ F, Erman M, el Sabagh A (2021) CRISPR-based genome editing tools: insights into technological breakthroughs and future challenges. Genes 12(6):797. https://doi.org/10.3390/genes12060797

Nidhi S, Anand U, Oleksak P, Tripathi P, Lal JA, Thomas G, Kuca K, Tripathi V (2021) Novel CRISPR–Cas systems: an updated review of the current achievements, applications, and future research perspectives. Int J Mol Sci 22(7):3327. https://doi.org/10.3390/ijms22073327

Piergentili R, Del Rio A, Signore F, Umani Ronchi F, Marinelli E, Zaami S (2021) CRISPR-Cas and its wide-ranging applications: from human genome editing to environmental implications, technical limitations, hazards and bioethical issues. Cells 10(5):969. https://doi.org/10.3390/cells10050969

Razzaq HA, Ijaz S, Haq IU, Khan IA (2022) Functional inhibition of the StERF3 gene by dual targeting through CRISPR/Cas9 enhances resistance to late blight in *Solanum tuberosum* L. Mol Biol Rep 49(12):11675–11684. https://doi.org/10.1007/s11033-022-07958-1

Simko I, Jia M, Venkatesh J, Kang BC, Weng Y, Barcaccia G, Lanteri S, Bhattarai G, Foolad MR (2021) Genomics and marker-assisted improvement of vegetable crops. Crit Rev Plant Sci 40(4):303–365. https://doi.org/10.1080/07352689.2021.1941605

Singh RK, Prasad A, Muthamilarasan M, Parida SK, Prasad M (2020) Breeding and biotechnological interventions for trait improvement: status and prospects. Planta 252(4):54. https://doi.org/10.1007/s00425-020-03465-4

Skendžić S, Zovko M, Živković IP, Lešić V, Lemić D (2021) The impact of climate change on agricultural insect pests. Insects 12(5):440. https://doi.org/10.3390/insects12050440

Swarup S, Cargill EJ, Crosby K, Flagel L, Kniskern J, Glenn KC (2021) Genetic diversity is indispensable for plant breeding to improve crops. Crop Sci 61(2):839–852. https://doi.org/10.1002/csc2.20377

Sweta, Singh B (2022) A review on heavy metal and metalloid contamination of vegetables: addressing the global safe food security concern. Int J Environ Anal Chem 104:1–22. https://doi.org/10.1080/03067319.2022.2115890

Thakur N, Nigam M, Mann NA, Gupta S, Hussain CM, Shah AA, Casini R, Elansary HO, Khan SA (2023) Host-mediated gene engineering and microbiome-based technology optimization for sustainable agriculture and environment. Funct Integr Genomics 23(1):57. https://doi.org/10.1007/s10142-023-00982-9

Tripathi AD, Mishra R, Maurya KK, Singh RB, Wilson DW (2019) Estimates for world population and global food availability for global health. In: The role of functional food security in global health, pp 3–24. https://doi.org/10.1016/b978-0-12-813148-0.00001-3

Vlčko J, Ohnoutková L (2020) Allelic variants of CRISPR/Cas9 induced mutation in an inositol trisphosphate 5/6 kinase gene manifest different phenotypes in barley. Plants 9(2):195. https://doi.org/10.3390/plants9020195

Wada N, Ueta R, Osakabe Y, Osakabe K (2020) Precision genome editing in plants: state-of-the-art in CRISPR/Cas9-based genome engineering. BMC Plant Biol 20(1):234. https://doi.org/10.1186/s12870-020-02385-5

Wang D, Samsulrizal NH, Yan C, Allcock NS, Craigon J, Blanco-Ulate B, Ortega-Salazar I, Marcus SE, Bagheri HM, Fons LP, Fraser PD, Foster T, Fray R, Knox JP, Seymour GB (2019) Characterization of CRISPR mutants targeting genes modulating pectin degradation in ripening tomato. Plant Physiol 179(2):544–557. https://doi.org/10.1104/pp.18.01187

Wang T, Zhang C, Zhang H, Zhu H (2021) CRISPR/Cas9-mediated gene editing revolutionizes the improvement of horticulture food crops. J Agric Food Chem 69(45):13260–13269. https://doi.org/10.1021/acs.jafc.1c00104. Epub 2021 Mar 18. PMID: 33734711

Yahia EM, García-Solís P, Celis MEM (2019) Contribution of fruits and vegetables to human nutrition and health. In: Postharvest physiology and biochemistry of fruits and vegetables. ScienceDirect, pp 19–45. https://doi.org/10.1016/b978-0-12-813278-4.00002-6

Zhu H, Li C, Gao C (2020) Applications of CRISPR–Cas in agriculture and plant biotechnology. Nat Rev Mol Cell Biol 21(11):661–677. https://doi.org/10.1038/s41580-020-00288-9

Zulfiqar F, Navarro M, Ashraf M, Akram NA, Munné-Bosch S (2019) Nanofertilizer use for sustainable agriculture: advantages and limitations. Plant Sci 289:110270. https://doi.org/10.1016/j.plantsci.2019.110270

Part VI

Postharvest Technology, Processing and Value Addition

23 Postharvest Storage Life Extension and Quality Management of Vegetables

Sajid Ali, Muhammad Hassan, Muhammad Akbar Anjum, Safina Naz, and Shaghef Ejaz

Abstract

Vegetables are an important group of horticultural crops that have considerably higher dietary significance for the humans. Nevertheless, these are perishable commodities and must be handled with extreme care in order to extend their storage life and conserve overall eating quality for the consumers. Postharvest management practices of vegetables generally depend upon the type of vegetables as some of these are less perishable (i.e., onion, potato, sweet potato, cassava) while others are easily decayable (i.e., spinach, cilantro, fenugreek, lettuce, tomato, etc.) in nature. Numerous types of postharvest approaches have been reported for postharvest shelf-life extension and quality management of vegetables such as use of cold storage, controlled atmosphere storage, packaging, edible coatings with plant extracts and essential oils, etc. In this chapter, various postharvest treatments and handling approaches have been summarized for the quality management of harvested vegetables.

Keywords

Edible coatings · Packaging · Postharvest quality · Shelf-life · Vegetable crops

23.1 Introduction

Vegetables considerably vary in their composition. For example, some vegetables contain a significantly higher percentage of water (spinach, cilantro, fenugreek, lettuce, cucumber, peppers, etc.) and require special management practices during

S. Ali (✉) · M. Hassan · M. A. Anjum · S. Naz · S. Ejaz
Department of Horticulture, Bahauddin Zakariya University, Multan, Pakistan
e-mail: sajidali@bzu.edu.pk

Z. Iqbal et al. (eds.), *Sustainable and Innovative Vegetable Production in times of Climate Change*, https://doi.org/10.1007/978-981-96-6283-8_23

postharvest in order to extend their availability and minimize loss of nutrients. On the other side, some vegetables do not contain as much higher water percentage as the aforementioned types of vegetables but these need to be cured and then handled with care and stored at appropriate conditions. These include onion, potato, sweet potato, and cassava. Nevertheless, these also need proper postharvest management and handling in order to reduce the extent of quantitative and qualitative losses after their harvest.

In general, the storage life of harvested vegetables depends upon different practices starting from field to market. Major factors that affect postharvest storability of harvested vegetable crops include higher moisture percentage, respiration rate, prevailing environmental conditions at harvest, bruise susceptibility, mechanical injuries, rough transportation (mainly in developing countries), microbial infestations, and postharvest diseases during storage thereby compromising global food security. Therefore, on the whole, harvested vegetables are generally considered perishable due to higher water content and microbial-based infestations thereby exhibiting short shelf-life (De Corato 2020). Hence, these should be handled suitably to retain their natural nutrients, keep them fresh looking with restored eating quality and extended storage life so that these could be available for broader marketable windows for the consumers (El-Ramady et al. 2015).

Different types of approaches have been detailed for management of postharvest storage life and quality of vegetable crops. These include sorting, grading, precooling, disinfectant-based treatments, proper packaging, improved transportation, and provision of cold storage facilities during transportation and postharvest storage. All these practices ultimately intend to delay rapid senescence and extend their storage life potential with preserved eating quality (El-Ramady et al. 2015). Most common practices now being used include removal of field heat by pre-cooling treatment (Dirapan et al. 2021), packaging (Nasrin et al. 2022), including modified atmosphere packaging (Jalali et al. 2017), and application of edible coatings (Bahri and Rashidi 2009). The objective of all these aforementioned approaches is to extend storage life and to maintain postharvest eating quality of harvested vegetables.

23.2 Types of General Storage Systems

23.2.1 Cold Storage

It is a technique to preserve vegetables at a temperature of 0–4 °C for a certain period of time to delay ripening/senescence or to conserve produce quality for later use and reduce food wastage. It not only conserves surplus produce but also helps to feed when vegetables are in short supply and their season of availability is ended. It also helps vegetable growers economically as farmers can store their produce for good price after bulk lots get short in the market (El-Ramady et al. 2015).

To avoid low temperature damages during cold storage, chemical treatment (as with methyl jasmonate, etc.), waxing, or packaging is done prior to storage. These practices help to reduce losses of stored vegetables. Frozen injured vegetables are more likely to spoil, so different vegetables must be stored at separate storage temperatures according to their nature (Raju et al. 2010). Different vegetables were stored at 4 °C for a period of a fortnight. Leafy vegetables showed more phenolic acid content as compared to other ones. Cold storage conserves nutritional constituents of the commodities that are essential for human health (Galani et al. 2017). Broccoli vegetable was tested in cold and hot conditions for certain duration of time; 1 h for cold (0 °C) and 30 s for hot conditions (55 °C). There was no considerable difference in both the treatments in enzymatic activities; both had a positive impact on storage life. However, results were different as longer shelf-life was noticed at <10 °C compared with other storage temperatures (Zhang et al. 2009).

23.2.2 Controlled Atmosphere (CA) Storage

It is a type of storage in which different gases like oxygen (O_2), carbon dioxide (CO_2), and nitrogen (N_2) are used separately or in combinations to preserve horticultural commodities. This is an effective way of keeping vegetables particularly during transportation. Different vegetables like root, leafy, and legumes have more shelf-life under CA storage freights (Ikegaya et al. 2023). These gases (CO_2, O_2, and N_2) are commonly used to reduce respiration rate and ethylene production ultimately delaying ripening and senescence. Use of nitric oxide (NO_2) as preservative gas has a positive effect on the ripening of fruits and vegetables by delaying production of ethylene and increasing their shelf-life during CA storage. NO_2 reduces the chances of chilling injuries and browning during CA storage by activating the enzymatic defense system in fruits and vegetables (Steelheart et al. 2019; Liu et al. 2023).

Improvement in CA has been made by the addition of 1-methylcyclopropene (1-MCP) which increases efficiency of the combinational treatment (CA + 1-MCP) in delaying the quality deterioration of the treated produce (Prange et al. 2005). Lettuce stored in 1-MCP-based atmosphere reduced reproductive growth and retained vitamin C content more as compared to that just kept in cold storage (Tay and Perera 2004). In a controlled atmosphere system, it is possible because low O_2, CO_2, and NO_2 concentrations reduce risk of chilling injuries and facilitate long-term storage. Different storage environments (CA, MAP) were tested to find out more sustainable one for green onions. Packaging preserved the produce for 15 days at 4 °C, while vacuum packaging under CA environment gave more promising results of 30 days of storage by maintaining their quality during storage (Zudaire et al. 2020).

23.3 Packaging of Vegetables

23.3.1 Vacuum Packaging

In vacuum packing, the produce is placed in a plastic film package. Then air is removed from inside and the package is sealed to stop entry of air. It is a good method to enhance the shelf-life of vegetables and other food products. Asparagus was stored at 4 °C under cold storage with the addition of vacuum packaging to increase its shelf-life. Cold temperature and vacuumed packaging resulted in increased shelf-life by slowing respiration rate. Moreover, lignin-producing enzyme was suppressed that maintained good texture and quality of asparagus (Lwin et al. 2020). Onions with green stacks were stored in different environments (MAP and vacuum packaging) at a temperature of 10 °C for a period of 28 days. Different parameters like weight loss, color change, and disease incidence were studied. Vacuum packaging was more successful as compared to MAP with more shelf-life and better produce quality (Hong and Kim 2004). Red chilies were stored as vacuum packaged and in jute bags at storage temperatures of 25 °C (room temperature) and 4 °C (cold environment) and at 10% and 12% RH under dark and light conditions. The produce was stored for a period of 24 months and quality parameters like changes in moisture, capsaicin, and color were studied after every 03 months. Chilies kept in vacuum packaging at 12% RH under cold storage had more positive effect on quality of produce (Chetti et al. 2014).

23.3.2 Modified Atmosphere Packaging

Modified atmosphere packaging (MAP) is a technique that involves modifying internal atmosphere of the package containing perishable produce or food product. It prolongs shelf-life, improves visual appearance, preserves freshness and food nutrients, and retains original favor of the produce. Usually, levels of gases (CO_2, O_2, and N_2) are manipulated inside the package. The amount of gases varies according to the kind of fresh produce being packaged (Kargwal et al. 2020). However, NO_2 is prohibited in modified atmosphere packaging (MAP) for fruits and vegetables conservation.

Technique of MAP for vegetables and fruits is now widely practiced to protect from postharvest diseases and other contaminants, maintain good quality, prolong their shelf-life as reliability for consumer's satisfaction (Werner and Hotchkiss 2005) and feed increasing population. Lettuce, a leafy vegetable, was stored by using active packaging, inserting a combination of O_2, CO_2, and N_2 gases, after dipping and cleansing with chlorine solution. This procedure hindered browning effectively as compared to control treatment (Horev et al. 2012). Turnip stored in MAP showed encouraging results in the form of better color retention and less weight loss. A comparison was carried out regarding the oxygen concentrations of 05% and 10% in MAP. Storage in less O_2 concentration, i.e., 05% was more helpful in enhancing shelf-life of turnip (Helland et al. 2016).

Three types of vegetables viz. broccoli, spinach, and asparagus were stored after different postharvest treatments like dipping in ozone water, chilled water, packaging in ordinary polythene bag, and MAP at 13 °C. MAP resulted in more shelf-life, green color, and enzymes as compared to other treatments. MA packaging may serve as a substitute for vegetables if refrigeration is unavailable (Chiebao et al. 2018). Pointed gourd was subjected to different packaging techniques, i.e., polypropylene coating with MAP and stored at ambient and refrigerated conditions. MAP with polypropylene film under refrigeration condition was the most suitable among all the treatments in extending shelf-life of gourd up to 16 days (Sahoo et al. 2015). Thin polyethylene film was used as coating to bell pepper fruits along with $KMnO_4$ pockets to absorb ethylene produced by respiration in MAP storage conditions at 10 °C. Control treatment was kept without absorbent. Results indicated that film-coated vegetables had least <1%, while control treatment had 15.1% weight loss. Ethylene absorber had positive effect on biochemical attributes (Cerit and Demirkol 2020).

23.3.3 Active Packaging

Active packaging, a type of packaging which uses ethylene scavengers is presently one of the major technologies used to conserve the quality and safety of fresh produce. Active packaging generally interacts with produce and/or environment to maintain the food quality significantly better in contrast with the conventional type of packaging (Kaewklin et al. 2018). The active MAP reached stable values of 4.8% O_2 and 7.1% CO_2 on the fourth day of storage, whereas passively modified packages failed to get stable condition until the completion of 12-day storage period. Different parameters like weight loss, firmness, color stability, phenolics content, and antioxidant capacity were studied. The shelf-life of actively packed fruits was increased to 28 days versus 12 days for passively packaged fruits. Thus, active atmosphere packaging was found helpful in delaying decay of fruits and thus prolonging their usefulness (Devgan et al. 2019). Use of antimicrobial agents in active packaging delivery systems has a crucial role in preserving produce quality. Active packaging was equipped with addition of natural substances that act as preserving agents to fruits and vegetables. Different sensory, physical and biochemical evaluations disclosed that addition of natural antimicrobial agents to active packaging system resulted in better quality of the produce ultimately ensuring consumers acceptance (Ayala-Zavala et al. 2008).

23.3.4 Passive Packaging

Passive modified atmosphere packaging (PMAP) is relatively a cheaper method to store fresh horticultural produce as it does not require any specialized equipment. In this type of MAP, the produce kept inside develops the internal gaseous composition on the basis of its respiratory metabolism by itself. In general, the compositions of O_2 and CO_2 change on the basis of nature of the stored produce, i.e., climacteric or non-climacteric. It has shown a good potential for prolonging postharvest life and enhancing quality of fresh horticultural produce when stored under non-optimum temperature conditions.

Cherry tomatoes were stored unpackaged (control) and in perforated and non-perforated packaging of polyethylene and polypropylene at a temperature of 7 °C for 21 days. The studied parameters included weight loss, color, and disease incidence. Tomato fruits with perforated polyethylene packaging showed more promising results in the form of extended shelf-life (21 days) as compared to other treatments (Paulsen et al. 2019). The efficacy of liquid ozone and chlorine applications on the postharvest shelf-life of green chilies packaged in PMAP was investigated. When contrasting with samples with no treatment (control), ozone (30 mg) treatment proved to be far better in preserving the postharvest quality of chilies. Ozone treatment also postponed aging and kept chilies firm during storage. When kept at 8 °C, MAP improved the value of peppers and extended their storage life. This combined treatment (ozone + MAP) further enhanced color retention in green chilies during storage and increased their shelf-life to 36 days, compared to 28 or 22 days for chlorine dipped and control ones, respectively (Chitravathi et al. 2015).

23.4 Edible Coatings

Coatings on fruits and vegetables serve as barriers to moisture loss, gases exchange, and entry of other substances, thus prolong their shelf-life. A number of edible coatings like polysaccharides, lipids, and essential oils are commonly used to conserve vegetables and extend their postharvest life. Consumable coating technique is one of the promising methods with higher demand, as it does not have any harmful effects on human health. Coatings derived from natural resources also contain numerous useful ingredients, such as anti-microbial and antioxidative components, thus when applied on freshly harvested horticultural produce, these improve their quality and prolong shelf-life. Different coatings like polysaccharides, carboxymethyl cellulose (CMC), starch-based, calcium based, etc. are used to prohibit microbial growth on fruits and vegetables.

23.4.1 Polysaccharide-Based Coatings

Polysaccharides are organic compounds which are easily degradable and biologically safe to use with no harmful effects on humans. These are gaining attention now-a-days as the produce remains safe, with good quality and longer shelf-life (Bilal et al. 2021). Different polysaccharide-based coatings were applied to zucchini vegetable (summer squash) which increased its storage ability and improved quality parameters. Different parameters like weight loss, chilling stress, and other biochemical attributes were studied. Additionally, the use of olive oil in coatings resulted in more promising results by improving quality parameters (Castro-Cegrí et al. 2023). Different types of film coatings were used on minimally processed carrots to uphold their quality deterioration. Different parameters like water loss, texture, respiration rate, acidity, total phenolic content, and activity of polyphenol oxidase, etc. were studied. Carboxymethyl cellulose (CMC) showed good results, reducing spoilage of carrots and extending their shelf-life up to 21 days when stored at 5 °C (Kowalczyk et al. 2020). CMC coating in combination with zinc oxide (ZnO) nanoparticles prepared from pineapple peel extract was prepared and applied on tomato fruits. Coated fruits had reduced weight loss and respiration rate, and increased fruit firmness and antioxidant compounds during storage period as compared to uncoated control fruits. In inoculated tomato fruits, CMC + ZnO nanoparticles coating commendably delayed disease incidence (Saekow et al. 2019).

Chitosan, a sugar extracted from the outer skeleton of shellfish, crab, lobster, and shrimp can be used as an alternative to chemicals as it is organic in nature and presents safe vegetables to consumers which are healthy and fresh. United States Food and Drug Administration (USFDA) has permitted chitosan for consumption as it is safe for human beings and also environment-friendly. Further, when used as coating, it is harmless for both fruits and vegetables (USFDA 2013). Chitosan has good potential for its applications in food industry as it is non-toxic, biodegradable, biocompatible with human tissues, and has good antifungal and antimicrobial activities. Chitosan is used profusely in agricultural sector as edible coating to minimize postharvest losses. Chitosan preserves vegetables by reducing rate of gaseous exchange and acting as a barrier to moisture loss or gain (Aider 2010). The effects of chitosan coatings were studied on broccoli and lettuce to minimize microbial activity and decay, improve safety of these vegetables, and maintain their quality traits. Chitosan with the addition of clove oil significantly increased the shelf-life of these vegetables (Alvarez et al. 2013). Chitosan coatings were tested against gray mold problem in stored tomato fruits. Coatings decreased polyphenol oxidase and increased phenolics and protein contents. Disease prevalence decreased with increase in chitosan concentration in coatings (Badawy and Rabea 2009). Similarly, chitosan coatings augmented with bioactive compounds and essential oils were tested for their antimicrobial properties on broccoli. Incorporation of bioactive compounds in chitosan enhanced antimicrobial action without any adverse effects on sensory parameters of broccoli (Alvarez et al. 2013). Chitosan and *Aloe vera* gel coatings were applied to green Capsicum (bell pepper) to study their effect on shelf-life and postharvest deterioration. Chitosan at concentration of 1% and 2% gave prominent results by inhibiting microbial activities and conserving biochemical features of fruits. The *Aloe vera* gel 20% had inhibitory effect on bacterial and fungal growth (Manoj et al. 2016).

23.4.2 Protein-Based Coatings

Zein protein from corn may act as barrier against postharvest losses by slowing respiration rate, ethylene production, and weight loss. Zein was applied to tomatoes at color breaker and pink stage to evaluate its effect in increasing shelf-life and conserving fruits stored at 21 °C. Coating significantly increased shelf-life to 6 days by delaying loss of firmness and color change, and conserving other quality attributes (Park et al. 1994). Two types of edible coatings viz. Alginate and Zein were applied to tomato fruits to overcome color development and faster senescence. Coated fruits had lower respiration rate, ethylene concentration, and decreased rate of color development as compared with uncoated fruits. Further, coating significantly delayed weight loss, softening, and color changes. Thus, quality of the coated fruits remained high as assessed by sensory evaluation and biochemical attributes, increasing their market value and resulting in better customer acceptance (Zapata et al. 2008).

23.4.3 Edible Gums-Based Coatings

Gums are organic compounds having adhesive nature, produced by trees and shrubs. Gums when applied to fruits and vegetables function as protective films against pathogenic organisms sustaining their quality. Gums are more beneficial as compared to the chemical-based coatings as these are eco-friendly, have no residual effects, and recommended as safe for consumers by FAO (Tahir et al. 2019). Incorporation of glycerol 20% to edible coatings containing gum (1%) applied on different vegetables and fruits (including tomatoes) minimizes weight loss, slows down rate of ethylene production, and preserves biochemical attributes, thus increases postharvest life of produce (Salehi 2020).

Gum Arabic was tested on mature green (breaker stage) tomato fruits in different concentrations (5–20%) stored at a temperature of 20 °C and RH 80–90% for 20 days. Tomatoes coated with gum Arabic (10%) resulted in delayed weight loss, color development, firmness, ripening process, disease incidence, and changes in biochemical attributes. Overall quality based on sensory evaluation was also superior in 10% gum Arabic coated tomatoes (Ali et al. 2010). Cucumber fruits were treated with various concentrations of gum Arabic (5–20%) and kept for 16 days at cold (10 °C) and ambient (25 °C) temperatures for 16 days. Gum Arabic coatings substantially decreased weight reduction and delayed softening of fruits at both temperatures during the storage period. The firmness of uncoated fruits dropped considerably at both storage conditions. The color, flavor, texture, appearance, and general acceptance of coated fruits were significantly better when stored at 10 or 25 °C as compared to untreated ones (Al-Juhaimi et al. 2012). The effect of different edible coatings on quality and storage life of bell pepper cv. "Yolo Wonder" kept at 8 °C with 90–95% RH for 24 days was studied. Bell pepper fruits were coated by immersing them in different concentrations of gum Arabic (6, 9, and 12%), *Aloe vera* gel (4, 5, and 6%), and *Cinnamomum verum* oil (0.5, 0.75, and 1%). All coating treatments enhanced biochemical behaviors and improved the quality of pepper, however 12% gum Arabic treatment greatly decreased weight reduction, membrane's leaking, cold injury, and disease incidence with lesser rise in pH, TSS, and sugar contents (Ullah et al. 2017). The suitability of *Aloe vera* gel and gum tragacanth as consumable coatings for bell pepper was studied and their impact on variations in biochemical characteristics was investigated when stored at different temperatures (4, 10, 15, and 23 °C) for 1 month. *A. vera* gel and gum tragacanth-coated fruits were significantly better as compared to uncoated control fruits (Mohebbi et al. 2014).

23.4.4 *Aloe vera* Gel-Based Coating

Use of eatable coatings is an efficient way, to maintain freshness of fruits and vegetables, prolong their postharvest life, and even serve as an option for modified atmosphere packing employed in traditional packaging. *Aloe vera* gel is an organic product made up primarily of polysaccharides. It is an edible and active component that can be used to make consumable coatings. It can serve as a semi-permeable shield for gases and water vapors, slowing respiration and ethylene generation from the produce and ripening process, and preserving its texture and biochemical components. Because of its antioxidative and anti-microbial properties, it is also a potential product for extending shelf-life of fruits and vegetables (Nicolau-Lapeña et al. 2021). It additionally serves to keep the coated produce fresh, extend shelf-life, and also offer some extra beneficial effects due to its curative properties. Films of *Aloe vera* gel, those are easy to make and use on the desired produce, have demonstrated their effectiveness in ensuring hygiene and nutritional value of different edible items (Maan et al. 2021). *Aloe vera* is also an important ingredient in the production of pharmaceuticals, beauty items, hair gels, and food coatings.

The impact of *Aloe vera* treatment was examined on chilies stored for 4 weeks at 10 °C. In treated chilies, weight loss, disease frequency, color development, rate of respiration, electrolytic leaking, and H_2O_2 decreased as compared to untreated fruits. The gel coating retained widespread acceptance in terms of its biochemical attributes, aesthetic value, and market value. Application of food-grade coating is a sustainable way of slowing aging and preserving the nutritional value of green chilies after harvest for up to 4 weeks (Hasan et al. 2021). *Aloe vera* gel as separate and in additive to other materials is an effective coating in extending shelf-life and lowering disease incidence in fresh produce. An edible coating was made by mixing purified shellac in a basic solution and combining it with *Aloe vera* gel. The addition of gel into shellac film increased the permeability of coating layer to air, CO_2, and moisture. In coated tomatoes, the films postponed aging which was characterized by limited alterations in air exchange and production of ethylene throughout the storage period. When evaluated in the sense of its texture and color, the fruits revealed only minor differences when compared to control. Maximum increase in shelf-life (up to 12 days) was found in composite coating of *Aloe vera* gel and shellac (Chauhan et al. 2015).

Aloe vera gel (50%) and ascorbic acid (1%) were applied alone and in combination to lotus roots slices before storing at 20 °C to preserve their quality attributes. Different parameters like weight loss, browning, electrolytic leakage, and physical appearance along with many biochemical parameters were studied. Composite coating of *Aloe vera* gel and ascorbic acid delayed surface browning, and maintained quality and market acceptability of lotus root slices (Ali et al. 2020). Salama and Aziz (2021) studied the effect of alginate coatings in combination with *Alo evera* gel and frankincense oil on green bell peppers to enhance their shelf-life as well as quality attributes. Sodium alginate containing 67% gel and 6% oil made a difference in the sense of lower weight loss and disease incidence of green bell pepper fruits as compared to control treatment, proving that the coating can be applied to other fruits and vegetables to increase their shelf-life.

23.4.5 Starch-Based Coating

Starch-based coatings are also being used for preservation of fruits and vegetables while minimizing their postharvest losses. These maintain quality of produce by reducing respiration rate and color change and prolonging shelf-life (Sapper and Chiralt 2018). Herbal edible coating containing corn starch and mint extract (as an antimicrobial agent) was applied on freshly harvested cucumber fruits and stored at 25 °C (room temperature) and 10 °C (low temperature) to prolong their shelf-life and maintain quality. The coated fruits had significantly less weight loss and better shelf-life as compared to uncoated control fruits. Further, appearance, color, texture, taste, flavor, and overall acceptability were better in herbal edible coated fruits than uncoated fruits (Raghav and Saini 2018).

Rice starch was tested as coating against biochemical changes and postharvest issues of tomato. Coating was prepared by addition of coconut oil and tea leaf extract. Tomato fruits at green mature stage were stored at room temperature for 20 days. Coatings significantly showed reduction in weight loss, and increase in titratable acidity, TSS, and other biochemical properties as compared to control ones indicating the effectiveness of coatings (Das et al. 2013).

23.5 Plant Extracts

Water and ethanol leaf extracts of leather bergenia were used with pullulan coating at different concentrations on bell pepper to investigate their antimicrobial characteristics. Pullulan coating in combination with 20% leather bergenia leaf extract effectively controlled postharvest decay, increased shelf-life, and also maintained turgidity and physicochemical and organoleptic attributes (Kraśniewska et al. 2015). *Flourensia cernua* extract was made and added to mixtures of alginate and chitosan. Tomato fruits treated with nanolaminate coating, and coating with addition of the extract, along with untreated fruits were stored at 20 °C for 15 days. Physicochemical studies, gas ratios of O_2 and CO_2, and ethylene generation of treated produce were all determined. The nanolaminate coverings reduced permeability to water and O_2. Nanolaminate + *F. cernua* extract coating was most effective treatment which prolonged shelf-life of tomatoes by lowering reduction in weight and growth of microbes, decreasing gaseous exchange and ethylene generation, and maintaining the rigidity and color of the tomato fruits (Salas-Méndez et al. 2019).

23.6 Essential Oils

These are natural compounds having antimicrobial ability. These are considered safe for consumption if recommended as coating on any vegetable. Bell pepper was coated with chitosan (0.5 or 1% solution) in addition to lemongrass essential oil (0.5 and 1%) to increase shelf-life of fruits. The fresh coated fruits were stored for 3 weeks at room temperature. Application of 1% chitosan and 0.5% essential oil coating resulted in fascinating results reducing anthracnose issue of bell pepper. Different parameters like weight loss, total soluble salts, and soluble sugars were studied. The coating increased shelf-life while also maintaining the quality of stored fruits (Ali et al. 2015). Persian hogweed oil used with chitosan on bell pepper also gave significant results as compared to control treatment by extending shelf-life up to 24 days under storage at 9 °C with acceptable flavor and appearance (Taheri et al. 2020).

Cinnamon oil was used in chitosan-based coating and applied on sweet pepper fruits, which were then kept under cold storage conditions (8 °C). Coated peppers were evaluated for disease incidence, sensory quality, and activities of antioxidant enzymes. After 35 days of storage, chitosan-cinnamon oil-coated fruits had good sensory quality, while untreated fruits lost quality and acceptability to consumers. In coated fruits, electrolyte leakage and MDA (malondialdehyde) content also increased but still lower as compared to uncoated fruits. On 35th day of storage, enzymatic activities (SOD, POD, and CAT) were also higher in coated fruits, indicating the potential of chitosan-cinnamon oil coating in maintaining quality attributes of sweet peppers (Xing et al. 2011). Cumin essential oil and basil mucilage were tested alone or in combination as coating material on tomato fruits to enhance their shelf-life. Treated and untreated (control) tomatoes were stored at 20 °C with relative humidity of 80–90% for 9 days. Coating significantly reduced decay and increased shelf-life, also preserved fruits from decay and faster weight loss as compared to control ones (Tabaestani et al. 2013). *Eruca vesicaria* (arugula) is a leafy vegetable having short postharvest life due to high rate of ethylene production that makes it yellow and un-consumable. By adding chitosan, its shelf-life was increased to 3–7 days as compared to untreated sample. Lemon essential oil has good antimicrobial properties so when lemon oil (1%) was added in chitosan (0.5%) coating, shelf-life increased to 10–14 days, not spoiling its taste and aroma, and conserving turgidity and green color (Sessa et al. 2015).

23.7 Chemical-Based Treatments

Food-borne diseases are a common problem in different countries including United States of America (USA). About more than 1300 persons die in the USA every year due to these illnesses, which necessitate formulating food safety policies and necessary interventions (Scallan et al. 2011). Chlorine dioxide (ClO_2) has proved to be a good decontaminator for fruits and vegetables reducing postharvest decay and food-borne diseases, which result in economic loss and health hazards. However, ClO_2 in gaseous form has more potential against postharvest problems rather than in liquid state; as in gaseous form it is less corrosive, easy to mix with air, and can rapidly diffuse and penetrate in porous surfaces and coverings (Sun et al. 2019). ClO_2 helps to minimize problem of contamination on potato tubers which deteriorates their quality making them unfit for consumption. Experiments showed that ClO_2 effectively controlled microorganisms, without affecting appearance and quality. It had no residual effects as it dissipated making tubers harmless to eat (Wu and Rioux 2010). Roma tomatoes were treated with ClO_2 and stored for 4 weeks at ambient temperature. ClO_2 treatment was effective in preventing pathogen growth, thus reducing economic loss. There were no residues; hence produce was safe for consumption (Trinetta et al. 2010).

Calcium chloride treatment on radish cv. Kadirli in cold storage environment (4 °C) exhibited positive results in the form of more cell turgidity, less hollowness, and less weight loss as compared to control treatment thereby increasing shelf-life. The treatment based on $CaCl_2$ with starch and glycerol combination performed better and ultimately increased shelf-life, as compared to control (Öz and Akyol 2020).

23.8 Ozone Application

Ozone is a gas being adopted recently as postharvest treatment to minimize microbial spoilage and extend postharvest life of fruits and vegetables. Iceberg lettuce and Romaine lettuce were treated with aqueous ozone (0.5 mg L^{-1}) and aqueous benzalkonium (120 mg L^{-1}) alone and in combination, and stored at 4 °C for 1 week. Results suggested that collective treatment of ozone and benzalkonium performed better for lettuce as it reduced microbial growth and preserved produce quality (Tooryan et al. 2023). Application of ozone in liquid and gaseous form was compared for its efficacy to disinfect tomato cv. FA-180 fruits at breaker stage. Ozone in liquid form was more effective than gaseous form. Treatment with ozone solution (1 mg L^{-1}) for 15 min was effective to adequately disinfect the tomatoes. Exposure to ozone treatment during storage extended shelf-life and preserved sensory quality of tomatoes (Venta et al. 2010). Ozone application to different vegetables was tested against microbial contamination. Treatment to onion and red pepper reduced disease incidence, while treated tomatoes showed minimal changes in color development, sugars, and antioxidants (Carletti et al. 2013). Cabbage (fresh-cut) was treated with ozone (1.4 mg L^{-1} of water) for different durations (1, 5, and 10 min) and kept in cold storage (4 °C) for 12 days to study its effect on the growth of microorganisms and quality characteristics of the produce. The ozone treatment for 5 and 10 min significantly reduced the growth of microbes during storage. The treatment also decreased ethylene generation and improved the quality of the produce (Liu et al. 2021).

23.9 Ultraviolet Radiation Technology

Ultraviolet (UV) radiation treatment is another way to reduce postharvest losses in vegetables, due to its germicidal activities by affecting the genetic material of microorganisms. Radiation is a form of physical treatment that not only inhibits microbes' growth and conserves the produce but also improves its biochemical nature (Ribeiro et al. 2012). UV irradiation preserves fruits and vegetables by decreasing the rate of ethylene production, delaying ripening and senescence processes, and reducing microbial spoilage and freezing injury. Radiation treatments increase antioxidants and phenolics activities, enhance shelf-life of commodities, and promote defensive mechanism against microbial activities. Three distinct types of fluorescent lamps were used to improve the flavonoid content of harvested onions. The application of radiation through these lamps led to the synthesis of quercetin in onions. Therefore, these outcomes imply that postharvest UV radiation can augment the quality of onions (Higashio et al. 2005). UV (A-range, 320–400 nm) irradiation effect was studied on physicochemical attributes and antioxidant properties of tomato fruit. UV irradiation did not cause any adverse effect on the physicochemical attributes of tomatoes. However, total phenolics, carotenoids, and flavonoids contents increased, and hence shelf-life was prolonged (Dyshlyuk et al. 2020). African leafy vegetables, nightshade, and amaranth were exposed to UV (C-range, 254 nm) irradiation to control microbial growth and prolong the shelf-life of vegetables. These vegetables were kept under controlled temperatures of 5 °C with 85% RH and 20 °C with 65% RH for 20 and 4 days, respectively. Radiation reduced weight loss, decreased microbial counts, increased health-promoting secondary metabolites, and enhanced shelf-life of both vegetables as compared to non-treated ones (Gogo et al. 2017). Cucumber fruits exhibited positive changes such as increased glow, turgidity, and variations in electrical properties following UV-C (253.7 nm) irradiation. These findings suggest that UV-C irradiation may contribute in extending shelf-life of cucumbers (Imaizumi et al. 2018).

23.10 Conclusion

Numerous types of postharvest approaches have been reported for postharvest shelf-life extension and quality management of vegetables such as cold storage, controlled atmosphere, modified atmosphere packaging, use of natural elicitors, edible coatings, and essential oils. The efficacy of the applied postharvest handling approach generally depends upon its nature and time of application from harvesting to retail market conditions. This is important that the combined use of some of the technologies has been found more effective in preserving postharvest quality and extending storage/shelf-life potential in contrast to their alone application. Postharvest handling protocols and treatments should be optimized crop by crop and even on a cultivar-to-cultivar basis in general.

References

Aider M (2010) Chitosan application for active bio-based films production and potential in the food industry. LWT-Food Sci Technol 43(6):837–842

Ali A, Maqbool M, Ramachandran S, Alderson PG (2010) Gum Arabic as a novel edible coating for enhancing shelf-life and improving postharvest quality of tomato (*Solanum lycopersicum* L.) fruit. Postharvest Biol Technol 58(1):42–47

Ali A, Noh NM, Mustafa MA (2015) Antimicrobial activity of chitosan enriched with lemongrass oil against anthracnose of bell pepper. Food Pack Shelf Life 3:56–61

Ali S, Anjum MA, Nawaz A, Naz S, Hussain S, Ejaz S, Sardar H (2020) Effect of pre-storage ascorbic acid and Aloe vera gel coating application on enzymatic browning and quality of lotus root slices. J Food Biochem 44(3):e13136

Al-Juhaimi F, Ghafoor K, Babiker EE (2012) Effect of gum arabic edible coating on weight loss, firmness and sensory characteristics of cucumber (*Cucumis sativus* L.) fruit during storage. Pak J Bot 44(4):1439–1444

Alvarez MV, Ponce AG, Moreira MDR (2013) Antimicrobial efficiency of chitosan coating enriched with bioactive compounds to improve the safety of fresh cut broccoli. LWT-Food Sci Technol 50(1):78–87

Ayala-Zavala JF, del Toro-Sánchez L, Alvarez-Parrilla E, Soto-Valdez H, Martín-Belloso O, Ruiz-Cruz S, González-Aguilar G (2008) Natural antimicrobial agents incorporated in active packaging to preserve the quality of fresh fruits and vegetables. Stewart Postharvest Rev 4(3):1–9

Badawy ME, Rabea EI (2009) Potential of the biopolymer chitosan with different molecular weights to control postharvest gray mold of tomato fruit. Postharvest Biol Technol 51(1):110–117

Bahri MH, Rashidi M (2009) Effects of coating methods and storage periods on some qualitative characteristics of carrot during ambient storage. Int J Agric Biol 11(4):443–447

Bilal M, Gul I, Basharat A, Qamar SA (2021) Polysaccharides-based bio-nanostructures and their potential food applications. Int J Biol Macromol 176:540–557

Carletti L, Botondi R, Moscetti R, Stella E, Monarca D, Cecchini M, Massantini R (2013) Use of ozone in sanitation and storage of fresh fruits and vegetables. J Food Agric Environ 11(3–4):585–589

Castro-Cegrí A, Ortega-Muñoz M, Sierra S, Carvajal F, Santoyo-Gonzalez F, Garrido D, Palma F (2023) Application of polysaccharide-based edible coatings to improve the quality of zucchini fruit during postharvest cold storage. Sci Hortic 314:111941

Cerit I, Demirkol O (2020) Effects of modified atmosphere packaging conditions and ethylene absorber on the quality of red bell pepper. J Food Nutr Res 59(1):35–43

Chauhan OP, Nanjappa C, Ashok N, Ravi N, Roopa N, Raju PS (2015) Shellac and Aloe vera gel-based surface coating for shelf life extension of tomatoes. J Food Sci Technol 52:1200–1205

Chetti MB, Deepa GT, Antony RT, Khetagoudar MC, Uppar DS, Navalgatti CM (2014) Influence of vacuum packaging and long-term storage on quality of whole chili (*Capsicum annuum* L.). J Food Sci Technol 51:2827–2832

Chiebao HP, Jenott JR, Unruh DA, Gragg SE, Rivard CL, Pliakoni ED (2018) Postharvest treatments to improve quality and safety of locally-grown vegetables stored at non-optimum temperatures. Acta Hortic 1194:1353–1360

Chitravathi K, Chauhan OP, Raju PS, Madhukar N (2015) Efficacy of aqueous ozone and chlorine in combination with passive modified atmosphere packaging on the postharvest shelf-life extension of green chilies (*Capsicum annuum* L.). Food Bioprocess Technol 8:1386–1392

Das DK, Dutta H, Mahanta CL (2013) Development of a rice starch-based coating with antioxidant and microbe-barrier properties and study of its effect on tomatoes stored at room temperature. LWT-Food Sci Technol 50(1):272–278

De Corato U (2020) Improving the shelf-life and quality of fresh and minimally-processed fruits and vegetables for a modern food industry: a comprehensive critical review from the traditional technologies into the most promising advancements. Crit Rev Food Sci Nutr 60(6):940–975

Devgan K, Kaur P, Kumar N, Kaur A (2019) Active modified atmosphere packaging of yellow bell pepper for retention of physico-chemical quality attributes. J Food Sci Technol 56:878–888

Dirapan P, Boonyakiat D, Poonlarp P (2021) Improving shelf life, maintaining quality, and delaying microbial growth of broccoli in supply chain using commercial vacuum cooling and package icing. Horticulturae 7(11):506

Dyshlyuk L, Babich O, Prosekov A, Ivanova S, Pavsky V, Chaplygina T (2020) The effect of postharvest ultraviolet irradiation on the content of antioxidant compounds and the activity of antioxidant enzymes in tomato. Heliyon 6(1):e03288

El-Ramady HR, Domokos-Szabolcsy É, Abdalla NA, Taha HS, Fári M (2015) Postharvest management of fruits and vegetables storage. Sustain Agric Rev 15:65–152

Galani JH, Patel JS, Patel NJ, Talati JG (2017) Storage of fruits and vegetables in refrigerator increases their phenolic acids but decreases the total phenolics, anthocyanins and vitamin C with subsequent loss of their antioxidant capacity. Antioxidants 6(3):59

Gogo EO, Opiyo AM, Hassenberg K, Ulrichs C, Huyskens-Keil S (2017) Postharvest UV-C treatment for extending shelf life and improving nutritional quality of African indigenous leafy vegetables. Postharvest Biol Technol 129:107–117

Hasan MU, Malik AU, Anwar R, Khan AS, Haider MW, Riaz R, Ali S, Rehman RNU, Ziaf K (2021) Postharvest Aloe vera gel coating application maintains the quality of harvested green chilies during cold storage. J Food Biochem 45(4):e13682

Helland HS, Leufvén A, Bengtsson GB, Skaret J, Lea P, Wold AB (2016) Storage of fresh-cut swede and turnip in modified atmosphere: effects on vitamin C, sugars, glucosinolates and sensory attributes. Postharvest Biol Technol 111:150–160

Higashio H, Hirokane H, Sato F, Tokuda S, Uragami A (2005) Effect of UV irradiation after the harvest on the content of flavonoid in vegetables. Acta Hortic 682:1007–1012

Hong SI, Kim D (2004) The effect of packaging treatment on the storage quality of minimally processed bunched onions. Int J Food Sci Technol 39(10):1033–1041

Horev B, Sela S, Vinokur Y, Gorbatsevich E, Pinto R, Rodov V (2012) The effects of active and passive modified atmosphere packaging on the survival of *salmonella enterica* serotype Typhimurium on washed romaine lettuce leaves. Food Res Int 45(2):1129–1132

Ikegaya A, Yamazaki S, Yamaga I, Kosugi T, Toyoizumi T, Nagafuji A, Saika T, Arai E (2023) Controlled atmosphere maintains the quality of certain fresh produce in mixed cargo shipments. J Food Process Eng 46(4):e14297. https://doi.org/10.1111/jfpe.14297

Imaizumi T, Yamauchi M, Sekiya M, Shimonishi Y, Tanaka F (2018) Responses of phytonutrients and tissue condition in persimmon and cucumber to postharvest UV-C irradiation. Postharvest Biol Technol 145:33–40

Jalali A, Seiiedlou S, Linke M, Mahajan P (2017) A comprehensive simulation program for modified atmosphere and humidity packaging (MAHP) of fresh fruits and vegetables. J Food Eng 206:88–97

Kaewklin P, Siripatrawan U, Suwanagul A, Lee YS (2018) Active packaging from chitosan-titanium dioxide nanocomposite film for prolonging storage life of tomato fruit. Int J Biol Macromol 112:523–529

Kargwal R, Garg MK, Singh VK, Garg R, Kumar N (2020) Principles of modified atmosphere packaging for shelf life extension of fruits and vegetables: an overview of storage conditions. Int J Chem Stud 8(3):2245–2252

Kowalczyk D, Skrzypek T, Łupina K (2020) Effect of carboxymethyl cellulose/candelilla wax edible coating incorporated with ascorbic acid on the physicochemical and sensory qualities of prepackaged minimally processed carrots (*Daucus carota* L.) during cold storage. J Food Process Preserv 44(9):e14713. https://doi.org/10.1111/jfpp.14713

Kraśniewska K, Gniewosz M, Synowiec A, Przybył JL, Bączek K, Węglarz Z (2015) The application of pullulan coating enriched with extracts from *Bergenia crassifolia* to control the growth of food microorganisms and improve the quality of peppers and apples. Food Bioprod Process 94:422–433

Liu C, Chen C, Jiang A, Zhang Y, Zhao Q, Hu W (2021) Effects of aqueous ozone treatment on microbial growth, quality, and pesticide residue of fresh-cut cabbage. Food Sci Nutr 9(1):52–61

Liu Y, Chen T, Tao N, Yan T, Wang Q, Li Q (2023) Nitric oxide is essential to keep the postharvest quality of fruits and vegetables. Horticulturae 9(2):135

Lwin WW, Pongprasert N, Boonyaritthongchai P, Wongs-Aree C, Srilaong V (2020) Synergistic effect of vacuum packaging and cold shock reduce lignification of asparagus. J Food Biochem 44(11):e13479

Maan AA, Ahmed ZFR, Khan MKI, Riaz A, Nazir A (2021) Aloe vera gel, an excellent base material for edible films and coatings. Trends Food Sci Technol 116:329–341

Manoj H, Sreenivas K, Shankarappa T, Krishna H (2016) Studies on chitosan and Aloe vera gel coatings on biochemical parameters and microbial population of bell pepper (*Capsicum annuum* L.) under ambient condition. Int J Curr Microbiol Appl Sci 5(1):399–405

Mohebbi M, Hasanpour N, Ansarifar E, Amiryousefi MR (2014) Physicochemical properties of bell pepper and kinetics of its color change influenced by Aloe vera and gum tragacanth coatings during storage at different temperatures. J Food Process Preserv 38(2):684–693

Nasrin TAA, Yasmin L, Arfin MS, Rahman MA, Molla MM, Sabuz AA, Afroz M (2022) Preservation of postharvest quality of fresh cut cauliflower through simple and easy packaging techniques. Appl Food Res 2(2):100125

Nicolau-Lapeña I, Colas-Meda P, Alegre I, Aguiló-Aguayo I, Muranyi P, Viñas I (2021) Aloe vera gel: an update on its use as a functional edible coating to preserve fruits and vegetables. Prog Org Coat 151:106007

Öz AT, Akyol B (2020) Effects of calcium chloride plus coating in modified-atmosphere packaging storage on whole-radish postharvest quality. J Sci Food Agric 100(10):3942–3949

Park HJ, Chinnan MS, Shewfelt RL (1994) Edible coating effects on storage life and quality of tomatoes. J Food Sci 59(3):568–570

Paulsen E, Barrios S, Lema P (2019) Ready-to-eat cherry tomatoes: passive modified atmosphere packaging conditions for shelf life extension. Food Pack Shelf Life 22:100407

Prange RK, DeLong JM, Daniels-Lake BJ, Harrison PA (2005) Innovation in controlled atmosphere technology. Stewart Postharvest Rev 1(3):1–11

Raghav PK, Saini M (2018) Development of mint (*Mentha viridis* L.) herbal edible coating for shelf life enhancement of cucumber (*Cucumis sativus*). Int J Green Herbal Chem 7(2):379–391

Raju PS, Chauhan OP, Bawa AS (2010) Postharvest handling systems and storage of vegetables. In: Sinha NK (ed) Handbook of vegetables and vegetable processing. Wiley-Blackwell, Ames, pp 185–198

Ribeiro C, Canada J, Alvarenga B (2012) Prospects of UV radiation for application in postharvest technology. Emirates J Food Agric 24(6):586–597

Saekow M, Naradisorn M, Tongdeesoontorn W, Hamauzu Y (2019) Effect of carboxymethyl cellulose coating containing ZnO-nanoparticles for prolonging shelf life of persimmon and tomato fruit. J Food Sci Agric Technol 5:41–48

Sahoo NR, Bal LM, Pal US, Sahoo D (2015) Effect of packaging conditions on quality and shelf-life of fresh pointed gourd (*Trichosanthes dioica* Roxb.) during storage. Food Pack Shelf Life 5:56–62

Salama HE, Aziz MSA (2021) Development of active edible coating of alginate and Aloe vera enriched with frankincense oil for retarding the senescence of green capsicums. LWT 145:111341

Salas-Méndez EJ, Vicente A, Pinheiro AC, Ballesteros LF, Silva P, Rodríguez-García RD, Hernández-Castillo FD, Díaz-Jiménez ML, Flores-López ML, Villarreal-Quintanilla JA, Peña-Ramos FM, Carrillo-Lomelí DA, de Rodríguez DJ (2019) Application of edible nanolaminate coatings with antimicrobial extract of *Flourensia cernua* to extend the shelf-life of tomato (*Solanum lycopersicum* L.) fruit. Postharvest Biol Technol 150:19–27

Salehi F (2020) Edible coating of fruits and vegetables using natural gums: a review. Int J Fruit Sci 20(Sup 2):S570–S589

Sapper M, Chiralt A (2018) Starch-based coatings for preservation of fruits and vegetables. Coatings 8(5):152

Scallan E, Hoekstra RM, Angulo FJ, Tauxe RV, Widdowson MA, Roy SL, Jones JL, Griffin PM (2011) Foodborne illness acquired in the United States – major pathogens. Emerg Infect Dis 17(1):7–15

Sessa M, Ferrari G, Donsì F (2015) Novel edible coating containing essential oil Nano emulsions to prolong the shelf life of vegetable products. Chem Eng Trans 43:55–60

Steelheart C, Alegre ML, Bahima JV, Senn ME, Simontacchi M, Bartoli CG, Grozeff GE (2019) Nitric oxide improves the effect of 1-methylcyclopropene extending the tomato (*Lycopersicum esculentum* L.) fruit postharvest life. Sci Hortic 255:193–201

Sun X, Baldwin E, Bai J (2019) Applications of gaseous chlorine dioxide on postharvest handling and storage of fruits and vegetables – a review. Food Control 95:18–26

Tabaestani HS, Sedaghat N, Pooya ES, Alipour A (2013) Shelf life improvement and postharvest quality of cherry tomato (*Solanum lycopersicum* L.) fruit using basil mucilage edible coating and cumin essential oil. Int J Agron Plant Prod 4(9):2346–2353

Taheri A, Behnamian M, Dezhsetan S, Karimirad R (2020) Shelf life extension of bell pepper by application of chitosan nanoparticles containing Heracleum persicum fruit essential oil. Postharvest Biol Technol 170:111313

Tahir HE, Xiaobo Z, Mahunu GK, Arslan M, Abdalhai M, Zhihua L (2019) Recent developments in gum edible coating applications for fruits and vegetables preservation: a review. Carbohydr Polym 224:115141

Tay SL, Perera CO (2004) Effect of 1-methylcyclopropene treatment and edible coatings on the quality of minimally processed lettuce. J Food Sci 69(2):fct135

Tooryan F, Naghashan M, Anvar N, Mortazavian A, Shariatifar N, Sohrabvandi S (2023) Effect of different treatments of aqueous benzalkonium chloride (BAC) and aqueous ozone (O3) on microbiological and sensory qualities of Iceberg lettuce and Romaine lettuce. Preprint submitted for publication. Available at SSRN: https://ssrn.com/abstract=4332277 or https://doi.org/10.2139/ssrn.4332277. Accessed 14 Aug 2023

Trinetta V, Morgan MT, Linton RH (2010) Use of high-concentration-short-time chlorine dioxide gas treatments for the inactivation of salmonella enterica spp. inoculated onto Roma tomatoes. Food Microbiol 27(8):1009–1015

Ullah A, Abbasi NA, Shafique M, Qureshi AA (2017) Influence of edible coatings on biochemical fruit quality and storage life of bell pepper cv. "Yolo Wonder". J Food Qual 2017:2142409

USFDA (2013) GRAS Notice (GRN) No. 397. Chitosan from Aspergillus Niger. GRAS notice inventory. US Food and Drug Administration, Silver Spring, MD 20993 USA

Venta MB, Broche SSC, Torres IF, Pérez MG, Lorenzo EV, Rodriguez YR, Cepero SM (2010) Ozone application for postharvest disinfection of tomatoes. Ozone Sci Eng 32(5):361–371

Werner BG, Hotchkiss JH (2005) Modified atmosphere packaging. In: Sapers GM, Gorny JR, Yousef AE (eds) Microbiology of fruits and vegetables. CRC Press, Boca Raton, pp 437–460

Wu VC, Rioux A (2010) A simple instrument-free gaseous chlorine dioxide method for microbial decontamination of potatoes during storage. Food Microbiol 27(1):179–184

Xing Y, Li X, Xu Q, Yun J, Lu Y, Tang Y (2011) Effects of chitosan coating enriched with cinnamon oil on qualitative properties of sweet pepper (Capsicum annuum L.). Food Chem 124(4):1443–1450

Zapata PJ, Guillén F, Martínez-Romero D, Castillo S, Valero D, Serrano M (2008) Use of alginate or zein as edible coatings to delay postharvest ripening process and to maintain tomato (*Solanum lycopersicon* Mill) quality. J Sci Food Agric 88(7):1287–1293

Zhang Z, Nakano K, Maezawa S (2009) Comparison of the antioxidant enzymes of broccoli after cold or heat shock treatment at different storage temperatures. Postharvest Biol Technol 54(2):101–105

Zudaire L, Viñas I, Abadias M, Lafarga T, Bobo G, Simó J, Aguiló-Aguayo I (2020) Effects of long-term controlled atmosphere storage, minimal processing, and packaging on quality attributes of calçots (*Allium cepa* L.). Food Sci Technol Int 26(5):403–412

Postharvest Management of Vegetable Seeds

24

Tanveer Ahmad, Muhammad Anwar-ul-Haq, Anam Noor, Sajjad Akhtar, and Taki Demir

Abstract

High-quality seeds are the foundation of successful vegetable production. This chapter ventures beyond the harvest, exploring the critical postharvest practices which ensure that vegetable seeds reach growers with optimal viability and performance. It delves into meticulous procedures for cleaning, drying, and storage, emphasizing the techniques that maximize seed longevity and minimize losses. Seed health treatments are also discussed to safeguard seed quality and prevent the spread of diseases. This chapter integrates these crucial elements (meticulous postharvest management and quality enhancement) to underscore their combined role in ensuring the profitability and sustainability of the vegetable seed industry as well as the growers, contributing to successful vegetable production and a robust food system.

T. Ahmad (✉)
Department of Horticulture, MNS University of Agriculture, Multan, Pakistan
e-mail: tanveer.ahmad@mnsuam.edu.pk

M. Anwar-ul-Haq
Department of Horticulture, MNS University of Agriculture, Multan, Pakistan

Department of Horticulture, Bahauddin Zakariya University, Multan, Pakistan

Department of Life Science, University of Modena and Reggio Emilia, Reggio Emilia, Italy

A. Noor
Department of Horticulture, Bahauddin Zakariya University, Multan, Pakistan

S. Akhtar
Rachna Agribusiness, Gujranwala, Pakistan

T. Demir
Department of Horticulture, Sakarya University of Applied Sciences, Sakarya, Türkiye

Z. Iqbal et al. (eds.), *Sustainable and Innovative Vegetable Production in times of Climate Change*, https://doi.org/10.1007/978-981-96-6283-8_24

Keywords

Seed aging and deterioration · Seed processing · Seed quality · Seed storage · Seed testing

24.1 Introduction

Vegetable seed production, a crucial aspect of global food security, operates at the nexus of agriculture, technology, and business. As the demand for high-quality seeds escalates, the postharvest management of vegetable seeds has emerged as a pivotal factor in sustaining agricultural productivity. Recent studies have underscored the crucial role of postharvest management in preserving seed quality. Jarrar et al. (2023) have demonstrated the impact of drying techniques on seed viability, highlighting the need for precise drying methods to mitigate quality degradation during storage. Seed processing technologies have accentuated the role of modern machinery in ensuring seed cleaning and grading efficiency and effectiveness (Dadlani 2023). Quality control at the postharvest level is a pressing research theme. Rana et al. (2022) have emphasized the importance of appropriate seed testing and certification protocols to uphold vegetable seeds' genetic purity and viability. This aligns with seed certification agencies' broader objectives to establish and sustain seed germination and purity standards. Recent seed coating and treatment innovations reveal promising avenues for enhancing seed performance and resistance to environmental stressors such as temperature, humidity, and light. Additionally, sustainable packaging solutions are minimizing the ecological footprint of seed distribution (Eissenberger et al. 2023; Pagano et al. 2023).

24.1.1 Importance of Postharvest Management in Vegetable Seed Industry

Postharvest management in the vegetable seed industry is paramount globally. Efficient postharvest practices are essential for preserving vegetable seeds' genetic purity, viability, and overall quality, ensuring their successful propagation during subsequent planting seasons. Any contamination or mixing of seeds during harvesting, processing, or storage can lead to adverse genetic variations, affecting the performance and characteristics of the following crops (Engels and Ebert 2021). Besides, proper drying, processing, and storage are also essential for seed viability and longevity. Hence, well-managed postharvest practices help in preventing premature seed germination, ensuring seeds to retain vitality over extended periods and facilitating their distribution to diverse regions and climates. Appropriate postharvest management reduces seed wastage, ensuring that a considerable proportion of seeds produced are suitable for planting.

Vegetables constitute a significant portion of agribusiness output. Effective postharvest practices directly impact the agricultural sector's economic viability, influencing many farmers' income and livelihoods. Proper postharvest management ensures a stable supply of high-quality seeds, which, in turn, supports the production of healthy and abundant vegetables, contributing to food security. Under the current climate change scenario, effective postharvest management practices, such as seed coating for stress tolerance, can enhance vegetable crops' resilience to adverse environmental conditions, contributing to sustained agricultural productivity (Kamal et al. 2022).

24.1.2 Postharvest Practices and Seed Quality

Seed quality is directly influenced by the procedures implemented during and after harvesting. Understanding and optimizing postharvest practices is essential for preserving vegetable seeds' genetic integrity, viability, and overall health. Here are key points to consider when exploring the relation between postharvest practices and seed quality.

- *Genetic Purity Preservation*: It includes measures to prevent the mixing of different varieties or species of vegetables (Yousaf et al. 2016).
- *Seed Viability:* Proper drying techniques help to lower seed moisture content to a level conducive to storage without compromising viability. Inadequate drying can lead to fungal growth and seed deterioration (Kameswara Rao et al. 2017; Osei et al. 2020).
- *Seed Health:* Thorough cleaning and sorting during postharvest processing removes debris, damaged seeds, and potential contaminants. This minimizes the risk of disease transmission and ensures that only healthy seeds are selected for storage and distribution (Sundareswaran et al. 2023).
- *Preventing Physical Damage:* Seeds are susceptible to physical damage during harvesting and processing. Gentle handling practices, including careful threshing and separation methods, prevent bruising or cracking of seeds, preserving their integrity and viability (Osei et al. 2020).
- *Storage Conditions:* Maintaining optimal temperature and humidity levels in storage facilities is essential for preventing seed deterioration. Proper ventilation and pest protection contribute to maintain seed quality over time (Dyck et al. 2023).
- *Quality Control Measures:* Seed testing evaluates quality factors such as germination rates, genetic purity, and seed health, providing valuable information about the overall quality of the seed lot (Kameswara Rao et al. 2017; Sundareswaran et al. 2023).
- *Technological Advancements:* Postharvest technologies, such as seed coating and treatment, enhance seed performance and resilience. Coating technologies may include protective layers that protect against environmental stressors and pathogens (Kameswara Rao et al. 2017).
- *Traceability and Documentation:* Maintaining comprehensive records of postharvest practices, including seed source, harvest conditions, and processing methods, facilitate traceability. This information is critical for quality assurance (Osei et al. 2020; Dyck et al. 2023).

- *Education and Training:* Educating farmers and industry professionals about best practices in postharvest management is crucial. Training programs can improve awareness and encourage adoption of postharvest practices for maintaining seed quality (Dyck et al. 2023).

The relationship between postharvest practices and seed quality is complex and multifaceted. By prioritizing genetic purity, seed viability, and overall health, the vegetable seed industry can ensure the production and distribution of high-quality seeds, contributing to agriculture's success and sustainability.

24.2 Seed Harvesting

Seed harvesting is a critical stage in seed production. Moisture content and physiological maturity are crucial factors that help in determining seed harvesting time. Different harvesting techniques (manual and mechanical) are used nowadays for seed harvesting. Hand-picking is the most used method for harvesting open-pollinated and hybrid vegetable seeds. After hand-picking, pods/fruits are dried under the sun, threshed by hand, and the seed is cleaned by winnowing. Mechanical harvesting is used for large-scale seed production.

24.2.1 Optimal Harvesting Time for Different Vegetable Crops

The right time for harvesting different vegetable seeds depends on their physiological maturity. The physiological maturity of some important vegetables for seed harvesting is as under:

Lady's finger	Pod color changes from green to brown and pods dry.
Tomato	Fruit color changes from green to orange or red.
Capsicum and chili	Fruit color turns green to red.
Brinjal or aubergine	Fruit color changes from green to yellow or dull purple.
Bitter gourd	Fruit color changes from green to yellow orange.
Ridge gourd	Fruit will dry on the vine with cracked skin.
Cucumber	Fruit color changes from greenish to pale yellow or golden.
Bottle gourd	Mature fruit sounds hollow when tapped. Fruit stalk color changes from green to brown.
Squash	Mature fruits' color changes from green to yellow or orange and golden yellow to straw yellow. The rind becomes harder.
Pumpkin	Fruit color changes from green to yellow, and the stalk becomes dry.
Radish	Mature pods have a pearl brown color.
Beans	Pods color changes from green to yellow.
Carrot	Seed heads turn green to brown color.
Coriander	Mature seeds turn yellowish green and become harder.

24.3 Postharvest Handling to Minimize Seed Damage

Compared to dry grains and legumes, vegetable seeds may have more moisture content and smaller size and are more difficult to manage (Rehman et al. 2019). Thus, extra care must be taken during processing and storage to minimize the danger of seed deterioration.

24.3.1 Seed Drying and Processing

Drying of seed is the process of lowering the moisture content of seeds to a level advised for storage, while ensuring that it does not negatively impact the seeds. Seeds have different moisture contents depending on the kind of seed, maturity stage, harvesting techniques, relative humidity, and seasonal variations. Higher moisture in fresh seeds promotes invasion of insects and fungi, and germination of seed embryos.

Seed drying to an optimal moisture level enables long-term seed storage (McCormack 2004; Rao et al. 2006). Previous research indicates that drying techniques may affect chemical and biological properties of seeds, as significant variations in chemical composition are noted across various drying techniques (Chua et al. 2019).

The speed at which moisture evaporates from a seed's surface determines how quickly it dries. This, in turn, is determined by the drying air's temperature and relative humidity. There is a link between initial seed moisture content and safe drying temperature. Drying seeds at temperatures of 40, 37, and 32 °C is advised for agricultural crops with moisture levels of no more than 18, 10–18, and less than 10%, respectively. Drying of seeds at temperatures higher than 40 °C reduces their viability. The seed is pushed to cool down after drying before being stored. The recommended moisture content range for certain crop seeds to be stored safely is 10–14% (Rao et al. 2006).

24.3.2 Seed Drying Techniques

24.3.2.1 Natural Sun Drying

When moist seed is dried naturally, the air in the atmosphere circulates on the fields or floors. Farmers have long used the sun and wind to dry little amounts of seed by spreading the threshed seed thinly on a floor. There is no additional cost or need for specific equipment for sun drying. The two main drawbacks of sun drying are the potential for mechanical admixtures and weather damage. Due to hot temperatures and UV radiation caused by direct sunlight, seed germination can also be negatively impacted, particularly if the seed has high moisture content. Seed should not be exposed to the sun for an extended period. Sun drying is quite effective for drying lesser amounts of seed and it takes only 3–4 days to attain a desirable seed moisture level, i.e., 12%. Weather, day and night temperatures, and humidity are crucial in sun drying.

24.3.2.2 Mechanical Drying

Mechanical drying is the best option if the weather is not good for drying seeds in the sun. Seed dryers are used for this purpose. Depending upon the dryer's capacity, about 12–28 tons of seed is placed in it. Warm (40 °C) air is passed through the wet seeds and these dry to 12% moisture within 70–110 h. Dry seed is packed in jute bags and stored in field until seed cleaning and grading are started.

Artificial drying, commonly called forced air drying or dehydration, involves a mechanical device forcing hot or cold air through a dryer. A layer of seeds is blown through either with natural air or air that has been heated until the drying process is completed. The two types of driers employed are continuous flow and batch driers.

A batch drier blows dry air through a seed layer until it dries to a desired moisture level. The seed is then removed and replaced with a new batch of seed. The process is easy to use and works well with little seed. In continuous flow dryers, the seed passes through a heated air stream vertically or horizontally before entering a chilling chamber. This method is used to dry large amounts of seed continuously in a factory setting.

24.3.3 Seed Cleaning and Grading

When the seeds are harvested from the field, they frequently contain contaminants (such as dirt, stones, plant waste, weeds' seeds, and the seeds of other plants), which should be removed to produce pure, viable crop seeds for sowing. To enhance and improve seed quality, cleaning involves removing inert materials, seeds of weeds, different crops, and also of the same crop that are cracked, broken, degraded, or infected (Rao et al. 2006). Cleaning eliminates damp and plant residue that could result in heating in storage and lowers the quality of the seeds to be stored. Various sets of machines and machine-combination sequences are required for multiple crops. Cleaning often involves different steps including basic cleaning, grading, and conditioning depending upon the crop.

Pre-cleaning, a partial cleaning process, is the removal of debris that is coarse enough for screens to separate. Pre-cleaning involves removing the bulkiest material and the trash most likely to clog conveyors and sieves using a combination of big mesh screens or cylinders and an air blast. Therefore, it makes easier for the seeds to pass through the machine during further cleaning steps. Seed-sized particles can flow through vibrating screens or rotating cylinders, while larger particles such as chaff, pods, stems, and leaves are retained and swept to the side.

Basic cleaning is a second cleaning step that may be used for any seed and involves vibrating screens and air blasts. It takes cleaning procedures one step further. Most basic cleaning is done using an air/screen cleaner apparatus. A sequence of screens separates particles bigger and smaller than the crop seeds, which might be broken or weed seeds, and the air blast eliminates lighter material. Certain seed lots need additional cleaning after the basic cleaning process to remove the contaminants which are too similar in size and shape to the pure seeds.

Depending on the kinds of crops, there are two different cleaning techniques, i.e., wet cleaning and dry cleaning (McCormack 2004). Wet cleaning can clean seeds in damp flesh of seeds. After being removed from the flesh of a mature fruit, the seeds are placed in a container and thoroughly scrubbed with coarse sand to eliminate any surrounding flesh. Then, the seeds are placed in a strainer and frequently rinsed under running water to remove the little pieces of flesh and mucilage. Seeds (e.g., tomato, muskmelon, watermelon, cucumber, etc.) are dried for 10 days after such washing before being stored. Dry cleaning involves cleaning developed seeds in dry capsules or pods. Dry pods or pod-bearing plants are harvested at one time, dried in shade, and then threshed to collect the seeds. After threshing, seeds are winnowed and stored. Examples include corn, peas, beans, okra, and radish (Lavezzo 2016).

During grading, broken, damaged, or faulty seeds are removed as they may have less vigor and capacity to germinate. Grading involves eliminating seeds bigger or smaller than the appropriate size. Based on consistent size and shape, most seed cleaning machines concurrently split the seed into first, second, etc., grades. More accurate sizing and grading methods can provide seeds of a consistent size and shape, which is necessary for mechanized planting.

24.4 Seed Storage and Preservation

In terms of storage, seeds fall into two categories: orthodox and recalcitrant seeds (Hartmann et al. 1997). Orthodox seeds can be dried to extremely low moisture levels and stored at very low temperatures without deteriorating. All legumes and most small-seeded vegetables had orthodox seeds. Recalcitrant seeds, due to their inability to dry, cannot be frozen or kept at below-freezing temperatures. Examples include *mango, lychee, avocado, mangosteen,* etc. The International Plant Genetic Resource Institute (IPGRI) does not advocate prolonged preservation conditions for recalcitrant species. Currently, no medium- or long-term seed storage techniques work for recalcitrant seeds (Ellis et al. 1985).

24.4.1 Factors Affecting Seed Storage and Longevity

Many factors including field conditions prior to harvesting (such as high temperature, humidity, and moisture), maturity at harvest, damage from machinery, fungi, insects, pests (such as rodents and birds), and treatment of seeds (fumigation) influence seed storage (Justice and Bass 1978). Seeds that have been damaged mechanically are less storable as they degenerate quickly and are also more vulnerable to attack of storage fungi. As the storage temperature and seed moisture content rise, the life span of seeds diminishes. According to Harrington and Douglas (1970), for seeds with a 5–14% moisture level, every 1% decrease in moisture level doubles the seed's storage life, while every 5 °C rise in storage temperature makes half the seed's life span.

Aspergillus and *Penicillium* species are the most common storage fungi and can damage seeds kept at relative humidity levels between 65% and 100% (Justice and Bass 1978). Storage fungi can develop quickly in conditions with high seed moisture content and/or high relative humidity in the atmosphere. Insects that feed on endosperm and embryos seriously harm seeds that are kept at 9% or higher moisture content. Pests such as rodents and birds consume seeds, destroy bags, and disperse, mix, and infect seeds (Justice and Bass 1978).

Certain seed treatment chemicals can occasionally negatively impact seeds that are very wet, mechanically damaged, or have been kept for extended periods of time. Certain agricultural seeds may not survive certain treatments, and a high treatment rate can also harm those seeds. Under some circumstances, some fumigants harm seed viability; the risk of damage increases with seed moisture content, temperature, fumigant dosage, and exposure duration (Justice and Bass 1978).

The characteristics of seeds determine how quickly they deteriorate during storage. According to Ellis and Hong (2006), orthodox seeds may withstand drying up to 0.03–0.07 g H_2O g^{-1} DW (dry weight) and storage at around −10 °C. Recalcitrant seeds have a tolerance limit of 0.2–0.3 g H_2O g^{-1} DW and cannot withstand extreme drying or long-term storage. For this reason, it is challenging to maintain them with conventional seed bank techniques.

24.4.2 Ageing and Deterioration in Vegetable Seeds

Seed longevity is a time period over which it retains its viability. It is a complex characteristic that varies among different species and even seed lots of the same species (Nagel and Börner 2010). The seed longevity is influenced by genotype, seed production environment, and storage conditions.

Longevity or storage life of different vegetable seeds under optimum storage conditions is given in Table 24.1. Seed shape and oxygen permeability of seed coat also impact this. Seed longevity can be increased by slowing down biochemical deterioration, especially oxidative damage. Elevated temperature, relative humidity, and oxygen content enhance the pace at which seeds deteriorate while being stored (Groot et al. 2012). Traditional artificial seed aging involves incubating them at high temperatures and high relative humidity levels, which raises the equilibrium seed moisture content. These artificial aging tests, which simulate seed degeneration in a comparatively short amount of time, are typically used in experiments to predict the lifespan of seeds. These include controlled deterioration assays and accelerated aging assays; both involve incubating seeds at elevated temperatures and humidity levels to produce elevated equilibrium seed moisture content (Ellis and Hong 2006).

A decrease in germination speed also indicates that seeds have first faced a loss of seed vigor before experiencing a loss of viability. Aging reduces the quality of crop seeds, first showing up as decreased vigor (germination performance) and eventually as decreased viability (seed death). Aging diminishes a seed's capacity to generate healthy seedlings in addition to vigor and viability loss (Groot et al. 2012; Schausberger et al. 2019). Understanding the molecular mechanism underlying the

Table 24.1 Longevity (years) of vegetable seeds under optimal storage conditions

Crop Name	Seed longevity				References
	Less than 1 year	1–2 years	3–4 years	More than 4 years	
Onion	✓				Manna et al. (2020)
Asparagus			✓		Mahaney (2014) Winterfeldt (2020) Ells et al. (2020) Iannotti (2022) Snively (2024)
Beet				✓	
Broccoli			✓		
Beans			✓		
Leek		✓			
Parsnip	✓				
Carrot		✓			
Celery			✓		
Cucumber				✓	
Endive				✓	
Okra		✓			
Kale			✓		
Pea			✓		
Pepper		✓			
Spinach			✓		
Cabbage			✓		Maynard and Hochmuth (2007) Ells et al. (2020)
Rutabaga			✓		
Cauliflower			✓		
Lettuce				✓	
Muskmelon				✓	
Pumpkin			✓		
Tomato			✓		
Turnip				✓	Burke (2024)
Watermelon			✓		Ziaf et al. (2017)
Eggplant			✓		Doijode (2010) Lindgren and Browning (2011) Ly (2015)
Parsley	✓				
New Zealand spinach			✓		
Squash			✓		
Radish				✓	
Swiss chard			✓		
Chicory			✓		
Chinese cabbage			✓		

decline in seed vigor and viability was made possible using traditional artificial aging experiments. The oxidation of macromolecules during seed storage is linked to seed aging (Sano et al. 2016). Auto-oxidation processes like Amadori and Maillard reactions, lipid peroxidation, or protein carbonylation cause progressive changes in cell constituents such as proteins, lipids, nucleic acids, sugars, etc. (Sano et al. 2016; Schausberger et al. 2019). The biochemical, biomechanical, and micromorphological characteristics of the layers that protect the seed, as well as capacity of antioxidant systems, DNA ligases, and O-methyltransferases to repair accumulated oxidative damage during seed imbibition, are critical factors in determining the quality of the seeds (Sano et al. 2016). Various crop species have shown that mechanical damage, such as threshing-induced cracks in the seed coat or drying-induced shrinkage, results in a decrease in seed viability and quality (Salimi and Boelt 2019). Permeability of the seed coverings to gases and surrounding oxygen concentration determines the internal oxygen concentration of a seed (Schwember and Bradford 2011). For vegetable seeds with a short lifespan, such as those for onions, these factors are particularly crucial (Schwember and Bradford 2011; Selvi and Saraswathy 2018).

As seeds age, oxidative damage that needs oxygen is increased and accompanied by a loss of vigor and viability. Therefore, the ideal strategy for storing seeds for an extended period at low oxygen concentrations is to combine cool, dry conditions with vacuum packaging (Schwember and Bradford 2011; Groot et al. 2015). On the other hand, accelerated deterioration of lettuce and cabbage seeds, as well as reduced percentage of normal seedlings, were seen when seeds were dry stored at elevated partial pressure of oxygen for 7 weeks (Groot et al. 2012).

24.4.3 Types of Seed Storage Facilities

According to Copeland and McDonald (1995), there are four main categories of seed storage, as described below.

24.4.3.1 Conditioned Storage

Most species' seeds may be kept for several years without risk if the temperature and relative humidity are carefully managed. Such circumstances may be important for conserving germplasm and some high-value seed stocks. Conditioned storage entails keeping seeds for lengthy periods in a cold, dry atmosphere. High oxygen environments during long-term storage harm seed viability, but high carbon dioxide environments are beneficial. Most of the seeds with a moisture level of 12–13%, kept for a year at standard storage conditions, germinate and grow satisfactorily. If a longer storage period is required, the temperature should not exceed 20 °C and the seed moisture level should be less than 11%. Similarly, seeds from most crops grown in temperate zones may be safely kept for a year at 10–11% moisture content at room temperature. Equipment for dehumidification and refrigeration are required to reduce relative humidity and temperature of the seed storage chamber and often chosen to provide adequate storage conditions in conditioned storage. While most vegetable crop seed lots may be too expensive to store using this strategy, however, high-value seed stocks can be preserved to great advantage.

24.4.3.2 Cryogenic Storage

In cryogenic preservation, seeds are kept in liquid nitrogen at −196 °C and can be stored for longer periods. However, for security and ease of working with, the seeds are kept in the gaseous form of liquid nitrogen at around −150 °C. The capability of this approach is restricted to the space of room for storage provided in the cryogenic tanks, although it is safer and less expensive than conditioned storage. Although preserving important germplasm over an extended length of time is useful, but not feasible for most commercial seeds. Liquid nitrogen is quite inexpensive in comparison to the expense of keeping conditioned store rooms (Pritchard and Nadarajan 2008).

24.4.3.3 Hermetic Storage

The term "hermetic storage" describes storing seeds in moisture-resistant or hermetically sealed containers. These containers are used to keep seeds at safe moisture content during storage and used to store seeds for at least 10 years. One of the main advantages of hermetic storage is the capacity to exclude ambient air from seeds and substitute them with certain gases that are known to extend seed storage life, e.g., reduced oxygen and elevated carbon dioxide levels. The materials' performance is closely related to their capacity to withstand moisture. While different lamination and polyethylene materials are effective, traditional paper and cloth containers are least effective. Seed moisture can be perfectly maintained in metal cans (Chala and Bekana 2017).

24.4.3.4 Containerized Storage

Complete humidity management in seed storage houses requires a large expenditure on specially designed spaces and dehumidifiers. However, in a confined container, chemical desiccants can control moisture without sophisticated equipment. Usually, when storing seeds in containers, the moisture content is controlled inside a confined space using chemical desiccants such as sulfuric acid, saturated salt solutions, or cobalt chloride-treated silica gel. The cobalt chloride acts as an indicator of color, changing from blue to pink when the relative humidity reaches 45%. The desiccant storage technique is less expensive and resistant to moisture, rodents, and insects. Because the seeds are kept at 45% relative humidity, they are not harmed by storage fungi.

24.5 Quality Control in Postharvest Management

Farmers usually need high-quality seeds for effective stand establishment, which may boost crop yields by up to 30% (Afzal et al. 2016). Numerous biotic and abiotic constraints can affect seeds, causing their stand performance to decline (Zinsmeister et al. 2020). Safeguarding and improving their performance is possible by applying chemical, biochemical, and biological seed treatments (Sharma et al. 2015). Postharvest treatments that increase seed germination and promote seedling growth are called seed enhancers. Seed coating, i.e., applying biostimulants, plant nutrients (such as inoculants), and other materials to seeds can help them withstand biotic and abiotic challenges that arise after planting (Taylor 2020).

24.5.1 Seed Testing and Quality Assessment

Growers and farmers throughout the world have two distinct requirements: first, the species or variety should correspond with what they have bought; and second, the seed should produce a crop free of weeds, growing consistently and without the possibility of illnesses brought on by seed-borne infection. The International Seed Testing Association (ISTA) has developed and standardized seed quality testing procedures to ensure seed quality for planting. These tests can be carried out using certain techniques. It's crucial to remember that samples from the seed lot (the population to be sown) are used for all testing. Viability and vigor are two important aspects that influence seed quality. Germination tests are frequently used to evaluate a seed sample. A tetrazolium test can be employed to assess its viability or potential for germination rapidly.

24.5.1.1 Variety Testing

Variety testing involves two aspects. In the first step, confirmation of the sample of the desired variety is done. The second step is to ensure that seeds of other varieties do not contaminate it.

Varietal purity means checking whether a variety is, for example, completely F1 variety as claimed or whether a conventional variety is contaminated by genetically modified (GM) seeds or vice versa. Morphological techniques are one type of traditional variety testing procedure. These techniques allow for the comparison of the seed properties. Variations in seed color, for instance, might indicate that several types have been blended. Alternatively, one may study various plant or fruit qualities in the field or observe the traits of seedlings in the field or a laboratory.

Variety testing has changed throughout time, and current approaches use several molecular and biological strategies. Electrophoresis is used in biochemical procedures to analyze the protein reserves of the seeds. This method is helpful when comparing F1 hybrids and parental lines. However, molecular techniques make use of molecular markers. These techniques include extracting deoxyribonucleic acid (DNA) and replicating the chosen DNA using polymerase chain reaction (PCR). Utilizing microsatellites/simple sequence repeats (SSRs) is an additional strategy.

In recent years, there has been much discussion and interest in seed testing, especially in genetically modified seed material detection. Numerous techniques have been employed, including DNA-based techniques like endpoint and real-time PCR, and protein-based techniques like enzyme-linked immunosorbent assay (ELISA) and bioassays.

24.5.1.2 Analytical Purity

The degree of contamination of the seed sample with undesired material is determined by analyzing the physical or analytical purity of the seed. Using lenses and microscopes, purity analysis is done by dividing seed samples into smaller seed lots. Blowers are used to remove light components from seed samples, such as trash and empty florets, while sieves are used to sort out trash, dirt, small fragments of seed, and other small particles. The percentage of pure seed by weight in a sample, along with the other recognized seeds and inert material, is shown by the results of analytical purity tests.

24.5.1.3 Viability/Tetrazolium Test

The tetrazolium test is a convenient and rapid method for determining a seed lot's viability. It is particularly helpful when a germination test takes too long, such as when seeds need to be planted right away after harvest or when they exhibit sluggish germination due to dormancy. The test can also assess the viability of seeds, which may be latent but not dead and that may not germinate after a germination test. Further, it can identify any damage, such as mechanical injury, which could have happened during harvesting and processing.

The tetrazolium test is a quick biochemical test for determining seed viability. In this test, 2, 3, 5-triphenyl tetrazolium chloride stain evaluates how much of the seed embryo's tissue is alive. The seed tissue's active respiratory enzymes react with this dye to color it crimson, which indicates that it is alive. The degree to which distinct tissues in the seeds must be alive to germinate and develop into a normal seedling may vary and minute quantities of dead tissue in certain portions of the seed are acceptable depending on the species. However, the embryo and surrounding tissue of a viable seed must be stained. When the test is performed on the entire seed sample, it gives viability for a seed lot in percentage. Specific details about seed preparation, stain concentration, staining period, and temperature are provided in the ISTA Rules. Additional specifics are found in the ISTA Working Sheets on Tetrazolium Testing (ISTA 2007).

24.5.1.4 Vigor Tests

The main method for determining a seed's ability to germinate and grow in the field is to conduct germination tests. Although conventional germination test results usually give a decent idea of how well the seeds will emerge under ideal circumstances, it may not provide an accurate picture of how the seeds would perform under less ideal circumstances, such as cold and damp soils. Seed lots with the same high laboratory germination rates may exhibit notable variations in field emergence.

Particularly under unfavorable field circumstances, the laboratory germination test does not usually predict accurate field emergence. Thus, there is another physiological component to seed quality that is referred to as seed vigor. Healthy and uniform seedlings are produced by high-vigor seeds. In contrast, low-vigor seeds germinate slowly over an extended period to produce a variety of seedling sizes. Furthermore, low-vigor seeds lose their capacity to germinate more quickly during storage, whereas high-vigor seeds have exceptional storage potential.

As aging is the primary driver of vigor changes, two tests have been designed to detect variations in seed vigor that are currently included in the ISTA Rules (ISTA 2009). Seeds' solute leakage is measured by the electrical conductivity test; old and low-vigor seeds exhibit higher amounts of leakage than high-vigor seeds. With the accelerated aging test, a seed sample is aged for an extra amount of time to ascertain the seed's initial position on the survival curve and, therefore, its vigor.

24.5.1.5 Electrical Conductivity Test

The electrical conductivity test measures the germination potential and vigor of seed batches. The cell membranes of dried seeds become disorganized and might sustain injuries while being stored. Because of this, during the first stages of germination, when seeds absorb water, parts of their cells, such as sugars, may leak out into the environment. This may affect seed performance in the field by providing soil-borne pathogens a feeding source. This may result in weak seedlings or a poor stand due to seedling deterioration or damping off. High-vigor seeds are more resilient to these circumstances; they leak less and remain in the soil for longer periods before becoming infected.

The electrical conductivity of the soak water is calculated after the seeds are soaked for a minimum of 24 h. One seed or a large sample of seeds can have its conductivity measured with a conductivity meter. High conductivity is typically linked to poor-quality seeds. The test is easy to administer and the results are ready in a day or two. There are a few problems with this test. Two important variables that must be controlled are soaking water's temperature and seed's initial moisture content. Leakage may arise because of cracked seed coatings. When using the bulk technique, laboratories may get a "false" high reading from a single faulty seed. Even though they have low conductivity, seeds kept at frigid temperatures may have poor vigor, as shown by various vigor tests.

24.5.1.6 Accelerated Aging Test

One of the most widely used vigor tests is the accelerated aging test. In this test, seeds are subjected to a conventional germination test after exposure to high temperatures and humidity for 3–4 days. High-vigor seeds have a limit to how much stress they can tolerate before their germination capacity and vitality start to decline. During the stressful storage of 3–4 days, low-vigor seeds quickly degrade and experience a reduction in germination. The test takes about 10 days to complete. This prolonged period could be a disadvantage when prompt judgments require a speedy test. A farmer can learn more about a seed's potential to perform in various soil conditions based on the results of a vigor test.

24.5.2 Standards for Seed Germination

A germination test is used to estimate germination percentage and adjust seed rate per unit area so that full potential of the seed can be exploited. Depending on the species, germination substrate, temperature, and planting time should be optimal for germination. Germination substrates include sand, organic medium, or a layer of paper below, between, or on top of seeds.

Constant or alternating temperatures are used for germination, where one temperature is administered for a certain amount of time. Then another temperature is applied for the remaining 24 h. In conclusion, the germination period in vegetable seeds varies considerably. The ISTA Rules for Seed Testing (ISTA 2009) provide the germination criteria for seeds of approximately 320 agricultural and vegetable species, 190 trees and shrub species, and 350 floral species.

Seed dormancy is a crucial factor to consider when conducting a germination test. In several plant species, dormant seeds require a particular environmental signal to germinate under the right circumstances. This implies that the existence of dormancy can impact a germination test's outcomes. This feature is a survival tactic that spreads germination over time and guarantees that the seed will only germinate when the environmental circumstances are favorable for seedling development and plant establishment. Although Baskin et al. (2000) thoroughly explained several forms of dormancy; these may be summed up as belonging to one of two categories: physiological or physical.

Certain pre-treatments are required in addition to the usual germination test criteria to break the dormancy of many species. For species with a short dormant period, dry storage disrupts physiological dormancy. Mucilaginous pre-chilling is applied to most vegetable and agricultural seeds at 5–10 °C. Additional choices include preheating, exposing to light, and supplying gibberellic acid or potassium nitrate to facilitate germination.

Physical dormancy may be caused by stiffness of seed coat. Mechanical scarification, acid scarification, or water immersion for 24–48 h can break this hard-seededness.

A seed is only deemed to have successfully germinated after a germination test if it has progressed to the point where the seedling's appearance indicates that it can produce a decent plant under favorable field circumstances. A normal seedling is the name given to this kind of seedling. Conversely, if normal seedling is not generated, the seedling is deemed abnormal and is not anticipated to germinate in the field. A germination test's outcome is expressed as a proportion of dead, fresh (i.e., wet yet firm), and hard (un-imbibed) seeds, and unusual and regular seedlings (ISTA 2009).

24.5.3 Advances in Seed Coatings and Treatments

Seed coatings and treatments employ a variety of materials. Based on their composition and place of origin, these materials are classified as synthetic chemicals, natural products or goods derived from natural products, biological agents, and minerals taken from the ground. The role of seed treatments and coatings (as liquids or solid particulates) further defines their characteristics.

The primary objective of using active substances for seed coatings and treatments is to safeguard and enhance seed germination, and seedling growth and development. The active ingredient's mechanism of action determines how it functions regarding enhancement or protection. Seed protectants are the most popular class of substances for managing pests and diseases during seeding. Protectants include insecticides, nematicides, fungicides, and bactericides. Commercial applications of certain beneficial microbes include plant protection and nitrogen fixation inoculants (Rocha et al. 2019). Certain biological agents and synthetic products can be used for seed coating to mitigate abiotic challenges caused by drought stress or salinized soil conditions that may arise after sowing. According to Tayyab et al. (2020), elicitors are considered as active ingredients for drought stress and insect control.

Active components added to seeds must remain attached to them throughout storage and until they are planted. Further, it must be simple to identify the treated seeds. Colorants, which comprise around 60% of the coating ingredient components, are often used to indicate that the seeds have been treated, and are added after the coating process in case of seed pelleting (Pedrini et al. 2017). Colorants also give a visual assessment of application consistency and cosmetic attractiveness. Low-viscosity liquids work best for atomization, and water is a universal carrier for liquids atomized onto seeds during the coating process. Most chemically active components in seed treatments have limited solubility in water, necessitating the use of adjuvants to develop aqueous seed treatment formulations. Surfactants have the potential to function as an active ingredient.

Binders for seed coatings function as adhesives and help the coating materials to stick with seeds. The coating integrity is provided by the binder both during and after drying. They keep coating material from breaking and dusting off during handling and sowing. Polyvinyl alcohol (Ryu et al. 2006), polyvinyl acetate (Chen and Turnblad, 2002), methylcellulose (Lopisso et al. 2017), and carboxymethyl cellulose (Kangsopa et al. 2018) are common binders used to preserve the physical integrity of seeds. Plant starches or maltodextrins (Kimmelshue et al. 2019) and gum Arabic (Javed and Afzal 2020) are frequently utilized for organic seed coatings. Polymer is the most frequent term used to describe most binders (Danielson and Gaul, 2011). While making binders in water, the solution viscosity must be low for the liquid to atomize onto the seeds completely; ideally less than 100 centipoises (cP) or 0.1 Pascal-second (Pa-s).

In seed coating technologies, solid particles serve as bulking ingredients, and when dried, create the physical coating (Chen and Turnblad 2002). Binders can also be solid particles. When water is added during the coating process, solid particle binders hydrolyze, which are coated as fine powders. To create a mixture for seed coating, fillers (fine powders) can be combined with solid particle binders. For seed pelleting, the optimization and selection of the best filler materials that don't impede germination, are essential.

According to Taylor (2003), filler materials are often non-toxic, affordable, readily accessible, and create a homogeneous coating surface texture that doesn't obstruct radicle emergence. Diatomaceous earth, limestone, gypsum, bentonite, vermiculite, talc, zeolite, silica sand, and barium sulfate are filler materials used in seed pelleting. Most of these fillers are made of mineral components extracted from the soil and ground into a fine powder and employed in seed coating.

To start the germination metabolism without radical protrusion, seeds are controlled and hydrated in water or a solution with low osmotic potential. This process is known as seed priming. According to several research, seed priming increases field crop production and stand establishment (Farooq et al. 2005, 2010). Hydropriming, halopriming, osmopriming, solid matrix priming, biopriming, nutripriming, and seed priming using hormones, plant growth regulators, and other organic sources, are among the several forms of seed priming.

24.6 Conclusion

Maintenance of vegetable seed quality is a major issue in developing countries. There is a dire need to maintain seed quality for better-quality crops. Advanced, economical, and productive methods for postharvest handling and storage of seeds need to be established and developed.

To meet the international standards, the latest technologies with special attention on seed response to environmental and other factors must be adopted. In conclusion, seed storage is an important aspect, thus understanding seed viability, harvesting and collecting seeds carefully, and choosing the right storage methods, are keys to success.

References

Afzal I, Rehman HU, Naveed M, Basra SMA (2016) Recent advances in seed enhancements. In: Susana A, Balestrazz A (eds) New challenges in seed biology: basic and translational research driving seed technology. Intech Open Press, London, pp 48–74

Baskin JM, Baskin CC, Li X (2000) Taxonomy, ecology, and evolution of physical dormancy in seeds. Plant Species Biol 15:139–152

Burke N (2024) Can seeds expire or go bad? https://www.gardenary.com/blog/can-seeds-expire-or-go-bad. Accessed 22 Feb 2025

Chala M, Bekana G (2017) Review on seed process and storage condition in relation to seed moisture and ecological factor. J Nat Sci Res 7:84–90

Chen Y, Turnblad KM (2002) Insecticidal seed coating. U.S. Patent 0,177,526 A1

Chua LYW, Chong CH, Chua BL (2019) Influence of drying methods on the antibacterial, antioxidant and essential oil volatile composition of herbs: a review. Food Bioproc Technol 12:450–476

Copeland LO, McDonald MB (1995) Principles of seed science and technology. Springer, New York

Dadlani M (2023) Emerging trends and promising technologies. In: Dadlani M, Yadava DK (eds) Seed science and technology. Springer, Singapore, pp 415–430

Danielson B, Gaul A (2011) Category 4, seed treatment-Iowa commercial pesticide applicator manual, CS16. Iowa State University Extension and Outreach, Ames, p 40

Doijode SD (2010) Long-term seed storage studies in radish (*Raphanus sativus* L.). J Hortl Sci 5(1):61–63

Dyck G, Hawley E, Hildebrand K, Paliwal J (2023) Digital twins: a novel traceability concept for post-harvest handling. Smart Agric Technol 3:100079

Eissenberger K, Ballesteros A, De Bisschop R, Bugnicourt E, Cinelli P, Defoin M, Demeyer E, Fürtauer S, Gioia C, Gomez L, Hornberger R (2023) Approaches in sustainable, bio based multilayer packaging solutions. Polymers 15:1184

Ellis RH, Hong TD (2006) Temperature sensitivity of the low moisture content limit to negative seed longevity moisture content relationships in hermetic storage. Ann Bot 97:785–791

Ellis RH, Hong TD, Roberts EH (1985) Handbook of seed technology for gene banks. Volume 1: principles and methodology. International Board for Plant Genetic Resources, IT, Rome

Ells JE, Bass LN, Whiting D (2020) Storing vegetable and flower seeds. Colorado State University Extension. https://extension.colostate.edu/topic-areas/yard-garden/storing-vegetable-and-flower-seeds-7-221. Accessed 27 Feb 2025

Engels JM, Ebert AW (2021) A critical review of the current global ex situ conservation system for plant agrobiodiversity. II. Strengths and weaknesses of the current system and recommendations for its improvement. Plan Theory 10:1904

Farooq M, Basra SMA, Hafeez K, Ahmad N (2005) Thermal hardening: a new seed vigor enhancement tool in rice. J Integr Plant Biol 47:187–193

Farooq M, Basra SMA, Wahid A, Ahmad N (2010) Changes in nutrient-homeostasis and reserves metabolism during rice seed priming: consequences for germination and seedling growth. Agric Sci China 9:101–108

Groot SPC, Surki AA, De Vos RC, Kodde J (2012) Seed storage at elevated partial pressure of oxygen, a fast method for analyzing seed ageing under dry conditions. Ann Bot 110:1149–1159

Groot SPC, De Groot L, Kodde J, van Treuren R (2015) Prolonging the longevity of ex situ conserved seeds by storage under anoxia. Plant Genet Resour Charact Util 13:18–26

Harrington JF, Douglas JE (1970) Seed storage and packaging. National Seeds Cropland Rockefeller Foundation, New Delhi, p 222

Hartmann HT, Kester DE, Davies FT, Geneve RL (1997) Plant propagation: principles and practices. Prentice-Hall Inc., Upper Saddle River

Iannotti M (2022) How long do seeds last and are they still good? https://www.thespruce.com/how-long-do-vegetable-seeds-last-1403089. Accessed 27 Feb 2025

ISTA (2007) International rules for seed testing. International Seed Testing Association, Bassersdorf

ISTA (2009) International rules for seed testing. International Seed Testing Association, Bassersdorf

Jarrar H, El-Keblawy A, Ghenai C, Abhilash PC, Bundela AK, Abideen Z, Sheteiwy MS (2023) Seed enhancement technologies for sustainable dryland restoration: coating and scarification. Sci Total Environ 904:166150

Javed T, Afzal I (2020) Impact of seed pelleting on germination potential, seedling growth and storage of tomato seed. Acta Hortic 1273:417–424

Justice OL, Bass LN (1978) Principles and practices of seed storage. Agriculture Handbook, No. 506. US Government Printing Office, Washington, p 289

Kamal AB, Sheikh MK, Azhar B, Munir M, Baig MB, Reed MR (2022) Role of agriculture extension in ensuring food security in the context of climate change: state of the art and prospects for reforms in Pakistan. In: Behnassi M, Baig MB, Sraïri MT, Alsheikh AA, Abu Risheh AWA (eds) Food security and climate-smart food systems. Springer, Cham, pp 189–218

Kameswara Rao N, Dulloo ME, Engels JM (2017) A review of factors that influence the production of quality seed for long-term conservation in gene banks. Genet Resour Crop Evol 64:1061–1074

Kangsopa J, Hynes RK, Siri B (2018) Lettuce seeds pelleting: a new bilayer matrix for lettuce (*Lactuca sativa*) seeds. Seed Sci Technol 46:521–531

Kimmelshue CH, Goggi AS, Cademartiri R (2019) The use of biological seed coatings based on bacteriophages and polymers against *Clavibacter michiganensis* subsp. *nebraskensis* in maize seeds. Sci Rep 9:17950

Lavezzo A (2016) Seed cleaning techniques for seed saving, sow true seed. Available at: https://sowtrueseed.com/blogs/seed-saving/seed-cleaning-techniques-for-seed-saving. Accessed 25 May 2024

Lindgren TD, Browning SJ (2011) Vegetable garden seed storage and germination requirements. University of Nebraska – Lincoln Extension, Institute of Agriculture and Natural Resources, pp 1–4

Lopisso DT, Kuhlmann V, Siebold M (2017) Potential of soil-derived fungal biocontrol agents applied as a soil amendment and a seed coating to control Verticillium wilt of sugar beet. Biocontrol Sci Tech 27:1019–1037

Ly L (2015) How long do old seeds last? https://gardenbetty.com/how-long-do-seeds-really-last-plus-a-cheat-sheet-on-seed-storage-life/. Accessed 27 Feb 2025

Mahaney E (2014) Seed viability (or, will those really old seeds in my Garage Germinate?) https://ucanr.edu/blogs/blogcore/postdetail.cfm?postnum=12950#. Accessed 27 Feb 2025

Manna D, Maity TK, Basu AK (2020) Comparison on condition and packaging materials for better storage of onion seeds. Asian J Hort 15(1):15–25

Maynard ND, Hochmuth GJ (2007) Knott's handbook for vegetable growers. Wiley, Hoboken

McCormack JH (2004) Principles and practices of seed harvesting, processing, and storage: an organic seed production manual for seed growers in the mid-Atlantic and southern U.S. Creative Commons, Stanford, p 28

Nagel M, Borner A (2010) The longevity of crop seeds stored under ambient conditions. Seed Sci Res 20:1–12

Osei MK, Adjebeng-Danquah J, Egbadzor KF, Osuman AS, Adjei EA, Abebrese SO, Melomey LD, Agyare RY (2020) Vegetable seed production: prospects and challenges: the case of Ghana. Adv Seed Prod Manag 1:91–117

Pagano A, Macovei A, Balestrazzi A (2023) Molecular dynamics of seed priming at the crossroads between basic and applied research. Plant Cell Rep 42:657–688

Pedrini S, Merritt DJ, Stevens J, Dixon K (2017) Seed coating: science or marketing spin? Trends Plant Sci 22:106–116

Pritchard HW, Nadarajan J (2008) Cryopreservation of orthodox (desiccation tolerant) seeds. In: Reed BM (ed) Plant cryopreservation: a practical guide. Springer, New York, pp 485–501

Rana AW, Gill S, Akram I (2022) PACE policy paper: seed system of Pakistan: policy challenges and prospects. International Food Policy Research Institute, Washington

Rao NK, Hanson J, Dulloo ME, Ghosh K, Nowell D, Larinde M (2006) Manual of seed handling in Genebanks. Handbooks for Genebanks No. 8. Biodiversity International, Rome, Italy

Rehman A, Ali H, Fatima Z, Ahmad S (2019) Fundamentals of seed production and processing of agronomic crops. In: Hasanuzzaman M (ed) Agronomic crops: production technologies. Springer, Singapore, pp 623–653

Rocha IDS, Ma Y, Souza-Alonso P, Vosatka M, Freitas H, Oliveira RS (2019) Seed coating: a tool for delivering beneficial microbes to agricultural crops. Front Plant Sci 10:1357

Ryu CM, Kim J, Choi O, Kim SH, Park CS (2006) Improvement of biological control capacity of *Paenibacillus polymyxa* E681 by seed pelleting on sesame. Biol Control 39:282–289

Salimi Z, Boelt B (2019) Classification of processing damage in sugar beet (*Beta vulgaris*) seeds by multispectral image analysis. Sensors 19:2360

Sano N, Rajjou L, North HM, Debeaujon I, Marion-Poll A, Seo M (2016) Staying alive: molecular aspects of seed longevity. Plant Cell Physiol 57:660–674

Schausberger C, Roach T, Stoggl W, Arc E, Finch-Savage WE, Kranner I (2019) Abscisic acid-determined seed vigour differences do not influence redox regulation during ageing. Biochem J 476:965–974

Schwember AR, Bradford KJ (2011) Oxygen interacts with priming, moisture content and temperature to affect the longevity of lettuce and onion seeds. Seed Sci Res 21:175–185

Selvi DT, Saraswathy S (2018) Seed viability, seed deterioration and seed quality improvements in stored onion seeds: a review. J Hortic Sci Biotechnol 93:1–7
Sharma KK, Singh US, Sharma P, Kumar A, Sharma L (2015) Seed treatments for sustainable agriculture – a review. J Appl Nat Sci 7:521–539
Snively T (2024) How long are Broccoli seeds good for? https://toddsseeds.com/how-long-are-broccoli-seeds-good-for/. Accessed 27 Feb 2025
Sundareswaran S, Ray Choudhury P, Vanitha C, Yadava DK (2023) Seed quality: variety development to planting – an overview. In: Dadlani M, Yadava DK (eds) Seed science and technology. Springer, Singapore, pp 1–16
Taylor AG (2003) Seed treatments. In: Thomas BDJ, Murphy BG (eds) Encyclopedia of applied plant sciences. Elsevier Academic Press, Cambridge, pp 1291–1298
Taylor AG (2020) Seed storage, germination, quality and enhancements. In: Wien HC, Stutzel H (eds) The physiology of vegetable crops. CAB International, Wallingford, pp 1–30
Tayyab N, Naz R, Yasmin H, Nosheen A, Keyani R, Sajjad M, Hassan MN, Roberts TH (2020) Combined seed and foliar pre-treatments with exogenous methyl jasmonate and salicylic acid mitigate drought-induced stress in maize. PLoS One 15:232269
Winterfeldt S (2020) Do seeds expire? A guide to seed longevity and proper storage. Park Seed. https://www.parkseed.com/blog/guide-to-seed-longevity-and-proper-storage
Yousaf Z, Saleh N, Ramazan A, Aftab A (2016) Post harvesting techniques and maintenance of seed quality. In: Araujo S, Balestrazzi A (eds) New challenges in seed biology: basic and translational research driving seed technology. Intech Open Publishers, London, pp 114–135
Ziaf K, Amjad M, Anjum MA (2017) Production and handling of vegetable and flower seeds. In: Khan AS, Ziaf K (eds) Horticulture: science and technology. University of Agriculture, Faisalabad, pp 102–131
Zinsmeister J, Leprince O, Buitink J (2020) Molecular and environmental factors regulating seed longevity. Biochem J 477:305–323

Prospects of Value Addition in Veggies: A Tool to Minimize Postharvest Losses

25

Muhammad Nadeem, Zafar Iqbal,
Tahir Mahmood Qureshi, Mian Nadeem Riaz,
and Muhammad Adnan Shahid

Abstract

The philosophy of value addition in agriculture can be described as revolution which transforms the quality as well as value of the actual products and more effectively addresses the numerous aspects of the issues in the sector. With effective use of technology that embraces cold chain systems, smart packaging, and digital tools, stakeholders should be able to have efficient and sustainable ways of delivering value so as to have competitive advantage in both domestic and international markets. The multiplier effects of value addition include increased income of the farmers, higher employment, and more export opportunities. These advantages reflect personal income growth and support the general economic progress and social uplift of small farmers, who are particularly prominent in developing nations. The development of infrastructure faces many challenges

M. Nadeem (✉)
Institute of Food Science and Nutrition, University of Sargodha, Sargodha, Pakistan

Horticultural Sciences Department, North Florida Research and Education Center, Institute of Food and Agricultural Sciences, University of Florida, Quincy, FL, USA
e-mail: nadeem.abdul@uos.edu.pk; nadeem.muhammad@ufl.edu

Z. Iqbal
Faculty of Agriculture, University of Sargodha, Sargodha, Pakistan

T. M. Qureshi
Department of Food Sciences, Cholistan University of Veterinary and Animal Sciences, Bahawalpur, Pakistan

M. N. Riaz
Department of Food Science and Technology, Texas A&M University, College Station, TX, USA

M. A. Shahid
Horticultural Sciences Department, North Florida Research and Education Center, Institute of Food and Agricultural Sciences, University of Florida, Quincy, FL, USA

Z. Iqbal et al. (eds.), *Sustainable and Innovative Vegetable Production in times of Climate Change*, https://doi.org/10.1007/978-981-96-6283-8_25

that include high cost and lack of technical know-how and friendly policies. Resolving these issues must involve capacity enhancement interventions, infrastructure development, and the harmonization of existing regulations. The future of value addition depends on Public-Private Partnership (PPP), innovative research, and dissemination of the latest technologies to stakeholders. By adopting these opportunities, agricultural sector can sustain its growth, fulfill the food security needs of the people, and satisfy the changing consumers' behaviors.

Keywords

Vegetables · Value addition · Cold chain · Smart packaging · Digital tools

25.1 Introduction

Agriculture is an important industry that supports food security and economic growth, with a huge input towards income generation in many developed and developing countries. One cannot imagine human diet without vegetables which provide vitamins, minerals, fibers as well as antioxidants (Noort et al. 2022). The importance of vegetables in the human diet cannot be overemphasized given the fact that they help in the prevention of chronic illnesses, thereby improve the health condition of the body (Fardet and Rock 2014). However, due to their high rate of perishability, some biotic and abiotic stresses may inflict postharvest losses. Postharvest losses of vegetables are estimated to range between 30% and 50% by the Food and Agriculture Organization (FAO) particularly in the developing countries where there is poor infrastructure (Kaur and Watson 2024).

Vegetables are imperative for the overall great well-being of individuals, giving fundamental nutrients, for instance minerals, dietary fiber, and phytochemicals, and hence lessening hazards from hazardous sicknesses and other ailments. Vegetables are produced worldwide in around 200 countries. A world vegetable review demonstrated 392 types of vegetables produced around the world. Just about three-fourths of the global vegetables' production is found in Asia, especially in China, which produces over portion of the world's vegetables. The space of world land committed to vegetables has been expanding due to increased population and rising demand (Butnariu and Butu 2015).

Vegetables' losses occur in the postharvest process through many factors such as improper method of harvesting, lack of storage and transport facilities, and inefficient markets (Anand and Barua 2022). These losses also pose severe financial risk to farmers and other industries within the food chain (Urugo et al. 2024a, b).

Value addition is the incorporation of a person's (or persons) time and effort to form a product in order to meet the tastes/preferences of consumers. It refers to those processes and steps taken to improve the status and consumers' acceptance of any product specifically related to food and agriculture. It is a technique opted not only as a marketing strategy but also to produce food items with refined shelf life and storage properties. In the process of value addition, inputs (raw materials, machines, and labor) are converted into an output by consuming the time and technology for the aim of achieving good economic return and healthy product. Vegetables can be purchased from the market during their peak season when prices would be very low. Then the purchased vegetables can be preserved by different methods of processing, i.e., drying, pickling, vacuum packaging, etc. For example, when potatoes are in massive quantity in the market, they can be value added (converted) into potato flour, grits, cubes, etc. Similarly, tomatoes can be converted into paste, puree, and ketchup as value-added products. In this way, processing (value addition) leads to lower wastage of vegetables and increases the economic return (Kulshrestha 2018). We can also add value to the vegetables by coating them with edible films, as these commodities have short shelf life and are exposed to physical injury during handling and transportation like bruising or enzymatic browning (Baldwin et al. 1996), skin wrinkling, blemishes, moisture loss, and discoloration (Choi et al. 2002). Good packaging is also considered as value addition because without good packaging, food may be damaged during transportation to the market. Glass jars, metal cans, and plastic containers are best for packaging during transport. During transportation, good humidity and temperature maintain the desired quality of value-added food product (Featherstone et al. 2012).

Value-added vegetable products include canned products, juices, soups, sauces, dehydrated vegetables, and vegetable oils. The processing of vegetables started in the middle of nineteenth century after the development of European industries to allow them to be exported and eaten throughout the year (Verma et al. 2017). The global fruit and vegetable processing industry has grown well since 2011 and is expected to grow rapidly by 2025 due to industrial growth and rapid urbanization. This leads to high levels of income generation, significant food diversity, and proximity to premium-branded products. The pattern of consumption varies between the developing and developed countries. Consumption of vegetables largely depends on the individuals in the latter case (FAO 2015).

Canning is a process that destroys microorganisms and inactivates enzymes to enhance the safety and quality of the food. Microbial recontamination of food (which causes food spoilage or otherwise poses a health risk) is prevented by putting it in a jar and heating the jar to a specific temperature which kills the microbes. Canning is usually used for fruits and vegetables, processed ketchup and other tomato sauces, and preserved products like jam and jelly. Growth of microorganisms can be prevented by using oxygen removal method (using oxygen absorbers) in combination with some other effective methods (Lamb et al. 2019).

Important steps of handling include packaging, storage, and shipping. Thermally processed fruits and vegetables are usually packed in steel cans or glass, but aseptically processed products are usually wrapped in metal plate laminate, plastic, and paper (Lucci et al. 2016).

Vegetables like tomatoes, potatoes, onions, peppers, and cucumbers are mostly grown in Pakistan's diverse agro-ecological zone. Despite the fact that Pakistan obtains high yield, this sector suffers postharvest losses of 30–40% due to the deteriorative nature of some vegetables (Dousa and Veclavovic 2012).

Due to the lack of advanced technology and processing industry, agriculture sector bears losses of worth 32–40 billion annually. To mitigate this problem, conversion of food crops to value-added products is direly needed. It is a market strategy which is designed according to the need and preferences of consumers. The processing of vegetables increases the shelf life, reduces wastage, and ensures the availability of products even though during off season (Stancu et al. 2016). With the help of value addition, safe and high-quality food is provided to the consumers and as a result, mutual linkage among farmers and consumers is also developed (Picha 2007).

Among the processing/preservation methods, dehydration and canning have been considered as the most effective followed by freezing. Cold storage is also used as a processing technique (Abdullah and Dong 2019). Value addition not only increases value of product but it also plays a significant role in minimizing the losses. Insufficient number of processing units also slows down the process of value addition which in turn affects the overall export potential of a country (Nyamulinda et al. 2011).

25.2 Significance of Vegetables in Global Food Security

Being an essential component of food security and nutrition, vegetables are considered very important part of sustainable agriculture (Pawlak and Kołodziejczak 2020). As sources of micronutrients, including vitamins A, C, and K and certain minerals (potassium and magnesium), the place of vegetables in eradicating malnutrition and controlling diet-related diseases cannot be underestimated. According to the WHO, a healthy adult should consume not less than 400 grams of fruits and vegetables every day with a view of reducing the level of chronic diseases such as cardiovascular diseases and certain types of cancers (Wallace et al. 2020).

The demand for vegetables has increased due to rise in awareness of the positive impact of vegetables on health and increased uptake of vegan diets. China, India, and the USA were among the major producing countries around the world in 2022 and worldwide vegetable production was higher than 1.1 billion metric tons. However, vegetables are viewed as luxury foods in the developing world, especially in the low-income areas where variations in the quantities and qualities can be observed (Bharatkumar et al. 2023). Vegetables also have a role in sustainability of the food system. They do not need much water and space as compared to other field crops. Moreover, vegetables mature much faster as one can harvest several times in a year. Also, vegetable farming is friendly to biodiversity and may be incorporated into crop production cycling to help enrich the ground (Yadav et al. 2021).

Unfortunately, the role of vegetables in solving the problem of food security at the global level is hindered by a high level of postharvest losses (Tadesse et al. 2024). These losses challenge food availability through decreasing the amenable quantity of produce to communities. Improving the handling, storage, and processing of vegetables would enable supply of vegetables in fresh and processed form to the consumers (Kaur and Watson 2024). When vegetables are prioritized in food security frameworks, food insecurity would be overcome ensuring improved living standards in rural areas.

25.3 Overview of Postharvest Losses in Vegetables

25.3.1 Causes of Postharvest Losses

In vegetables, postharvest losses depend on a number of factors that occur at the time of harvest, transit, storage, and marketing (Elik et al. 2019). Following are the key causes of postharvest losses of vegetables.

25.3.1.1 Perishability of Vegetables

Vegetables are perishable foods because they contain a lot of water and undergo metabolic activities.

25.3.1.2 Improper Harvesting Techniques

The absence of information on when and how to tap these products correctly results in mechanical injury and degradation.

25.3.1.3 Inadequate Storage Facilities

Insufficient or wrong temperature storage increases decay and microbial infestation.

25.3.1.4 Poor Transportation Infrastructure

A lot of losses occur as a result of delays, exposure to unfavorable conditions, and rough handling during transit.

25.3.1.5 Market Gluts and Inefficiencies

This results in seasonality, and farmers end up throwing away many unsold vegetables.

25.3.1.6 Lack of Awareness and Training

Small farmers and handlers are sometimes not technically equipped to adopt good postharvest management practices.

25.3.2 Magnitude and Impact of Losses

The levels of postharvest losses in vegetables differ by region and crop type. FAO estimates that overall postharvest food losses of vegetables range from 20% to 50% depending on the conditions of the supply chain. It reveals that loss is significantly higher in developing countries due to the backwardness in overall infrastructure and technology (Dang and Pheng 2015).

25.3.2.1 Impacts of Postharvest Losses

25.3.2.1.1 Economic Losses

Farmers lose part of their income whereas nations lose part of their productivity in the form of precious agricultural produce (Hazell and Wood 2008).

25.3.2.1.2 Food Insecurity

Shortage of vegetables may lead to food insecurity adding to miseries of common and rural people.

25.3.2.1.3 Environmental Impact

The lost vegetables emit green gases after decomposition. Improvements in handling, storage, and processing of vegetables can make supply chain sustainable (Magalhães et al. 2022). Value addition comes out as the best approach to conquer loss, thereby increasing the revenue, and satisfying customers' needs.

25.4 Concept of Value Addition in Vegetables

Value addition may be defined as the processes or techniques used for the improvement of market value, quality or utility of raw vegetables or vegetable products (Rao et al. 2022). It involves a wide variety of work such as tidiness, separation, categorization, organization, classification, encasement, working over and reworking of work pieces, and changing the form of products into new articles. Generally, the goal of value addition is mainly to increase the shelf life and quality of produce and products to meet consumers' demands, and reduce the immense losses that occur after harvest (Onwude et al. 2020). Through increasing product value by turning raw produce into semi-processed or fully processed goods, value addition also creates avenues for getting into other markets for the generation of huge revenue.

Canned vegetables, frozen mixed vegetables, ready-to-cook vegetable kits, and other value-added products have taken root as more people shift toward convenience products (Ding and Liu 2024). These products support the busy lifestyles of the consumers and help in reducing the wastage of the excess produce. Value addition is also crucial in enhancing the standing of smallholder farmers, income generation, and growth of rural businesses (Materia et al. 2021).

25.4.1 Importance of Value Addition

Compared to raw commodity, value-added products can be preserved for longer time periods in a wide array of temperature and storage conditions (Insights 2011). Innovative and easily accessible items are produced commercially that fulfill the ever-changing consumer preferences. Value-added food products are generally considered as safe, healthy, and convenient (Sharangi and Dutta 2015). Food processing with value-addition is important because value addition increases shelf stability, brings forward more farming opportunities for new and experienced farmers, acts as a substitute for heavily taxed imported products, minimizes postharvest handling and transport losses, and adds commercial profit and nutritious diet (Augustin et al. 2016).

Vegetables in larger quantity have short shelf life and there is need to convert them into value-added products or sell in the market immediately. These value-added vegetable products have gained recognition in the defense sector of various countries (Sogi and Bhullar 2000). Frozen and minimally processed vegetables with high nutritive value also hold prominent position in the markets (Keukens et al. 1994). These products are nutritious and usually rich sources of minerals, vitamins, antioxidants, and fibers. Spinach, green beans, and smooth wrinkled peas are cooked and usually preserved by home-bottling or deep freezing, canning, freeze drying, or air-drying on a semi-technical scale with variable storage periods of up to 6 months (Van Wageningen et al. 2010). As vegetables have short shelf life, so dehydration and drying are two methods that usually enhance the shelf life and minimize deterioration. These also lessen the packaging cost and transportation process. The material to be transported is converted into a dry solid and the weight is usually reduced but volume remains the same (Sharangi and Datta 2015). Value addition of vegetables by edible packaging is also very important because it usually supports to reduce the waste and helps in the preservation of food. Ideal and lightweight packaging can also be reused or recycled many times. Food industries now move toward the biopolymer packaging which maintains the freshness and also preserves food products (Nandane and Jain 2011).

Waste vegetables can also be transformed into value-added products because they contain very useful substances which usually have a high value. By utilizing the vegetable residues, many useful products can be produced. Residues of vegetables can be used as bio absorbents for the wastewater treatment (Benu and Muditta 2017).

25.4.2 Role of Different Techniques in Value-Added Vegetables

Delayed consumption or processing of vegetables may lead to spoilage. To control the issues of food wastage, idea to preserve the food was first introduced by our ancestors to keep the food items edible and fresh. Depending on seasonal and climatic variations, some concepts of food preservation such as salting, sun drying, and pasteurization were introduced. Food preservation helps to prevent the microorganisms and control the rancidity caused by the reactions of fats. As a result of rapid industrialization, some processes like canning, thermal treatments, and freezing methods control the pathogens and ensure a long shelf-life of food items (Saravanan et al. 2023).

Now the concept of food preservation is expanding rapidly (Białkowska et al. 2020). Though there are some disadvantages of the freezing method and heat treatment such as loss of nutrients and food texture and huge changes in organic properties which ultimately lead to the loss of food product (Jayasena et al. 2015). Recently, some treatments involving the use of microbiological and chemical coatings, extracts of various polyphenolic plants, and different additives have also been used for effective solutions to preserve the food (Tsironi et al. 2020). Following are the benefits of various techniques to be used for value addition of vegetables.

- Timely processing of vegetables ensures safe preservation and high quality of products to be consumed for the whole year.
- The wastage of vegetables is minimized by preserving and storing them using different techniques (Miller-Cebert et al. 2016).
- Vegetables subjected to drying are further used in different food preparations, like soups and other kitchen recipes (Datta et al. 2015).
- Marketing of processed vegetables in national and international markets saves from risk of financial loss.
- The colorful presentation of processed vegetables' products on shelves in superstores attracts consumers.

25.4.2.1 Value Addition Through Canning

In France, Appert invented a process in 1804 for preserving the food in hermetically sealed containers after sterilizing the containers through heating. As an honor of inventor, the canning is also known as appertizing. In 1807, Saddington was the first in England who described the canning method for food. Another Englishman Peter Durand in 1810 was the first who obtained the British Patent to give canned foods in tin containers. In the USA, William Underwood was the one who introduced canned fruits on a commercial scale in 1817. During the season, a lot of vegetables are available which are canned and sold in off-seasons (Ronoh et al. 2020).

Canning is a process of preservation in which chances of recontamination caused by microorganisms in food are minimized by inserting it in a jar. This technique can be applied for vegetables, fruit juices, processed ketchup, vegetable juices, tomato sauce, jams, and jellies. Steam is used for air/oxygen removal which can be utilized to avoid microbial development that requires O_2 for functioning. Generally, canned veggies require more processing because some microorganisms are more heat-resistant (Sharangi et al. 2018).

25.4.2.2 Value Addition Through Pickling

Pickling is a sort of adding values in fruits as well as vegetables. It boosts the nutritional content of fruits and vegetable as well as increase the shelf life through preservation. Pickling is defined as the process of maintaining a food by either anaerobic fermentation in brine or dipping in vinegar. The consequential food is pickle. Pickling can also be done by dipping veggies in mustard oil. The term pickle comes from the Dutch word "Pekel" that signifies "brine". In Canada and the USA, various forms of pickle, i.e., pickled cucumber, pickled onion, pickled beets are being prepared (Park et al. 2011).

25.4.2.3 Value Addition Through Dehydration and Drying

Some fruits and vegetables have their own growing season and are delicate in nature. They are available in market at different prices. More losses occur during transportation as there is no proper availability of storage conditions. Preservation of such types of fruits and vegetables can be done by using different preservation techniques which enable their availability all the year round at varying prices (Seidu et al. 2012).

In this case, dehydration strategy can be beneficial since dehydrated veggies may be stored for prolonged time and are readily available even during off-season. Drying or dehydration of food is an easy and cheap solution to keep food away from contamination (Parveen et al. 2014). Due to the occasional accessibility, endeavors are made to handle the vegetables in huge amount to stretch out the time frame of realistic usability so that these can be used throughout the year (Deulkar and Barve 2018).

Dehydration is the process to remove moisture from food items while maintaining their intrinsic quality under induced temperature and relative humidity. People used to dry vegetables in the open fields in ancient times (Kaur et al. 2019). Fruits and vegetables that have been dried or dehydrated, can be used in a variety of ways like dairy and confectionery products, instant soups, and baking items (Savo et al. 2012). Vegetables can be converted into unique value-added products like powders, corn meal, chips, mash, puree, etc. (Hassan et al. 2021). Table 25.1 presents value-added products prepared from vegetables by drying.

25.4.2.4 Value Addition Through Packaging

Packaging for transporting, marketing, and handling of fresh produce includes cartons, crates, baskets, bulk bins, palletized containers, and bags. Maintaining suitable temperatures during their delivery and storage is another crucial aspect in extending the shelf life of vegetables. A variety of films produces a permeable package that retains gas levels at an acceptable level, extending shelf life exponentially. Before or after packaging, the items should be promptly pre-cooled to room temperature. Precooling provides the longest possible shelf life for any produce. Until they are conveyed to the broker, packed and precooled commodities must be retained at ideal storage temperatures.

25.4.2.5 Value Addition Through Processing

Food processing includes any value addition with single step or in combination that is used for the agricultural produce after harvesting of vegetables (Floros et al. 2010). Consumers desire high-quality, "fresh-like" foods even after the processing of vegetables. This type of value-addition should have a strong commitment to food safety standards and practices in order to ensure product quality (Augustin et al. 2016). Value-addition through processing of vegetables includes operations that change fresh vegetables to produce more value-added useful product for human consumption. It usually includes canned products, soups, sauces, and dehydrated vegetables. Food processing increases shelf life of food products, adds value, increases market opportunities, enhances farmer's returns on their produce,

Table 25.1 Value-added products prepared from vegetables by drying

Sr. no	Vegetables	Value-added products	Methods of drying
1	Potato	Papad, potato chips, potato flour	Remove the rind and trim into thick slices, then do frying till they become crispy.
2	Tomato	Dried tomato, tomato powder	Peel, quarter, or slice after dipping in boiling water so that their skin becomes soft. Then drying is done for 6–12 h until they are crispy.
3	Chili	Dried chili, chili powder	Sun drying or dehydration may be done.
4	Cauliflower	Dried cauliflower	Eliminate stalks, stems, and its leaves. Breakdown flowers into little chunks. Blanch for 5 min in water, then dry for 6–14 h.
5	Cabbage	Dried cabbage	Eliminate stalks, leaves, and shred it. Then do water blanching for specific time using 1% of sodium bicarbonate. Then do drying until specific time period.
6	Garden pea	Dried pea	Dry for 5–14 h until becomes breakable.
7	Carrot	Carrot powder, dried carrot pieces	Peeling, slicing, and shredding are all options. Dry for 6–12 h, or until the material is almost breakable.
8	Onion	Dehydrated onion, onion powder	Pretreat and dry for 6–12 h till crunchy after peeling and slicing into 3–5 mm thick slices.
9	Broccoli	Dried broccoli powder	For 4–10 h, cutting and drying.
10	Pumpkin	Pumpkin lather	Dry for 3–10 h until rubbery, after peeling and cutting into 1/4-inch slices.

improves perishability quality, and ensures food accessibility at household and community levels (Sharma et al. 2019).

25.4.2.6 Value Addition Through Cooling

Low-temperature preservation includes refrigeration (−1–8 °C) and freezing (−2 °C and below) (Galati et al. 2021). Cooling is a relatively simple technology for maintaining good quality and nutritional contents of products. Despite the packaging technology employed, maintaining proper temperatures during distribution and storage is a critical component of increasing shelf life.

25.4.2.6.1 Types of Cooling for Vegetable Preservation

Precooling

Although the phrase has been used to describe cooling at any moment before shipping, precooling is the practice of cooling a large number of produce prior to grading and packaging. Effective precooling is becoming obsolete as precooled products have the potential to heat up during subsequent processes.

Room Cooling
Keeping food in a refrigerated area for an unidentified time duration is called as room cooling. For many semi-perishable vegetable items like root crops, potatoes, and cabbage, the cooling pace of room cooling may be sufficient.

Hydro-Cooling
Flooding, spraying, or submerging freshly picked vegetables with enormous amounts of cold water to cool it is known as hydro-cooling. Hydro-cooling is faster and generally thorough process than other cooling processes. Hydro-cooling is only available for products that can withstand liquid water.

Icing
This approach can only be used for produce that is not damaged by ice.

25.4.3 Economic and Social Impacts of Value Addition

Today, neither the monetary nor wholesome force of vegetables is adequately figured out. To tap the monetary benefits from vegetables, governments should build their interest in ranch efficiency, sanitation, and market access. There is an ideal opportunity to focus on interests in vegetables, giving expanded monetary freedoms to smallholder ranchers and giving solid eating regimens to all (Syahza and Asmit 2019). Extension of food production is a conspicuous initial step as developing population and expanded wages, particularly in the metropolitan territories, are now making an ascent in market interest as consumers try to enhance their eating regimens.

Expanding vegetable farming would be beneficial in opening the doors for monetary rewards. Vegetable cultivation pays more to smallholder ranchers. Variety of vegetable yields, short developing cycles, and effective utilization of water system can lessen ranchers' weakness to environmental changes (Fanzo et al. 2020).

25.4.3.1 Reduction in Postharvest Losses

Value addition greatly reduces postharvest losses since it improves the preservation and quality of the final product. Where the raw materials are transformed into juices, jams, or dehydrated items, wastage due to excess produce is minimized (Urugo et al. 2024a, b).

25.4.3.2 Income and Profitability for Farmers

Since the limited increase in income of farmers does not necessarily ensure profitability. This suggests that any improvement that conforms to a given trend may not guarantee profitability to farmers. They may enhance marketing opportunities since value-added products are costly than raw produce in the market (Minot and Sawyer 2016). Farmers gain economic independence through secondary processing or direct marketing of processed products. The markets too offer improved returns on investment.

25.4.3.3 Employment Generation

The processing and value addition by the industries open up employment at different stages of production, i.e., handling, packaging, and marketing. Both skilled and unskilled manpower can be recruited which enhances employment opportunities in the rural and urban sectors (Al Hinai et al. 2022). Training programs also add to employment prospects as well as to the growth in economic status.

25.4.3.4 Export Opportunities Through Value Addition

Value addition of vegetables makes it easier for growers and exporters to penetrate into international markets since they adopt approved standard of quality and safety. Longer durability of the products and good packaging enhance exportability by attracting foreigners to purchase the products. Export market development provides foreign exchange earnings and, therefore, boosts the status of agricultural production (Leonidou et al. 2015).

25.4.4 Constraints Related to Food Processing and Value Addition

Following constraints are related to value-addition and food processing:

25.4.4.1 Rudimentary Technologies

Sometimes resources are not utilized effectively in terms of transforming skills and technologies to local people of a society due to following rudimentary technologies (Hasan et al. 2019).

Normally, conventional technologies are very cheap and easy to apply but may cause many problems in food processing and value addition (Sekulya et al. 2018). Improper use of packaging or packaging material, poor techniques of processing, and low-quality market for the value-added food are the factors that hinder the smooth progress of food value addition and processing (Van Gogh et al. 2017).

25.4.4.2 Safety and Quality Constraints

Modern food safety and quality management systems are not established in most of the processing facilities (Aworh 2020). There are certain techniques in the food processing that are used to inhibit the bacterial growth but these processes affect the quality of the food. For example, thermal processes used in food industry are well known as they effectively control the development of microbial population in food. But these processes also have some drawbacks as these result in destruction of food biochemical properties, ultimately causing the loss of sensory and nutritional quality of food (Wafula et al. 2016).

25.4.4.3 Poor Linkage Between Producers and Processors

The weak linkage between the producers and the processors is the main constraint for value addition. This results in high operational cost that processor passes to the consumer and consumer does not afford this production cost (Gil et al. 2015).

25.4.4.4 Lack of Innovativeness and Product Diversification

In the world of food science and technology, there is need to develop new and innovative food products by using latest tools and techniques for doing value addition. Lack of innovation in product development and food processing does not allow food industry to grow well (Leneveu-Jenvrin et al. 2020).

25.4.4.5 Shortage of Mechanical Professionals

By using old techniques, we cannot add value in products. Specific modules should be designed to train professionals and meet current needs of communities (Arah et al. 2015).

25.4.4.6 Inadequate Artisanal Skills

The existing food processing equipment are latest and unique. But local artisans do not have latest facilities and are unaware of the techniques to maintain or repair the present food processing tools and equipment. So, there is a need to know or learn the latest techniques and methods in the food industry to ensure advanced processing and value addition to form new products (Snyder and Worobo 2018).

25.4.4.7 Inadequate Supply of Appropriate Packaging

Packaging material procured from local sources is unpredictable in terms of its value, protection, and qualification standards. Most of the food losses in the food industry are due to improper or poor packaging. This is because of the limited supply of packaging material and machine which are used for packaging. Almost 40% of food is lost due to improper packaging of food items (Shinde et al. 2018).

25.4.4.8 Poor Processing Methods

Most of the food companies use poor processing and manufacturing methods that result in low levels of value addition.

25.4.4.9 Poor Market Analysis

Commonly, least effort is done for analyzing the expectation of consumers and the requirements of market. Ill planned launching of products does not conform to market demands. Consequently, the disliking of consumers results in the loss of business concerning to food processing (Kaptan and Kayısoglu 2015).

25.5 Conclusion

The value addition in vegetables is an emerging demand for postharvest management system. This goal can be achieved by effective planning and use of technology that embraces cold chain systems, smart packaging, and digital tools. The stakeholders must be aware of efficient and sustainable ways of delivering value so as to have competitive advantage in both domestic and international markets. Nevertheless, the utilization of value-addition practices is not devoid of constraints. Some of the challenges include lack of resources and proper planning. Solving these problems must involve capacity enhancement interventions, infrastructure development, and the harmonization of regulations. Thus, case studies and success stories from different countries may offer relevant recommendations and useful experience. Therefore, the future of value addition will depend on Public-Private Partnership (PPP), research and development, learning facilities, amicable policies, and provision of funds to stakeholders. By availing these opportunities, agricultural sector can sustain its growth, fulfill the food security needs of the people, and satisfy the changing consumer behaviors.

25.6 Future Prospects and Opportunities

25.6.1 Public-Private Partnership Programs

Investment, innovation, and capacity-building relating to value addition are best deliverable through Public-Private Partnership (PPP). The public and private sectors, being key players in the development of the transport infrastructure and market networks, can fill existing gaps more effectively, and would do so at a much faster pace. Based on the analysis of theoretical prerequisites, it is possible to conclude that provisions of policies that provide positive incentives in terms of subsidies, tax exemptions, and or affordable credit funding, compel key players in the value chains to engage in practices that support value addition initiatives for vegetables. The continued simplification of these procedures and the offering of technical support also contribute to the creation of excellent conditions for development.

25.6.1.1 Advancements in Research and Development

Constant innovation is crucial to introduce new solution-oriented technologies as well as efficient and eco-friendly approaches. The advancement in food processing, biotechnology, and packaging has the potential to change the dimension of value addition. Academic and industrial cooperation can achieve innovations with steady advancements and competitiveness that are indispensable for the international community.

References

Abdullah ZW, Dong Y (2019) Biodegradable and water-resistant poly (vinyl) alcohol (PVA)/starch (ST)/glycerol (GL)/halloysite nanotube (HNT) nanocomposite films for sustainable food packaging. Front Mater 6:58

Al Hinai A, Jayasuriya H, Pathare PB, Al Shukaili T (2022) Present status and prospects of value addition industry for agricultural produce – a review. Open Agric 7(1):207–216

Anand S, Barua M (2022) Modeling the key factors leading to postharvest loss and waste of fruits and vegetables in the agri-fresh produce supply chain. Comput Electron Agric 198:106936

Arah IK, Amaglo H, Kumah EK, Ofori H (2015) Preharvest and postharvest factors affecting the quality and shelf life of harvested tomatoes: a mini review. Int J Agron 2015(1):478041

Augustin MA, Riley M, Stockmann R, Bennett L, Kahl A, Lockett T, Cobiac L (2016) Role of food processing in food and nutrition security. Trends Food Sci Tech 56:115–125
Aworh OC (2020) Food safety issues in fresh produce supply chain with particular reference to sub-Saharan Africa. Food Control 107737
Baldwin EA, Nisperos MO, Chen XIUHUA, Hagenmaier RD (1996) Improving storage life of cut apple and potato with edible coating. Postharvest Biol Tech 9:151–163
Benu FL, Mudita IW (2017) Money matters for subsistence farmers: provision of a low-rate credit scheme by a local bank allowed mungbean farmers to adopt improved cultivation practices. Net J Agric Sci 5:71–78
Bharatkumar DP, Khatun P, Kumar C, Yadav AK (2023) Role of agriculture processing in export growth of agricultural products. J Current Res Food Sci 4(1):49–56
Białkowska A, Majewska E, Olczak A, Twarda-Clapa A (2020) Ice binding proteins: diverse biological roles and applications in different types of industry. Biomol Ther 10:274
Butnariu M, Butu A (2015) Chemical composition of vegetables and their products. In: Handbook of food chemistry, pp 627–692
Choi WY, Park HJ, Ahn DJ, Lee J, Lee CY (2002) Wettability of chitosan coating solution on 'Fuji'apple skin. J Food Sci 67:2668–2672
Dang G, Pheng LS (2015) Infrastructure investments in developing economies, vol 10. Springer Science Business Media, Singapore, pp 978–981
Datta S, Chatterjee R, Jana JC (2015) Value addition in vegetable crops. In: Value addition of horticultural crops: recent trends and future directions. Springer, New Delhi, pp 43–57
Deulkar SS, Barve SS (2018) An automated tomato quality grading using clustering-based support vector machine. In: 3rd international conference on communication and electronics systems (ICCES), pp 1128–1133
Ding H, Liu M (2024) From root to seed: Unearthing the potential of carrot processing and comprehensive utilization. Food Sci Nutr 12(11):8762–8778
Dousa J, Vaclavovic P (2012) GOP Analysis Center Report 2012–2013. IGS Central Bureau, 65
Elik A, Yanik DK, Istanbullu Y, Guzelsoy NA, Yavuz A, Gogus F (2019) Strategies to reduce post-harvest losses for fruits and vegetables. Strategies 5(3):29–39
Fanzo J, Covic N, Dobermann A, Henson S, Herrero M, Pingali P, Staal S (2020) A research vision for food systems in the 2020s: defying the status quo. Glob Food Sec 26:100397
FAO (2015) Fruit and vegetable products. Accessed 19 Oct 2015
Fardet A, Rock E (2014) Toward a new philosophy of preventive nutrition: from a reductionist to a holistic paradigm to improve nutritional recommendations. Adv Nutr 5(4):430–446
Featherstone D, Ince A, Mackinnon D, Strauss K, Cumbers A (2012) Progressive localism and the construction of political alternatives. Trans Inst Br Geogr 37:177–182
Floros JD, Newsome R, Fisher W, Barbosa-Cánovas GV, Chen H, Dunne CP, Ziegler GR (2010) Feeding the world today and tomorrow: the importance of food science and technology: an IFT scientific review. Compr Rev Food Sci Food Saf 9:572–599
Galati S, Gullì M, Giannelli G, Furini A, DalCorso G, Fragni R, Visioli G (2021) Heavy metals modulate DNA compaction and methylation at CpG sites in the metal hyperaccumulator Arabidopsis halleri. Environ Mol Mutagen 62:133–142
Gil MI, Selma MV, Suslow T, Jacxsens L, Uyttendaele M, Allende A (2015) Pre-and postharvest preventive measures and intervention strategies to control microbial food safety hazards of fresh leafy vegetables. Crit Rev Food Sci Nutr 55:453–468
Hasan MU, Malik AU, Ali S, Imtiaz A, Munir A, Amjad W, Anwar R (2019) Modern drying techniques in fruits and vegetables to overcome postharvest losses: a review. J Food Process Preserv 43:e14280
Hassan AK, Issa IM, Munishi EJ (2021) Constraints to intermediaries' role in value addition and quality improvement for fresh fruits and vegetables supply chain in dar es salaam city, Tanzania: intermediaries, value addition, fruits and vegetables, supply chain, impediments. Bus Edu J 10(2)
Hazell P, Wood S (2008) Drivers of change in global agriculture. Philos Trans R Soc B Biol Sci 363(1491):495–515

Insights IM (2011) Innova's food & beverage database. Published online at http://www.innovadatabase.com/home/index.rails. Innova Market Insights, Duiven, Netherlands

Jayasena DD, Kim HJ, Yong HI, Park S, Kim K, Choe W, Jo C (2015) Flexible thin-layer dielectric barrier discharge plasma treatment of pork butt and beef loin: effects on pathogen inactivation and meat-quality attributes. Food Microbiol 46:51–57

Kaptan B, Kayısoglu S (2015) Consumers' attitude towards food additives. Am J Food Sci Nutr Res 2:21–25

Kaur R, Watson JA (2024) A scoping review of postharvest losses, supply chain management, and technology: implications for produce quality in developing countries. J ASABE 67(5):1103–1131

Kaur P, Ghoshal G, Jain A (2019) Bio-utilization of fruits and vegetables waste to produce β-carotene in solid-state fermentation: characterization and antioxidant activity. Process Biochem 76:155–164

Keukens EA, Hop ME, Jongen WM (1994) Rapid high-performance liquid chromatographic method for the quantification of alpha-tomatine in tomato. J Agri Food Chem 42:2475–2477

Kulshrestha K (2018) Horticultural crops value addition for nutritional security. Int J Res-Granthaalayah 6:110–120

Lamb FC, Farrow RP, Elkins ER (2019) Effect of processing on nutritive value of food: canning. In: Handbook of nutritive value of processed food. CRC Press, pp 11–30

Leneveu-Jenvrin C, Charles F, Barba FJ, Remize F (2020) Role of biological control agents and physical treatments in maintaining the quality of fresh and minimally-processed fruit and vegetables. Crit Rev Food Sci Nutr 60:2837–2855

Leonidou LC, Fotiadis TA, Christodoulides P, Spyropoulou S, Katsikeas CS (2015) Environmentally friendly export business strategy: its determinants and effects on competitive advantage and performance. Int Bus Rev 24(5):798–811

Lucci P, Pacetti D, Loizzo MR, Frega NG (2016) Canning: Impact on food products quality attributes. In: Food processing technologies. CRC Press, pp 41–60

Magalhães VS, Ferreira LMD, Silva C (2022) Prioritizing food loss and waste mitigation strategies in the fruit and vegetable supply chain: a multi-criteria approach. Sustain Prod Consum 31:569–581

Materia VC, Linnemann AR, Smid EJ, Schoustra SE (2021) Contribution of traditional fermented foods to food systems transformation: value addition and inclusive entrepreneurship. Food Secur 13(5):1163–1177

Miller-Cebert RL, Boateng J, Cebert E, Shackelford L, Verghese M (2016) Chemo preventive potential of canola leafy greens and other cruciferous vegetables on Azoxymethane (AOM)-induced Colon Cancer in Fisher-344 male rats. Food Nutr Sci 7:964

Minot N, Sawyer B (2016) Contract farming in developing countries: Theory, practice, and policy implications. Successes and challenges, Innovation for inclusive value chain development, pp 127–155

Nandane AS, Jain R (2011) Value addition of fruits and vegetables by edible packaging: scope and constraints. In Proceedings of the national symposium on emerging innovative technologies for assurance of quality and safety in processed food, Kharagpur, India, p 24–25

Noort MW, Renzetti S, Linderhof V, du Rand GE, Marx-Pienaar NJ, de Kock HL, Magano N, Taylor JR (2022) Towards sustainable shifts to healthy diets and food security in sub-Saharan Africa with climate-resilient crops in bread-type products: a food system analysis. Food Secur 11(2):135

Nyamulinda B, Bizoza A, Rukazambuga D, Wanjiku C, Buruchara R, Mugabo J, Tenywa M (2011) Agricultural postharvest Innovative technologies and access to niche market: experience from Gataraga IP, Rwanda. Learn Public J Agric Environ Stud 2:1–23

Onwude DI, Chen G, Eke-Emezie N, Kabutey A, Khaled AY, Sturm B (2020) Recent advances in reducing food losses in the supply chain of fresh agricultural produce. PRO 8(11):1431

Park J, Kim S, Moon B (2011) Changes in carotenoids, ascorbic acids, and quality characteristics by the pickling of paprika (*Capsicum annuum* L.) cultivated in Korea. J Food Sci 76:C1075–C1080

Parveen S, Bushra I, Humaira K, Shazia S, Azhar MA (2014) Value addition: a tool to minimize the postharvest losses in horticultural crops. Greener J Agri Sci 4:195–198
Pawlak K, Kołodziejczak M (2020) The role of agriculture in ensuring food security in developing countries: considerations in the context of the problem of sustainable food production. Sustain For 12(13):5488
Picha MS (2007) Enhanced oil recovery by hot CO2 flooding. In: SPE Middle East oil and gas show and conference, Manama. https://doi.org/10.2118/105425-MS
Rao KS, Haran RH, Rajpoot VS (2022) Value addition: a novel strategy for quality enhancement of medicinal and aromatic plants. J Appl Res Med Aromat Plants 31:100415
Ronoh P, Furlong C, Kansiime F, Mugambe R, Brdjanovic D (2020) Are there seasonal variations in faecal contamination of exposure pathways? An assessment in a low–income settlement in Uganda. Int J Environ Res Public Health 17:6355
Saravanan D, Khatoon S, Sabarathinam S (2023) A narrative review on the impact of processed foods/junk foods/preserved foods: why special attention is required to prevent metabolic syndrome? Obesity Med 42:100507
Savo I, Dragan M, Miladin B (2012) Specificities of fruit freeze drying and product prices. Економика пољопривреде 59(3):461–471
Seidu JM, Bobobee EYH, Kwenin WKJ, Frimpong R, Kubge SD, Tevor WJ, Mahama AA (2012) Preservation of indigenous vegetables by solar drying. ARPN J Agri Biol Sci 7:407–415
Sekulya S, Nandutu A, Namutebi A, Ssozi J, Masanza M, Jagwe JN, Hedwig A (2018) Effect of postharvest handling practices, storage technologies and packaging material on postharvest quality and antioxidant potential of *Solanum Aethiopicum* (Shum) leafy vegetable. Am J Food Sci Technol 6(4):167–180
Sharangi AB, Datta S (eds) (2015) Value addition of horticultural crops: recent trends and future directions. Springer, New York, pp 1–342
Sharangi AB, Bhutia PH, Raj AC, Sreenivas M (2018) Underexploited spice crops: present status, agrotechnology, and future research directions. Apple Academic Press
Sharma P, Srivastava P, Seth A, Tripathi PN, Banerjee AG, Shrivastava SK (2019) Comprehensive review of mechanisms of pathogenesis involved in Alzheimer's disease and potential therapeutic strategies. Prog Neurobiol 174:53–89
Shinde R, Rodov V, Krishnakumar S, Subramanian J (2018) Active and intelligent packaging for reducing postharvest losses of fruits and vegetables. Postharvest Biol Nanotechnol 171–189
Snyder AB, Worobo RW (2018) The incidence and impact of microbial spoilage in the production of fruit and vegetable juices as reported by juice manufacturers. Food Control 85:144–150
Sogi DS, Bhullar JK (2000) Shelf life studies and refining of tomato seed oil. J Food Sci Technol 37:542–544
Stancu V, Haugaard P, Lähteenmäki L (2016) Determinants of consumer food waste behavior: two routes to food waste. Appetite 96:7–17
Syahza A, Asmit B (2019) Regional economic empowerment through oil palm economic institutional development. Manag Environ Qual Int J 30:1256–1278
Tadesse KT, Demise R, Mulat F, Zerfu A (2024) The role of postharvest management in ensuring food security in a changing world: review article. J Clin Res Case Stud 2(3):1–14
Tsironi T, Houhoula D, Taoukis P (2020) Hurdle technology for fish preservation. Aquac Fish 5:65–71
Urugo MM, Teka TA, Gemede HF, Mersha S, Tessema A, Woldemariam HW, Admassu H (2024a) A comprehensive review of current approaches on food waste reduction strategies. Compr Rev Food Sci Food Saf 23(5):e70011
Urugo MM, Yohannis E, Teka TA, Gemede HF, Tola YB, Forsido SF, Tessema A, Suraj M, Abdu J (2024b) Addressing postharvest losses through agro-processing for sustainable development in Ethiopia. J Agri Food Res 18:101316
Van Gogh B, Boerrigter H, Noordam M, Ruben R, Timmermans T (2017) Postharvest loss reduction: a value chain perspective on the role of postharvest management in attaining economically and environmentally sustainable food chains (No. 1751) Wageningen Food & Biobased Research

Van Wageningen S, Kemmeren P, Lijnzaad P, Margaritis T, Benschop JJ, De Castro IJ, Holstege FC (2010) Functional overlap and regulatory links shape genetic interactions between signaling pathways. Cell 143:991–1004

Verma DK, Kapri M, Billoria S, Mahato DK, Prem PS (2017) Effects of thermal processing on nutritional composition of green leafy vegetables: a review. In: Engineering interventions in foods and plants. As part of book series on "innovations in agricultural and biological engineering". Apple Academic Press, Palm Bay

Wafula EN, Franz CM, Rohn S, Huch M, Mathara JM, Trierweiler B (2016) Fermentation of African indigenous leafy vegetables to lower postharvest losses, maintain quality and increase product safety. Afr J Hort Sci 9:1–13

Wallace TC, Bailey RL, Blumberg JB, Burton-Freeman B, Chen CO, Crowe-White KM, Drewnowski A, Hooshmand S, Johnson E, Lewis R (2020) Fruits, vegetables, and health: a comprehensive narrative, umbrella review of the science and recommendations for enhanced public policy to improve intake. Crit Rev Food Sci Nutr 60(13):2174–2211

Yadav AN, Kour D, Kaur T, Devi R, Yadav A, Dikilitas M, Abdel-Azeem AM, Ahluwalia AS, Saxena AK (2021) Biodiversity, and biotechnological contribution of beneficial soil microbiomes for nutrient cycling, plant growth improvement and nutrient uptake. Biocatal Agric Biotech 33:102009

Recent Advances in Fresh-Cut Vegetables

26

Anwar Ali, Joanna Trafialek,
and Muhammad Faisal Manzoor

Abstract

Fresh-cut vegetables, also known as minimally processed vegetables, are becoming more popular recently due to their ease of utilization, freshness, and potential health benefits. Their industrial processing involves several steps, including washing, sorting, cutting, packaging, cooling, storage, and distribution. Recent advances in fresh-cut vegetables have focused on improving the processing methods' safety, quality, and sustainability. These emphasize to use alternative disinfectants, such as peroxyacetic acid and ozone, to reduce reliance on chlorine, develop novel packaging materials and technologies to extend shelf life, and reduce environmental impacts. Additionally, advancements in processing equipment, automation, and data analytics have increased efficiency, traceability, and quality control in fresh-cut vegetables' processing. However, there are also challenges and limitations in the data related to recent advances, including gaps in knowledge regarding consumer safety and acceptability. Further research, collaboration among stakeholders, and regulatory guidelines are needed to fully evaluate the benefits and limitations of these advances in fresh-cut vegetables' processing. Overall, recent advances in fresh-cut vegetables have contributed to improved safety, quality, and sustainability of these convenient and popular food products, with the potential for continued advancements in the future.

A. Ali · J. Trafialek
Institute of Human Nutrition Sciences, Warsaw University of Life Sciences–SGGW, Warsaw, Poland

M. F. Manzoor (✉)
Guangdong Provincial Key Laboratory of Intelligent Food Manufacturing, School of Food Science and Engineering, Foshan University, Foshan, China

Faculty of Sciences and Technology, ILMA University, Karachi, Pakistan
e-mail: faisaluos26@gmail.com

Z. Iqbal et al. (eds.), *Sustainable and Innovative Vegetable Production in times of Climate Change*, https://doi.org/10.1007/978-981-96-6283-8_26

Keywords

Fresh-cut vegetables · Fruits · Recent advances · Processing · Food safety

26.1 Introduction

Fresh-cut vegetables are becoming increasingly popular due to their convenience, long shelf life, and the growing demand for healthy and nutritious food options. In recent years, there have been several advances in producing and introducing fresh-cut vegetables (Giannakourou and Tsironi 2021). Advances in packaging technology have allowed for the development of packaging materials that help maintain the freshness and quality of fresh-cut vegetables for extended periods. These include modified atmosphere packaging, which involves controlled gas mixtures to slow down the rate of respiration and microbial growth, and vacuum packaging, which removes the air from the package to slow down oxidation (Qu et al. 2022). There has been a growing demand for a broader range of fresh-cut vegetables beyond traditional options such as carrots, lettuce, and cucumbers. Producers now offer more exotic and diverse options such as kale, bok choy, and rainbow carrots to cater to consumers' changing tastes and preferences (Bumbudsanpharoke and Ko 2022). Advances in processing technology have made it possible to produce fresh-cut vegetables that are more visually appealing and retain their nutritional value for extended periods. For example, some processors now use high-pressure processing (HPP) to preserve fresh-cut vegetables' color, texture, and flavor while eliminating harmful bacteria. There is an increasing focus on sustainability in the fresh-cut vegetables industry, and producers are adopting activities based on minimizing waste, using biodegradable packaging materials, and sourcing produce from local and organic farms (Yousuf et al. 2021).

Fresh vegetables and fruits are minimally processed to meet the consumer demands of convenience and lifestyle modifications. The definition of fresh-cut produce is: "any vegetable or fruit or mixture of both that has been reformed physically (by slicing, peeling, shredding, tearing, chopping, etc.) from raw to modified form after harvesting from the fields without blanching or cooking" (USFDA 2018). During the last two decades, the demand for fresh-cut produce has been urged worldwide due to the advanced lifestyle culture, health-boosting potential, sustainability, and convenience (Fan and Wang 2021). The rise in fresh-cut produce is on peak mostly in developing countries and will grow even faster. Organic nature, accessibility, and sustainability are the main factors in the growth of the European fruit and vegetable market (CBI 2023). The market demand for ready-to-eat salads is higher in Europe, with 50–70% shares compared to fresh-cut produce. According to a survey, it is estimated that each year, 1/3 of the population of developed countries may be infected by foodborne illnesses, which will be much higher for underdeveloped countries (Raffo and Paoletti 2022).

The population in developing countries mostly lacks information about food safety (Aworh 2021). The safety of fresh produce usually depends on factors like fertilizers, chemical contaminants, including heavy metals, chemical/industrial waste, and fungal attacks on fresh by-products. Pathogenic contamination is crucial for fresh vegetables and fruits (Mostafidi et al. 2020).

Concerns about food safety and security have recently increased. Both of these are directly linked to the processing of fresh-cut produce. This chapter will provide an up-to-date overview of fresh-cut vegetables' latest developments, innovations, and technologies. It includes advancements in processing methods, food safety practices, quality control, packaging, storage, distribution, alternative disinfectants, packaging materials, and technologies to enhance fresh-cut vegetables' safety, quality, and sustainability.

26.2 Nutritional Value of Fresh-Cut Produce

The health-boosting potentials of fresh-cut vegetables are retained due to minimal processing operations of freshly harvested produce without any nutrient loss. Previously, several studies proved that processing and storage conditions impact mineral, vitamin, and bioactive profiles based on factors such as cultivars and harvesting/growing conditions (De Corato 2020). Studies revealed that the varietal and ecological differences had a major influence on the level of processing impacts on the nutrient contents of fresh-cut vegetables. Moreover, contradictory results of effect of processing conditions on phytochemical and vitamin profiles have been reported within the same species of vegetables (Sumner and Peters 2023). These results did not depict industrial processing conditions as experiments have been conducted in laboratories. Studies showed that the health-boosting potential of fresh-cut vegetables is not only limited to phytochemicals and vitamin profile, but a major contribution comes from minerals and fiber constituents of fresh produce, which remains unaffected by processing and storage conditions. Previous studies establish that dietary fiber protects against various chronic diseases. Fresh vegetables and fruits are low-calorie foods and physicochemical properties of plant-based foods play a vital role in weight management; meanwhile, overweight and excessive body fat percentage are major risk factors in the onset of chronic health issues (Saravanakumar et al. 2021). So, based on all the previous data, it can be assumed that there is no substantial difference in the nutritional profile of fresh-cut produce compared to non-processed produce. Meanwhile, a significant difference in food safety issues between fresh-cut vegetables and non-processed fresh produce has been noticed that is influenced directly by the processing phase (Fallik and Ilić 2021).

26.3 Health-Boosting Potential of Fresh-Cut Produce

Fresh-cut produce significantly impacts by providing essential vitamins, minerals, and other beneficial plant compounds important for human health. Further, these provide a meager amount of dietary fiber, which improves digestion and prevents constipation (Agriopoulou et al. 2020). The high nutrient content and beneficial plant compounds found in fresh-cut produce shield the human body against chronic diseases. The vitamins and minerals in fresh-cut produce improve immune system and protect from infections and illnesses. High vitamin and antioxidant contents of fresh-cut produce improve skin health and slow down the aging process. Consuming fresh-cut produce can significantly impact overall physical and mental health, making it essential for a healthy and balanced diet (Singla et al. 2020).

Fresh-cut vegetables contain many phytochemicals like polyphenols that possess excellent antioxidant properties. Sulforaphane and indole compounds, found in cruciferous vegetables, have been linked to reducing cancer risk and improving heart health (Shi et al. 2022). Overall, fresh-cut vegetables contain a diverse range of phytochemicals with various health-promoting benefits. Individuals can ensure that they receive a wide range of these beneficial compounds by consuming various fresh-cut vegetables.

26.4 Processing of Fresh-Cut Vegetables: The Fundamental Step of Washing/Disinfection

During the last decade, several studies revealed that among all the processing steps involved in fresh-cut vegetables, the most critical is washing or disinfection due to its significant impact on fresh-cut quality, safety, and environmental sustainability. Various machines are used in this step; apart from being designed differently, the general mechanism involved is subjecting fresh vegetables to an upward flow of pressurized water or air bubbles with or without a disinfecting agent in a water tank (Mahnot et al. 2020). In the next step, vegetables are removed from the water tank and rinsed with clean water if exposed to any chemical disinfectant. The washing step in the fresh-cut industry is crucial due to the difference between fresh-cut vegetables and non-processed produce. Firstly, the vegetable is peeled in the cutting operation, and the outer skin is removed. It naturally prevents the deterioration of vegetables by preventing any tissue damage due to microbial invasion and channeling the juice leakage from the inside vegetable to the outer part (Mahnot et al. 2020). These cutting steps briefly explain the reason for significant microbial growth in fresh-cut vegetables compared to intact counterpart produce.

Generally, a microbiological load of fresh-cut vegetables is 5–9 logarithmic units, while 1 log unit reduction in microbial load is achieved merely by pressurized water flow inserting mechanical force (Kaczmarek et al. 2019). However, washing fresh vegetables with portable water has limitations as organic matter quickly accumulates in the bottom of the tank, thus promoting microbial growth. Hence, fresh-cut vegetables conveyed into the water tank could cross-contaminate unless continuously renewed (Fig. 26.1). Studies have shown that excessive water is required for this purpose, which requires huge investment and is difficult to manage. Besides, some studies revealed that this method is not cost-effective, whereas others showed that even excessive water would not prevent this risk (Mieszczakowska-Frąc et al. 2021).

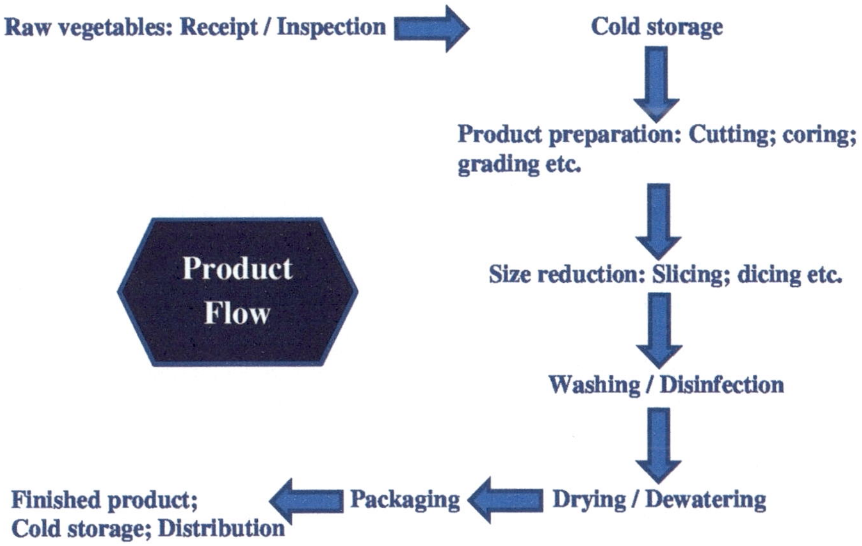

Fig. 26.1 Product flow diagram for the fresh-cut vegetable industry

26.4.1 Disinfecting Agents: Chlorine

At the industrial level, chlorine is utilized as a disinfectant agent for the washing. Chlorinated water is applied for fresh-cut industry, in which chlorine is used in any of the three forms, such as liquid sodium hypochlorite (NaOCl) or solid calcium hypochlorite [$Ca(OCl)_2$], or gas chlorine (Cl_2) or chlorine dioxide (ClO_2). Chlorine is utilized extensively as a processing water disinfecting agent due to its cost-effectiveness, non-reactive nature, antimicrobial capacity, and ease of usage at the industrial level (Sharma et al. 2019). The steps to use chlorine as a disinfecting agent for fresh-cut vegetables at an industrial level include the preparation of a solution of chlorine by adding the appropriate amount of chlorine bleach to water based on the concentration recommended by local regulatory agency (Zhang et al. 2020).

The concentration typically ranges from 50 to 200 parts per million (ppm). Fresh-cut vegetables are normally washed using specialized equipment such as flumes or conveyor belts to remove visible dirt or debris. The vegetables are then submerged in the prepared chlorine solution for a specific amount, typically ranging from 1 to 5 min, depending on the type of vegetable and the concentration of the chlorine solution. After soaking, the vegetables are rinsed with clean water to remove residual chlorine (Julien-Javaux et al. 2019). The vegetables are then drained and dried before further processing or packaging. Appropriate concentration of chlorine solution is important to follow food safety guidelines when washing fresh-cut vegetables at an industrial level to ensure product safety. The concentration of the chlorine should be regularly monitored to ensure its recommended range (Balali et al. 2020).

26.4.2 Chlorine: Environmental Issues

While chlorine is an effective disinfecting agent for fresh-cut vegetables, environmental issues are associated with its use. A potential environmental issue related to the use of chlorine includes chlorine gas emissions. Chlorine gas can be released into the atmosphere during the production, transport, and use of chlorine, which can contribute to air pollution and climate change. Disposal of chlorine solutions can lead to contamination of waterways and soil, which can harm aquatic life and damage ecosystems (Cuggino et al. 2020).

Accidental chlorine spills can occur during transport or handling, posing a risk to human health and the environment. Alternative disinfecting agents and methods are being developed and implemented to address these environmental issues. For example, some processors use ozone or ultraviolet light as an alternative to chlorine for disinfecting fresh-cut vegetables. Additionally, best management practices such as proper handling, storage, and disposal of chlorine can help reduce the risk of environmental harm (Tudela et al. 2019b).

26.4.3 Food Safety Risks Associated with Chlorine

While chlorine is an effective disinfecting agent for fresh-cut vegetables, there are potential food safety risks associated with its application if not used properly. If the concentration of residual chlorine is too high on fresh-cut vegetables after disinfection, it can cause a strong odor or taste and, in some cases, harm human health (Botondi et al. 2021). If the chlorine concentration is too low or the exposure time is too short, the fresh-cut vegetables may not be adequately disinfected, increasing the risk of foodborne illness. If the equipment used to wash or disinfect fresh-cut vegetables is not adequately cleaned or sanitized, it can lead to cross-contamination with harmful bacteria (López-Gálvez et al. 2021).

26.4.4 Food Quality Risks of Chlorine

While chlorine is commonly used as a disinfectant for fresh-cut vegetables, there are also potential food quality risks.

26.4.4.1 Development of Off-Flavor and Odors

Organic compounds present on the surface of fresh-cut vegetables may react with chlorine and cause off-flavors and odors, which can affect the taste and quality of the final product. This reaction can occur when the chlorine concentration is too high or the exposure time is too long (Simpson and Mitch 2022). The specific off-flavors and odors that may develop can vary depending on the type of vegetable, the concentration of chlorine, and the exposure time. For example, chlorophenol compounds may form in lettuce and spinach, giving the vegetables a medicinal or disinfectant-like taste. Additionally, it is essential to rinse the vegetables thoroughly after disinfection to remove any residual chlorine and organic compounds that may have reacted with the chlorine (Chain et al. 2020).

26.4.4.2 Discoloration

Chlorine can also cause discoloration of fresh-cut vegetables, especially if exposed to high chlorine concentrations or too long (Maringgal et al. 2020).

26.4.4.3 Loss of Nutrients

Exposure to chlorine can lead to a loss of nutrients in fresh-cut vegetables, especially vitamin C, which is sensitive to oxidation. Chlorine can impact the nutrient content of fresh-cut vegetables by oxidizing and degrading specific vitamins and phytochemicals. This oxidation can occur when the chlorine concentration is too high or the exposure time is too long. Vitamin C is an important antioxidant that helps protect cells from damage and is also involved in collagen synthesis and immune function. To minimize chlorine's impact on fresh-cut vegetables' nutrient content, it is important to use the appropriate concentration of chlorine solution and maintain the correct exposure time. Additionally, it is important to rinse the vegetables thoroughly after disinfection to remove any residual chlorine and prevent further oxidation. Other disinfectants or sanitizers may also be used as an alternative to chlorine, depending on the specific needs and requirements of the food processing industry (Bhargava et al. 2021).

26.4.4.4 Formation of Disinfection by-Products

Organic matter on the surface of fresh produce in the water may react with chlorine that results in the formation of trihalomethanes (THMs) and haloacetic acids (HAAs), which are source of off-flavor of the final product. To minimize these food quality risks, it is important to use the appropriate concentration of chlorine solution, maintain contact time, and properly rinse the fresh-cut vegetables after disinfection to remove any residual chlorine. Additionally, it is important to properly store and handle fresh-cut vegetables to prevent exposure to excessive chlorine or other disinfectants (Chen et al. 2020).

26.4.5 Legal Issues Associated with Chlorine Usage

Government agencies regulate chlorine as a disinfectant for fresh-cut vegetables to ensure that it is used safely and effectively. Government agencies may establish MRLs for chlorine in fresh-cut vegetables to ensure they do not contain excessive chlorine residues. These limits may vary depending on the type of vegetable, the intended use of the vegetable, and the specific regulations of the country or region (Mahendran et al. 2019). The use of chlorine in the food industry is subject to health and safety regulations to ensure that it is used safely and does not pose a risk to workers or consumers. These regulations may cover workers' protection, safe handling and storage, and disposal of chlorine-containing solutions (Botondi et al.

2021). Chlorine use in the food industry is also subject to environmental regulations to prevent contamination of waterways and other natural resources. These regulations may cover areas such as wastewater treatment, discharge limits, and disposal of chlorine-containing solutions (Giannakourou and Tsironi 2021).

26.4.6 Strategies to Minimize Chlorine Usage as a Disinfectant

Several strategies can be used to minimize chlorine as a disinfectant for fresh-cut vegetables. One strategy is to use alternative disinfectants, such as hydrogen peroxide, ozone, or UV light. These disinfectants can be as effective as chlorine in reducing microbial contamination and may have fewer negative impacts on the quality and safety of fresh-cut vegetables. Several alternative disinfectants can be used instead of chlorine to sanitize fresh-cut vegetables (Wong et al. 2021). Acidic electrolyzed water is a solution that is generated by passing an electrical current through a saltwater solution.

Another strategy is to implement good agricultural practices (GAPs) to reduce the likelihood of microbial contamination in the first place. GAPs may include measures such as using clean water for irrigation, ensuring proper worker hygiene, and minimizing contact between vegetables and soil or other potential sources of contamination. GAPs are principles and practices designed to minimize the risk of microbial contamination in producing and handling fresh produce (Jelsma et al. 2019). By implementing GAPs, growers can reduce the amount of microbial contamination on fresh produce, thus, reducing the amount of chlorine needed to disinfect the produce.

Another source of microbial contamination on fresh produce is the poor hygiene of workers. By implementing good hygiene practices, such as hand washing and clean gloves and clothing, growers can reduce the amount of microbial contamination on their produce and, as a result, reduce the amount of chlorine needed to disinfect the produce. Soil can also be a source of microbial contamination of fresh produce. If the product is over-rinsed, the amount of chlorine used for disinfection may be diluted to the point where it is no longer effective (Turner et al. 2020).

To optimize the use of chlorine and ensure the safety and quality of fresh-cut vegetables, it is important to follow established best practices for rinsing. It typically involves using the appropriate amount of water to effectively rinse the produce without over-diluting the disinfectant. Additionally, it is important to monitor chlorine concentration in the rinsing water and adjust it as needed to ensure its effectiveness (Table 26.1).

26.4.7 Development of Chlorine-Free Processing Technologies

Researchers are also exploring new processing technologies that can eliminate the need for chlorine, such as high-pressure processing or nonthermal plasma technology. The development of chlorine-free processing technologies can significantly

Table 26.1 Summary of latest studies in optimizing chlorine levels for washing water disinfection

Aim of the study	Type of fresh-cut produce	Experimental setting	Measured parameter	Free chlorine concentration (mg L^{-1}) in washing water	References
Efficacy study of water, SH, and PAA against *Salmonella*	Fresh-cut iceberg lettuce	Laboratory scale	Physicochemical and microbial	25 mg/L and 80 mg/L of free chlorine	Cuggino et al. (2023)
Efficacy study of ozone and chlorine on the quality of fresh-cut produce during storage	Fresh-cut onion	Laboratory scale	Physicochemical and microbial	Levels: 100	Aslam et al. (2022)
Efficacy of chlorine and peracetic acid (PAA) against pathogens	Lettuce, baby leaves, carrots	Industrial scale	Physicochemical and microbial	Levels: 80	Petri et al. (2021)
Efficacy of peroxy acid solution composed of organic acids and chlorine	Cherry tomatoes	Laboratory scale	Microbial	100 mL of the 8 ppm of free chlorine solution	Gurtler et al. (2022)
Efficacy of gaseous chlorine dioxide against pathogens	Tomatoes, carrots, blueberries, baby-cut almonds, peppercorns	Commercial scale	Microbial	Levels: 360	Tan et al. (2023)
Efficacy of chlorine dioxide against pathogens	Iceberg lettuce	Laboratory scale	Microbial	20 and 30 mg L^{-1}	Hassenberg et al. (2021)
Efficacy of gaseous chlorine dioxide against *Salmonella enterica, E. coli* O157:H7, and 2 *Listeria monocytogenes*	Strawberries and blueberries	Laboratory scale	Microbiological	3–5 mg/L	Luu et al. (2021)

reduce the use of chlorine in fresh-cut vegetable processing. These technologies use alternative methods for disinfection that do not require the use of chlorine or other chemicals.

Example of chlorine-free processing technologies includes ultraviolet (UV) light implementation which is an effective method for disinfecting fresh-cut vegetables. This technology uses a specific wavelength of light to destroy microorganisms on the surface of the produce (Lv et al. 2021). Pulsed electric fields (PEF) is a nonthermal technology that uses short bursts of high-voltage electricity to disrupt the cell membranes of microorganisms, resulting in their inactivation (Manzoor et al. 2019, 2021). Ozone is a powerful oxidant used to disinfect fresh-cut vegetables. This technology involves exposing the produce to ozone gas, which destroys microorganisms on the surface of the produce. Plasma is a novel technology that uses ionized gas to disinfect fresh-cut vegetables. This technology effectively reduces the microbial load on fresh-cut vegetables (Arshad et al. 2020).

As discussed earlier, several alternative disinfectants, such as UV light, PEF, ozone, and plasma, can replace chlorine. Specific natural antimicrobial agents, such as essential oils, plant extracts, and organic acids, effectively reduce the microbial load on fresh-cut vegetables (Lepaus et al. 2020). These agents can be used in place of chlorine to provide adequate disinfection. Certain processing conditions, such as temperature, pH, and water activity, can be modified to reduce the microbial load on fresh-cut vegetables. For example, low-temperature washing effectively reduces the microbial load on fresh-cut vegetables (Tudela et al. 2019a). MAP involves modifying the atmosphere surrounding fresh-cut vegetables to minimize microbial growth. Adopting these strategies makes it possible to replace chlorine as a disinfectant for fresh-cut vegetables while ensuring their safety and quality (Nicolau-Lapeña et al. 2019) (Table 26.2).

26.4.7.1 Peroxyacetic Acid (PAA)

PAA is a colorless liquid with a strong, pungent odor. It is a powerful oxidizing agent and disinfectant used in various industrial and agricultural applications, including as a sanitizer for food and beverage processing equipment, a disinfectant for medical and dental equipment, and a bleaching agent for textiles and paper. PAA is composed of acetic acid (the main component of vinegar) and hydrogen peroxide, which react together to produce a highly effective oxidizing and disinfecting agent. PAA is highly effective against many microorganisms, including bacteria, viruses, fungi, and spores. It is also biodegradable and breaks down into harmless by-products (Cheng et al. 2020).

PAA is a safer alternative to traditional disinfectants such as chlorine and quaternary ammonium (QACs) because it does not produce harmful disinfection by-products or leave toxic residues on treated surfaces. It is also effective at lower concentrations than chlorine, making it a more economical choice for disinfection applications. The food industry commonly uses PAA as a sanitizer for fresh produce, meat, and poultry. It disinfects food contact surfaces, such as cutting boards, processing equipment, and food storage containers (Cano et al. 2021).

Table 26.2 Advantages and disadvantages of alternative disinfecting methods applied in the fresh-cut industry

Disinfecting method	Advantages	Disadvantages	References
Synthetic methods			
Peroxyacetic acid	Simple management, ease of implementation, economical, efficacy independent of physiochemical properties of water, GRAS substance, low tendency to form DBPs and not corrosive (<80 mg/L)	Low antimicrobial efficacy at the permitted level and not allowed for organic products	Min et al. (2023)
Ozone	Possesses antimicrobial activity even at low concentration and short time, independent from pH, GRAS substance, and no formation of chlorinated DBPs	Toxic if inhaled, in situ generation needed, corrosive, formation of brominated DBPs, high initial cost, and not recommended for organic products	Sun et al. (2022)
Electrolyzed water	Based on the use of salt and water, no need to handle chlorine solution, reduced concentration of chlorine in wash water, and lower corrosiveness	High cost and limited availability of equipment, organic matter level dependent efficacy in wash water, little evidence of efficacy at the commercial level, and not allowed for organic matter	Ahmed et al. (2022)
Hydrogen peroxide	Easy to apply, economical, non-corrosive, and no formation of chlorinated DBPs	Fast disinfectant with slow kinetics, negative impact on product quality, phytotoxic, and not recommended for organic products	Zhou et al. (2023)
Organic acids (citric, malic, lactic acid)	Allowed for organic products, no formation of DBPs, and easy to apply	Low efficacy, long contact time, not suitable for industrial line, impact on product sensory attributes, and high COD and BOD of wastewater	Bangar et al. (2022)
Biologically based methods			
Bacteriocins, bacteriophages, enzymes	Environmentally friendly, harmless to humans, specificity of antimicrobial activity, and reduced impacts on product sensory attributes	Poorly suitable for industrial processing, long contact time, low antimicrobial efficacy, high cost, need for mixtures of disinfecting agents, and consumer acceptance to be verified	Rendueles et al. (2022)

(continued)

Table 26.2 (continued)

Disinfecting method	Advantages	Disadvantages	References
Physically based methods			
Ultraviolet radiation (UV)	Strong antimicrobial efficacy, low cost of energy, easy to implement, no formation of toxic by-products, and useful application of wastewater	Negative impact on product sensory attributes, long treatment times, low efficacy at high organic matter levels, and complex standardization at the commercial scale	Abadias et al. (2021)
Ultrasound	No formation of toxic by-products, environmentally friendly Useful in combination with other disinfecting technologies	Consumer acceptance to be verified, limited antimicrobial efficacy, dependent on organic matter, water hardness, dissolved gasses, and increased water temperature during application	Khaire et al. (2022)
Combination of disinfecting methods			
O_3 + UV radiation O_3 + calcium lactate O_3 + citric acid O_3 + lactic acid Sonication+UV $Na_2HCO_3 + H_2O_2$ PAA+ UV radiation	Wider antimicrobial action, increased antimicrobial efficacy, and reduced chlorine dosage (when applied with chlorine)	Complex standardization at commercial levels and consumer acceptance to be verified	Ruan et al. (2022), Sun et al. (2022)

PAA is highly effective against many microorganisms, including bacteria, viruses, fungi, and spores. PAA works quickly, with a disinfection time of only a few minutes. PAA breaks down into harmless by-products, which means it does not leave any harmful residues on treated surfaces. PAA is biodegradable and does not persist in the environment. PAA is effective at lower concentrations than chlorine, which makes it a more economical choice for disinfection applications. Unlike other disinfectants, its use also has disadvantages. PAA can be corrosive to some materials, including metals and plastics, limiting its use in specific applications. PAA can irritate the skin and eyes, so workers must take precautions when handling it. PAA is a hazardous chemical and requires proper handling, storage, and disposal to minimize potential risks to workers and the environment. PAA has limited stability over time, so it must be prepared fresh and used immediately to ensure maximum efficacy. PAA has a strong, pungent odor that can be unpleasant to some people and may require ventilation or other measures to control the odor in the workplace. PAA can be more expensive than other disinfectants, hindering its adoption in some industries (Donatti Leão Alvarenga et al. 2020).

26.4.7.2 Ozone (O_3)

Ozone is a colorless gas with a pungent odor formed when oxygen molecules (O_2) are exposed to ultraviolet (UV) radiation or electrical discharges. Ozone is a powerful oxidizing agent that can react with a wide range of organic and inorganic substances, making it an effective disinfectant and sanitizer. Ozone chemistry is complex and involves a series of reactions that depend on the specific conditions and substances. When ozone reacts with organic matter, it can break down complex molecules into simpler compounds, such as carbon dioxide, water, and smaller organic molecules. This process is known as oxidation, and it helps to destroy bacteria, viruses, and other microorganisms by disrupting their cell membranes and interfering with their metabolic processes (Epelle et al. 2023).

Ozone is highly reactive and unstable, so it has a relatively short half-life and cannot be stored or transported easily. It also requires specialized equipment to generate and distribute, making it more expensive and difficult to use than other disinfectants. Ozone is effective against many microorganisms, including bacteria, viruses, fungi, and spores. Ozone breaks down into harmless by-products, which means it does not leave any harmful residues on treated surfaces. Ozone is biodegradable and does not persist in the environment (Niveditha et al. 2021). Ozone does not produce toxic fumes that can be harmful to workers. Ozone is a safer alternative to traditional disinfectants such as chlorine and quaternary ammonium compounds (QACs) because it does not produce harmful disinfection by-products.

26.4.7.3 Electrolyzed Water (EW)

It is a type of water that has been electrically activated or ionized, resulting in a solution with strong antimicrobial properties. The electrolysis process involves passing an electric current through a saltwater solution, which creates two separate solutions viz. acidic and alkaline. The acidic solution is often called "anolyte" and contains hypochlorous acid, while the alkaline solution is called "catholyte" and contains sodium hydroxide. Both solutions have antimicrobial properties, but anolyte is typically used as a disinfectant and sanitizer in the food industry. Anolyte is a non-toxic and environmentally friendly disinfectant, making it safe for food processing and handling. Anolytes are effective against many microorganisms, including bacteria, viruses, fungi, and spores (Zhao et al. 2021).

The acidic solution acts quickly, killing microorganisms within seconds or minutes of contact. Unlike some other disinfectants, it can provide residual protection against recontamination. The cost of producing anolyte is relatively low compared to other disinfectants, and the technology required to generate it is relatively simple. However, anolyte has limited stability and can lose effectiveness over time or when exposed to light or heat (Ampiaw et al. 2021). Anolyte can be corrosive to specific materials, such as aluminum, copper, and brass, damaging equipment and infrastructure over time. The buildup of minerals in the anolyte generator can lead to scaling, reducing the system's effectiveness.

26.5 Alternative Approaches for the Organic Sector

For a few decades, the organic sector has been exploring alternative strategies to reduce the use of synthetic disinfectants and ensure food safety. Organic farmers can follow good agricultural practices (GAPs) to minimize the risk of contamination in the first place. It can include regular soil and water testing, using compost and manure safely, and following proper harvest and handling practices. Physical interventions, such as washing, brushing, and air drying, can remove pathogens from fresh produce. These interventions are particularly effective when combined with other strategies, such as good agricultural practices and proper storage and handling (Ali et al. 2023a, b; Manzoor et al. 2023).

High-pressure processing (HPP) is a nonthermal technology that uses high pressure to inactivate pathogens. This technology is particularly effective for preserving fresh produce's nutritional and sensory properties (Fam et al. 2021). Ultraviolet-C (UV-C) light is a nonthermal treatment that can inactivate pathogens on surfaces. This technology is effective against many microorganisms and is being explored as a disinfection method for fresh produce. Natural antimicrobials, such as essential oils, organic acids, and bacteriophages, can control pathogens in fresh organic produce (Koutchma 2019). These substances are effective against various microorganisms and can be used as an alternative to synthetic disinfectants. Modified atmosphere packaging (MAP) is a packaging technique that involves modifying the atmosphere around the product to inhibit the growth of microorganisms. This technology can be particularly effective for extending the shelf life of fresh produce. Overall, the organic sector is exploring a range of strategies to reduce the use of synthetic disinfectants and ensure food safety. These strategies focus on preventing contamination in the first place and using alternative technologies and natural antimicrobials to control pathogens (Tabassum and Khan 2020).

26.6 Major Data Gaps for the Use of Disinfectants in Fresh Produce

There is a need to study the long-term environmental impacts of disinfectant use, particularly concerning the accumulation of disinfectant residues in soil and water. There is a need to examine the potential health effects of disinfectant residues on consumers, farm workers, and the environment. There is a need to understand consumer perceptions and acceptability of disinfectants used in fresh produce. It includes understanding consumer attitudes towards different types of disinfectants and their preferences for organic or conventionally grown produce (Masotti et al. 2019).

There is a need for comparative studies to evaluate the effectiveness of different disinfectants and disinfection methods for controlling pathogens in fresh produce. It's wise to assess the economic feasibility of using alternative disinfectants and disinfection methods, particularly for small-scale and organic farmers. There is also need to develop standards and guidelines for using disinfectants in fresh produce, particularly for organic farming systems. Addressing these data gaps is essential for ensuring the safe and sustainable use of disinfectants in fresh produce and for promoting alternative strategies to reduce reliance on synthetic disinfectants (Pedreira et al. 2021).

26.7 Advanced Conceptual Tools for a Comprehensive Evaluation of All the Processing Impacts on Fresh Produce

Life Cycle Assessment (LCA) is a tool that evaluates the environmental impacts of a product or process throughout its entire life cycle, from raw material extraction to disposal. LCA can be used to assess the environmental effects of different disinfectants and disinfection methods and the impacts of other processing steps, such as packaging and transportation (Yang et al. 2022). Multi-Criteria Decision Analysis (MCDA) is a tool that helps decision-makers evaluate different options based on multiple criteria or objectives. Risk assessment can be used to evaluate the risks associated with different disinfectants and disinfection methods and identify strategies for reducing those risks (Bolaji et al. 2021). Environmental Impact Assessment (EIA) is a tool that evaluates the potential environmental impacts of a project or process. EIA can evaluate the environmental effects of disinfectants, disinfection methods, and other processing steps, such as packaging and transportation.

26.8 Limitations

The data has several limitations related to recent advances in fresh-cut vegetables. A limited amount of data is available on using alternative disinfectants, such as peracetic acid and electrolyzed water, and their impacts on food safety, quality, and shelf life. There is a lack of standardization in the methods used to evaluate the efficacy of disinfectants and their effects on fresh-cut vegetables. It makes, comparing the results of different studies and drawing general conclusions, difficult. There is limited data on consumers' acceptability of fresh-cut vegetables treated with alternative disinfectants.

It is important to consider when evaluating the feasibility of using these disinfectants at an industrial level. There is high variability in the types of fresh-cut vegetables available in the market, which can impact the efficacy of different disinfection methods and their impacts on food safety and quality. Many studies on using alternative disinfectants for fresh-cut vegetables are short-term and do not evaluate the long-term effects on food safety and quality. These limitations highlight the need for more standardized and long-term studies on alternative disinfectants for fresh-cut vegetables, consumer acceptability, and the variability of fresh-cut vegetable products.

26.9 Conclusion

Recent advances in the processing and disinfection of fresh-cut vegetables have shown great potential to improve food safety, extend shelf life, and reduce the use of harmful chemicals such as chlorine. Alternative disinfectants such as peracetic acid and electrolyzed water effectively reduce microbial contamination while maintaining fresh-cut vegetables' nutritional and sensory quality. However, there are still gaps in our understanding of the long-term impacts of these disinfectants on food safety, quality, and consumer acceptability. Further research is needed to address these gaps and ensure that the benefits of these advances can be fully realized. Overall, the development of new technologies and best practices in the fresh-cut vegetable industry offers a promising avenue for improving the safety and quality of our food supply.

References

Abadias M, Colás-Medà P, Viñas I, Bobo G, Aguiló-Aguayo I (2021) Application of an innovative water-assisted ultraviolet C light technology for the inactivation of microorganisms in tomato processing industries. Food Microbiol 94:103631

Agriopoulou S, Stamatelopoulou E, Sachadyn-Król M, Varzakas T (2020) Lactic acid bacteria as antibacterial agents to extend the shelf life of fresh and minimally processed fruits and vegetables: quality and safety aspects. Microorganisms 8(6):952

Ahmed S, Akther S, Alam SS, Ahiduzzaman M, Islam MN, Azam MS (2022) Individual and combined effects of electrolyzed water and ultrasound treatment on microbial decontamination and shelf life extension of fruits and vegetables: a review of potential mechanisms. J Food Process Preserv 46(8):e16765

Ali M, Cheng JH, Tazeddinova D, Aadil RM, Zeng XA, Goksen G, Manzoor MF (2023a) Effect of plasmaactivated water and buffer solution combined with ultrasound on fungicide degradation and quality of cherry tomato during storage. Ultrason Sonochem 97:106461

Ali M, Manzoor MF, Goksen G, Aadil RM, Zeng XA, Iqbal MW, Lorenzo JM (2023b) High-intensity ultrasonication impact on the chlorothalonil fungicide and its reduction pathway in spinach juice. Ultrason Sonochem 94:106303

Ampiaw RE, Yaqub M, Lee W (2021) Electrolyzed water as a disinfectant: a systematic review of factors affecting the production and efficiency of hypochlorous acid. J Water Proc Eng 43:102228

Arshad RN, Abdul-Malek Z, Munir A, Buntat Z, Ahmad MH, Jusoh YM, Aadil RM (2020) Electrical systems for pulsed electric field applications in the food industry: An engineering perspective. Trends Food Sci Technol 104:1–13

Aslam R, Alam MS, Kaur P (2022) Comparative study on efficacy of sanitizing potential of aqueous ozone and chlorine on keeping quality and shelf-life of minimally processed onion (*Allium Cepa* L.). Ozone Sci Eng 44(2):196–207

Aworh OC (2021) Food safety issues in fresh produce supply chain with particular reference to sub-Saharan Africa. Food Control 123:107737

Balali GI, Yar DD, Afua Dela VG, Adjei-Kusi P (2020) Microbial contamination, an increasing threat to the consumption of fresh fruits and vegetables in today's world. Intl J Microbiol 2020:1

Bangar SP, Suri S, Trif M, Ozogul F (2022) Organic acids production from lactic acid bacteria: a preservation approach. Food Biosci 46:101615

Bhargava N, Mor RS, Kumar K, Sharanagat VS (2021) Advances in application of ultrasound in food processing: a review. Ultrason Sonochem 70:105293

Bolaji I, Nejad B, Billham M, Mehta N, Smyth B, Cunningham E (2021) Multi-criteria decision analysis of agri-food waste as a feedstock for biopolymer production. Resour Conserv Recycl 172:105671

Botondi R, Barone M, Grasso C (2021) A review into the effectiveness of ozone technology for improving the safety and preserving the quality of fresh-cut fruits and vegetables. Food Secur 10(4):748

Bumbudsanpharoke N, Ko S (2022) Packaging technology for home meal replacement: innovations and future prospective. Food Control 132:108470

Cano C, Meneses Y, Chaves BD (2021) Application of peroxyacetic acid for decontamination of raw poultry products and comparison to other commonly used chemical antimicrobial interventions: a review. J Food Prot 84(10):1772–1783

CBI (2023) Which trends offer opportunities or pose threats on the European fresh fruit and vegetables market? https://www.cbi.eu/market-information/fresh-fruit-vegetables/trends. Accessed 25 Jan 2025

Chain et al (2020) Risk assessment of chlorinated paraffins in feed and food. EFSA J 18(3):e05991

Chen YQ, Cheng JH, Sun DW (2020) Chemical, physical and physiological quality attributes of fruit and vegetables induced by cold plasma treatment: mechanisms and application advances. Crit Rev Food Sci Nutr 60(16):2676–2690

Cheng C, Li H, Wang J, Wang H, Yang X (2020) A review of measurement methods for peracetic acid (PAA). Front Environ Sci Eng 14:1–10

Cuggino SG, Bascon-Villegas I, Rincón F, Pérez MA, Posada-Izquierdo G, Marugan J, Carro CP, Pérez-Rodríguez F (2020) Modelling the combined effect of chlorine, benzyl isothiocyanate, exposure time and cut size on the reduction of *Salmonella* in fresh-cut lettuce during washing process. Food Microbiol 86:103346

Cuggino SG, Posada-Izquierdo G, Villegas IB, Theumer MG, Pérez-Rodríguez F (2023) Effects of chlorine and peroxyacetic acid wash treatments on growth kinetics of *Salmonella* in fresh-cut lettuce. Food Res Int 167:112451

De Corato U (2020) Improving the shelf-life and quality of fresh and minimally-processed fruits and vegetables for a modern food industry: a comprehensive critical review from the traditional technologies into the most promising advancements. Crit Rev Food Sci Nutr 60(6):940–975

Donatti Leão Alvarenga P, Mileib Vasconcelos C, de São José JFB (2020) Application of ultrasound combined with acetic acid and peracetic acid: microbiological and physicochemical quality of strawberries. Molecules 26(1):16

Epelle EI, Macfarlane A, Cusack M, Burns A, Okolie JA, Mackay W, Rateb M, Yaseen M (2023) Ozone application in different industries: a review of recent developments. Chem Eng J 454:140188

Fallik E, Ilić Z (2021) The influence of physical treatments on phytochemical changes in fresh produce after storage and marketing. Agronomy 11(4):788

Fam SN, Khosravi-Darani K, Massoud R, Massoud A (2021) High-pressure processing in food. Biointerf Res Appl Chem 11(4):11553–11561

Fan X, Wang W (2021) Quality of fresh and fresh-cut produce impacted by nonthermal physical technologies intended to enhance microbial safety. Crit Rev Food Sci Nutr 62(2):362–382

Giannakourou MC, Tsironi TN (2021) Application of processing and packaging hurdles for fresh-cut fruits and vegetables preservation. Food Secur 10(4):830

Gurtler JB, Dong X, Zhong B, Lee R (2022) Efficacy of a mixed peroxyorganic acid antimicrobial wash solution against *Salmonella*, *Escherichia coli* O157: H7, or *Listeria monocytogenes* on cherry tomatoes. J Food Prot 85(5):773–777

Hassenberg K, Praeger U, Herppich WB (2021) Effect of chlorine dioxide treatment on human pathogens on iceberg lettuce. Food Secur 10(3):574

Jelsma I, Woittiez LS, Ollivier J, Dharmawan AH (2019) Do wealthy farmers implement better agricultural practices? An assessment of implementation of good agricultural practices among different types of independent oil palm smallholders in Riau, Indonesia. Agric Syst 170:63–76

Julien-Javaux F, Gérard C, Campagnoli M, Zuber S (2019) Strategies for the safety management of fresh produce from farm to fork. Curr Opin Food Sci 27:145–152

Kaczmarek M, Avery SV, Singleton I (2019) Microbes associated with fresh produce: sources, types and methods to reduce spoilage and contamination. In: Advances in applied microbiology, vol 107. Elsevier, pp 29–82

Khaire RA, Thorat BN, Gogate PR (2022) Applications of ultrasound for food preservation and disinfection: a critical review. J Food Process Preserv 46(10):e16091

Koutchma T (2019) Ultraviolet light in food technology: principles and applications, vol 2. CRC Press

Lepaus BM, Rocha JS et al (2020) Organic acids and hydrogen peroxide can replace chlorinated compounds as sanitizers on strawberries, cucumbers and rocket leaves. Food Sci Technol 40:242–249

López-Gálvez F, Allende A, Gil MI (2021) Recent progress on the management of the industrial washing of fresh produce with a focus on microbiological risks. Curr Opin Food Sci 38:46–51

Luu P, Chhetri VS, Janes ME, King JM, Adhikari A (2021) Efficacy of gaseous chlorine dioxide in reducing *Salmonella enterica, E. coli* O157: H7, and *Listeria monocytogenes* on strawberries and blueberries. Lwt 141:110906

Lv R, Liu D, Zhou J (2021) Bacterial spore inactivation by non-thermal technologies: resistance and inactivation mechanisms. Curr Opin Food Sci 42:31–36

Mahendran R, Ramanan KR, Barba FJ, Lorenzo JM, López-Fernández O, Munekata PE, Roohinejad S, Sant'Ana AS, Tiwari BK (2019) Recent advances in the application of pulsed light processing for improving food safety and increasing shelf life. Trends Food Sci Technol 88:67–79

Mahnot NK, Siyu L-P, Wan Z, Keener KM, Misra N (2020) In-package cold plasma decontamination of fresh-cut carrots: microbial and quality aspects. J Phys D Appl Phys 53(15):154002

Manzoor MF, Zeng XA, Rahaman A, Siddeeg A, Aadil RM, Ahmed Z, Niu D (2019) Combined impact of pulsed electric field and ultrasound on bioactive compounds and FT-IR analysis of almond extract. J Food Sci Technol 56:2355–2364

Manzoor MF, Ahmed Z, Ahmad N, Karrar E, Rehman A, Aadil RM, Zeng XA (2021) Probing the combined impact of pulsed electric field and ultra-sonication on the quality of spinach juice. J Food Process Preserv 45(5):e15475

Manzoor MF, Ali M, Aadil RM, Ali A, Goksen G, Li J, Proestos C (2023) Sustainable emerging sonication processing: Impact on fungicide reduction and the overall quality characteristics of tomato juice. Ultrason Sonochem 94:106313

Maringgal B, Hashim N, Tawakkal ISMA, Mohamed MTM (2020) Recent advance in edible coating and its effect on fresh/fresh-cut fruits quality. Trends Food Sci Technol 96:253–267

Masotti F, Cattaneo S, Stuknytė M, De Noni I (2019) Airborne contamination in the food industry: an update on monitoring and disinfection techniques of air. Trends Food Sci Technol 90:147–156

Mieszczakowska-Frąc M, Celejewska K, Płocharski W (2021) Impact of innovative technologies on the content of vitamin C and its bioavailability from processed fruit and vegetable products. Antioxidants 10(1):54

Min A, Hossain MI, Jung S, Yeo D, Wang Z, Song M, Zhao Z, Park S, Choi C (2023) Evaluation of the efficacy of ethanol, peracetic acid, and quaternary ammonium compounds against murine norovirus using carrier and suspension tests. Food Control 143:109325

Mostafidi M, Sanjabi MR, Shirkhan F, Zahedi MT (2020) A review of recent trends in the development of the microbial safety of fruits and vegetables. Trends Food Sci Technol 103:321–332

Nicolau-Lapeña I, Lafarga T, Viñas I, Abadias M, Bobo G, Aguiló-Aguayo I (2019) Ultrasound processing alone or in combination with other chemical or physical treatments as a safety and quality preservation strategy of fresh and processed fruits and vegetables: a review. Food Bioprocess Technol 12:1452–1471

Niveditha A, Pandiselvam R, Prasath VA, Singh SK, Gul K, Kothakota A (2021) Application of cold plasma and ozone technology for decontamination of *Escherichia coli* in foods-a review. Food Control 130:108338

Pedreira A, Taşkın Y, García MR (2021) A critical review of disinfection processes to control SARS-CoV-2 transmission in the food industry. Food Secur 10(2):283

Petri E, Virto R, Mottura M, Parra J (2021) Comparison of Peracetic acid and chlorine effectiveness during fresh-cut vegetable processing at industrial scale. J Food Prot 84(9):1592–1602

Qu P, Zhang M, Fan K, Guo Z (2022) Microporous modified atmosphere packaging to extend shelf life of fresh foods: a review. Crit Rev Food Sci Nutr 62(1):51–65

Raffo A, Paoletti F (2022) Fresh-cut vegetables processing: environmental sustainability and food safety issues in a comprehensive perspective. Front Sustain Food Syst 5:681459

Rendueles C, Duarte AC, Escobedo S, Fernández L, Rodríguez A, García P, Martínez B (2022) Combined use of bacteriocins and bacteriophages as food biopreservatives. A review. Int J Food Microbiol 368:109611

Ruan J, Zhou Y, Yan J, Zhou M, Woo S-H, Weng W, Cheng J, Zhang K (2022) Tartary buckwheat: an under-utilized edible and medicinal herb for food and nutritional security. Food Rev Intl 38(4):440–454

Saravanakumar K, Sathiyaseelan A, Mariadoss AVA, Chelliah R, Shin S, Park S, Oh D-H, Wang M-H (2021) Slightly acidic electrolyzed water combination with antioxidants and fumaric acid treatment to maintain the quality of fresh-cut bell peppers. LWT 147:111565

Sharma P, Shehin V, Kaur N, Vyas P (2019) Application of edible coatings on fresh and minimally processed vegetables: a review. Int J Veg Sci 25(3):295–314

Shi M, Gu J, Wu H, Rauf A, Emran TB, Khan Z, Mitra S, Aljohani AS, Alhumaydhi FA, Al-Awthan YS (2022) Phytochemicals, nutrition, metabolism, bioavailability, and health benefits in lettuce – a comprehensive review. Antioxidants 11(6):1158

Simpson AM-A, Mitch WA (2022) Chlorine and ozone disinfection and disinfection byproducts in postharvest food processing facilities: a review. Crit Rev Environ Sci Technol 52(11):1825–1867

Singla G, Chaturvedi K, Sandhu PP (2020) Status and recent trends in fresh-cut fruits and vegetables. In: Fresh-cut fruits and vegetables. Elsevier, pp 17–49

Sumner SS, Peters DL (2023) Microbiology of vegetables. In: Processing vegetables. Routledge, pp 87–114

Sun Y, Wu Z, Zhang Y, Wang J (2022) Use of aqueous ozone rinsing to improve the disinfection efficacy and shorten the processing time of ultrasound-assisted washing of fresh produce. Ultrason Sonochem 83:105931

Tabassum N, Khan MA (2020) Modified atmosphere packaging of fresh-cut papaya using alginate based edible coating: quality evaluation and shelf life study. Sci Hortic 259:108853

Tan JN, Hwang CA, Huang L, Wu VC, Hsiao HI (2023) A pilot-scale evaluation of using gaseous chlorine dioxide for decontamination of foodborne pathogens on produce and low-moisture foods. J Food Saf 43(2):e12937

Tudela JA, Lopez-Galvez F, Allende A, Gil MI (2019a) Chlorination management in commercial fresh produce processing lines. Food Control 106:106760

Tudela JA, López-Gálvez F, Allende A, Hernández N, Andújar S, Marín A, Garrido Y, Gil MI (2019b) Operational limits of sodium hypochlorite for different fresh produce wash water based on microbial inactivation and disinfection by-products (DBPs). Food Control 104:300–307

Turner ER, Luo Y, Buchanan RL (2020) Microgreen nutrition, food safety, and shelf life: a review. J Food Sci 85(4):870–882

USFDA (2018) Guide to minimize food safety hazards of fresh-cut produce: draft guidance for industry. https://www.fda.gov/media/117526/download. Accessed 25 Jan 2025

Wong JX, Ramli S, Desa S, Chen SN (2021) Use of *Centella asiatica* extract in reducing microbial contamination and browning effect in fresh cut fruits and vegetables during storage: a potential alternative of synthetic preservatives. LWT 151:112229

Yang N, Li F, Liu Y, Dai T, Wang Q, Zhang J, Dai Z, Yu B (2022) Environmental and economic life-cycle assessments of household food waste management systems: a comparative review of methodology and research progress. Sustain14(13):7533

Yousuf B, Wu S, Siddiqui MW (2021) Incorporating essential oils or compounds derived thereof into edible coatings: effect on quality and shelf life of fresh/fresh-cut produce. Trends Food Sci Technol 108:245–257

Zhang H, Yamamoto E, Murphy J, Locas A (2020) Microbiological safety of ready-to-eat fresh-cut fruits and vegetables sold on the Canadian retail market. Int J Food Microbiol 335:108855

Zhao L, Li S, Yang H (2021) Recent advances on research of electrolyzed water and its applications. Curr Opin Food Sci 41:180–188

Zhou B, Luo Y, Nou X, Mwangi E, Poverenov E, Rodov V, Demokritou P, Fonseca JM (2023) Effects of a novel combination of gallic acid, hydrogen peroxide and lactic acid on pathogen inactivation and shelf-life of baby spinach. Food Control 143:109284

27 Innovative Use of Nanotechnology in Vegetable Processing

Rimsha Gulzar, Amna Shehbaz, Tuba Mustafa, Muhammad Afzaal, and Hanisah Kamilah

Abstract

Nanotechnology is gaining attention across various food and pharmaceutical industries, and agriculture. In the food industry, it contributes a significant role in food processing, preservation, and packaging. The global fruit and vegetable processing industries extensively use nanotechnology to improve the taste, flavor, and bioavailability of different compounds. High temperatures and various processing conditions in vegetable processing can affect nutrient availability. However, the use of nanotechnology in vegetable processing has helped retain nutrients and extend the quality of vegetables. Nanotechnology is also crucial in intelligent packaging and extending the shelf life of processed vegetables. Therefore, this chapter focuses on various aspects of nanotechnology in food processing, safety, and packaging, including classifications and safety concerns.

Keywords

Nanotechnology in vegetable processing · intelligent packaging · Eustressors · Nano sensors · Nano fertilizers

R. Gulzar · A. Shehbaz · T. Mustafa · M. Afzaal (✉)
Safety & Biotechnology Lab, Department of Food Science, Government College University, Faisalabad, Pakistan
e-mail: muhammadafzaal@gcuf.edu.pk

H. Kamilah
Department of Crop Science, Faculty of Agricultural and Forestry Sciences, Universiti Putra Malaysia, Bintulu Saralu Sarawak Campus, Bintulu, Malaysia

Z. Iqbal et al. (eds.), *Sustainable and Innovative Vegetable Production in times of Climate Change*, https://doi.org/10.1007/978-981-96-6283-8_27

27.1 Introduction

The creation, manufacturing, and utilization of systems, devices, and structures through manipulation of material size and shape at the nanoscale (10^{-9} of a meter) level, where special phenomena allow for new applications, is known as nanotechnology (Ravichandran and Sasikala 2006). Numerous industries, including computer electronics, communications, energy generation, medical, and the food business, have already benefited from the use of nanoscience and nanotechnology. Proteins, DNA, membranes, and other natural biomolecules are examples of nanoscale devices that are produced with the intention of mimicking the nanodevices present in nature (German et al. 2006). Furthermore, biogenic nanomaterials are essential to both the cutting-edge scientific applications and contemporary agriculture (Chellamuthu et al. 2019).

Development in rural areas is primarily driven by nanoparticle (NP) agriculture. A robust agricultural economy boosts income, employment, and productivity, which ultimately strengthens and advances rural community (Pouratashi and Iravani 2012). Nanotechnology is playing an excellent role in increasing food availability. It is also revolutionizing different fields viz. food safety, green processing, clean energy, electronics, diagnostics and water treatment. It is a rapidly expanding and developing region with novel and unique uses in food production and agriculture (Sadeghi et al. 2017). Latest development in science and technology supports the agriculture sector by providing it with new ideas and solutions for resolving difficult issues. Since nanotechnology has advanced, more effective nanoformulations for sustainable agriculture have been created on a regular basis. This is because these formulations include less impurities (Fraceto et al. 2016). The application of nanomaterials in agriculture has yielded excellent outcomes, including more output, better-quality yields with improved nutritional value, and efficient growth. A strategy to improve crops under varying climate conditions is called agri-nanotechnology. Metal oxide nanoparticles (NPs) have excellent benefits in agriculture, such as efficient growth, increased yield, and high-quality nutrition (Kolenčík et al. 2019).

To achieve fantabulous living standards and food security, newer technologies that boost production and decrease food waste are crucial. Foods with exceptional quality can be produced using nanotechnology in a much more practical manner, increasing the bioavailability of nutrients. The increased use of nanotechnology in crop production and food processing is the subject of numerous research studies (Abobatta 2018). In addition to the aforementioned output, nanotechnology also holds the potential to lower pesticide usage. The greatest solution may involve increasing production and lowering postharvest costs through improved results from more recent technical studies using food-related nanotechnology and biotechnology (Yadollahi et al. 2010).

Applications of nanotechnology include the processing and storage of food, the creation of insect-resistant plant varieties through the transfer of genes or DNA via nanoparticles, the addition of nanofood additives, and the extension of products' shelf life. Nanotechnology is the study of materials at the nanoscale (size range of 1–100 nm), which have improved optical, magnetic, physicochemical, and biocompatibility features. When it comes to reducing postharvest losses, nano-based materials including nanoparticles, nanoemulsions, nanocomposites, and nanofilms have proved to be more helpful than traditional methods (Onyszko et al. 2022). The specific properties of nanomaterials, such as their smaller size (nano-scale), increased surface area, stability, and biological activity, can extend the shelf stability of food in two ways: first, by preserving the food's natural nutritional qualities and visual appeal, and second, by providing protection from diverse pathogenic microorganisms.

One of the keys to influencing customer acceptability in the agri-food sector is increasing awareness of nanotechnology such as intelligent packaging, intelligent monitoring, smart feed, and food ingredients. This chapter provides an overview of cutting-edge and unconventional methods that are now being employed to improve the quality of agricultural goods after harvest and reduce production losses in the best interests of all the stakeholders including farmers, distributors, and consumers.

27.2 Improvement of Resistance, Production and Quality of Vegetable Crops

Stresses on plants are divided into two categories: "external," or originating from outside the plant, and "internal," or coming from within. According to Kranner et al. (2010), external and internal strains are typically referred to as "stress factors" and "stressors," respectively. "Any unfavorable condition or substance that affects or blocks a plant's metabolism or development" is the definition of plant stress given by Lichtenthaler (1996). Stress elements in horticultural and agricultural operations across the globe restrict yield, growth, quality, and performance. They can be classified as biotic (such as infections and insect pests) or abiotic (drought, severe temperatures, salt, etc.) depending on where they came from. The duration of the stressors also affects their impact, in addition to their dosage. Accordingly, plant sensitivity to stress and its response have a positive (eustress) or negative (distress) effect on plant metabolism. For instance, low temperatures for a brief period of time in the morning (referred to as a "cool morning pulse" or cold hardiness) cause hardening and tend to improve growth in some unfavorable conditions (Kalberer et al. 2006); however, low temperatures below freezing point cause distress for vascular plants vegetative tissues and are fatal for many plant species (Schwarz et al. 2010). A number of other chemical or physical non-biological stresses can also cause a plant to respond favorably, leading to increased growth or improved quality (Demkura and Ballaré 2012). According to current definitions, hormesis is a biphasic response in which a toxic agent (stressor) at high doses can cause inhibition, or stress, and at low doses, a beneficial effect, or eustress; this mild overcompensation response is referred to as "hormetic low dose stimulation" (Vargas-Hernandez et al. 2017). Nanotechnology has the potential to enhance vegetable tolerance to various biotic and abiotic stresses. Use of nanoparticles (NPs) strengthens cell wall, improves nutrient uptake, enhances photosynthesis, and modulates gene expression ultimately increasing plant resistance to stressors. NPs improve water retention and

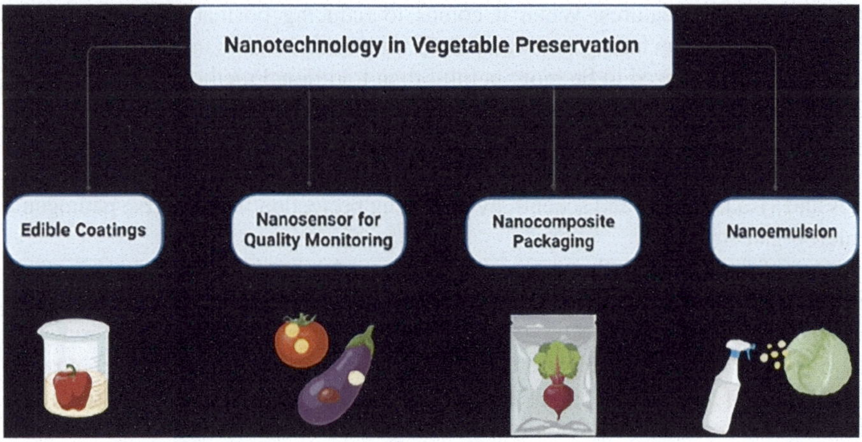

Fig. 27.1 Applications of nanotechnology for vegetable preservation

promote uptake of essential nutrients under severe drought conditions. They reduce oxidative stress, improve soil fertility and promote plant growth. They also reduce salt stress by mitigating the toxic effects of high salt concentrations viz. ion toxicity, decreased water potential and nutritional imbalances which eventually result in low crop production. No doubt, adoption of good pre-harvest practices and wise post-harvest processing and preservation protocols using nanotechnology (Fig. 27.1) may fetch good economic returns for both growers and entrepreneurs.

27.3 Reduction of Postharvest Spoilage of Vegetables by Nanosensor

It has been found that about 40% of various types of food is lost during handling in developed countries (FAO 2019). Pest damages, particularly those induced by fungi during storage, can result in even greater losses in fruits and vegetables' yield and quality than losses in the field. These losses are particularly pronounced when produce is not transported and stored under specific cold conditions (Batta 2007). Nanomaterials are used to preserve the quality and safety of food that is meant for human consumption. Many different methods are used to identify substances that jeopardize quality and safety (Andre et al. 2018). During food manufacturing and processing, the nanosensors can aid in ensuring food security by identifying bacteria that might cause food spoilage. Nanosensors are essential for the detection and reduction of organic molecules, contaminants, other substances, and pathogenic propagules of microscopic sizes. Devices with quick reaction times, high sensitivity, potential, and recovery are known as nanosensors. One possible technology for the current market is the application of nanosensors to food packaging to ensure product quality for the ultimate customer and monitor quality during the logistical process stages (Batista et al. 2018).

By utilizing nanotechnology, food may be quickly and easily analyzed by nanosensors. Since abzymes are employed in nanosensors, specific warning indicators are produced in the event as specific hazardous compounds are discovered. Consequently, the nanosensors have a high degree of accuracy when identifying potentially harmful ingredients in food (Li and Sheng 2014).

In horticulture, nanosensors are used to detect crop pests and determine nutrient requirements, pesticide residues, and soil moisture levels (Rana et al. 2021). The ability to detect gases, particularly ethylene gas, a plant hormone that regulates a variety of fruit-related physiological processes, is one of the nanosensor's features. Climacteric fruits may experience negative storage conditions that result in a reduction in fruit quality (Martínez-Romero et al. 2007). Figure 27.2 presents the applications of nanosensor in crop sustainability.

Fruits and vegetables can be tested for the presence of pesticides on their surface using nanosensors. It has also been stated that some nanosensors can identify food products that contain carcinogens (Meetoo 2011). The created nanosensor may be used to detect chlorpyrifos. Chlorpyrifos in genuine samples, such as fruits and vegetables, may be quantitatively detected using a nanosensor up to 10 nM for apples and cabbage and 50 nM for pomegranates. The current biosensor surpasses the sensitivity and specificity constraints linked to previous biosensor development (Talan et al. 2018). Compounds of interest can be widely detected. Potato storage-related volatile organic compounds (VOCs) have been connected to fungi and bacteria that cause soft rot (*Pythium ultimum* and *Pectobacterium carotovorum* subsp. *carotovorum*, respectively) (Ratiu et al. 2019; Tietel et al. 2021; Sangjan et al. 2022).

Crop productivity can be greatly increased by utilizing biosensors in conjunction with advanced technology in the fields of nanomaterials, molecular biology, and microfluidics. They can also be used to forecast the likelihood of soil infections and track the activity of microorganisms in the soil. The main goal of analyzing soil with a biosensor is to ascertain how differently positive and negative microorganisms in the soil behave, based on the amount of oxygen they take in while breathing. They also present numerous prospects for the use of novel nanomaterial characteristics in the detection of pollutants and their obstructions. Methyl salicylate and azelaic acid biosensors for nitrate concentration detection in plants and marker detection for infected plant identification have been reported (Griesche and Baeumner 2020).

Smart delivery systems and nanosensors are used in precision farming to help with the economical use of natural agricultural resources such as pesticides, fertilizers, and water. Satellite monitoring, geographic information systems, and remote sensing techniques that may detect pests on crops or indicators of stress like drought, are a few examples of these tools. For nanotechnology-assisted instruments, it is thought to be essential to use independent sensors connected to real-time GPS monitoring (Davari et al. 2017). Nanosensors can be installed across the field to monitor crop development and soil properties.

Microorganisms' attack can cause swift degradation of recently harvested high moisture produce. Nanotechnology is one of the most recent and sophisticated technologies that can aid in reducing postharvest losses. By creating functional packaging materials with the fewest amounts of bioactive components, better gas and

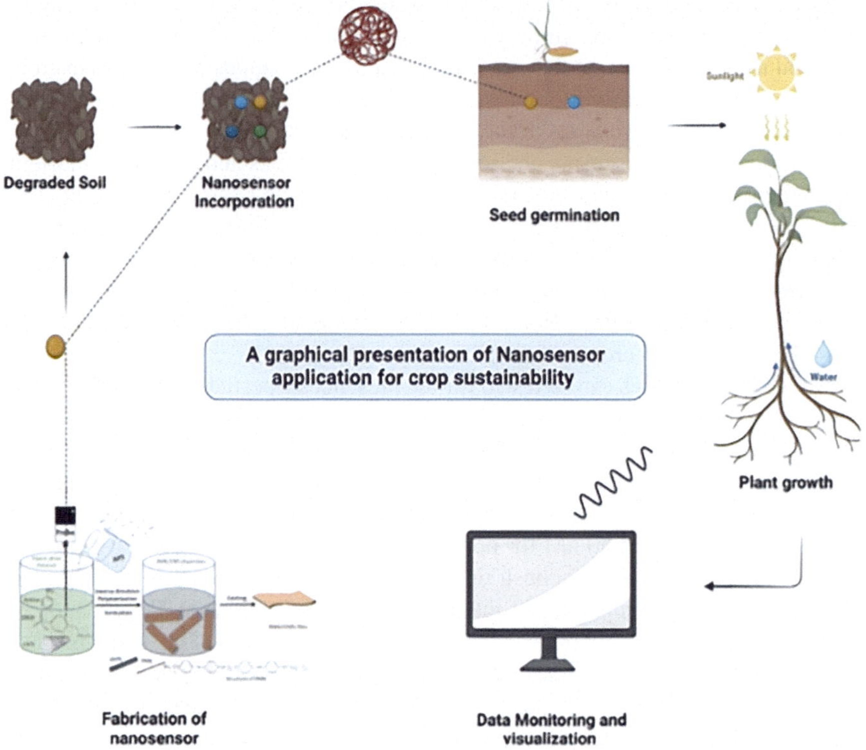

Fig. 27.2 Nanosensor applications for crop sustainability

mechanical qualities, and less of an impact on the sensing qualities of fruits and vegetables, nanotechnology can be used to reduce postharvest losses (Flores-López et al. 2016). The use of nanosensors in food packaging to guarantee product quality for the final consumer and track quality throughout the logistical process stages is one potential technology for the present market (Kuswandi et al. 2012). In nanosensors, for the capture of ethylene, the adsorption of gas (adsorbate) occurs on the surface of the solid material (adsorbent). The amount of material adsorbed will depend on temperature, pressure, and adsorbate concentration.

Certain ethylene capture platforms have an oxidizing agent adsorbed on their surface, which, when in contact with ethylene, produces an oxidation-reduction reaction that changes the nanosensor's color. Depending on the chemicals being employed, multiple capture processes, supports, and even color changes, are the basis of current colorimetric platforms for ethylene capture. Many of the substitutes for the colorimetric systems currently in use rely on metallic compounds to oxidize gases. For example, to capture external ethylene, palladium sulfate catalyst and polymolybdate (IV) are utilized as oxidizing agents. The ethylene oxidation produces gas that is then trapped, changing the coordination of the molybdenum metal center from Mo (VI) to Mo(V). This results in noticeable alterations that turn the light-yellow compound into a dark-blue color. Alteration of pH of the ammonium molybdate solution being used can change the sensor's sensitivity (Lang and Hübert 2012).

27.4 Use of Nanotechnology in Vegetable Crops

Scientists are attempting to create fertilizers that can increase both the effectiveness of plant-nutrient uses and crop yields while lowering contamination from environment because the long history of using conventional fertilizers has led to major environmental problems throughout the world (Salama et al. 2019; Chandre et al. 2019). Currently, one problem is the creation of innovative methods that can lower the amount of nutrients that leak into the surrounding environment. As a result, numerous practical measures should be implemented to reduce this nutrient loss (Chen et al. 2018; Mariano et al. 2019). In order to do this, nutrients in the form of nanoparticles called plant fertilizers release those nutrients gradually over an extended period of time, minimizing soil loss and groundwater and soil contamination (Shebl et al. 2019; Guru et al. 2015).

Nanotechnology has a broad range of applications for smart packaging, medication delivery, electronics, tissue engineering, and food science and safety (Shebl et al. 2020; Youssef et al. 2019a). It has recently been extensively employed in contemporary agriculture to realize the idea of sustainable agriculture and recognized as potential technological innovation to reenergize food and agricultural sectors. When it comes to agriculture, nanotechnology (NT) is essential for food production, crop enhancement, plant defense, food manufacturing, food wrapping, water refinement, food safety, and environmental treatment (El-Mohamedy et al. 2019).

Plants stimulated genetically by nanomaterials, chemicals transferred to cellular molecular grades in plants, and stress-adapted plants supplemented with genetic material based on nanoarrays can all increase agricultural output (Pramanik et al. 2020). Stated differently, due to their small size, unique optical properties, physicochemical qualities, and surface area to volume ratio, nanomaterials have uses in the administration of agricultural activities (Ghormade et al. 2011). Nanomaterials come in a variety of morphologies, including nanopowders, nanotubes, nano-sheets, nanofibers, nanoclusters, and nanocrystals, with one or more dimensions falling between 2 and 100 nm (Abou Hammad et al. 2019).

Moreover, nano-coating slows down the pace at which fertilizers dissolve and release, and prolonged release from coated fertilizers results in greater absorption through plant roots. In order to minimize fertilizer depletion and lessen environmental contamination, slow-release fertilizers have recently gained popularity as a new method (Duhan et al. 2017). Furthermore, to enhance soil health and activity, micronutrients in nanoform can be sprayed on plants or made available to the soil for root uptake (Peteu et al. 2010; Duhan et al. 2017).

Vegetables' output and quality have enhanced with application of nanotechnology in vegetable crop production over time for a variety of reasons, including good seed germination, plant growth and development, and the detailed description of biotic and abiotic stress detection and management.

27.4.1 Seed Germination

According to Almutairi and Alharbi (2015), it has been discovered that the application of nanoparticles has a beneficial impact on the procedure of germination of seeds because they aid in the seeds' absorption of water, help the seed's antioxidant system, lessen the amount of peroxides and superoxide radicles in the environment, boost enzyme and catalase activity, enhance seed qualities and vigor of seedlings, and eventually raise the germination rate.

27.4.2 Plant Growth and Development

Depending on the kind of nanoparticle, dosage, length of treatment, species of plant, and developmental stage, the impacts on plant growth and development might be either positive or negative (Hendel et al. 2016). According to Mikkelsen (2018), there are various kinds of nano fertilizers, including carbon nanotubes, zinc nanofertilizer, boron nanofertilizer, nanoporous zeolite, nanoaptamers, and nanoherbicide. Different techniques, including seed priming, soil, foliar and root dip, aerosol dusting, and fertigation using a system of drip irrigation, can be used to apply these nanofertilizers (Kalia and Kaur 2019). According to Kalikeri et al. (2015), fertilizers based on nanomaterials have a number of benefits over conventional fertilizers, including the ability to improve crops more precisely, a three-fold increase in nutrient use efficiency, a steady release of nutrients, increased yield of crop, and reduced toxicity of Eco. Table 27.1 describes different nanoparticles enhancing the quality of crops.

27.4.3 Disease Detection and Management

According to Alghuthaymi et al. (2015), quantum dots, which are densely packed semiconductor crystals made of millions of atoms, can also be used to detect diseases. Nanobiosensors, on the other hand, can be used to assess crop health and identify microbial or chemical contamination. Lastly, nanofabrication can be used to study the behavior and process of pathogen infection inside host plants. Nanobiosensors have been used to identify and diagnose diseases brought on by a variety of microorganisms (viruses, bacteria, and fungus) contaminated in the soil, by quantifying the amount of differential oxygen used by beneficial and harmful microorganisms of the soil (Rai et al. 2012).

27.4.4 Applications of Nanotechnology in the Preservation of Vegetables

Nanotechnology has vast applications for preserving vegetables including packaging, coatings, and sensors as detailed in Table 27.2.

27.5 Opportunities and Challenges for Food Industry

Nanotechnology is anticipated to be a novel technology that will aid in the advancement of genetically modified crops, fisheries and livestock production input, chemical pesticides, and precision agricultural methods. One of the most crucial methods for increasing crop mass output is precision farming, which monitors environmental factors and takes suitable action (Chen and Yada 2011). The use of nanotechnology in animal husbandry has increased the nutritional value and effectiveness of animal feeds (enriched with antimicrobial additives, detoxifying nanomaterials, and nanosupplements), as well as the diagnostic capabilities of nanobiosensors for animal diseases. Few commercially available products that incorporate nanosized additives that are designed specifically for animal feed exist. Examples of this include mycotoxin binders, like nanoclay, which shield animals against mycotoxicosis (Handford et al. 2014). In their 2011 review, Chaudry and Castle examined the advantages and disadvantages of using nanotechnology in food for developing nations. They discovered that developing nations benefited from the new developments brought about by nanotechnology in the food sector. Resultantly, this improved quality and safety of food and feed, and caused reduction in the likelihood of food-borne illnesses. Nonetheless, food quality can be enhanced by nanotechnologies, leading to a more nutrient-dense and healthful diet. Cushen et al. (2012) examined recent advancements in nanotechnologies in the food industry, covering food-related

Table 27.1 Nanoparticles to improve crop quality

Nanoparticles	Quantity	Crop	Trait improved	References
Applying zinc oxide nanoparticles topically	20 mg/mL	Tomato	Better maturation & production of biomass	Panwar et al. (2012)
Foliar application of nano-chitosan NPK fertilizer	10%	Potato	Modification of chemical composition and improvement of growth and yield	Elshamy et al. (2019)
Foliar application of liquid nano NPK	6 mL and 9 mL/plant	Cucumber	Improved vegetative and qualitative parameters	Merghany et al. (2019)
Applying lithovit nanofertilizer as a foliar spray	0.75 g/liter	Okra	Boosting of total yield and increase in number of pods per plant	El-Hamd and Abd-Elwahed (2017)
Foliar spray of zinc oxide nanoparticle	20 and 30 μg/mL	Onion	Early flowering by 12–14 days; improved growth	Laware and Raskar (2014)

Table 27.2 Nanocomposites in packaging of fresh vegetables

Biopolymer used	Fruit/ vegetable	Nanocomposite type	Outcomes	References
Soy protein isolate + nano-SiOx	Apple	Film coatings	SPI film coatings with good dispersion of nano-SiOx particles helped in fruit preservation	Liu et al. (2017)
PVA + cellulose + apple Peel extract	Fresh cherry tomatoes and potatoes	Films	The PVA/CNC/ TiO2/APE film exhibited excellent antioxidant/ antimicrobial properties	Van Nguyen and Lee (2023)
Apple Peel powder + carboxymethylcellulose	Fruit/ vegetable	Novel nanocomposite films	Can be used as a novel biodegradable packaging material	Shin et al. (2014)
Soy protein + cellulose nanofibril	Fruit/ vegetable	Nanocomposite films	Showed excellent antioxidant, antimicrobial, and light barrier capacities	Yu et al. (2019)
Konjac glucomannan + carboxymethyl chitosan + gliadin + casein	Grapes	Packaging films	Possess great potential of GCNPs as functional nanofillers in enhancing the physicochemical properties of KC film	Bu et al. (2023)
Gelatin + iron oxide	Grapes	Nanocomposite films	Extend the shelf life of grapes and can be used as a replacement to non-biodegradable packaging	Mehmood et al. (2020)
Chitosan + hydroxypropyl methylcellulose	Grapes	Nanocomposite films	Extend up to 10 days with good sensory and textural qualities compared with other films	Shanmuga Priya et al. (2014)
Chitosan + cellulose acetate phthalate	Black grape fruit	Films	Increase the shelf life of black grape fruits	Indumathi et al. (2019)
Chitosan + nettle leaf extract	Guava	Films	Extend postharvest shelf life	Kalia et al. (2021)

(continued)

Table 27.2 (continued)

Biopolymer used	Fruit/ vegetable	Nanocomposite type	Outcomes	References
Soybean protein isolate + PVA + Nanocellulose	Cherry tomatoes	Films	Excellent packaging performance	Zhao et al. (2020)
Zein + konjac glucomannan	Cherry tomatoes	Nanocomposite films	Showed high potential of KNZ film for application in cherry tomatoes preservation.	Xiang et al. (2021)
Chitosan + cellulose + beta-cyclodextrin	Cherry	Edible films	Preserve cherries under cold storage	Bizymis et al. (2023)
Gum Arabic + cellulose nanocrystals	Strawberry	Bio-nanocomposite films	Delay the deterioration of strawberry quality during storage	Kang et al. (2021)
Potato peel starch + ZnO-starch	Strawberry	Coatings	Prolong the shelf life of strawberries for longer at lower temperatures	Chitena, (2022)

industries such as food packaging as well as the food matrix. Food additives derived from nanotechnology offer their better uses. Natural food ingredients that are organic include fat, protein, and carbohydrates. These substances can be produced with important goals in mind, such as encapsulating nutrients to improve the consistency, flavor, and texture of food, or they can be utilized to cover up an off-putting taste or smell. Though the science underlying the creation of food ingredients derived from nanotechnology is still in its infancy, it already seems more promising in terms of the potential to improve product functionality without sacrificing the product's quality and safety.

Some important issues need to be resolved before employing nanotechnology in the food sector. The primary challenge is to create effective and safe delivery methods for better use of people. These systems require processing that is both efficient and smart. The leaching and migration of nanoparticles from packaging into food is an interest for food safety. When users begin to leave a system, some of the NSMs that were either directly or indirectly introduced to it may isolate themselves. Materials act considerably differently at the nanoscale, and our understanding of how to investigate this phenomenon is still quite limited. A solid understanding of nanomaterial toxicity and nanoscale functionalities will be critical to nanotechnology's practical applications and safety needs. It is critical to recognize the possible hazards, toxicological difficulties, and environmental concerns related to nanoparticles. It is well known that nanoparticles can enter various tissues and organs and cross the biological barrier. Numerous chemical techniques can be used to create nanoparticles, and each one carries the risk of yielding unfavorable byproducts that seriously contaminate the environment. It is critical to conduct a complete threat assessment program, review regulatory frameworks and biosecurity, and address public concerns before manufacturing, packaging, and consuming nano-based food products for human consumption. Furthermore, in vitro and in vivo investigations on nanoparticle interfaces with live beings are required before commercial application and production of antibacterial nanoparticles with environmentally friendly properties.

27.6 Conclusion

The food business will undoubtedly benefit more from nanotechnology as new advancements in this field continue to emerge. The majority of incremental nanotechnology's benefits are anticipated to improve product choice and quality, and they will be viewed as gradual modifications to established practices in technology. Crops treated with nanoparticles exhibit increased resilience to weeds, insects, pests, and fungi, in addition to greater growth and production. Nanotechnology is also necessary for safe food manufacturing, food processing, food modification, stability, sensing, longer shelf life, and less food loss. In conclusion, nanotechnology can be used to make products with higher yields and quality. It can be used to protect the crop as a nanofungicide, nanoherbicide, or nanopesticide.

References

Abobatta WF (2018) Nanotechnology application in agriculture. Acta Sci Agric 2(6):99–102

Abou Hammad AB et al (2019) A novel electromagnetic biodegradable nanocomposite based on cellulose, polyaniline, and cobalt ferrite nanoparticles. Carbohydr Polym 216:54–62

Alghuthaymi MA et al (2015) Myconanoparticles synthesis and their role in phytopathogens management. Biotechnol Biotechnol Equip 29:221–236

Almutairi ZM, Alharbi A (2015) Effect of silver nanoparticles on seed germination of the crop plants. J Adv Agric 4:280–285

Andre RS, Sanfelice RC et al (2018) Hybrid nanomaterials designed for volatile organic compounds sensors: a review. Mater Des 156:154–166

Batista PF et al (2018) Bioactive compounds and antioxidant activity in tropical fruits grown in the lower-middle São Francisco Valley. Rev Ciênc Agron 49:616–623

Batta YA (2007) Control of postharvest diseases of fruit with an invert emulsion formulation of *Trichoderma harzianum* Rifai. Postharvest Biol Technol 43(1):143–150

Bizymis AP, Kalantzi S, Mamma D, Tzia C (2023) Addition of silver nanoparticles to composite edible films and coatings to enhance their antimicrobial activity and application to cherry preservation. Food Secur 12(23):4295

Bu N, Zhou N et al (2023) Konjac glucomannan/carboxymethyl chitosan film embedding gliadin/casein nanoparticles for grape preservation. Int J Biol Macromol 249:126131

Chandre MS, Lal M, Naresh R et al (2019) Role of polymer coated fertilizers (PCFs) as an advanced technology for improving nutrient use efficiency and crop productivity: a review. Int J Chem Stud 7:2667–2679

Chellamuthu P, Tran F et al (2019) Engineering bacteria for biogenic synthesis of chalcogenide nanomaterials. Microb Biotechnol 12(1):161–172

Chen H, Yada RY (2011) Nanotechnologies in agriculture: new tools for sustainable development. Trends Food Sci Technol 22(11):585–594

Chen J, Lü S et al (2018) Environmentally friendly fertilizers: a review of materials used and their effects on the environment. Sci Total Environ 613-614:829–839

Chitena L (2022) Investigation of the efficacy of a ZnO-starch nanocomposite packaging to prolong the shelf life of *Fragaria ananassa* (strawberries). Thesis (MSc of science in biology and biotechnology), Botswana International University of Science and Technology. https://repository.biust.ac.bw/handle/123456789/554

Cushen M, Kerry JM, Morris M, Cruz-Romero M, Cummins E (2012) Nanotechnologies in the food industry—recent developments, risks and regulation. Trends Food Sci Technol 24(1):30–46. https://doi.org/10.1016/j.tifs.2011.10.006

Davari MR, Bayat Kazazi S et al (2017) Nanomaterials: implications on agroecosystem. In: Nanotechnology: an agricultural paradigm, pp 59–71

Demkura PV, Ballare CL (2012) UVR8 mediates UV-B-induced Arabidopsis defense responses against *Botrytis cinerea* by controlling sinapate accumulation. Mol Plant 5(3):642–652

Duhan JS, Kumar R et al (2017) Nanotechnology: the new perspective in precision agriculture. Biotechnol Rep 15:11–23

El-Hamd ASA, Abd-Elwahed AHM (2017) Improving the growth and yield of okra plants (*Abelmoschus esculentus* L.) using Lithovit fertilizer. Acad J Sci Res 6:65–71

El-Mohamedy R et al (2019) Antifungal activity of chitosan nanoparticles against some plant pathogenic fungi in vitro. Agric Eng Int CIGR J 21:201–209

Elshamy MT, El-Khallal SM et al (2019) Application of nano-chitosan NPK fertilizer on growth and productivity of potato plant. J Sci Res 36:424–441

FAO (2019) The state of food and agriculture 2019. Moving forward on food loss and waste reduction. FAO, Rome, pp 2–13

Flores-López ML et al (2016) Perspectives on utilization of edible coatings and nano-laminate coatings for extension of postharvest storage of fruits and vegetables. Food Eng Rev 8:292–305

Fraceto LF, Grillo R et al (2016) Nanotechnology in agriculture: which innovation potential does it have? Front Environ Sci 4:20

German JB, Smilowitz J et al (2006) Lipoproteins: when size really matters. Curr Opin Colloid Interface Sci 11(2–3):171–183

Ghormade V, Deshpande MV et al (2011) Perspectives for nano-biotechnology enabled protection and nutrition of plants. Biotechnol Adv 29:792–803

Griesche C, Baeumner AJ (2020) Biosensors to support sustainable agriculture and food safety. TrAC Trends Anal Chem 128:115906

Guru T, Veronica N et al (2015) Crop nutrition management with nano fertilizers. Int J Environ Sci Technol 1:4–6

Handford C, Dean M et al (2014) Implications of nanotechnology for the agri-food industry: opportunities, benefits and risks. Trends Food Sci Technol 40(2):226

Hendel AM, Gawecki R et al (2016) Diverse influence of nanoparticles on plant growth with a particular emphasis on crop plants. Acta Agrobot 69:1694

Indumathi MP, Sarojini KS, Rajarajeswari GR (2019) Antimicrobial and biodegradable chitosan/cellulose acetate phthalate/ZnO nano composite films with optimal oxygen permeability and hydrophobicity for extending the shelf life of black grape fruits. Int J Biol Macromol 132:1112–1120

Kalberer SR, Wisniewski M, Arora R (2006) Deacclimation and reacclimation of cold-hardy plants: current understanding and emerging concepts. Plant Sci 171(1):3–16

Kalia A, Kaur H (2019) Nanofertilizers: an innovation towards new generation fertilizers for improved nutrient use efficacy (NUE) and environmental sustainability. In: Singh B (ed) Emerging trends in nanobiomedicine. Nanoagroceuticals & Nanophytochemicals, pp 1–17

Kalia A, Kaur M et al (2021) Nettle-leaf extract derived ZnO/CuO nanoparticle-biopolymer-based antioxidant and antimicrobial nanocomposite packaging films and their impact on extending the postharvest shelf life of guava fruit. Biomol Ther 11(2):224

Kalikeri S, Shiralgi Y et al (2015) Fe2O3 magnetic nanoparticles to enhance *S. lycopersicum* (tomato) plant growth and their biomineralization. Appl Nanosci 6:1–9

Kang S, Xiao Y, Guo X, Huang A, Xu H (2021) Development of gum Arabic-based nanocomposite films reinforced with cellulose nanocrystals for strawberry preservation. Food Chem 350:129199

Kolenčík M, Ernst D et al (2019) Effect of foliar spray application of zinc oxide nanoparticles on quantitative, nutritional, and physiological parameters of foxtail millet (*Setaria italica* L.) under field conditions. Nano 9(11):1559

Kranner I, Minibayeva FV et al (2010) What is stress? Concepts, definitions and applications in seed science. New Phytol 188(3):655–673

Kuswandi B, Restyana A et al (2012) A novel colorimetric food package label for fish spoilage based on polyaniline film. Food Control 25(1):184–189

Lang C, Hübert T (2012) A colour ripeness indicator for apples. Food Bioprocess Technol 5:3244–3249

Laware SL, Raskar S (2014) Influence of zinc oxide nanoparticles on growth, flowering and seed productivity in onion. Int J Curr Microbiol Appl Sci 3:874–881

Li Z, Sheng C (2014) Nanosensors for food safety. J Nanosci Nanotechnol 14(1):905–912

Lichtenthaler HK (1996) Vegetation stress: an introduction to the stress concept in plants. J Plant Physiol 148(1–2):4–14

Liu R, Liu D, Liu Y et al (2017) Using soy protein SiOx nanocomposite film coating to extend the shelf life of apple fruit. Int J Food Sci Technol 52(9):2018–2030

Mariano E, de Sant Ana Filho CR et al (2019) Ammonia losses following surface application of enhanced-efficiency nitrogen fertilizers and urea. Atmos Environ 203:242–251

Martínez-Romero D, Bailén G et al (2007) Tools to maintain postharvest fruit and vegetable quality through the inhibition of ethylene action: a review. Crit Rev Food Sci Nutr 47(6):543–560

Meetoo DD (2011) Nanotechnology and the food sector: from the farm to the table. Emirates J Food Agric 23:387–403

Mehmood Z, Sadiq MB, Khan MR (2020) Gelatin nanocomposite films incorporated with magnetic iron oxide nanoparticles for shelf-life extension of grapes. J Food Saf 40(4):Article e12814

Merghany MM, Shahein MM et al (2019) Effect of nano-fertilizers on cucumber plant growth, fruit yield and its quality. Plant Arch 19:165–172

Mikkelsen R (2018) Nanofertilizer and nanotechnology: a quick look. Better Crops Plant Food 102:18–19

Onyszko M, Markowska-Szczupak A et al (2022) The cellulose fibers functionalized with star-like zinc oxide nanoparticles with boosted antibacterial performance for hygienic products. Sci Rep 12(1):1321

Panwar J, Jain N, Bhargava A et al (2012) Positive effect of zinc oxide nanoparticles on tomato plants: a step towards developing "Nano-fertilizers". In: 3rd international conference on environmental research and technology (ICERT). University of Sains, Pinang. https://doi.org/10.13140/2.1.2697.8889

Peteu SF, Oancea F et al (2010) Responsive polymers for crop protection. Polymers 2:229–251

Pouratashi M, Iravani H (2012) Farmers' knowledge of integrated pest management and learning style preferences: implications for information delivery. Int J Pest Manag 58(4):347–353

Pramanik P, Krishnan P et al (2020) Application of nanotechnology in agriculture. In: Dasgupta N, Ranjan S, Lichtfouse E (eds) Environmental Nanotechnology. Springer, Cham, pp 317–348

Rai V, Acharya S, Dey N (2012) Implications of nanobiosensors in agriculture. J Biomater Nanobiotechnol 3:315–324

Rana RA, Siddiqui MN et al (2021) Prospects of nanotechnology in improving the productivity and quality of horticultural crops. Horticulturae 7(10):332

Ratiu IA, Ligor T et al (2019) GC-MS application in determination of volatile profiles emitted by infected and uninfected human tissue. J Breath Res 13(2):026003

Ravichandran R, Sasikala P (2006) Nanoscience and nanotechnology: perspectives and overview. J Sci Educ 44(2):43–49. http://61.2.46.60:8088/jspui/bitstream/123456789/1852/1/science_science_june06.pdf#page=42

Sadeghi R, Rodriguez RJ et al (2017) Advances in nanotechnology as they pertain to food and agriculture: benefits and risks. Annu Rev Food Sci Technol 8:467–492

Salama DM, Osman SA et al (2019) Effect of zinc oxide nanoparticles on the growth, genomic DNA, production and the quality of common dry bean (*Phaseolus vulgaris*). Biocatal Agric Biotechnol 18:101083

Sangjan W, Marzougui A et al (2022) Identification of volatile biomarkers for high-throughput sensing of soft rot and Pythium leak diseases in stored potatoes. Food Chem 370:130910
Schwarz D, Rouphael Y et al (2010) Grafting as a tool to improve tolerance of vegetables to abiotic stresses: thermal stress, water stress and organic pollutants. Sci Hortic 127(2):162–171
Shanmuga Priya D et al (2014) Chitosan-incorporated different nanocomposite HPMC films for food preservation. J Nanopart Res 16:1–16
Shebl A, Hassan AA et al (2019) Green synthesis of nanofertilizers and their application as a foliar for *Cucurbita pepo* L. J Nanomater 2019:3476347
Shebl A, Hassan AA et al (2020) Template-free microwave-assisted hydrothermal synthesis of manganese zinc ferrite as a nanofertilizer for squash plant (*Cucurbita pepo* L). Heliyon 6:e03596
Shin SH, Kim SJ et al (2014) Apple peel and carboxymethylcellulose-based nanocomposite films containing different nanoclays. J Food Sci 79(3):E342–E353
Talan A, Mishra A et al (2018) Ultrasensitive electrochemical immuno-sensing platform based on gold nanoparticles triggering chlorpyrifos detection in fruits and vegetables. Biosens Bioelectron 105:14–21
Tietel Z, Melamed S et al (2021) Characterization of volatile signatures of *Pectobacterium* and *Dickeya* spp. as biomarkers for early detection and identification: a major tool in potato blackleg and tuber soft rot management. LWT 144:111236
Van Nguyen S, Lee BK (2023) Multifunctional nanocomposite based on polyvinyl alcohol, cellulose nanocrystals, titanium dioxide, and apple peel extract for food packaging. Int J Biol Macromol 227:551–563
Vargas-Hernandez M et al (2017) Plant hormesis management with biostimulants of biotic origin in agriculture. Front Plant Sci 8:1762
Xiang F, Xia Y et al (2021) Preparation of konjac glucomannan based films reinforced with nanoparticles and its effect on cherry tomatoes preservation. Food Packag Shelf Life 29:100701
Yadollahi A, Arzani K, Khoshghalb H (2010) The role of nanotechnology in horticultural crops postharvest management. Southeast Asia Symposium on Quality and Safety of Fresh and Fresh-Cut Produce. ISHS ActaHorticulturae 875:49–56
Youssef AM, Abd El-Aziz ME et al (2019a) Rational design and electrical study of conducting bionanocomposites hydrogel based on chitosan and silver nanoparticles. Int J Biol Macromol 140:886–894
Youssef AM, Assem FM et al (2019b) Development of bionanocomposite materials and its use in coating of Ras cheese. Food Chem 270:467–475
Yu Z, Dhital R et al (2019) Development of multifunctional nanocomposites containing cellulose nanofibrils and soy proteins as food packaging materials. Food Packag Shelf Life 21:100366
Zhao X, Xu L, Zhang H et al (2020) Ultrasonic treatment for improved properties of soybean protein nanocomposite packaging film and its application in preservation of cherry tomatoes. Shipin Kexue/Food Sci 41(19):230–237

Vegetables' Juices, Blends, and Smoothies: Preparation and Processing

28

Hafsa Ashraf, Shakeela Khalid, Waseem Khalid, Muhammad Nadeem, and Muhammad Usman Tahir

Abstract

Vegetable-rich diets offer numerous health advantages. Adding blended vegetables and fruit juices in diet can retain fiber which is a useful nutrient for heart and digestive health. Smoothies are basically a blended mixture of vegetables and fruits often mixed with nuts, yogurt, and crushed ice. This type of mixture is quite palatable and nutritious as it incorporates healthy types of vegetables and fruits into the diet. These blended mixtures are also a good option for kids of all ages. Moreover, these liquids contain rich amount of minerals, trace elements, vitamins, carbohydrates, flavonoids, and phenols. Cold plasma, inert gas treatments, ultrasonic, radiation, high-pressure processing, non-thermal electrical therapies, and membrane processing are some of the useful methods to preserve food value. This chapter details comparison of the dietetic superiority of juices

H. Ashraf · M. U. Tahir
Department of Food Science, Government College University, Faisalabad, Pakistan

S. Khalid
Department of Zoology, University of Sargodha, Sargodha, Pakistan

W. Khalid (✉)
Faculty of Chemical Sciences and Technologies, Department of Organic Chemistry, University of Castilla La Mancha, Ciudad Real, Spain

University Institute of Food Science and Technology, The University of Lahore, Lahore, Pakistan
e-mail: Waseem.khalid@uifst.uol.edu.pk

M. Nadeem
Institute of Food Science and Nutrition, University of Sargodha, Sargodha, Pakistan

Horticultural Sciences Department, North Florida Research and Education Center, Institute of Food and Agricultural Sciences, University of Florida, Quincy, FL, USA

Z. Iqbal et al. (eds.), *Sustainable and Innovative Vegetable Production in times of Climate Change*, https://doi.org/10.1007/978-981-96-6283-8_28

of various vegetables and fruits including carrot, orange, spinach, and sweet lime, processed through novel processing approaches.

Keywords

Smoothies · Non-thermal technologies · Thermal pasteurization · Antioxidant activity · Total phenolic content

28.1 Introduction

Most of the fresh and eatable lots of herbaceous plants are known as vegetables, containing significant nutritional contents that are effective in repairing and strengthening the body's mechanisms. Vegetables are available in several forms and varieties. They could be edible fruits, seeds, stalks, roots, or leaves. Each form has a unique contribution toward diet. Vegetables are invaluable due to having high contents of nutrients, vitamins, minerals, and antioxidants. The body's alkaline reserve can be maintained by consuming vegetables (Hanif et al. 2006).

Fresh juice mixes are a prevalent diurnal meal enjoyed by individuals of all ages and socioeconomic backgrounds due to their affordable cost and delicious taste. Despite their reasonable cost, these juices are highly nutritious due to their high beta-carotene content and the presence of minerals like calcium, potassium, magnesium, and iron (Tabikha et al. 2010). Liquids made solely of fruits and vegetables are a great way to receive essential nutrients, trace elements, vitamins, and soluble fiber (only present in liquids that have "bits"). Juices can also contain a variety of phytochemicals, including flavonoids and polyphenols. These have been allied to the protection of cardiovascular diseases and cancer. Fruit and vegetable juices lower low-density lipoprotein (LDL) levels, according to a health claim documented by the US Food and Drug Administration, but only if the liquids "contain fiber" (Ruxton et al. 2006). Vegetable juice consumption has been reported to reduce the incidence of Alzheimer's disease. In a population of more than 3000 elderly people, high vegetable consumption was linked to a reduction in overall cognitive deterioration during a period of up to 6 years of observation. Fresh vegetable juices have, therefore, emerged as a novel functional food for a healthy diet (Amal et al. 2008).

Smoothies are ready-to-drink beverage that offers a practical means of cohering different foods, such as fruits, vegetables, nuts, and milk. Smoothies consumption might be a better way to attain the synergistic effects of different phytochemicals that have been linked to the reported health benefits. Smoothies are a high-nutrition beverage that keeps the body hydrated and gives it a boost of readily available antioxidants (Aderinola 2018).

28.2 Types of Vegetables Juices

One of the main categories of processed vegetable products is vegetable juices and blends. Juicing vegetables is a nutritious and delicious method to incorporate concentrated collection of nutrients in a daily diet. These are liquid meals made mostly using veggies as the primary ingredient. Juices can be divided into three categories, which are based upon the amount and size of unsolvable materials present in them: pure juice, cloudy juice, and pulpy juice. There are no insoluble particles in clear juice. Translucent, cloudy juice contains uniformly suspended, insoluble particles. Coarse particles can precipitate to the bottom, float on the surface, or be suspended in the pulpy juice.

Some important vegetable juices which are quite beneficial, refreshing, and flavorful and possess health-promoting properties are detailed below.

28.2.1 Cucumber Juice

Cucumbers belong to the family Cucurbitaceae, which comprises 825 species and 118 genera. Members of this family are mostly found in tropical and subtropical regions of the world, despite their Asian ancestry. The benefits of cucumber juice for skin, hair, and nails are immense. Frequent sun exposure causes the skin to produce free radicals, which cause oxidative stress and inflammatory reactions in the dermal layer of the connective tissues, resulting in aging and damaging of biological molecules and cell membranes (Uthpala et al. 2020). As a vegetable crop, cucumber is high in polyphenols and other phytochemicals that contain a diversity of biological bioactive compounds, together with anti-inflammatory, anti-hyperglycemic, anti-inflammatory, anti-carcinogenic, antioxidant, and analgesic properties. Cucumber is the best ingredient for skin care products as it contains Vitamin C and folic acid. Its juice, containing 95% water, has the property to hydrate and moisturize the skin. Its cooling and soothing effect can calm discomfort and irritation on the skin. Its antioxidant content helps to fight free radicals.

28.2.2 Beetroot Juice

The consumption of beetroot juice (BRJ) by athletes, seeking to enhance their athletic performance, has attained much popularity. BRJ contains high levels of nitrate that can be changed into nitric oxide (NO) after ingestion. The human body uses NO for a number of purposes, one of which is the vasodilator action, which reduces hypertension and increases the nutritive supply and oxygen to different body parts. The results suggest that BRJ might be suitable in the management and prevention of heart disease. Additionally, BRJ's consumption affects muscle endurance, tolerance, and efficiency as well as the transport of oxygen to the skeletal muscles. These effects may improve sports performances (Zamani et al. 2021).

28.2.3 Carrot Juice

The root vegetable known as carrot (*Daucus carota* subsp. *sativus*) comes in a variety of hues, including purple, black, red, white, and yellow. Polyacetylene, flavonoids, carotenoids, minerals, and vitamins are all present in carrots and offer a variety of dietetic and fitness benefits. The vitamins, polyphenols, and carotenoids present in carrots serve as immune system boosters, anti-carcinogens, and antioxidants, proving the ancient saying that carrots are beneficial for the eyesight. Carrots have an abundance of potassium. Therefore, while maintaining the moisture content of human skin, drinking carrot juice may help to keep skin healthy and fresh. Carotenoids and antioxidants found in carrots protect and condition skin, boosting its solar resistance and promoting sunburn repair. Carrot juice actually acts as a natural sunscreen in the summer (Varshney and Mishra 2022).

28.2.4 Tomato Juice

One of the supreme popular fruits of the world is tomato. Though it is classified as a vegetable, on a scientific basis, it is the fruit grown on a vine. Tomato is enriched with iron that provides suitable nourishment. In tomato, more iron is balanced by ascorbic acid. Ten to twenty percent iron needs of a person can be fulfilled by consuming one glass of tomato juice daily (Nasir et al. 2015).

28.2.5 Green Leafy Vegetables

28.2.5.1 Kale Juice

Juice from green leafy vegetables is a practical way to receive the recommended daily allowance of vegetables. In Korea, the most favorite green leafy juice is of kale (*Brassica oleracea acephala*), which is enriched with flavonoids, minerals, and vitamins. An in vitro study proves that kale extract can control cholesterol metabolism through bile acid production and HMG-CoA inhibition (Kim et al. 2008).

28.2.5.2 Spinach Juice

Spinach juice contains vitamins A, C, D, and K and nutrients like calcium, potassium, and magnesium in addition to antioxidants viz. quercetin, kaempferol, and lutein. It strengthens bones, helps in digestion, and reduces the risk of chronic diseases.

28.2.5.3 Swiss Chard Juice

Swiss chard contains Vitamin A, B_6, C, and D and nutrients like calcium, iron, magnesium, potassium, and cobalamin. Alpha-lipoic acid in Swiss chard helps to manage blood sugar levels. Its antioxidants viz. beta-carotene, lutein, and zeaxanthin protect cells from damage. Swiss chard also improves brain function and supports cardiovascular system. It can be mixed to some other juices like lemon or ginger can be added to it.

28.2.5.4 Parsley Juice

In mild climates, parsley is a biennial herbaceous plant and in subtropical and tropical areas, it is an annual herb. Iron and vitamins A, B, and C, are abundantly present in parsley. An oil called apiol is extracted from the seeds of parsley, which helps to treat kidney stones, urinary tract infections, and some related health problems. Juice made from parsley and cilantro is commonly used as a detox beverage. Consuming parsley juice or tea on a regular basis is also acknowledged to expedite the process of uric acid clearance, which otherwise affects human joints. Parsley juice can also be used to treat ear infections and toothaches. Parsley juice or tea is recommended for people who feel trouble in absorbing iron supplements (Ajmera et al. 2019).

Celery contains Vitamins A, B12, C, E, K and antioxidants like caffeic acid, kaempferol, and ferulic acid. It also contains calcium, iron, magnesium, phosphorus, potassium, zinc, copper, manganese, selenium, and folate (USDA 2019). Juice of celery possesses health-promoting properties. It reduces inflammation, improves digestion, and reduces blood pressure. Its antioxidants protect from cell damage.

28.2.5.5 Collard Greens Juice

Collard greens are a good source of nutrients like calcium, potassium, magnesium, cobalamin, iron, and vitamins A, B_6, C, D, and K. Collard greens' juice promotes strong bones and lowers risk of different types of cancer. Its fiber content aids in digestion and reduces the risk of cardiovascular disease.

28.2.6 Broccoli Juice

Broccoli is rich in potassium, calcium, and vitamins A, B_6, C, and K. Broccoli juice provides numerous health benefits. It controls inflammation, supports immune system, lowers cholesterol and blood pressure, and reduces risk of heart disease. Vitamin A and antioxidants in broccoli juice help to maintain eye health. Sulforaphane in broccoli juice restricts the growth of cancer cells. Vitamin B12 and riboflavin minimize the chances of cancer. Iron in broccoli helps produce red blood cells.

28.2.7 Turnips Juice

Turnip juice has many health benefits. It is high in fiber and so it helps in digestion. It also supports to strengthen bones. It helps maintain normal blood pressure, reduces risk of heart disease, and improves brain function. Glucosinolates in turnip help in cancer prevention.

28.2.8 Ginger Juice

Ginger contains nutrients like potassium, magnesium, calcium, and cobalamin, and vitamins viz. B_6, C, and D. Ginger juice helps to reduce inflammation, relieve pain, and improve digestion. It lowers blood pressure and cholesterol level. It soothes sore muscles and prevents the formation of blood clots.

28.3 Blends of Various Vegetables

Vegetable-rich diets offer numerous health advantages. A latest study explains that almost 30% of infants and toddlers don't consume any vegetable part on a daily basis. Previous studies reveal that infant foods in the United States contain no single vegetable options made up of dark green vegetables, as babies tend to reject bitter veggies more frequently. Dark green vegetables typically appear less in the dietary list than other fruits and vegetables. Infant food is typically combined with orange and red fruit and sugar-coated vegetables. Given that sweetness can effectively disguise bitterness so blending darker green veggies with sweeter fruits or vegetables may yield a more palatable product (Bakke et al. 2020).

Reducing the risk of osteoporotic fractures requires prevention of low bone mass. Consumption of vegetables, fruits, and their juices and blends is linked to higher bone mineral density (BMD) in humans as these foods buffer the metabolic acid that is considered to disintegrate bone.

Researchers found that patients with hypercholesterolemia (lipid disorder) using a canned mixed fruit beverage and green vegetables including broccoli and cabbage as the principal ingredients, showed a decrease in lipid level of blood (Suido et al. 2002).

Flexitarian diet is considered as one of the best nutritious diets in the world. Fruits, nuts, vegetables, legumes, grains, and olive oil are all part of the Flexitarian diet. But the quality of this diet may also be significantly influenced by the herbs and spices that are a part of this diet (Bower et al. 2016).

El-Malky et al. (2010) describe a dehydrated vegetable mix that includes squash, peas, green beans, potatoes, carrots, tomatoes, onions, green, red, and yellow peppers, cauliflower, and eggplant. The mix is free of herbs and is thought to be a good source of antioxidants and minerals. It is nutrient dense, well-palatable, and safe to store for 3 months at room temperature.

Every vegetable has a different phenolic makeup with different antioxidant power. Consumers and food industries may find it useful to know each vegetable's bioactive content and antioxidant capacity profile when choosing which leaves or parts to use to create salad combinations or blends that are high in nutritional and functional value. Incorporating these substances into the diet on a regular basis is significant since they can prevent certain oxidative stress-related chronic degenerative illnesses.

28.4 Vegetable-Based Smoothies

Low fruit and vegetable consumption leads to deficiency of micronutrients, which are mostly obtained from the intake of vegetables and fruits. The amount of fruits and vegetables consumed by most communities does not align with the WHO's recommended daily intake of 400 g. Latest research indicates that even while fruits and vegetables possess healing, curative, and preventive properties, people of many countries especially developing ones still do not consume enough of them (Dhandevi and Rajesh 2015).

There are numerous factors that can cause or contribute to low daily consumption. Though rates of consumption range among various groups, differences in tastes and looks have been identified as contributing factors to youth's low intake of fruits and vegetables. Keeping in view the taste, flavor, and aesthetic preferences, smoothies are the best choice for people of almost all age groups.

They are suggested as affordable and practical meals that provide comparable nutrients (Rehm and Drewnowski 2016). This is a popular, convenient, and economical way to consume beneficial nutrients of fruits and vegetables even in the wake of the worsening global economic situation.

Smoothies are made from frozen or fresh vegetables and fruits, mixed, and commonly served with crushed ice (Rehm and Drewnowski 2016). They are non-alcoholic drinks that can be drunk right away. Certain smoothies frequently contain additional ingredients like milk, yogurt, ice cream, tea, or lemonade. They are thicker than slush drinks and have a consistency similar to a milkshake (Rodríguez-Veráestegui et al. 2016). According to current studies, people's health may significantly improve if they regularly consume green smoothies (Maeda 2013).

The primary problem with smoothies processing is that these products have a short shelf life and their quality may deteriorate (Buzrul et al. 2008). To the detriment of associated health-promoting qualities, heat treatments may lower the phytochemical nutrients of smoothies, including antioxidants. So, processors must employ minimal heat treatments to extend the shelf life of smoothies without compromising value (Di Cagno et al. 2011).

Research on how heat treatment and subsequent storage affect bioactive components and quality variations in fresh vegetable smoothie is extremely rare as there are currently no published studies on green smoothies.

28.5 Processing of Vegetable Juices, Blends, and Smoothies

As of right now, heat processing is the most economical way to check the enzyme deactivation and microbiological safety in fruit and vegetable juices and drinks. Thermal treatments, however, may result in a number of physical and chemical alterations that compromise the taste, color, and texture of food and lower the nutritional content or bioavailability of specific nutritive components; these effects are typically highly reliant on the dietary medium (Tiwari and Cummins 2013; Petruzzi et al. 2017).

Cold plasma, microwave processing, high pressure processing, and ultrasonication are some of the useful methods to preserve food value (Bevilacqua et al. 2018).

28.5.1 Cold Plasma Processing

Cold plasma (CP) processing has attracted considerable interest as a revolutionary decontamination tool in food and biological sciences research in the recent years. Cold plasma processing is particularly useful in the food industry for in-package decontamination of juices since it eliminates post-processing contamination. In cold plasma processing noble gas or air is used to generate plasma (Misra et al. 2014). The mechanism of cold plasma involves ionizing gas (carbon dioxide, argon, nitrogen, helium, oxygen, or air) with active particles such as electrons, ions, free radicals, and atoms in both ground and excited states at room temperature (Niemira 2012). The process variables include voltage, frequency, flow rate, and plasma generating method, and are mostly determined by the kind of gas employed (Pankaj et al. 2014).

28.5.2 Microwave Processing

Microwaves, employed in ovens for food processing, generate wavelength of <1 cm with a frequency range of 2.5 GHz and are basically the non-ionizing radiations that do not possess the ability to break covalent bonds (Anantheswaran and Ramaswamy 2001). Microwaves may induce changes in food products either by thermal effect or non-thermal effect (Marani and Feirabend 1994). Thermal effect arises due to the arrangement of food constituents (water, ions, salts, etc.) to adjust their dipole with that of microwaves that results in the friction among constituents resulting in heat generation while non-thermal effects are the observable changes that do not depend upon temperature. In most of the foods along with beverages/juices, water is the major constituent that has polarity. The absorption of microwaves by water molecules results in the disruption of hydrogen bonds generating heat. The presence of salts and charged ions also increases the heating effect (Gomma 2010).

28.5.3 High-pressure Processing

High-pressure processing involves Le Chatelier's principle where high pressure may induce changes in the properties of a food system that are prone to change in volume. High pressure has the tendency to reduce the volume that in turn affects the hydrogen bonding, ionic interactions, and hydrophobic linkages in a food system as these are sensitive to pressure. High pressure does not affect the low molecular weight components that mostly define the nutritional and sensory properties of food rather affect high molecular weight components (Carlez et al. 1994). In batch process, after specific time interval, the pressure chamber is depressurized and processed liquid food is transferred to sterile tanks for storage or shipment. The pressure ranges from 3000 to 8000 bars with a temperature range from −20 to 80 °C (Anon 2007). In a continuous process, the food is continuously pressurized with the help of hold vessel (Anon 2001).

28.5.4 Ultrasound or Sonication Processing

A new and practical method called sonication is used in place of traditional thermal processing. Ultrasound refers to sound waves of frequency greater than 20 kHz. When ultrasound waves pass through a medium, they generate different phenomena viz. mechanical vibration, acoustic streaming, and acoustic cavitation. Mechanical vibrations have the potential to severely damage any solid particle structure while acoustic streaming has the potential of mass transfer through a medium and could be used in cleaning and extraction application in industries (Tho et al. 2007). The collapse of bubbles produces high localized temperatures ranging between 2000 and 5000 K along with pressure (Ashokkumar and Mason 2007). Besides all these physical forces, cavitation also produces some chemical effects. The physical and chemical effects are frequency dependent as low-frequency ultrasounds (20 kHz) produce higher physical and lower chemical effects, while high-frequency ultrasounds (300–500 kHz) generate low physical and high chemical effects (Ashokkumar and Mason 2007). These specified effects produced by specified frequencies, broaden the range of choice for different desired processing techniques. According to Anaya-Esparza et al. (2017), it can lessen the effect on the dietary value of vegetable and fruit juices, boost microbiological and enzymatic denaturation rates, and increase product storage life.

In a related study, the nutritional value of juices of various vegetables and fruits including carrot, orange, spinach, and sweet lime, was compared employing novel approaches like ultrasound and ultraviolet irradiation. The antioxidant activity, total phenolic content, carbs, vitamin C, and other components of the treated juices were determined. When compared to all other methods utilized in the work, juices processed using ultrasound technology preserved the majority of the nutritious components to a greater degree. When compared to the thermal approach, the grouping of ultraviolet irradiations and ultrasound, effectively decontaminated liquids (suggested 5-log drop of bacteria) and maintained nutrition to an advanced degree (Khandpur and Gogate 2015).

28.6 Advantages of Non-thermal Processing

The non-thermal methodologies cause slight change in the functional, nutritional, and sensory qualities of beverages and, therefore, vegetable juice processing has attracted increasing attention in recent years. The benefits of non-thermal processing rely heavily on the composition or matrix of the meal. Therefore, to evaluate the technical standards and generate value-added food items, it is imperative to have a grasp of the special effects that enhance the qualities of drinks and juices. Food manufacturers are focusing on application of diverse methods for preparation of beverages, juices, and smoothies to strengthen the retaining capacity of nutritious contents without altering the physical qualities (Markowski et al. 2017). They also continue to struggle for mitigation of microbial risks in different food products in order to extend the storage life. To address safety and quality issues, they are looking for appropriate treatments, such as nonthermal or thermal methods, that will boost product's acceptability without compromising its sensory qualities or health benefits. The food scientists are always searching for an appropriate treatment that uses latest strategy to prolong the shelf life while retaining more bioactive chemicals during processing.

28.7 Conclusion

Mostly people look for organic or minimally processed and additive-free products. Conventional thermal and non-thermal treatments affect the color, taste, and flavor of food products including juices and smoothies. Modern technologies like cold plasma processing, microwave processing, high-pressure processing, and sonication provide alternatives for food processing that guarantee the consistency and quality of final food products. So, there is a dire need that food manufacturers should focus on applying advanced methods in the preparation of diverse products to enhance the retaining capacity of nutritious contents without altering the physical qualities.

References

Aderinola TA (2018) The impacts of lemon juice on the physicochemical and antioxidant properties of smoothies. Ann Food Sci Technol 19(4):675–683

Ajmera P, Kalani S, Sharma L (2019) Parsley: benefits & side effects on health. Int J Physiol Nutr Phys Educ 4(1):1236–1242

Amal ME, Hareedy LA, El-Malky WA (2008) Production and evaluation of nutraceutical vegetable juice blends. Arab Univ J Agric Sci 16(2):389–397

Anantheswaran RC, Ramaswamy HS (2001) Bacterial destruction and enzyme inactivation during microwave heating. In: Datta AK, Anantheswaran RC (eds) Handbook of microwave technology for food applications. Marcel Dekker, New York, pp 191–213

Anaya-Esparza LM et al (2017) Thermosonication: an alternative processing for fruit and vegetable juices. Trends Food Sci Technol 61:26–37

Anonymous (2001) Technical elements of new and emerging non-thermal food technologies. Fao.org/Ag/ags/Agsi/Nonthermal/nonthermal_1. Accessed Aug 2006

Anonymous (2007) Technical elements of new and emerging non-thermal food technologies. Invenia.es/tech:05_fr_soaa_0d6b. Accessed Mar 2008

Ashokkumar M, Mason T (2007) Sonochemistry, Kirk-Othmer encyclopedia of chemical technology. Wiley, Chichester, pp 1–34

Bakke AJ, Carney EM et al (2020) Blending dark green vegetables with fruits in commercially available infant foods makes them taste like fruit. Appetite 150:104652

Bevilacqua A, Petruzzi L et al (2018) Nonthermal technologies for fruit and vegetable juices and beverages: overview and advances. Compr Rev Food Sci Food Saf 17(1):2–62

Bower A, Marquez S, de Mejia EG (2016) The health benefits of selected culinary herbs and spices found in the traditional Mediterranean diet. Crit Rev Food Sci Nutr 56(16):2728–2746

Buzrul S, Alpas H, Largeteau A, Demazeau G (2008) Inactivation of *Escherichia coli* and *Listeria innocua* in kiwifruit and pineapple juices by high hydrostatic pressure. Int J Food Microbiol 124(3):275–278. https://doi.org/10.1016/j.ijfoodmicro.2008.03.015

Carlez A, Rosec JP, Richard N, Cheftel JC (1994) Bacterial growth during chilled storage of pressure-treated minced meat. LWT – Food Sci Technol 27(1):48–54

Dhandevi P, Rajesh J (2015) Fruit and vegetable intake: benefits and progress of nutrition education interventions—narrative review article. Iran J Public Health 44(10):1309–1321

Di Cagno R, Minervini G et al (2011) Effect of lactic acid fermentation on antioxidant, texture, color and sensory properties of red and green smoothies. Food Microbiol 28(5):1062–1071. https://doi.org/10.1016/j.fm.2011.02.011

El-Malky WA et al (2010) Production of low caloric dehydrated vegetable mixtures. Afr J Food Sci 9(5):326–330

Gomma AI (2010) An investigation of effects of microwave treatment on the structure, enzymatic hydrolysis, and nutraceutical properties of β-lactoglobulin. PhD thesis. Food Science Department, McGill University
Hanif R, Iqbal Z, Iqbal M, Hanif S, Rasheed M (2006) Use of vegetables as nutritional food: role in human health. J Agric Biol Sci 1(1):18–22
Khandpur P, Gogate PR (2015) Effect of novel ultrasound-based processing on the nutrition quality of different fruit and vegetable juices. Ultrason Sonochem 27:125–136
Kim SY, Yoon S, Kwon SM et al (2008) Kale juice improves coronary artery disease risk factors in hypercholesterolemic men. Biomed Environ Sci 21(2):91–97
Maeda E (2013) The effects of green smoothie consumption on blood pressure and health-related quality of life: a randomized controlled trial. Dissertations and theses. Portland State University
Marani E, Feirabend H (1994) Future perspectives in microwave applications in life sciences. Eur J Morphol 32(2–4 Suppl):330–334
Markowski J, Celejewska K et al (2017) Impact of different thermal preservation technologies on the quality of apple-based smoothies. LWT – Food Sci Technol. Available from: https://doi.org/10.1016/j.lwt.2017.01.004
Misra NN, Moiseev T, Patil S, Pankaj SK, Bourke P, Mosnier JP et al (2014) Cold plasma in modified atmospheres for postharvest treatment of strawberries. Food Bioprocess Technol 7:3045–3054
Nasir MU, Hussain S, Jabbar S (2015) Tomato processing, lycopene and health benefits: a review. Sci Lett 3(1):1–5
Niemira BA (2012) Cold plasma decontamination of foods. Ann Rev Food Sci Technol 3:125–142
Pankaj SK, Bueno-Ferrer C, Misra NN, Milosavljević V, O'Donnell CP, Bourke P et al (2014) Applications of cold plasma technology in food packaging. Trends Food Sci Technol 35(1):5–17
Petruzzi L, Campaniello D et al (2017) Thermal treatments for fruit and vegetable juices and beverages: a literature overview. Compr Rev Food Sci Food Saf 16(4):668–691
Rehm CD, Drewnowski A (2016) Dietary and economic effects of eliminating shortfall in fruit intake on nutrient intakes and diet costs. BMC Pediatr 16:83. https://doi.org/10.1186/s12887-016-0620-z
Rodríguez-Veráestegui LL et al (2016) Bioactive compounds and enzymatic activity of red vegetable smoothies during storage. Food Bioprocess Technol 9:137–146. https://doi.org/10.1007/s11947-015-1609-6
Ruxton CH, Gardner EJ, Walker D (2006) Can pure fruit and vegetable juices protect against cancer and cardiovascular disease too? A review of the evidence. Int J Food Sci Nutr 57(3–4):249–272
Suido H, Tanaka T, Tabei T et al (2002) A mixed green vegetable and fruit beverage decreased the serum level of low-density lipoprotein cholesterol in hypercholesterolemic patients. J Agric Food Chem 50(11):3346–3350
Tabikha MMM, El-Shehawy SM, Helal DMA (2010) Changes in chemical and nutritional quality during cold storage of some fruit and vegetable juice blends. J Food Dairy Sci 1(4):181–191
Tho P, Manasseh R, Ooi A (2007) Cavitation microstreaming patterns in single and multiple bubble systems. J Fluid Mech 576:191–233
Tiwari U, Cummins E (2013) Fruit and vegetables. In: Handbook of plant food phytochemicals. Wiley-Blackwell, Chichester, pp 107–137
USDA (2019) Celery, Raw. https://fdc.nal.usda.gov/food-details/169988/nutrients. Accessed 19 Jan 2025
Uthpala TGG, Marapana RAUJ et al (2020) Nutritional bioactive compounds and health benefits of fresh and processed cucumber (*Cucumis sativus* L.). Sumer J Biotechnol 3(9):75–82
Varshney K, Mishra K (2022) An analysis of health benefits of carrot. Int J Innov Res Eng Manag 9:211–214
Zamani H, De Joode MEJR et al (2021) The benefits and risks of beetroot juice consumption: a systematic review. Crit Rev Food Sci Nutr 61(5):788–804

The manufacturer's authorised representative in the EU is Springer Nature Customer Service Centre GmbH, Europaplatz 3, 69115 Heidelberg, Germany. If you have any concerns regarding our products, please contact ProductSafety@springernature.com

Printed and bound by CPI Group (UK) Ltd, Croydon, CR0 4YY
01/10/2026
02225921-0020